Springer Series in Computational Mathematics

25

AF181226

Vidar Thomée

Galerkin Finite Element Methods for Parabolic Problems

Second Edition

Springer

Vidar Thomée

Department of Mathematics
Chalmers University of Technology
S-41296 Göteborg
Sweden
email: thomee@math.chalmers.se

Mathematics Subject Classification (2000): 65M60, 65M12, 65M15

ISSN 0179-3632

ISBN 978-3-642-06967-3 e-ISBN 978-3-540-33122-3

Springer is a part of Springer Science+Business Media
springer.com

Softcover reprint of the hardcover 2nd edition 2006

Cover design: *design & production* GmbH, Heidelberg

Preface

My purpose in this monograph is to present an essentially self-contained account of the mathematical theory of Galerkin finite element methods as applied to parabolic partial differential equations. The emphases and selection of topics reflects my own involvement in the field over the past 25 years, and my ambition has been to stress ideas and methods of analysis rather than to describe the most general and farreaching results possible. Since the formulation and analysis of Galerkin finite element methods for parabolic problems are generally based on ideas and results from the corresponding theory for stationary elliptic problems, such material is often included in the presentation.

The basis of this work is my earlier text entitled *Galerkin Finite Element Methods for Parabolic Problems,* Springer Lecture Notes in Mathematics, No. 1054, from 1984. This has been out of print for several years, and I have felt a need and been encouraged by colleagues and friends to publish an updated version. In doing so I have included most of the contents of the 14 chapters of the earlier work in an updated and revised form, and added four new chapters, on semigroup methods, on multistep schemes, on incomplete iterative solution of the linear algebraic systems at the time levels, and on semilinear equations. The old chapters on fully discrete methods have been reworked by first treating the time discretization of an abstract differential equation in a Hilbert space setting, and the chapter on the discontinuous Galerkin method has been completely rewritten.

The following is an outline of the contents of the book:

In the introductory Chapter 1 we begin with a review of standard material on the finite element method for Dirichlet's problem for Poisson's equation in a bounded domain, and consider then the simplest Galerkin finite element methods for the corresponding initial-boundary value problem for the linear heat equation. The discrete methods are based on associated weak, or variational, formulations of the problems and employ first piecewise linear and then more general approximating functions which vanish on the boundary of the domain. For these model problems we demonstrate the basic error estimates in energy and mean square norms, in the parabolic case first for the semidiscrete problem resulting from discretization in the spatial variables only, and then also for the most commonly used fully discrete schemes obtained by discretization in both space and time, such as the backward Euler and Crank-Nicolson methods.

In the following five chapters we study several extensions and generalizations of the results obtained in the introduction in the case of the spatially semidiscrete approximation, and show error estimates in a variety of norms. First, in Chapter 2, we formulate the semidiscrete problem in terms of a more general approximate solution operator for the elliptic problem in a manner which does not require the approximating functions to satisfy the homogeneous boundary conditions. As an example of such a method we discuss a method of Nitsche based on a nonstandard weak formulation. In Chapter 3 more precise results are shown in the case of the homogeneous heat equation. These results are expressed in terms of certain function spaces $\dot{H}^s(\Omega)$ which are characterized by both smoothness and boundary behavior of its elements, and which will be used repeatedly in the rest of the book. We also demonstrate that the smoothing property for positive time of the solution operator of the initial value problem has an analogue in the semidiscrete situation, and use this to show that the finite element solution converges to full order even when the initial data are nonsmooth. The results of Chapters 2 and 3 are extended to more general linear parabolic equations in Chapter 4. Chapter 5 is devoted to the derivation of stability and error bounds with respect to the maximum-norm for our plane model problem, and in Chapter 6 negative norm error estimates of higher order are derived, together with related results concerning superconvergence.

In the next six chapters we consider fully discrete methods obtained by discretization in time of the spatially semidiscrete problem. First, in Chapter 7, we study the homogeneous heat equation and give analogues of our previous results both for smooth and for nonsmooth data. The methods used for time discretization are of one-step type and rely on rational approximations of the exponential, allowing the standard Euler and Crank-Nicolson procedures as special cases. Our approach here is to first discretize a parabolic equation in an abstract Hilbert space framework with respect to time, and then to apply the results obtained to the spatially semidiscrete problem. The analysis uses eigenfunction expansions related to the elliptic operator occurring in the parabolic equation, which we assume positive definite. In Chapter 8 we generalize the above abstract considerations to a Banach space setting and allow a more general parabolic equation, which we now analyze using the Dunford-Taylor spectral representation. The time discretization is interpreted as a rational approximation of the semigroup generated by the elliptic operator, i.e., the solution operator of the initial-value problem for the homogeneous equation. Application to maximum-norm estimates is discussed. In Chapter 9 we study fully discrete one-step methods for the inhomogeneous heat equation in which the forcing term is evaluated at a fixed finite number of points per time stepping interval. In Chapter 10 we apply Galerkin's method also for the time discretization and seek discrete solutions as piecewise polynomials in the time variable which may be discontinuous at the now not necessarily equidistant nodes. In this *discontinuous Galerkin* procedure the forcing term enters in integrated form rather than at a finite number of points. In Chapter 11 we consider multistep backward difference methods. We first study such methods with constant time steps of order at most 6, and show stability as well as smooth and nonsmooth data error estimates, and then discuss the second order backward difference method with variable time steps. In Chapter 12 we study the incomplete iterative solution of the finite dimensional linear systems of algebraic equations which need to be solved at each level of the time stepping procedure, and exemplify by the use of a V-cycle multigrid algorithm.

The next two chapters are devoted to nonlinear problems. In Chapter 13 we discuss the application of the standard Galerkin method to a model nonlinear parabolic equation. We show error estimates for the spatially semidiscrete problem as well as the fully discrete backward Euler and Crank-Nicolson methods, using piecewise linear finite elements, and then pay special attention to the formulation and analysis of time stepping procedures based on these, which are linear in the unknown functions. In Chapter 14 we derive various results in the case of semilinear equations, in particular concerning the extension of the analysis for nonsmooth initial data from the case of linear homogenous equations.

In the last four chapters we consider various modifications of the standard Galerkin finite element method. In Chapter 15 we analyze the so called lumped mass method for which in certain cases a maximum-principle is valid. In Chapter 16 we discuss the H^1 and H^{-1} methods. In the first of these, the Galerkin method is based on a weak formulation with respect to an inner product in H^1 and for the second, the method uses trial and test functions from different finite dimensional spaces. In Chapter 17, the approximation scheme is based on a mixed formulation of the initial boundary value problem in which the solution and its gradient are sought independently in different spaces. In the final Chapter 18 we consider a singular problem obtained by introducing polar coordinates in a spherically symmetric problem in a ball in $\mathbf{R}^3$ and discuss Galerkin methods based on two different weak formulations defined by two different inner products.

References to the literature where the reader may find more complete treatments of the different topics, and some historical comments, are given at the end of each chapter.

A desirable mathematical background for reading the text includes standard basic partial differential equations and functional analysis, including Sobolev spaces; for the convenience of the reader we often give references to the literature concerning such matters.

The work presented, first in the Lecture Notes and now in this monograph, has grown from courses, lecture series, summer-schools, and written material that I have been involved in over a long period of time. I wish to thank my students and colleagues in these various contexts for the inspiration and support they have provided, and for the help they have given me as discussion partners and critics. As regards this new version of my work I particularly address my thanks to Georgios Akrivis, Stig Larsson, and Per-Gunnar Martinsson, who have read the manuscript in various degrees of detail and are responsible for many improvements. I also want to express my special gratitude to Yumi Karlsson who typed a first version of the text from the old lecture notes, and to Gunnar Ekolin who generously furnished me with expert help with the intricacies of TEX.

Göteborg
July 1997

Vidar Thomée

Preface to the Second Edition

I am pleased to have been given the opportunity to prepare a second edition of this book. In doing so, I have kept most of the text essentially unchanged, but after correcting a number or typographical errors and other minor inadequacies, I have also taken advantage of this possibility to include some new material representing work that I have been involved in since the time when the original version appeared about eight years ago.

This concerns in particular progress in the application of semigroup theory to stability and error analysis. Using the theory of analytic semigroups it is convenient to reformulate the stability and smoothing properties as estimates for the resolvent of the associated elliptic operator and its discrete analogue. This is particularly useful in deriving maximum-norm estimates, and has led to improvements for both spatially semidiscrete and fully discrete problems. For this reason a somewhat expanded review of analytic semigroups is given in the present Chapter 6, on maximum-norm estimates for the semidiscrete problem, where now resolvent estimates for piecewise linear finite elements are discussed in some detail. These changes have affected the chapter on single step time stepping methods, expressed as rational approximation of semigroups, now placed as Chapter 9. The new emphasis has led to certain modifications and additions also in other chapters, particularly in Chapter 10 on multistep methods and Chapter 15 on the lumped mass method.

I have also added two chapter at the end of the book on other topics of recent interest to me. The first of these, Chapter 19, concern problems in which the spatial domain is polygonal, with particular attention given to noncovex such domains. rather than with smooth boundary, as in most of the rest of the book. In this case the corners generate singularites in the exact solution, and we study the effect of these on the convergence of the finite element solution.

The second new chapter, Chapter 20, considers an alternative to time stepping as a method for discretization in time, which is based on representing the solution as an integral involving the resolvent of the elliptic operator along a smooth curve extending into the right half of the complex plane, and then applying an accurate quadrature rule to this integral. This reduces the parabolic problem to a finite set of elliptic problems that may be solved in parallel. The method is then combined with finite element discretization in the spatial variable. When applicable, this method gives very accurate approximations of the exact solution in an efficient way.

I would like to take this opportunity to express my warm gratitude to Georgios Akrivis for his generous help and support. He has critically read through the new material and made many valuable suggestions.

Göteborg *Vidar Thomée*
March 2006

Table of Contents

1. The Standard Galerkin Method

In this introductory chapter we shall study the standard Galerkin finite element method for the approximate solution of the model initial-boundary value problem for the heat equation,

$$(1.1)\qquad \begin{aligned} u_t - \Delta u &= f \quad \text{in } \Omega, \qquad \text{for } t > 0,\\ u &= 0 \quad \text{on } \partial\Omega, \quad \text{for } t > 0, \quad \text{with } u(\cdot,0) = v \quad \text{in } \Omega, \end{aligned}$$

where Ω is a domain in $\mathbb{R}^d$ with smooth boundary $\partial\Omega$, and where $u = u(x,t)$, u_t denotes $\partial u/\partial t$, and $\Delta = \sum_{j=1}^d \partial^2/\partial x_j^2$ the Laplacian.

Before we start to discuss this problem we shall briefly review some basic relevant material about the finite element method for the corresponding stationary problem, the Dirichlet problem for Poisson's equation,

$$(1.2)\qquad -\Delta u = f \quad \text{in } \Omega, \quad \text{with } u = 0 \quad \text{on } \partial\Omega.$$

Using a variational formulation of this problem, we shall define an approximation of the solution u of (1.2) as a function u_h which belongs to a finite-dimensional linear space S_h of functions of x with certain properties. This function, in the simplest case a continuous, piecewise linear function on some partition of Ω, will be a solution of a finite system of linear algebraic equations. We show basic error estimates for this approximate solution in energy and least square norms.

We shall then turn to the parabolic problem (1.1) which we first write in a weak form. We then proceed to discretize this problem, first in the spatial variable x, which results in an approximate solution $u_h(\cdot,t)$ in the finite element space S_h, for $t \geq 0$, as a solution of an initial value problem for a finite-dimensional system of ordinary differential equations. We then define the fully discrete approximation by application of some finite difference time stepping method to this finite dimensional initial value problem. This yields an approximate solution $U = U_h$ of (1.1) which belongs to S_h at discrete time levels. Error estimates will be derived for both the spatially and fully discrete solutions.

For a general $\Omega \subset \mathbb{R}^d$ we denote below by $\|\cdot\|$ the norm in $L_2 = L_2(\Omega)$ and by $\|\cdot\|_r$ that in the Sobolev space $H^r = H^r(\Omega) = W_2^r(\Omega)$, so that for real-valued functions v,

$$\|v\| = \|v\|_{L_2} = \Big(\int_\Omega v^2\,dx\Big)^{1/2},$$

and, for r a positive integer,

$$\|v\|_r = \|v\|_{H^r} = \Big(\sum_{|\alpha|\le r} \|D^\alpha v\|^2\Big)^{1/2}, \tag{1.3}$$

where, with $\alpha = (\alpha_1, \dots, \alpha_d)$, $D^\alpha = (\partial/\partial x_1)^{\alpha_1} \cdots (\partial/\partial x_d)^{\alpha_d}$ denotes an arbitrary derivative with respect to x of order $|\alpha| = \sum_{j=1}^d \alpha_j$, so that the sum in (1.3) contains all such derivatives of order at most r. We recall that for functions in $H_0^1 = H_0^1(\Omega)$, i.e., the functions v with $\nabla v = \operatorname{grad} v$ in L_2 and which vanish on $\partial\Omega$, $\|\nabla v\|$ and $\|v\|_1$ are equivalent norms (Friedrichs' lemma, see, e.g., [42] or [51]), and that

$$c\|v\|_1 \le \|\nabla v\| \le \|v\|_1, \quad \forall v \in H_0^1, \qquad \text{with } c > 0. \tag{1.4}$$

Throughout this book c and C will denote positive constants, not necessarily the same at different occurrences, which are independent of the parameters and functions involved.

We shall begin by recalling some basic facts concerning the Dirichlet problem (1.2). We first write this problem in a weak, or variational, form: We multiply the elliptic equation by a smooth function φ which vanishes on $\partial\Omega$ (it suffices to require $\varphi \in H_0^1$), integrate over Ω, and apply Green's formula on the left-hand side, to obtain

$$(\nabla u, \nabla\varphi) = (f, \varphi), \quad \forall \varphi \in H_0^1, \tag{1.5}$$

where we have used the L_2 inner products,

$$(v, w) = \int_\Omega vw\,dx, \quad (\nabla v, \nabla w) = \int_\Omega \sum_{j=1}^d \frac{\partial v}{\partial x_j}\frac{\partial w}{\partial x_j}\,dx. \tag{1.6}$$

A function $u \in H_0^1$ which satisfies (1.5) is called a variational solution of (1.2). It is an easy consequence of the Riesz representation theorem that a unique such solution exists if $f \in H^{-1}$, the dual space of H_0^1. In this case (f, φ) denotes the value of the functional f at φ. Further, since we have assumed the boundary $\partial\Omega$ to be smooth, the solution u is smoother by two derivatives in L_2 than the right hand side f, which may be expressed in the form of the *elliptic regularity* inequality

$$\|u\|_{m+2} \le C\|\Delta u\|_m = C\|f\|_m, \quad \text{for any } m \ge -1. \tag{1.7}$$

In particular, using also Sobolev's embeddning theorem, this shows that the solution u belongs to C^∞ if f belongs to C^∞. We refer to, e.g., Evans [96] for such material.

We remark for later reference that, for $m = -1, 0$, (1.7) holds also in the case of a convex polygonal domain Ω, but that this is not true for nonconvex polygonal domains.

As a preparation for the definition of the finite element solution of (1.2), we consider briefly the approximation of smooth functions in Ω which vanish on $\partial\Omega$. For concreteness, we shall exemplify by piecewise linear functions in a convex plane domain.

Let thus Ω be a convex domain in the plane with smooth boundary $\partial\Omega$, and let $\mathcal{T}_h$ denote a partition of Ω into disjoint triangles τ such that no vertex of any triangle lies on the interior of a side of another triangle and such that the union of the triangles determine a polygonal domain $\Omega_h \subset \Omega$ with boundary vertices on $\partial\Omega$.

Let h denote the maximal length of the sides of the triangulation $\mathcal{T}_h$. Thus h is a parameter which decreases as the triangulation is made finer. We shall assume that the angles of the triangulations are bounded below by a positive constant, independently of h, and sometimes also that the triangulations are *quasiuniform* in the sense that the triangles of $\mathcal{T}_h$ are of essentially the same size, which we express by demanding that the area of $\tau \in \mathcal{T}_h$ is bounded below by ch^2, with $c > 0$, independent of h.

Let now S_h denote the continuous functions on the closure $\bar{\Omega}$ of Ω which are linear in each triangle of $\mathcal{T}_h$ and which vanish outside Ω_h. Let $\{P_j\}_{j=1}^{N_h}$ be the interior vertices of $\mathcal{T}_h$. A function in S_h is then uniquely determined by its values at the points P_j and thus depends on N_h parameters. Let Φ_j be the *pyramid function* in S_h which takes the value 1 at P_j but vanishes at the other vertices. Then $\{\Phi_j\}_{j=1}^{N_h}$ forms a *basis* for S_h, and every χ in S_h admits a unique representation

$$\chi(x) = \sum_{j=1}^{N_h} \alpha_j \Phi_j(x), \quad \text{with } \alpha_j = \chi(P_j).$$

A given smooth function v on Ω which vanishes on $\partial\Omega$ may now be approximated by, for instance, its interpolant $I_h v$ in S_h, which we define as the function in S_h which agrees with v at the interior vertices of $\mathcal{T}_h$, i.e.,

$$I_h v(x) = \sum_{j=1}^{N_h} v(P_j)\Phi_j(x). \tag{1.8}$$

Using this notation in our plane domain Ω, the following error estimates for the interpolant defined in (1.8) are well known (see, e.g., [42] or [51]), namely, for $v \in H^2 \cap H_0^1$,

$$\|I_h v - v\| \le Ch^2 \|v\|_2 \quad \text{and} \quad \|\nabla(I_h v - v)\| \le Ch\|v\|_2. \tag{1.9}$$

They may be derived by showing the corresponding estimate for each $\tau \in \mathcal{T}_h$ and then taking squares and adding. For an individual $\tau \in \mathcal{T}_h$ the proof is achieved by means of the Bramble-Hilbert lemma (cf. [42] or [51]), noting that $I_h v - v$ vanishes on τ for v linear.

We shall now return to the general case of a domain Ω in $\mathbb{R}^d$ and assume that we are given a family $\{S_h\}$ of finite-dimensional subspaces of H_0^1 such that, for some integer $r \geq 2$ and small h,

$$\inf_{\chi \in S_h} \{\|v - \chi\| + h\|\nabla(v - \chi)\|\} \leq Ch^s\|v\|_s, \quad \text{for } 1 \leq s \leq r, \tag{1.10}$$

when $v \in H^s \cap H_0^1$. The number r is referred to as the order of accuracy of the family $\{S_h\}$. The above example of piecewise linear functions in a plane domain corresponds to $d = r = 2$. In the case $r > 2, S_h$ often consists of piecewise polynomials of degree at most $r - 1$ on a triangulation $\mathcal{T}_h$ as above. For instance, $r = 4$ in the case of piecewise cubic polynomial subspaces. Also, in the general situation estimates such as (1.10) may often be obtained by exhibiting an *interpolation operator* $I_h : H^r \cap H_0^1 \to S_h$ such that

$$\|I_h v - v\| + h\|\nabla(I_h v - v)\| \leq Ch^s\|v\|_s, \quad \text{for } 1 \leq s \leq r. \tag{1.11}$$

When $\partial\Omega$ is curved and $r > 2$ there are difficulties in the construction and analysis of such operators near the boundary, but this may be accomplished, in principle, by mapping a curved triangle onto a straight-edged one (isoparametric elements). We shall not dwell on this here, but return in Chapter 2 to this problem.

We remark for later reference that if the family $\{S_h\}$ is based on a family of *quasiuniform* triangulations $\mathcal{T}_h$ and S_h consists of piecewise polynomials of degree at most $r - 1$, then one may show the *inverse inequality*

$$\|\nabla\chi\| \leq Ch^{-1}\|\chi\|, \quad \forall \chi \in S_h. \tag{1.12}$$

This follows by taking squares and adding from the corresponding inequality for each triangle $\tau \in \mathcal{T}_h$, which in turn is obtained by a transformation to a fixed reference triangle, and using the fact that all norms on a finite dimensional space are equivalent, see, e.g., [51].

The optimal orders to which functions and their gradients may be approximated under our assumption (1.10) are $O(h^r)$ and $O(h^{r-1})$, respectively, and we shall now construct approximations to these orders of the solution of the Dirichlet problem (1.2) by the finite element method. The approximate problem is then to find a function $u_h \in S_h$ such that, cf., (1.5),

$$(\nabla u_h, \nabla\chi) = (f, \chi), \quad \forall \chi \in S_h. \tag{1.13}$$

This way of defining an approximate solution in terms of the variational formulation of the problem is referred to as Galerkin's method, after the Russian applied mathematician Boris Grigorievich Galerkin (1871-1945).

Note that, as a result of (1.5) and (1.13),

$$(\nabla(u_h - u), \nabla\chi) = 0, \quad \forall \chi \in S_h, \tag{1.14}$$

that is, the error in the discrete solution is orthogonal to S_h with respect to the Dirichlet inner product $(\nabla v, \nabla w)$.

In terms of a basis $\{\Phi_j\}_1^{N_h}$ for the finite element space S_h, our discrete problem may also be formulated: Find the coefficients α_j in $u_h(x) = \sum_{j=1}^{N_h} \alpha_j \Phi_j(x)$ such that

$$\sum_{j=1}^{N_h} \alpha_j (\nabla \Phi_j, \nabla \Phi_k) = (f, \Phi_k), \quad \text{for } k = 1, \dots, N_h.$$

In matrix notation this may be expressed as

$$\mathcal{A}\alpha = \widetilde{f},$$

where $\mathcal{A} = (a_{jk})$ is the *stiffness matrix* with elements $a_{jk} = (\nabla \Phi_j, \nabla \Phi_k)$, $\widetilde{f} = (f_k)$ the vector with entries $f_k = (f, \Phi_k)$, and α the vector of unknowns α_j. The dimensions of all of these arrays then equal N_h, the dimension of S_h (which equals the number of interior vertices in our plane example above). The stiffness matrix $\mathcal{A}$ is a Gram matrix and thus in particular positive definite and invertible so that our discrete problem has a unique solution. To see that $\mathcal{A} = (a_{jk})$ is positive definite, we note that

$$\sum_{j,k=1}^{d} a_{jk} \xi_j \xi_k = \|\nabla \Big(\sum_{j=1}^{d} \xi_j \Phi_j \Big)\|^2 \geq 0.$$

Here equality holds only if $\nabla(\sum_{j=1}^{d} \xi_j \Phi_j) \equiv 0$, so that $\sum_{j=1}^{d} \xi_j \Phi_j = 0$ by (1.4), and hence $\xi_j = 0, \ j = 1, \dots, N_h$.

When S_h consists of piecewise polynomial functions, the elements of the matrix $\mathcal{A}$ may be calculated exactly. However, unless f has a particularly simple form, the elements (f, Φ_j) of $\widetilde{f}$ have to be computed by some quadrature formula.

We shall prove the following estimate for the error between the solutions of the discrete and continuous problems. Note that these estimates are of optimal order as defined by our assumption (1.10). Here, as will always be the case in the sequel, the statements of the inequalities assume that u is sufficiently regular for the norms on the right to be finite.

We remark that since we require $\partial\Omega$ to be smooth, according to the elliptic regularity estimate (1.7), the solution of (1.2) can be guaranteed to have any degree of smoothness required by assuming the right hand side f to be sufficiently smooth. In particular, $u \in H^r \cap H_0^1$ if $f \in H^{r-2}$, and the solution u belongs to C^∞ if $\partial\Omega \in C^\infty$ and $f \in C^\infty$.

Theorem 1.1 *Assume that* (1.10) *holds, and let* u_h *and* u *be the solutions of* (1.13) *and* (1.2), *respectively. Then, for* $1 \leq s \leq r$,

$$\|u_h - u\| \leq C h^s \|u\|_s \quad \textit{and} \quad \|\nabla u_h - \nabla u\| \leq C h^{s-1} \|u\|_s.$$

Proof. We start with the estimate for the error in the gradient. Since by (1.14) u_h is the orthogonal projection of u onto S_h with respect to the Dirichlet inner product, we have by (1.10)

$$\|\nabla(u_h - u)\| \le \inf_{\chi \in S_h} \|\nabla(\chi - u)\| \le Ch^{s-1}\|u\|_s. \tag{1.15}$$

For the error bound in L_2−norm we proceed by a duality argument. Let φ be arbitrary in L_2, take $\psi \in H^2 \cap H_0^1$ as the solution of

$$-\Delta\psi = \varphi \quad \text{in } \Omega, \qquad \text{with } \psi = 0 \quad \text{on } \partial\Omega, \tag{1.16}$$

and recall the fact that by (1.7) the solution ψ of (1.16) is smoother by two derivatives in L_2 than the right hand side φ, so that

$$\|\psi\|_2 \le C\|\Delta\psi\| = C\|\varphi\|. \tag{1.17}$$

For any $\psi_h \in S_h$ we have

$$\begin{aligned}(u_h - u, \varphi) &= -(u_h - u, \Delta\psi) = (\nabla(u_h - u), \nabla\psi)\\ &= (\nabla(u_h - u), \nabla(\psi - \psi_h)) \le \|\nabla(u_h - u)\|\,\|\nabla(\psi - \psi_h)\|,\end{aligned} \tag{1.18}$$

and hence, using (1.15) together with (1.10) with $s = 2$ and (1.7) with $m = 0$,

$$(u_h - u, \varphi) \le Ch^{s-1}\|u\|_s\, h\|\psi\|_2 \le Ch^s\|u\|_s\|\varphi\|.$$

Choosing $\varphi = u_h - u$ completes the proof. □

After these preparations we now turn to the initial-boundary value problem (1.1) for the heat equation. As in the elliptic case we begin by writing the problem in weak form: We multiply the heat equation by a smooth function φ which vanishes on $\partial\Omega$ (or $\varphi \in H_0^1$), integrate over Ω, and apply Green's formula to the second term, to obtain, with (v, w) and $(\nabla v, \nabla w)$ as in (1.6),

$$(u_t, \varphi) + (\nabla u, \nabla\varphi) = (f, \varphi), \quad \forall \varphi \in H_0^1,\ t > 0. \tag{1.19}$$

We say that a function $u = u(x,t)$ is a weak solution of (1.1) on $[0, \bar{t}]$ if (1.19) holds with $u \in L_2(0, \bar{t}; H_0^1)$ and $u_t \in L_2(0, \bar{t}; H^{-1})$, and if $u(\cdot, 0) = v$. Again, since the boundary $\partial\Omega$ is smooth, such a solution is smooth provided the data are smooth functions, and in this case also satisfy certain compatibility conditions at $t = 0$. Similarly to (1.7) this may be expressed by a *parabolic regularity* estimate such as, cf. [96], with $u^{(j)} = (\partial/\partial t)^j u$ and $C = C_{m,\bar{t}}$,

$$\sum_{j=0}^{m+1} \int_0^{\bar{t}} \|u^{(j)}\|^2_{2(m-j)+2} dt \le C\Big(\|v\|^2_{2m+1} + \sum_{j=0}^{m} \int_0^{\bar{t}} \|f^{(j)}\|^2_{2(m-j)} dt\Big), \tag{1.20}$$

for $m \ge 0$. The compatibility conditions required express that the different conditions imposed in (1.1) at $\partial\Omega$ are consistent with each other. The first such condition, required for $m = 0$, is that since $u(t) = 0$ on $\partial\Omega$ for $t > 0$, then $u(0) = v$ also has to vanish on $\partial\Omega$. Next, for $m = 1$, since $u_t(t) = 0$ on $\partial\Omega$ for $t > 0$, smoothness requires that $u_t(0) = g := \Delta v + f(0) = 0$ on $\partial\Omega$, and similarly for $u^{(m)}(0)$ with $m \geq 2$. Again we refer to, e.g., Evans [96] for details.

As indicated above it is convenient to proceed in two steps with the derivation and analysis of the approximate solution of (1.1). In the first step we approximate $u(x, t)$ by means of a function $u_h(x, t)$ which, for each fixed t, belongs to a finite-dimensional linear space S_h of functions of x of the type considered above. This function will be a solution of an h-dependent finite system of ordinary differential equations in time and is referred to as a *spatially discrete*, or *semidiscrete*, solution. As in the elliptic case just considered, the spatially discrete problem is based on a weak formulation of (1.1). We then proceed to discretize this system in the time variable to obtain produce a *fully discrete* approximation of the solution of (1.1) by a *time stepping* method. In our basic schemes this discretization in time will be accomplished by a finite difference approximation of the time derivative.

We thus first pose the spatially semidiscrete problem, based on the weak formulation (1.19), to find $u_h(t) = u_h(\cdot, t)$, belonging to S_h for $t \geq 0$, such that

$$(1.21)\quad (u_{h,t}, \chi) + (\nabla u_h, \nabla \chi) = (f, \chi), \quad \forall\, \chi \in S_h,\ t > 0, \quad \text{with } u_h(0) = v_h,$$

where v_h is some approximation of v in S_h.

In terms of the basis $\{\Phi_j\}_1^{N_h}$ for S_h, our semidiscrete problem may be stated: Find the coefficients $\alpha_j(t)$ in $u_h(x, t) = \sum_{j=1}^{N_h} \alpha_j(t)\Phi_j(x)$ such that

$$\sum_{j=1}^{N_h} \alpha_j'(t)(\Phi_j, \Phi_k) + \sum_{j=1}^{N_h} \alpha_j(t)(\nabla\Phi_j, \nabla\Phi_k) = (f, \Phi_k), \quad k = 1, \ldots, N_h,$$

and, with γ_j the components of the given initial approximation v_h, $\alpha_j(0) = \gamma_j$ for $j = 1, \ldots, N_h$. In matrix notation this may be expressed as

$$\mathcal{B}\alpha'(t) + \mathcal{A}\alpha(t) = \widetilde{f}(t), \quad \text{for } t > 0, \quad \text{with } \alpha(0) = \gamma,$$

where $\mathcal{B} = (b_{jk})$ is the *mass matrix* with elements $b_{jk} = (\Phi_j, \Phi_k)$, $\mathcal{A} = (a_{jk})$ the stiffness matrix with $a_{jk} = (\nabla\Phi_j, \nabla\Phi_k)$, $\widetilde{f} = (f_k)$ the vector with entries $f_k = (f, \Phi_k)$, $\alpha(t)$ the vector of unknowns $\alpha_j(t)$, and $\gamma = (\gamma_k)$. The dimension of all these items equals N_h, the dimension of S_h.

Since, like the stiffness matrix $\mathcal{A}$, the mass matrix $\mathcal{B}$ is a Gram matrix, and thus in particular positive definite and invertible, the above system of ordinary differential equations may be written

$$\alpha'(t) + \mathcal{B}^{-1}\mathcal{A}\alpha(t) = \mathcal{B}^{-1}\widetilde{f}(t), \quad \text{for } t > 0, \quad \text{with } \alpha(0) = \gamma,$$

and hence obviously has a unique solution for t positive.

Our first aim is to prove the following estimate in L_2 for the error between the solutions of the semidiscrete and continuous problems.

Theorem 1.2 *Let u_h and u be the solutions of* (1.21) *and* (1.1), *and assume $v = 0$ on $\partial\Omega$. Then*

$$\|u_h(t) - u(t)\| \le \|v_h - v\| + Ch^r\big(\|v\|_r + \int_0^t \|u_t\|_r\,ds\big), \quad \textit{for } t \ge 0.$$

Here as earlier we require that the solution of the continuous problem has the regularity implicitly assumed by the presence of the norms on the right. Note also that if (1.11) holds and $v_h = I_h v$, then the first term on the right is dominated by the second. This also holds if $v_h = P_h v$, where P_h denotes the orthogonal projection of v onto S_h with respect to the inner product in L_2, since this choice is the best approximation of v in S_h with respect to the L_2 norm, and thus at least as good as $I_h v$.

Another such optimal order choice for v_h is the so-called *elliptic* or *Ritz projection* R_h onto S_h which we define as the orthogonal projection with respect to the inner product $(\nabla v, \nabla w)$, so that

$$(1.22) \qquad (\nabla R_h v, \nabla\chi) = (\nabla v, \nabla\chi), \quad \forall\, \chi \in S_h, \quad \text{for } v \in H_0^1.$$

In view of (1.14), this definition may be expressed by saying that $R_h v$ is the finite element approximation of the solution of the corresponding elliptic problem with exact solution v. A pervading strategy throughout the error analysis in the rest of this book is to write the error in the parabolic problem as a sum of two terms,

$$(1.23) \quad u_h(t) - u(t) = \theta(t) + \rho(t), \quad \text{where } \theta = u_h - R_h u, \quad \rho = R_h u - u,$$

which are then bounded separately. The second term, $\rho(t)$, is thus the error in an elliptic problem and may be handled as such, whereas the first term $\theta(t)$ will be the main object of the analysis.

It follows at once from setting $\chi = R_h v$ in (1.22) that the Ritz projection is stable in H_0^1, or

$$(1.24) \qquad \|\nabla R_h v\| \le \|\nabla v\|, \quad \forall\, v \in H_0^1.$$

As an immediate consequence of Theorem 1.1 we have the following error estimate for $R_h v$.

Lemma 1.1 *Assume that* (1.10) *holds. Then, with R_h defined by* (1.22) *we have*

$$\|R_h v - v\| + h\|\nabla(R_h v - v)\| \le Ch^s\|v\|_s, \quad \textit{for } v \in H^s \cap H_0^1, \ 1 \le s \le r.$$

Proof of Theorem 1.2. We write the error according to (1.23) and obtain easily by Lemma 1.1 and obvious estimates

(1.25) $$\|\rho(t)\| \le Ch^r\|u(t)\|_r \le Ch^r\big(\|v\|_r + \int_0^t \|u_t\|_r\,ds\big).$$

In order to bound θ, we note that by our definitions

(1.26) $$\begin{aligned}(\theta_t,\chi) + (\nabla\theta,\nabla\chi) &\\ &= (u_{h,t},\chi) + (\nabla u_h,\nabla\chi) - (R_h u_t,\chi) - (\nabla R_h u,\nabla\chi)\\ &= (f,\chi) - (R_h u_t,\chi) - (\nabla u,\nabla\chi) = (u_t - R_h u_t,\chi),\end{aligned}$$

or

(1.27) $$(\theta_t,\chi) + (\nabla\theta,\nabla\chi) = -(\rho_t,\chi), \quad \forall \chi \in S_h,\ t>0,$$

where we have used the easily established fact that the operator R_h commutes with time differentiation. Since θ belongs to S_h, we may choose $\chi = \theta$ in (1.27) and conclude

(1.28) $$(\theta_t,\theta) + \|\nabla\theta\|^2 = -(\rho_t,\theta), \quad \text{for } t>0.$$

Here the second is nonnegative, and we obtain thus

$$\tfrac12 \frac{d}{dt}\|\theta\|^2 = \|\theta\|\,\frac{d}{dt}\|\theta\| \le \|\rho_t\|\,\|\theta\|,$$

and hence, after cancellation of one factor $\|\theta\|$ (the case that $\|\theta(t)\| = 0$ for some t may easily be handled), and integration,

(1.29) $$\|\theta(t)\| \le \|\theta(0)\| + \int_0^t \|\rho_t\|\,ds.$$

Here, using Lemma 1.1, we find

$$\|\theta(0)\| = \|v_h - R_h v\| \le \|v_h - v\| + \|R_h v - v\| \le \|v_h - v\| + Ch^r\|v\|_r,$$

and since

$$\|\rho_t\| = \|R_h u_t - u_t\| \le Ch^r\|u_t\|_r,$$

the desired bound for $\|\theta(t)\|$ now follows. □

In the above proof we have made use in (1.28) of the fact that $\|\nabla\theta\|^2$ is nonnegative, and simply discarded this term. By using it in a somewhat more careful way, one may demonstrate that the effect of the initial data upon the error tends to zero exponentially as t tends to ∞. In fact, with λ_1 the smallest eigenvalue of $-\Delta$, with Dirichlet boundary data, we have

(1.30) $$\|\nabla v\|^2 \ge \lambda_1\|v\|^2, \quad \forall v \in H_0^1,$$

and hence (1.28) yields

$$\frac{1}{2}\frac{d}{dt}\|\theta\|^2 + \lambda_1\|\theta\|^2 \le \|\rho_t\|\|\theta\|.$$

It follows as above that

$$\frac{d}{dt}\|\theta\| + \lambda_1\|\theta\| \le \|\rho_t\|,$$

and hence

$$\begin{aligned}\|\theta(t)\| &\le e^{-\lambda_1 t}\|\theta(0)\| + \int_0^t e^{-\lambda_1(t-s)}\|\rho_t(s)\|\,ds \\ &\le e^{-\lambda_1 t}\|v_h - v\| + Ch^r\Big(e^{-\lambda_1 t}\|v\|_r + \int_0^t e^{-\lambda_1(t-s)}\|u_t(s)\|_r\,ds\Big).\end{aligned} \tag{1.31}$$

Using the first part of (1.25) we conclude that with v_h appropriately chosen

$$\|u_h(t) - u(t)\| \le Ch^r\Big(e^{-\lambda_1 t}\|v\|_r + \|u(t)\|_r + \int_0^t e^{-\lambda_1(t-s)}\|u_t(s)\|_r\,ds\Big).$$

We shall not pursue the error analysis for large t below.

We shall now briefly look at another way of expressing the argument in the proof of Theorem 1.2, which consists in working with the equation for θ in operator form. We first recall that by Duhamel's principle, the solution of (1.1) may be written

$$u(t) = E(t)v + \int_0^t E(t-s)f(s)\,ds. \tag{1.32}$$

Here $E(t)$ is the solution operator of the homogeneous equation, the case $f \equiv 0$ of (1.1), i.e., the operator which takes the initial values $u(0) = v$ into the solution $u(t)$ at time t. This operator may also be thought of as the *semigroup* $e^{\Delta t}$ on L_2 generated by the Laplacian, considered as defined in $\mathcal{D}(\Delta) = H^2 \cap H_0^1$. We now introduce a *discrete Laplacian* $\Delta_h : S_h \to S_h$ by

$$(\Delta_h\psi, \chi) = -(\nabla\psi, \nabla\chi), \quad \forall \psi, \chi \in S_h; \tag{1.33}$$

this analogue of Green's formula clearly defines $\Delta_h\psi = \sum_{j=1}^{N_h} d_j\Phi_j$ by

$$\sum_{j=1}^{N_h} d_j(\Phi_j, \Phi_k) = -(\nabla\psi, \nabla\Phi_k), \quad \text{for } k = 1, \dots, N_h,$$

since the matrix of this system is the positive definite mass matrix encountered above. The operator $-\Delta_h$ is easily seen to be selfadjoint and positive definite in S_h with respect to $(\cdot,\cdot)$. Note that Δ_h is related to our other operators by

(1.34) $$\Delta_h R_h = P_h \Delta.$$

For, by our definitions,

$$(\Delta_h R_h v, \chi) = -(\nabla R_h v, \nabla \chi) = -(\nabla v, \nabla \chi) = (\Delta v, \chi) = (P_h \Delta v, \chi), \ \forall \chi \in S_h.$$

With this notation the semidiscrete equation takes the form

$$(u_{h,t}, \chi) - (\Delta_h u_h, \chi) = (P_h f, \chi), \quad \forall \chi \in S_h, \quad t > 0,$$

and thus, since the factors on the left are all in S_h, (1.21) may be written as

(1.35) $$u_{h,t} - \Delta_h u_h = P_h f, \quad \text{for } t > 0, \quad \text{with } u_h(0) = v_h.$$

Using (1.34) we hence obtain for θ

$$\begin{aligned}\theta_t - \Delta_h \theta &= (u_{h,t} - \Delta_h u_h) - (R_h u_t - \Delta_h R_h u) \\ &= P_h f + (P_h - R_h) u_t - P_h (u_t - \Delta u) = P_h (I - R_h) u_t = -P_h \rho_t,\end{aligned}$$

or

(1.36) $$\theta_t - \Delta_h \theta = -P_h \rho_t, \quad \text{for } t > 0, \quad \text{with } \theta(0) = v_h - R_h v.$$

We now denote by $E_h(t)$ the discrete analogue of the operator $E(t)$ introduced above, the solution operator of the homogeneous semidiscrete problem (1.35). The analogue of (1.32), together with (1.36), then shows

(1.37) $$\theta(t) = E_h(t)\theta(0) - \int_0^t E_h(t-s) P_h \rho_t(s)\, ds.$$

We now note that $E_h(t)$ is stable in L_2, or, more precisely, as in the proof of (1.31),

(1.38) $$\|E_h(t) v_h\| \le e^{-\lambda_1 t} \|v_h\| \le \|v_h\|, \quad \text{for } v_h \in S_h, \ t \ge 0.$$

Since obviously P_h has unit norm in L_2, (1.37) implies (1.29), from which Theorem 1.2 follows as above. The desired estimate for θ is thus a consequence of the stability estimate for $E_h(t)$ combined with the error estimate for the elliptic problem applied to $\rho_t = (R_h - I) u_t$.

In a similar way we may prove the following estimate for the error in the gradient.

Theorem 1.3 *Under the assumptions of Theorem 1.2 we have*

$$\begin{aligned}\|\nabla u_h(t) - \nabla u(t)\| &\le C \|\nabla v_h - \nabla v\| \\ &\quad + C h^{r-1} \Big(\|v\|_r + \|u(t)\|_r + \big(\int_0^t \|u_t\|_{r-1}^2\, ds \big)^{1/2} \Big), \quad \textit{for } t \ge 0.\end{aligned}$$

Proof. As before we write the error in the form (1.23). Here, by Lemma 1.1,

$$\|\nabla\rho(t)\| = \|\nabla(R_h u(t) - u(t))\| \le Ch^{r-1}\|u(t)\|_r.$$

In order to estimate $\nabla\theta$, we use again (1.27), now with $\chi = \theta_t$. We obtain

$$\|\theta_t\|^2 + \tfrac{1}{2}\frac{d}{dt}\|\nabla\theta\|^2 = -(\rho_t, \theta_t) \le \tfrac{1}{2}\|\rho_t\|^2 + \tfrac{1}{2}\|\theta_t\|^2,$$

so that $(d/dt)\|\nabla\theta\|^2 \le \|\rho_t\|^2$ or

$$\begin{aligned}\|\nabla\theta(t)\|^2 &\le \|\nabla\theta(0)\|^2 + \int_0^t \|\rho_t\|^2\,ds \\ &\le \big(\|\nabla(v_h - v)\| + \|\nabla(R_h v - v)\|\big)^2 + \int_0^t \|\rho_t\|^2\,ds.\end{aligned} \tag{1.39}$$

Hence, in view of Lemma 1.1,

$$\|\nabla\theta(t)\|^2 \le 2\|\nabla(v_h - v)\|^2 + Ch^{2r-2}\Big(\|v\|_r^2 + \int_0^t \|u_t\|_{r-1}^2\,ds\Big), \tag{1.40}$$

which completes the proof. □

Note that if $v_h = I_h v$ or $v_h = R_h v$, then, by Lemma 1.1 or (1.11), respectively, the first term on the right hand side in Theorem 1.3 is again bounded by the second.

In the case of a quasiuniform family of triangulations $\mathcal{T}_h$ of a plane domain, or, more generally, when the inverse estimate (1.12) holds, an estimate for the error in the gradient may also be obtained directly from the result of Theorem 1.2. In fact, we obtain then, for χ arbitrary in S_h,

$$\begin{aligned}\|\nabla u_h(t) - \nabla u(t)\| &\le \|\nabla(u_h(t) - \chi)\| + \|\nabla\chi - \nabla u(t)\| \\ &\le Ch^{-1}\|u_h(t) - \chi\| + \|\nabla\chi - \nabla u(t)\| \\ &\le Ch^{-1}\|u_h(t) - u(t)\| + Ch^{-1}(\|\chi - u(t)\| + h\|\nabla\chi - \nabla u(t)\|).\end{aligned} \tag{1.41}$$

Here, by our approximation assumption (1.10), we have, with suitable $\chi \in S_h$,

$$\|\chi - u(t)\| + h\|\nabla\chi - \nabla u(t)\| \le Ch^r\|u(t)\|_r,$$

and hence, bounding the first term on the right in (1.41) by Theorem 1.2, for the appropriate choice of χ,

$$\|\nabla u_h(t) - \nabla u(t)\| \le Ch^{r-1}\Big(\|v\|_r + \int_0^t \|u_t(s)\|_r\,ds\Big).$$

We make the following observation concerning the gradient of the term $\theta = u_h - R_h u$ in (1.23): Assume that we have chosen $v_h = R_h v$ so that $\theta(0) = 0$. Then, in addition to (1.40), we have from (1.39)

(1.42) $$\|\nabla\theta(t)\| \le C\Big(\int_0^t \|\rho_t\|^2\,ds\Big)^{1/2} \le Ch^r\Big(\int_0^t \|u_t\|_r^2\,ds\Big)^{1/2}.$$

Hence $\nabla\theta(t)$ is of order $O(h^r)$, whereas the gradient of the total error can only be $O(h^{r-1})$. Thus ∇u_h is a better approximation to $\nabla R_h u$ than is possible to ∇u. This is an example of a phenomenon which is sometimes referred to as *superconvergence* .

Because the formulation of Galerkin's method is posed in terms of L_2 inner products, the most natural error estimates are expressed in L_2-based norms. Error analyses in other norms have also been pursued in the literature, and for later reference we quote the following *maximum-norm error estimate*, for piecewise linear approximating functions in a plane domain Ω, see, e.g., [42]. Here we write $L_\infty = L_\infty(\Omega)$ and $W_\infty^r = W_\infty^r(\Omega)$, with

$$\|v\|_{L_\infty} = \sup_{x\in\Omega} |u(x)|, \quad \|v\|_{W_\infty^r} = \max_{|\alpha|\le r} \|D^\alpha v\|_{L_\infty}.$$

We note first that the error in the interpolant introduced above is second order also in maximum-norm, so that (cf. (1.9))

(1.43) $$\|I_h v - v\|_{L_\infty} \le Ch^2\|v\|_{W_\infty^2}, \quad \text{for } v \in W_\infty^2 \cap H_0^1.$$

The error estimate for the elliptic finite element problem is then the following.

Theorem 1.4 *Let $\Omega \subset \mathbb{R}^2$ and assume that S_h consists of piecewise linear finite element functions, and that the family $\mathcal{T}_h$ is quasiuniform. Let u_h and u be the solutions of* (1.13) *and* (1.2), *respectively. Then*

(1.44) $$\|u_h - u\|_{L_\infty} \le Ch^2\ell_h\|u\|_{W_\infty^2}, \quad \textit{where } \ell_h = \max(1, \log(1/h))$$

We note that, in view of (1.43), this error estimate is nonoptimal, but it has been shown, see Haverkamp [116], that the logarithmic factor in (1.44) cannot be removed. Note that although ℓ_h is unbounded for small h, it is of moderate size for realistic values of h.

Recall the definition (1.22) of the Ritz projection $R_h : H_0^1 \to S_h$, and its stability in H_0^1. When the family of triangulations is quasiuniform, this projection is known to have the almost maximum-norm stability property

(1.45) $$\|R_h v\|_{L_\infty} \le C\ell_h\|v\|_{L_\infty}.$$

The proof of this is relatively difficult, and will not be included here. We remark that in contrast to (1.24) and (1.45), R_h is not bounded in L_2. The error bound of Theorem 1.4 is now an easy consequence of this stability result and the interpolation error estimate of (1.43), since

$$\|R_h v - v\|_{L_\infty} \le \|R_h(v - I_h v)\|_{L_\infty} + \|I_h v - v\|_{L_\infty} \le Ch^2\ell_h\|v\|_{W_\infty^2}.$$

As a simple example of an application of the superconvergent order estimate (1.42), we shall indicate briefly how it may be used to show an essentially optimal order error bound for the parabolic problem in the maximum-norm. Consider thus the concrete situation described in the beginning of this chapter with Ω a plane smooth convex domain and S_h consisting of piecewise linear functions ($d = r = 2$) on quasiuniform triangulations of Ω. Then, by Theorem 1.4,

$$\|\rho(t)\|_{L_\infty} = \|R_h u(t) - u(t)\|_{L_\infty} \le Ch^2 \ell_h \|u(t)\|_{W^2_\infty}. \tag{1.46}$$

In two dimensions, Sobolev's inequality almost bounds the maximum-norm by the norm in H^1, and it may be shown (cf. Lemma 6.4 below) that for functions in the subspace S_h,

$$\|\chi\|_{L_\infty} \le C\ell_h^{1/2} \|\nabla\chi\|, \quad \forall \chi \in S_h.$$

Applied to θ this shows, by (1.42) (with $r = 2$), that

$$\|\theta(t)\|_{L_\infty} \le Ch^2 \ell_h^{1/2} \Big(\int_0^t \|u_t\|_2^2 \, ds \Big)^{1/2},$$

and we may thus conclude for the error in the parabolic problem that

$$\|u_h(t) - u(t)\|_{L_\infty} \le \|\rho(t)\|_{L_\infty} + \|\theta(t)\|_{L_\infty} \le C(t,u) h^2 \ell_h.$$

We now turn our attention to some simple schemes for discretization also with respect to the time variable. We introduce a time step k and the time levels $t = t_n = nk$, where n is a nonnegative integer, and denote by $U^n = U_h^n \in S_h$ the approximation of $u(t_n)$ to be determined.

We begin by the *backward Euler Galerkin method*, which is defined by replacing the time derivative in (1.21) by a backward difference quotient, or, if $\bar\partial U^n = (U^n - U^{n-1})/k$,

$$(\bar\partial U^n, \chi) + (\nabla U^n, \nabla \chi) = (f(t_n), \chi), \quad \forall \chi \in S_h, \ n \ge 1, \quad U^0 = v_h. \tag{1.47}$$

For U^{n-1} given this defines U^n implicitly from the equation

$$(U^n, \chi) + k(\nabla U^n, \nabla \chi) = (U^{n-1} + kf(t_n), \chi), \quad \forall \chi \in S_h,$$

which is the finite element formulation of an elliptic equation of the form $(I - k\Delta)u = g$. With matrix notation as in the semidiscrete situation, this may be written

$$(\mathcal{B} + k\mathcal{A})\alpha^n = \mathcal{B}\alpha^{n-1} + k\widetilde{f}(t_n),$$

where $\mathcal{B} + k\mathcal{A}$ is positive definite and hence, in particular, invertible.

We shall prove the following error estimate.

Theorem 1.5 *With U^n and u the solutions of* (1.47) *and* (1.1), *respectively, we have, if $\|v_h - v\| \le Ch^r\|v\|_r$ and $v = 0$ on $\partial\Omega$,*

$$\|U^n - u(t_n)\| \le Ch^r\big(\|v\|_r + \int_0^{t_n} \|u_t\|_r\,ds\big) + k\int_0^{t_n} \|u_{tt}\|\,ds, \quad \textit{for } n \ge 0.$$

Proof. In analogy with (1.23) we write

$$U^n - u(t_n) = (U^n - R_h u(t_n)) + (R_h u(t_n) - u(t_n)) = \theta^n + \rho^n,$$

and here $\rho^n = \rho(t_n)$ is bounded as claimed in (1.25). This time, a calculation corresponding to (1.26) yields

$$(1.48) \qquad (\bar\partial\theta^n, \chi) + (\nabla\theta^n, \nabla\chi) = -(\omega^n, \chi), \quad \forall \chi \in S_h,\ n \ge 1,$$

where

$$\omega^n = R_h\bar\partial u(t_n) - u_t(t_n) = (R_h - I)\bar\partial u(t_n) + (\bar\partial u(t_n) - u_t(t_n)) = \omega_1^n + \omega_2^n.$$

Choosing $\chi = \theta^n$ in (1.48), we have $(\bar\partial\theta^n, \theta^n) \le \|\omega^n\|\,\|\theta^n\|$, or

$$\|\theta^n\|^2 - (\theta^{n-1}, \theta^n) \le k\|\omega^n\|\,\|\theta^n\|,$$

so that

$$(1.49) \qquad \|\theta^n\| \le \|\theta^{n-1}\| + k\|\omega^n\|,$$

and, by repeated application,

$$(1.50) \qquad \|\theta^n\| \le \|\theta^0\| + k\sum_{j=1}^n \|\omega^j\| \le \|\theta^0\| + k\sum_{j=1}^n \|\omega_1^j\| + k\sum_{j=1}^n \|\omega_2^j\|.$$

Here, as before, $\theta^0 = \theta(0)$ is bounded as desired. We write

$$(1.51) \qquad \omega_1^j = (R_h - I)k^{-1}\int_{t_{j-1}}^{t_j} u_t\,ds = k^{-1}\int_{t_{j-1}}^{t_j} (R_h - I)u_t\,ds,$$

and obtain

$$k\sum_{j=1}^n \|\omega_1^j\| \le \sum_{j=1}^n \int_{t_{j-1}}^{t_j} Ch^r\|u_t\|_r\,ds = Ch^r\int_0^{t_n} \|u_t\|_r\,ds.$$

Further,

$$(1.52) \qquad k\,\omega_2^j = u(t_j) - u(t_{j-1}) - ku_t(t_j) = -\int_{t_{j-1}}^{t_j} (s - t_{j-1})u_{tt}(s)\,ds,$$

so that

$$k\sum_{j=1}^n \|\omega_2^j\| \le \sum_{j=1}^n \|\int_{t_{j-1}}^{t_j} (s - t_{j-1})u_{tt}(s)\,ds\| \le k\int_0^{t_n} \|u_{tt}\|\,ds.$$

Together our estimates complete the proof of the theorem. □

In order to show an estimate for $\nabla\theta^n$ we may choose instead $\chi = \bar\partial\theta^n$ in (1.48) to obtain $\bar\partial\|\nabla\theta^n\|^2 \le \|\omega^n\|^2$, or, if $\nabla\theta^0 = 0$,

$$\|\nabla\theta^n\|^2 \le k\sum_{j=1}^{n}\|\omega^j\|^2 \le Ch^{2s}\int_0^{t_n}\|u_t\|_s^2\,dt + Ck^2\int_0^{t_n}\|u_{tt}\|^2\,dt, \tag{1.53}$$

for $1 \le s \le r$. Together with the standard estimate for $\nabla\rho$ this shows, with $s = r-1$ in (1.53),

$$\|\nabla(U^n - u(t_n))\| \le C(u)(h^{r-1} + k).$$

If one uses Theorem 1.5 together with the inverse inequality (1.12) one now obtains the weaker estimate $\|\nabla(U^n - u(t_n))\| \le C(u)(h^{r-1} + kh^{-1})$. We also note that with $s = r$ in (1.53) one may conclude the maximum-norm estimate

$$\|U^n - u(t_n)\|_{L_\infty} \le C(u)\ell_h(h^r + k).$$

Note that because of the nonsymmetric choice of the discretization in time, the backward Euler Galerkin method is only first order in k. We therefore now turn to the *Crank-Nicolson Galerkin method.* Here the semi-discrete equation is discretized in a symmetric fashion around the point $t_{n-\frac12} = (n-\frac12)k$, which will produce a second order accurate method in time. More precisely, we set $\widehat{U}^n = \frac12(U^n + U^{n-1})$ and define $U^n \in S_h$ by

$$(\bar\partial U^n, \chi) + (\nabla\widehat{U}^n, \nabla\chi) = (f(t_{n-\frac12}), \chi), \quad \forall\chi \in S_h, \quad \text{for } n \ge 1, \tag{1.54}$$

with $U^0 = v_h$. Here the equation for U^n may be written in matrix form as

$$(\mathcal{B} + \tfrac12 k\mathcal{A})\alpha^n = (\mathcal{B} - \tfrac12 k\mathcal{A})\alpha^{n-1} + k\widetilde{f}(t_{n-\frac12}),$$

with a positive definite matrix $\mathcal{B} + \frac12 k\mathcal{A}$. Now the error estimate reads as follows.

Theorem 1.6 *Let U^n and u be the solutions of* (1.54) *and* (1.1), *respectively, and let $\|v_h - v\| \le Ch^r\|v\|_r$ and $v = 0$ on $\partial\Omega$. Then we have, for $n \ge 0$,*

$$\|U^n - u(t_n)\| \le Ch^r\Big(\|v\|_r + \int_0^{t_n}\|u_t\|_r\,ds\Big) + Ck^2\int_0^{t_n}(\|u_{ttt}\| + \|\Delta u_{tt}\|)\,ds.$$

Proof. With ρ^n bounded as above, we only need to consider θ^n. We have

$$(\bar\partial\theta^n, \chi) + (\nabla\widehat{\theta}^n, \nabla\chi) = -(\omega^n, \chi), \quad \text{for } \chi \in S_h,\ n \ge 1, \tag{1.55}$$

where now

$$\omega^n = (R_h - I)\bar\partial u(t_n) + (\bar\partial u(t_n) - u_t(t_{n-\frac12})) + \Delta\big(u(t_{n-\frac12}) - \tfrac12(u(t_n) + u(t_{n-1}))\big) = \omega_1^n + \omega_2^n + \omega_3^n. \tag{1.56}$$

Choosing this time $\chi = \widehat{\theta}^n$ in (1.55), we find

$$(\bar\partial\theta^n, \widehat{\theta}^n) \le \tfrac12\|\omega^n\|(\|\theta^n\| + \|\theta^{n-1}\|),$$

or

$$\|\theta^n\|^2 - \|\theta^{n-1}\|^2 \le k\|\omega^n\|(\|\theta^n\| + \|\theta^{n-1}\|),$$

so that, after cancellation of a common factor,

$$\|\theta^n\| \le \|\theta^{n-1}\| + k\|\omega^n\|, \quad \text{for } n \ge 1.$$

After repeated application this yields

$$\|\theta^n\| \le \|\theta^0\| + k\sum_{j=1}^n \big(\|\omega_1^j\| + \|\omega_2^j\| + \|\omega_3^j\|\big).$$

With θ^0 and ω_1^j estimated as before, it remains to bound the terms in ω_2^j and ω_3^j. We have

$$\begin{aligned} k\|\omega_2^j\| &= \|u(t_j) - u(t_{j-1}) - ku_t(t_{j-\frac12})\| \\ &= \tfrac12\Big\|\int_{t_{j-1}}^{t_{j-\frac12}} (s - t_{j-1})^2 u_{ttt}(s)\,ds + \int_{t_{j-\frac12}}^{t_j} (s - t_j)^2 u_{ttt}(s)\,ds\Big\| \\ &\le Ck^2\int_{t_{j-1}}^{t_j} \|u_{ttt}\|\,ds, \end{aligned}$$

and similarly,

$$k\|\omega_3^j\| = k\|\Delta\big(u(t_{j-\frac12}) - \tfrac12(u(t_j) + u(t_{j-1}))\big)\| \le Ck^2\int_{t_{j-1}}^{t_j}\|\Delta u_{tt}\|\,ds.$$

Altogether,

$$k\sum_{j=1}^n(\|\omega_2^j\| + \|\omega_3^j\|) \le Ck^2\int_0^{t_n}(\|u_{ttt}\| + \|\Delta u_{tt}\|)\,ds, \tag{1.57}$$

which completes the proof. □

Another way to attain second order accuracy in the discretization in time is to approximate the time derivative in the differential equation by a second order backward difference quotient. Setting

$$\bar{D}U^n = \bar{\partial}U^n + \tfrac{1}{2}k\bar{\partial}^2U^n = (\tfrac{3}{2}U^n - 2U^{n-1} + \tfrac{1}{2}U^{n-2})/k,$$

we have at once by Taylor expansion, for a smooth function u,

$$\bar{D}u(t_n) = u_t(t_n) + O(k^2), \quad \text{as } k \to 0.$$

We therefore pose the discrete problem

$$(\bar{D}U^n, \chi) + (\nabla U^n, \nabla\chi) = (f(t_n), \chi), \quad \forall\chi \in S_h,\ n \geq 2. \tag{1.58}$$

Note that for n fixed this equation employs three time levels rather than the two of our previous methods. We therefore have to restrict its use to $n \geq 2$, because we do not want to use U^n with n negative. With $U^0 = v_h$ given, we then also need to define U^1 in some way, and we choose to do so by employing one step of the backward Euler method, i.e., we set

$$(\bar{\partial}U^1, \chi) + (\nabla U^1, \nabla\chi) = (f(t_1), \chi), \quad \forall\chi \in S_h. \tag{1.59}$$

We note that in our earlier matrix notation, (1.58) may be written as

$$(\tfrac{3}{2}\mathcal{B} + k\mathcal{A})\alpha^n = 2\mathcal{B}\alpha^{n-1} - \tfrac{1}{2}\mathcal{B}\alpha^{n-2} + k\widetilde{f}(t_n), \quad \text{for } n \geq 2,$$

with the matrix coefficient of α^n again positive definite.

We have this time the following $O(h^r + k^2)$ error estimate.

Theorem 1.7 *Let U^n and u be the solutions of* (1.58) *and* (1.1)*, with $U^0 = v_h$ and U^1 defined by* (1.59)*. Then, if $\|v_h - v\| \leq Ch^r\|v\|_r$ and $v = 0$ on $\partial\Omega$, we have*

$$\begin{aligned}\|U^n - u(t_n)\| \leq Ch^r\Big(\|v\|_r + \int_0^{t_n} \|u_t\|_r\, ds\Big)& \\ + Ck\int_0^k \|u_{tt}\|\, ds + Ck^2\int_0^{t_n} \|u_{ttt}\|\, ds,& \quad \textit{for } n \geq 0.\end{aligned}$$

Proof. Writing again $U^n - u(t_n) = \theta^n + \rho^n$ we only need to bound θ^n, which now satisfies

$$\begin{aligned}(\bar{D}\theta^n, \chi) + (\nabla\theta^n, \nabla\chi) &= -(\omega^n, \chi), \quad \text{for } n \geq 2, \\ (\bar{\partial}\theta^1, \chi) + (\nabla\theta^1, \nabla\chi) &= -(\omega^1, \chi),\end{aligned} \tag{1.60}$$

where

$$\omega^n = \bar{D}R_hu^n - u_t^n = (R_h - I)\bar{D}u^n + (\bar{D}u^n - u_t^n) = \omega_1^n + \omega_2^n, \quad n \geq 2,$$
$$\omega^1 = (R_h - I)\bar{\partial}u^1 + (\bar{\partial}u^1 - u_t^1) = \omega_1^1 + \omega_2^1.$$

We shall show the inequality

$$\|\theta^n\| \le \|\theta^0\| + 2k\sum_{j=2}^{n} \|\omega^j\| + \tfrac{5}{2}k\|\omega^1\|, \quad \text{for } n \ge 1. \tag{1.61}$$

Assuming this for a moment, we need to bound the errors ω_1^j and ω_2^j. Using Taylor expansions with the appropriate remainder terms in integral form we find easily, for $j \ge 2$,

$$k\|\omega_1^j\| \le Ch^r k\|\bar{D}u^j\|_r \le Ch^r \int_{t_{j-2}}^{t_j} \|u_t\|_r\, ds, \quad k\|\omega_2^j\| \le Ck^2 \int_{t_{j-2}}^{t_j} \|u_{ttt}\|\, ds.$$

As for the backward Euler method we have

$$k\|\omega_1^1\| + k\|\omega_2^1\| \le Ch^r \int_0^k \|u_t\|_r\, ds + k\int_0^k \|u_{tt}\|\, ds,$$

and we hence conclude

$$k\sum_{j=1}^{n} \|\omega^j\| \le Ch^r \int_0^{t_n} \|u_t\|_r\, ds + k\int_0^k \|u_{tt}\|\, ds + Ck^2 \int_0^{t_n} \|u_{ttt}\|\, ds.$$

Together with our earlier estimate for θ^0, this completes the proof of the estimate for θ^n and thus of the theorem.

It remains to show (1.61). Introducing the difference operators $\delta_l\theta^n = \theta^n - \theta^{n-l}$ for $l = 1, 2$, we may write $k\bar{D}\theta^n = 2\delta_1\theta^n - \frac{1}{2}\delta_2\theta^n$. Since $2(\delta_l\theta^n, \theta^n) = \delta_l\|\theta^n\|^2 + \|\delta_l\theta^n\|^2$, we therefore have

$$k(\bar{D}\theta^n, \theta^n) = \delta_1\|\theta^n\|^2 - \tfrac{1}{4}\delta_2\|\theta^n\|^2 + \|\delta_1\theta^n\|^2 - \tfrac{1}{4}\|\delta_2\theta^n\|^2, \text{ for } n \ge 2.$$

Replacing n by j and then summing from 2 to n, we have

$$\sum_{j=2}^{n}(\delta_1\|\theta^j\|^2 - \tfrac{1}{4}\delta_2\|\theta^j\|^2) = \tfrac{3}{4}\|\theta^n\|^2 - \tfrac{1}{4}\|\theta^{n-1}\|^2 - \tfrac{3}{4}\|\theta^1\|^2 + \tfrac{1}{4}\|\theta^0\|^2,$$

and further, since $\delta_2\theta^n = \delta_1\theta^n + \delta_1\theta^{n-1}$, we obtain

$$\begin{aligned}\sum_{j=2}^{n}(\|\delta_1\theta^j\|^2 - \tfrac{1}{4}\|\delta_2\theta^j\|^2) &\ge \sum_{j=2}^{n}\big(\|\delta_1\theta^j\|^2 - \tfrac{1}{4}(\|\delta_1\theta^j\| + \|\delta_1\theta^{j-1}\|)^2\big)\\ &\ge \tfrac{1}{2}\sum_{j=2}^{n}(\|\delta_1\theta^j\|^2 - \|\delta_1\theta^{j-1}\|^2) = \tfrac{1}{2}(\|\delta_1\theta^n\|^2 - \|\delta_1\theta^1\|^2).\end{aligned}$$

Hence,

$$
\begin{aligned}
&k(\bar\partial\theta^1,\theta^1)+k\sum_{j=2}^{n}(\bar D\theta^j,\theta^j)\\
(1.62)\qquad &\ge \tfrac12\big(\|\theta^1\|^2-\|\theta^0\|^2+\|\delta_1\theta^1\|^2\big)+\tfrac12\big(\|\delta_1\theta^n\|^2-\|\delta_1\theta^1\|^2\big)\\
&\quad+\big(\tfrac34\|\theta^n\|^2-\tfrac14\|\theta^{n-1}\|^2-\tfrac34\|\theta^1\|^2+\tfrac14\|\theta^0\|^2\big)\\
&\ge \tfrac34\|\theta^n\|^2-\tfrac14\|\theta^{n-1}\|^2-\tfrac14\|\theta^1\|^2-\tfrac14\|\theta^0\|^2.
\end{aligned}
$$

But by (1.60) with $\chi=\theta^n$ we have

$$k(\bar\partial\theta^1,\theta^1)+k\sum_{j=2}^{n}(\bar D\theta^j,\theta^j)+k\sum_{j=1}^{n}(\nabla\theta^j,\nabla\theta^j)=-k\sum_{j=1}^{n}(\omega^j,\theta^j),$$

and by (1.62) this yields

$$\|\theta^n\|^2\le\tfrac13\big(\|\theta^{n-1}\|^2+\|\theta^1\|^2+\|\theta^0\|^2\big)+\tfrac43k\sum_{j=1}^{n}\|\omega^j\|\,\|\theta^j\|.$$

Suppose m is chosen so that $\|\theta^m\|=\max_{0\le j\le n}\|\theta^j\|$. Then

$$\|\theta^m\|^2\le\tfrac13\Big(\|\theta^m\|+\|\theta^1\|+\|\theta^0\|+4k\sum_{j=1}^{n}\|\omega^j\|\Big)\|\theta^m\|,$$

whence

$$\|\theta^n\|\le\|\theta^m\|\le\tfrac12(\|\theta^1\|+\|\theta^0\|)+2k\sum_{j=1}^{n}\|\omega^j\|.$$

Since, as follows from above, $\|\theta^1\|\le\|\theta^0\|+k\|\omega^1\|$, this completes the proof of (1.61) and thus of the theorem. □

In the above time discretization schemes we have used a constant time step k. We shall close this introductory discussion of fully discrete methods with an example of a variable time step version of the backward Euler method.

Let thus $0=t_0<t_1<\cdots<t_n<\cdots$ be a partition of the positive time axis and set $k_n=t_n-t_{n-1}$. We may then consider the approximation U^n of $u(t_n)$ defined by

$$(\bar\partial_n U^n,\chi)+(\nabla U^n,\nabla\chi)=(f(t_n),\chi),\quad\forall\chi\in S_h,\ n\ge1,\tag{1.63}$$

with $U^0=v_h$, where $\bar\partial_n U^n=(U^n-U^{n-1})/k_n$. We have the following error estimate which reduces to that of Theorem 1.5 for constant time steps.

Theorem 1.8 *Let U^n and u be the solutions of (1.63) and (1.1), with $U^0=v_h$ such that $\|v_h-v\|\le Ch^r\|v\|_r$ and $v=0$ on $\partial\Omega$. Then we have for $n\ge0$*

$$\|U^n-u(t_n)\|\le Ch^r\Big(\|v\|_r+\int_0^{t_n}\|u_t\|_r\,ds\Big)+\sum_{j=1}^{n}k_j\int_{t_{j-1}}^{t_j}\|u_{tt}\|\,ds.$$

Proof. This time we have for θ^n,

$$(\bar\partial_n\theta^n,\chi)+(\nabla\theta^n,\nabla\chi)=-(\omega^n,\chi),\quad \forall\chi\in S_h,\ n\geq 1,$$

where now

$$\omega^n=(R_h-I)\bar\partial_n u^n+(\bar\partial_n u^n-u_t^n)=\omega_1^n+\omega_2^n.$$

Referring to the proof of Theorem 1.5, (1.49) will be replaced by $\|\theta^n\|\leq\|\theta^{n-1}\|+k_n\|\omega^n\|$, and hence (1.50) by

$$\|\theta^n\|\leq\|\theta^0\|+\sum_{j=1}^n k_j(\|\omega_1^j\|+\|\omega_2^j\|).$$

Now

$$\sum_{j=1}^n k_j\|\omega_1^j\|\leq\sum_{j=1}^n\int_{t_{j-1}}^{t_j}Ch^r\|u_t\|_r\,ds=Ch^r\int_0^{t_n}\|u_t\|_r\,ds,$$

and, since (1.52) still holds, with k replaced by k_j,

$$\sum_{j=1}^n k_j\|\omega_2^j\|\leq\sum_{j=1}^n\Big\|\int_{t_{j-1}}^{t_j}(s-t_{j-1})u_{tt}(s)\,ds\Big\|\leq\sum_{j=1}^n k_j\int_{t_{j-1}}^{t_j}\|u_{tt}\|\,ds.$$

Together with the standard estimates for ρ^n and θ^0, this completes the proof of the theorem. □

We note that the form of the error bound in Theorem 1.8 suggests using shorter time steps when $\|u_{tt}\|$ is larger. We shall return to such considerations in later chapters.

We complete this introductory chapter with some short remarks about other initial boundary value problems for the heat equation than (1.1), and consider first a simple situation with Neumann rather than Dirichlet boundary conditions. Consider thus instead of (1.1) the initial boundary value problem

$$\begin{aligned}&u_t-\Delta u+u=f\quad\text{in }\Omega,\qquad\text{for }t>0,\\&\frac{\partial u}{\partial n}=0\quad\text{on }\partial\Omega,\quad\text{for }t>0,\quad u(\cdot,0)=v\quad\text{in }\Omega,\end{aligned}\tag{1.64}$$

where $\partial u/\partial n$ denotes the derivative in the direction of the exterior normal to $\partial\Omega$. The corresponding stationary problem is then

$$-\Delta u+u=f\quad\text{in }\Omega,\quad\text{with }\frac{\partial u}{\partial n}=0\quad\text{on }\partial\Omega.\tag{1.65}$$

In order to formulate this in variational form, we now multiply by $\varphi\in H^1$, thus without requiring $\varphi=0$ on $\partial\Omega$, integrate over Ω, and use Green's formula to obtain

$$(\nabla u, \nabla \varphi) + (u, \varphi) = (f, \varphi), \quad \forall \varphi \in H^1.$$

We note that if u is smooth, this in turn shows

$$(-\Delta u + u, \varphi) + \int_{\partial\Omega} \frac{\partial u}{\partial n} \varphi \, ds = (f, \varphi), \quad \forall \varphi \in H^1,$$

from which (1.65) follows since φ is arbitrary. In particular, the boundary condition is now a consequence of the variational formulation, in contrast to our earlier discussion when the boundary condition was enforced by looking for a solution in H_0^1. We therefore say that $\partial u/\partial n = 0$ is a *natural boundary condition*, whereas the Dirichlet boundary condition is referred to as an *essential boundary condition*. The lower order term in the differential equation was included to make (1.65) uniquely solvable; note that $\lambda = 0$ is an eigenvalue of $-\Delta$ under Neumann boundary conditions since $\Delta 1 \equiv 0$, whereas $-\Delta + I$ is positive definite.

From the above variational formulation it is natural to assume now that the approximating space S_h is a subspace of H^1, without requiring its elements to vanish on $\partial\Omega$, and safisfies (1.10) when $v \in H^s$. The discrete stationary problem is then

$$(\nabla u_h, \nabla \chi) + (u_h, \chi) = (f, \chi), \quad \forall \chi \in S_h,$$

and this may be analyzed as in Theorem 1.1. The corresponding spatially discrete version of (1.64) is

$$(u_{h,t}, \chi) + (\nabla u_h, \nabla \chi) + (u_h, \chi) = (f, \chi), \ \forall \chi \in S_h, \ t > 0, \quad u_h(0) = v_h,$$

and the analysis of this method, and also of corresponding fully discrete ones, follow the same lines as in the case of Dirichlet boundary conditions.

We also mention the time periodic boundary value problem

$$(1.66) \qquad \begin{aligned} u_t - \Delta u &= f \quad \text{in } \Omega, \qquad \text{for } 0 < t < \omega, \\ u &= 0 \quad \text{on } \partial\Omega, \quad \text{for } 0 < t < \omega, \quad u(\cdot, 0) = u(\cdot, \omega) \quad \text{in } \Omega, \end{aligned}$$

where $\omega > 0$ is the period. Setting $u(0) = v$ we have by Duhamel's principle for a possible solution

$$v = u(\omega) = E(\omega)v + \int_0^\omega E(\omega - s) f(s) \, ds,$$

and since $\|E(\omega)\| < 1$ by (1.38), this equation has a unique solution v. Once v is known, (1.66) may be solved as an initial value problem. Spatially semidiscrete and fully discrete versions of the problem may be formulated in obvious ways and analyzed by the techniques developed here.

The *finite element method* originated in the engineering literature in the 1950s, when structural engineers combined the well established framework analysis with *variational methods* in continuum mechanics into a discretization method in which a structure is thought of as consisting of *elements* with locally defined strains or stresses; a standard reference from the engineering literature is Zienkiewicz [249]. In the mid 1960s, a number of papers appeared independently in the numerical analysis literature which were concerned with the construction and analysis of *finite difference* schemes for elliptic problems by *variational principles*, e.g., Céa [45], Demjanovič [68], Feng [98], Friedrichs and Keller [101], and Oganesjan and Ruchovets [187]. By considering approximating functions as defined at all points rather than at meshpoints, the mathematical theory of finite elements then became established through contributions such as Birkhoff, Schultz and Varga [27], where the theory of splines was brought to bear on the development, and Zlámal [250], with the first stringent error analysis of more complicated elements. The duality argument for the L_2 error estimate quoted in Theorem 1.1 was developed independently by Aubin [7], Nitsche [179] and Oganesjan and Ruchovets [188], and later maximum-norm error estimates such as (1.44) were shown by Scott [214], Natterer [175], and Nitsche [182], see Schatz and Wahlbin [208]. The sharpness of this estimate, with the logarithmic factor, was shown in Haverkamp [116].

General treatments of the mathematics of the finite element method for *elliptic* problems can be found in textbooks such as, e.g., Babuška and Aziz [11], Strang and Fix [221], Ciarlet [51] and Brenner and Scott [42], and we shall sometimes quote these for background material.

The development of the theory of finite elements for *parabolic* problems started around 1970. At this time finite difference analysis for such problems had reached a high level of refinement after the fundamental 1928 paper by Courant, Friedrichs and Lewy [52], and became the background and starting point for the finite element analysis of such problems. Names of particular distinction in the development of finite differences in the 50s and 60s are, e.g., F. John, D. G. Aronson, H. O. Kreiss, O. B. Widlund, J. Douglas, Jr., and collaborators, Russian researchers such as Samarskii, etc. (cf. the survey paper Thomée [230]).

The material presented in this introductary chapter is standard; some early references are Douglas and Dupont [74], Price and Varga [196] and Fix and Nassif [99]. An important step in the development was the introduction and exploitation by Wheeler [246] of the Ritz projection, which made it possible to improve earlier suboptimal L_2-norm error estimates to optimal order ones. The nucleus of the present survey is Thomée [229]. Several of the topics that have been touched upon only lightly in this chapter will be developed in more detail in the rest of the book where we will consider both more general equations and wider classes of discretization methods, as well as more detailed investigations of the dependence of the error bounds on the regularity of the exact solutions of our problems. Concerning the discretization of the time-periodic problem mentioned at the end, see Carasso [44], Bernardi [26], and Hansbo [114].

For standard material concerning the mathematical treatment of elliptic and parabolic differential equations we refer to Evans [96], cf. also Lions and Magenes [156] and, for parabolic equations, Friedman [100].

2. Methods Based on More General Approximations of the Elliptic Problem

In our above discussion of finite element approximation of the parabolic problem, the discretization in space was based on using a family of finite-dimensional spaces $S_h \subset H_0^1 = H_0^1(\Omega)$, such that, for some $r \geq 2$, the approximation property (1.10) holds. The most natural example of such a family in a plane domain Ω is to take for S_h the continuous functions which reduce to polynomials of degree at most $r-1$ on the triangles τ of a triangulation $\mathcal{T}_h$ of Ω of the type described in the beginning of Chapter 1, and which vanish on $\partial\Omega$. However, for $r > 2$ and in the case of a domain with smooth boundary, it is not possible, in general, to satisfy the homogeneous boundary conditions exactly for this choice. This difficulty occurs, of course, already for the elliptic problem, and several methods have been suggested to deal with it. In this chapter we shall consider, as a typical example, a method which was proposed by Nitsche for this purpose. This will serve as background for our subsequent discussion of the discretization of the parabolic problem. Another example, a so called mixed method, will be considered in Chapter 17 below.

Consider thus, with Ω a plane domain with smooth boundary, the Dirichlet problem

$$-\Delta u = f \quad \text{in } \Omega, \quad \text{with } u = 0 \quad \text{on } \partial\Omega. \tag{2.1}$$

Let now the $\mathcal{T}_h = \{\tau_j\}_{j=1}^{M_h}$ belong to a family of quasiuniform triangulations of Ω, with $\max_j \operatorname{diam}(\tau_j) \leq h$, where the boundary triangles are allowed to have one curved edge along $\partial\Omega$, and let S_h denote the finite-dimensional linear space of continuous functions on $\bar{\Omega}$ which reduce to polynomials of degree $\leq r-1$ on each triangle τ_j, without any boundary conditions imposed at $\partial\Omega$, i.e.,

$$S_h = \{\chi \in \mathcal{C}(\bar{\Omega});\ \chi\big|_{\tau_j} \in \Pi_{r-1}\}, \tag{2.2}$$

where Π_s denotes the set of polynomials of degree at most s.

In addition to the inner product in $L_2 = L_2(\Omega)$ we set

$$\langle \varphi, \psi \rangle = \int_{\partial\Omega} \varphi\psi\, ds, \quad \text{and} \quad |\varphi| = \langle \varphi, \varphi \rangle^{1/2} = \|\varphi\|_{L_2(\partial\Omega)},$$

and introduce the bilinear form

(2.3) $$N_\gamma(\varphi,\psi) = (\nabla\varphi,\nabla\psi) - \langle\frac{\partial\varphi}{\partial n},\psi\rangle - \langle\varphi,\frac{\partial\psi}{\partial n}\rangle + \gamma h^{-1}\langle\varphi,\psi\rangle,$$

where γ is a positive constant to be fixed later and $\partial/\partial n$ denotes differentiation in the direction of the exterior normal to $\partial\Omega$.

Now let u be a solution of our Dirichlet problem (2.1). Then, using Green's formula, we have, since u vanishes on $\partial\Omega$,

(2.4) $$\begin{aligned} N_\gamma(u,\chi) &= (\nabla u,\nabla\chi) - \langle\frac{\partial u}{\partial n},\chi\rangle - \langle u,\frac{\partial\chi}{\partial n}\rangle + \gamma h^{-1}\langle u,\chi\rangle \\ &= -(\Delta u,\chi) = (f,\chi), \quad \text{for } \chi\in S_h. \end{aligned}$$

With this in mind we define Nitsche's method for (2.1) to find $u_h \in S_h$ satisfying the variational equation

(2.5) $$N_\gamma(u_h,\chi) = (f,\chi), \quad \forall\chi\in S_h.$$

We shall demonstrate below that if γ is appropriately chosen, then this problem admits a unique solution for which optimal order error estimates hold.

For our analysis we introduce, for φ appropriately smooth, the norm

$$|||\varphi||| = \big(\|\nabla\varphi\|^2 + h\Big|\frac{\partial\varphi}{\partial n}\Big|^2 + h^{-1}|\varphi|^2\big)^{1/2}.$$

We first note the following inverse property.

Lemma 2.1 *There is a constant C independent of h such that*

$$|||\chi||| \le Ch^{-1}\|\chi\|, \quad \forall\chi\in S_h.$$

Proof. Because of the quasiuniformity of the family of triangulations $\mathcal{T}_h$, $\nabla\chi$ is estimated by (1.12). Further,

(2.6) $$\Big|\frac{\partial\chi}{\partial n}\Big|^2 \le C_0h^{-1}\|\nabla\chi\|^2, \quad \forall\chi\in S_h.$$

This follows easily by using for each boundary triangle τ_j a linear transformation to map it onto a unit size reference triangle $\widetilde{\tau}_j$ with vertices $(0,0)$, $(1,0)$, and $(0,1)$, say, with the curved edge between $(0,1)$ and $(1,0)$, and noting that here $\|\eta\|_{L_2(\partial\widetilde{\tau}_j)} \le C\|\eta\|_{L_2(\widetilde{\tau}_j)}$ for $\eta = \partial\chi/\partial x_i$, since the right hand side is a norm on Π_{r-2}. Using the inverse of the linear transformation to map $\widetilde{\tau}_j$ back to τ_j, we obtain $\|\partial\chi/\partial x_i\|^2_{L_2(\partial\tau_j)} \le Ch^{-1}\|\partial\chi/\partial x_i\|^2_{L_2(\tau_j)}$, and (2.6) follows by summation over the boundary triangles. Using also (1.12) this bounds $\partial\chi/\partial n$ in the desired way. Finally, in the same way, $|\chi|^2 \le C_0h^{-1}\|\chi\|^2$ for $\chi\in S_h$. Together these estimates show the lemma. □

We now show that the bilinear form $N_\gamma(\cdot,\cdot)$ defined in (2.3) is continuous in terms of $|||\cdot|||$ and positive definite when restricted to S_h.

Lemma 2.2 *We have, for γ fixed and for φ, ψ appropriately smooth,*

$$|N_\gamma(\varphi, \psi)| \leq C|||\varphi|||\, C|||\psi|||$$

and, with S_h defined in (2.2), *there exist positive numbers γ_0 and c such that*

$$N_\gamma(\chi, \chi) \geq c|||\chi|||^2, \quad \forall \chi \in S_h, \quad \textit{for } \gamma \geq \gamma_0. \tag{2.7}$$

Proof. The first part of the lemma is obvious from our definitions. For the second part we use (2.6) to obtain

$$\begin{aligned}
N_\gamma(\chi, \chi) &= \|\nabla\chi\|^2 - 2\langle \frac{\partial\chi}{\partial n}, \chi\rangle + \gamma h^{-1}|\chi|^2 \\
&\geq \|\nabla\chi\|^2 - 2\Big|\frac{\partial\chi}{\partial n}\Big|\, |\chi| + \gamma h^{-1}|\chi|^2 \\
&\geq \|\nabla\chi\|^2 - \frac{h}{4C_0}\Big|\frac{\partial\chi}{\partial n}\Big|^2 - \frac{4C_0}{h}|\chi|^2 + \frac{\gamma}{h}|\chi|^2 \\
&\geq \frac{1}{2}\|\nabla\chi\|^2 + \frac{h}{4C_0}\Big|\frac{\partial\chi}{\partial n}\Big|^2 + \frac{\gamma - 4C_0}{h}|\chi|^2 \\
&\geq c|||\chi|||^2, \quad \text{if} \quad \gamma \geq \gamma_0 > 4C_0.
\end{aligned}$$

□

We shall now show an approximation property of our subspaces S_h with respect to $|||\cdot|||$.

Lemma 2.3 *With S_h defined in* (2.2), *we have*

$$\inf_{\chi \in S_h} |||v - \chi||| \leq Ch^{s-1}\|v\|_s, \quad \textit{for } 2 \leq s \leq r,\ v \in H^s. \tag{2.8}$$

Proof. Because the functions in S_h do not belong to $H^2(\Omega)$, even though they are in $H^2(\tau_j)$ for each j, we shall use the trianglewise defined norm

$$\|\varphi\|_{2,h} = \Big(\sum_{j=1}^{M_h} \|\varphi\|^2_{H^2(\tau_j)}\Big)^{1/2},$$

and show that, for appropriately smooth φ,

$$|||\varphi||| \leq Ch^{-1}(\|\varphi\| + h\|\varphi\|_1 + h^2\|\varphi\|_{2,h}). \tag{2.9}$$

Since it is easy to find a local interpolation operator I_h into S_h (using, e.g., Lagrange interpolation) such that

$$\|I_h v - v\| + h\|I_h v - v\|_1 + h^2\|I_h v - v\|_{2,h} \leq Ch^s\|v\|_s, \quad \text{for } 2 \leq s \leq r,$$

this will complete the proof of (2.8).

To show (2.9), we begin by bounding the term $|\varphi|$. Let τ_j be a boundary triangle of $\mathcal{T}_h$ and $(\partial\Omega)_j$ the corresponding part of $\partial\Omega$. Mapping τ_j onto the unit size reference triangle $\widetilde{\tau}_j$, we note that for $\widetilde{\varphi}$ defined in $\widetilde{\tau}_j$ we have the trace inequality (see., e.g., [42])

$$\|\widetilde{\varphi}\|_{L_2(\partial\widetilde{\tau}_j)}^2 \le C\|\widetilde{\varphi}\|_{L_2(\widetilde{\tau}_j)}\|\widetilde{\varphi}\|_{H^1(\widetilde{\tau}_j)}. \tag{2.10}$$

Transforming back to τ_j we find that for any $\varphi \in H^1(\tau_j)$,

$$\|\varphi\|_{L_2((\partial\Omega)_j)}^2 \le C\|\varphi\|_{L_2(\tau_j)}\big(\|\nabla\varphi\|_{L_2(\tau_j)} + h^{-1}\|\varphi\|_{L_2(\tau_j)}\big). \tag{2.11}$$

Hence

$$h^{-1}\|\varphi\|_{L_2((\partial\Omega)_j)}^2 \le C(h^{-2}\|\varphi\|_{L_2(\tau_j)}^2 + \|\varphi\|_{H^1(\tau_j)}^2),$$

and after summation this shows

$$h^{-1}|\varphi|^2 \le C(h^{-2}\|\varphi\|^2 + \|\varphi\|_1^2).$$

Similarly, considering now first $\varphi \in H^2(\tau_j)$, we have

$$h\Big|\frac{\partial\varphi}{\partial n}\Big|^2 \le C(\|\varphi\|_1^2 + h^2\|\varphi\|_{2,h}^2).$$

Since $\|\nabla\varphi\|$ is obviously bounded as desired, (2.9) follows, and the proof is complete. □

We assume from now on that γ is chosen so that the second estimate of Lemma 2.2 holds. Then, in particular, $N_\gamma(\cdot,\cdot)$ is positive definite on S_h and, consequently, our discrete Dirichlet problem has a unique solution. By subtraction we obtain at once from (2.5) and (2.4) the error equation

$$N_\gamma(u_h - u, \chi) = 0, \quad \forall \chi \in S_h, \tag{2.12}$$

which we shall use to prove the following error estimate.

Theorem 2.1 *Let S_h be defined in (2.2). Then, with u_h and u the solutions of (2.5) and (2.1), respectively, we have*

$$|||u_h - u||| \le Ch^{s-1}\|u\|_s, \quad \textit{for } 2 \le s \le r.$$

In particular, $\|\nabla(u_h - u)\| \le Ch^{r-1}\|u\|_r$.

Proof. We have, for any $\chi \in S_h$,

$$|||u_h - u||| \le |||u - \chi||| + |||\chi - u_h|||.$$

Now, by Lemma 2.2 and (2.12),

$$|||\chi - u_h|||^2 \le CN_\gamma(\chi - u_h, \chi - u_h) = CN_\gamma(\chi - u, \chi - u_h)$$

$$\le C|||\chi - u|||\ |||\chi - u_h|||.$$

Hence $|||\chi - u_h||| \leq C|||\chi - u|||$, and so, by Lemma 2.3,

$$|||u_h - u||| \leq (1 + C) \inf_{\chi \in S_h} |||\chi - u||| \leq Ch^{s-1}\|u\|_s, \quad \text{for } 2 \leq s \leq r,$$

which proves the theorem. □

We note that although the discrete solution u_h does not satisfy the boundary condition $u_h = 0$ on $\partial\Omega$, it is small on the boundary because, as a result of Theorem 2.1, we have

$$|u_h| = |u_h - u| \leq h^{1/2}|||u_h - u||| \leq Ch^{r-1/2}\|u\|_r.$$

We also have the following L_2-norm estimate.

Theorem 2.2 *Under the assumptions of Theorem 2.1, we have*

$$\|u_h - u\| \leq Ch^s\|u\|_s, \quad \textit{for} \ \ 2 \leq s \leq r.$$

Proof. We shall use the standard duality argument. Define ψ by

$$(2.13) \qquad -\Delta\psi = \varphi \quad \text{in } \Omega, \quad \text{with } \psi = 0 \quad \text{on } \partial\Omega,$$

and recall the elliptic regularity estimate (1.17). We have, for $e = u_h - u$,

$$(e, \varphi) = -(e, \Delta\psi) = (\nabla e, \nabla\psi) - \langle e, \frac{\partial\psi}{\partial n}\rangle = N_\gamma(e, \psi).$$

Now for ψ_h the approximate solution of our auxiliary problem (2.13) we have, using (2.12), Lemma 2.2, Theorem 2.1 with $s = 2$, and (1.17)

$$\begin{aligned} |(e, \varphi)| &= |N_\gamma(e, \psi - \psi_h)| \leq C|||e|||\,|||\psi - \psi_h||| \\ &\leq Ch\,\|\psi\|_2\,|||e||| \leq Ch\,\|\varphi\|\,|||e|||. \end{aligned}$$

Hence applying Theorem 2.1 once more to bound $|||e|||$ we obtain

$$|(e, \varphi)| \leq Ch^s\|u\|_s\|\varphi\|, \quad \text{for } 2 \leq s \leq r,$$

which shows the theorem. □

We now resume our discussion of the parabolic problem

$$(2.14) \qquad \begin{aligned} u_t - \Delta u &= f \quad \text{in } \Omega, \qquad \text{for } t > 0, \\ u &= 0 \quad \text{on } \partial\Omega, \quad \text{for } t > 0, \quad \text{with } u(\cdot, 0) = v \quad \text{in } \Omega. \end{aligned}$$

We shall present an alternative derivation of our previous L_2-norm error estimate which will be general enough to cover situations when $S_h \not\subset H_0^1$, as, for instance, in the case of Nitsche's method described above. We now allow $\Omega \subset \mathbb{R}^d$ with $d \geq 2$.

For motivation let us first recall the standard Galerkin method for the elliptic problem (2.1) with $S_h \subset H_0^1$, namely

$$(\nabla u_h, \nabla \chi) = (f, \chi), \quad \forall \chi \in S_h, \tag{2.15}$$

and define the linear operator $T_h : L_2 \to S_h$ by $T_h f = u_h$, so that $u_h = T_h f \in S_h$ is the approximate solution of (2.1). Letting $u = Tf$ be the solution of this problem, so that $T : L_2 \to H_0^1$ denotes the exact solution operator of (2.1), we have

$$T_h = R_h T, \tag{2.16}$$

where R_h is the elliptic projection operator defined in (1.22). In fact, by our definitions we have

$$(\nabla T_h f, \nabla \chi) = (f, \chi) = (\nabla T f, \nabla \chi) = (\nabla R_h T f, \nabla \chi), \quad \forall \chi \in S_h,$$

which shows (2.16).

Recalling that

$$\|R_h u - u\| + h\|\nabla(R_h u - u)\| \leq Ch^s \|u\|_s, \quad \text{for } u \in H^s \cap H_0^1, \quad 1 \leq s \leq r,$$

we obtain

$$\|T_h f - Tf\| = \|(R_h - I)Tf\| \leq Ch^s \|Tf\|_s.$$

By the elliptic regularity estimate, we have

$$\|u\|_s \leq C\|\Delta u\|_{s-2}, \quad \text{if } u = 0 \text{ on } \partial\Omega, \quad \text{for } s \geq 2,$$

or $\|Tf\|_s \leq C\|f\|_{s-2}$, so that thus

$$\|T_h f - Tf\| \leq Ch^s \|f\|_{s-2}, \quad \text{for } 2 \leq s \leq r, \quad \text{if } f \in H^{s-2}.$$

We also note that T_h is selfadjoint, positive semidefinite on L_2:

$$(f, T_h g) = (\nabla T_h f, \nabla T_h g) = (T_h f, g), \quad \forall f, g \in L_2.$$

In particular,

$$(T_h f, f) = \|\nabla T_h f\|^2 \geq 0. \tag{2.17}$$

In fact, T_h is positive definite on S_h, considered as an inner product space with respect to the L_2 inner product. For assume $f_h \in S_h$ is such that $(T_h f_h, f_h) = 0$. Then $T_h f_h = 0$ by (2.17) and hence

$$\|f_h\|^2 = (f_h, f_h) = (\nabla T_h f_h, \nabla f_h) = 0.$$

Recalling the definition (1.33) of the discrete Laplacian $\Delta_h : S_h \to S_h$ we have $T_h = (-\Delta_h)^{-1}$ on S_h. For

$$(f_h, \chi) = (\nabla(T_h f_h), \nabla\chi) = -(\Delta_h(T_h f_h), \chi), \quad \text{for } \chi \in S_h,$$

so that $-\Delta_h(T_h f_h) = f_h$ for $f_h \in S_h$. Note also that $T_h P_h = T_h$, since

$$(2.18) \qquad (\nabla(T_h P_h) f, \nabla\chi) = (P_h f, \chi) = (f, \chi) = (\nabla T_h f, \nabla\chi), \quad \forall \chi \in S_h.$$

We now recall the semidiscrete problem

$$(2.19) \qquad u_{h,t} - \Delta_h u_h = P_h f, \quad \text{for } t > 0, \quad \text{with } u_h(0) = v_h.$$

In view of the above definition of the discrete solution operator T_h, this may equivalently be written

$$T_h u_{h,t} + u_h = T_h P_h f = T_h f, \quad \text{for } t > 0, \quad \text{with } u_h(0) = v_h.$$

Similarly, for the continuous problem, we have

$$T u_t + u = T f, \quad \text{for } t > 0, \quad \text{with } u(0) = v.$$

For the same reason as for T_h, the operator T is selfadjoint and, in fact, positive definite on L_2. For $(f, \varphi) = (\nabla(Tf), \nabla\varphi)$ for $\varphi \in H_0^1$ shows $(f, Tf) = \|\nabla T f\|^2 \geq 0$, and clearly $Tf = 0$ implies $f = -\Delta(Tf) = 0$.

From now on, instead of defining the approximate solution of the elliptic problem by (2.15), we shall assume only that we are given an approximate solution operator T_h with the properties:

(i) *T_h is selfadjoint, positive semidefinite on L_2, and positive definite on S_h.*

(ii) *There is a positive integer $r \geq 2$ such that*

$$\|(T_h - T) f\| \leq C h^s \|f\|_{s-2}, \quad \text{for } 2 \leq s \leq r, \quad f \in H^{s-2}.$$

We may then pose the *semidiscrete* problem to find $u_h(t) \in S_h$ for $t \geq 0$ such that

$$(2.20) \qquad T_h u_{h,t} + u_h = T_h f, \quad \text{for } t > 0, \quad \text{with } u_h(0) = v_h,$$

which may be solved uniquely for $t \geq 0$ since T_h^{-1} exists on S_h by (i). Since T_h is positive definite on S_h, we may define $\Delta_h = -T_h^{-1} : S_h \to S_h$, and note that (2.20) may then also be written in the form (2.19).

As an example, we may consider Nitsche's method for the elliptic problem and define T_h by

$$(2.21) \qquad N_\gamma(T_h f, \chi) = (f, \chi), \quad \forall \chi \in S_h, \ f \in L_2.$$

Property (i) follows then essentially as for the standard Galerkin method, and property (ii) is the L_2 error estimate for Nitsche's method (Theorem 2.2). In this case $\Delta_h : S_h \to S_h$ may be defined by

$$-(\Delta_h \psi, \chi) = N_\gamma(\psi, \chi), \quad \forall \psi, \ \chi \in S_h.$$

Since (2.20) is equivalent to the variational formulation

$$N_\gamma(T_h u_{h,t}, \chi) + N_\gamma(u_h, \chi) = N_\gamma(T_h f, \chi), \quad \forall \chi \in S_h,$$

equation (2.21) shows that the semidiscrete problem is now equivalent to

$$(u_{h,t}, \chi) + N_\gamma(u_h, \chi) = (f, \chi), \quad \forall \chi \in S_h, \ t > 0, \quad \text{with } u_h(0) = v_h.$$

Note that this time we make no explicit assumption about the approximation properties of $\{S_h\}$, which are now instead implicitly contained in (ii). In fact, it follows from (ii) that

$$(2.22) \qquad \begin{aligned} \inf_{\chi \in S_h} \|v - \chi\| &\le \|v - T_h(-\Delta v)\| = \|(T - T_h)\Delta v\| \\ &\le C h^s \|\Delta v\|_{s-2} \le C h^s \|v\|_s, \quad \text{for } 2 \le s \le r. \end{aligned}$$

In particular, for P_h the orthogonal L_2-projection, we have

$$\|v - P_h v\| \le C h^s \|v\|_s, \quad \text{for } 2 \le s \le r,$$

and if we now introduce the *elliptic* projection $R_h = T_h(-\Delta) : H^2 \cap H_0^1 \to S_h$, (2.22) shows

$$(2.23) \qquad \|v - R_h v\| \le C h^s \|v\|_s, \quad \text{for } 2 \le s \le r.$$

For the standard Galerkin method the present projection coincides with the old elliptic projection, and for Nitsche's method we have by our definitions

$$N_\gamma((R_h - I)v, \chi) = 0, \quad \forall \chi \in S_h.$$

We note that by (i), we have

$$(2.24) \qquad T_h P_h = T_h.$$

In fact, since T_h is selfadjoint,

$$(T_h P_h f, g) = (P_h f, T_h g) = (f, T_h g) = (T_h f, g), \quad \forall f, g \in L_2,$$

from which (2.24) follows.

Under our new general assumptions we shall now prove an error estimate of the same form as in the special case of the standard Galerkin method shown earlier (Theorem 1.2).

Theorem 2.3 *Assume that T_h satisfies (i) and (ii) and let u_h and u be the solutions of* (2.20) *and* (2.14), *respectively. Then*

$$\|u_h(t) - u(t)\| \le \|v_h - v\| + Ch^r\big(\|v\|_r + \int_0^t \|u_t\|_r ds\big), \quad \textit{for } t \ge 0.$$

Proof. We have for the error $e = u_h - u$,

$$\begin{aligned} T_h e_t + e &= (T_h u_{h,t} + u_h) - (T_h u_t + u) \\ &= T_h f - (Tu_t + u) + (T - T_h)u_t = (T - T_h)(u_t - f) = (T - T_h)\Delta u, \end{aligned}$$

that is, using also (2.16),

$$T_h e_t + e = \rho, \quad \text{where } \rho = -(T_h - T)\Delta u = (R_h - I)u. \tag{2.25}$$

We multiply by $2e_t$ and integrate over Ω to find

$$2(T_h e_t, e_t) + \frac{d}{dt}\|e\|^2 = 2(\rho, e_t) = 2\frac{d}{dt}(\rho, e) - 2(\rho_t, e),$$

and hence, after integration with respect to t,

$$\begin{aligned} \|e(t)\|^2 &\le \|e(0)\|^2 + 2\|\rho(t)\|\,\|e(t)\| + 2\|\rho(0)\|\,\|e(0)\| + 2\int_0^t \|\rho_t\|\,\|e\|\,ds \\ &\le \sup_{s\le t}\|e(s)\|\Big(\|e(0)\| + 4\sup_{s\le t}\|\rho(s)\| + 2\int_0^t \|\rho_t\, ds\Big). \end{aligned}$$

Applying this with τ such that $\|e(\tau)\| = \sup_{s\le t}\|e(s)\|$, we have

$$\begin{aligned} \|e(t)\| \le \|e(\tau)\| &\le \|e(0)\| + 4\sup_{s\le t}\|\rho(s)\| + 2\int_0^t \|\rho_t\|\,ds \\ &\le \|e(0)\| + C\Big(\|\rho(0)\| + \int_0^t \|\rho_t\|\,ds\Big). \end{aligned} \tag{2.26}$$

Here $e(0) = v_h - v$, and

$$\|\rho(0)\| = \|(T_h - T)\Delta v\| \le Ch^r\|\Delta v\|_{r-2} \le Ch^r\|v\|_r,$$

and, similarly, $\|\rho_t\| = \|(T_h - T)\Delta u_t\| \le Ch^r\|u_t\|_r$, which completes the proof. □

Note from our discussion preceding Theorem 2.3 that if v_h is chosen as $v_h = P_h v$ or $v_h = R_h v$, then the first term on the right in the error estimate is bounded by the second.

We remark for later reference that the essence of the proof is the following.

Lemma 2.4 *Let* $T_h e_t + e = \rho$ *for* $t \geq 0$ *where* T_h *is nonnegative* ($(T_h f, f) \geq 0$) *with respect to the (semi-)inner product* $(\cdot,\cdot)$. *Then for the corresponding (semi-)norm* $\|\cdot\| = (\cdot,\cdot)^{1/2}$,

$$\|e(t)\| \leq \|e(0)\| + C\Big(\|\rho(0)\| + \int_0^t \|\rho_t\|\,ds\Big).$$

For the standard Galerkin method we saw that optimal order error estimates for the gradient could be derived from the weak formulation of the parabolic problem; a similar argument would give estimates for the gradient also when the semidiscrete parabolic problems is based on Nitsche's method. We also saw that such estimates could be derived using an inverse property, and this still applies under our present more general assumptions:

Theorem 2.4 *Assume that (i), (ii) hold together with the inverse property* (1.12) *and the approximation property*

$$\inf_{\chi\in S_h}\{\|v-\chi\| + h\|\nabla(v-\chi)\|\} \leq Ch^r\|v\|_r. \tag{2.27}$$

Let u_h *and* u *be the solutions of* (2.20) *and* (2.14), *with* v_h *chosen so that* $\|v_h - v\| \leq Ch^r\|v\|_r$. *Then we have*

$$\|\nabla(u_h(t)-u(t))\| \leq Ch^{r-1}\Big(\|v\|_r + \int_0^t \|u_t\|_r\,ds\Big), \quad \text{for } t \geq 0.$$

Proof. This follows exactly as in Chapter 1 from (1.41), (1.12), and (2.27). □

We shall end this chapter with an estimate of the error using the mean square norm also in time. Note that the error bound does not contain the time derivative of the solution.

Theorem 2.5 *Assume that* T_h *satisfies (i) and (ii) and let* u_h *and* u *be the solutions of* (2.20) *and* (2.14), *with* $v_h = P_h v$. *Then*

$$\Big(\int_0^t \|u_h - u\|^2\,ds\Big)^{1/2} \leq Ch^r\Big(\int_0^t \|u\|_r^2\,ds\Big)^{1/2}, \quad \text{for } t \geq 0.$$

Proof. We take the L_2-inner product of the error equation (2.25) by e and observe that since T_h is selfadjoint, $2(T_h e_t, e) = \frac{d}{dt}(T_h e, e)$. Hence

$$\frac{d}{dt}(T_h e, e) + 2\|e\|^2 = (\rho, e) \leq \|e\|^2 + \|\rho\|^2.$$

After integration this shows

$$(T_h e(t), e(t)) + \int_0^t \|e\|^2\,ds \leq (T_h e(0), e(0)) + \int_0^t \|\rho\|^2\,ds.$$

We now note that $T_h e(0) = 0$ for $v_h = P_h v$. For

$$(T_h e(0), w) = (P_h v - v, T_h w) = 0, \quad \forall w \in L_2, \tag{2.28}$$

since $T_h w \in S_h$. Hence

$$\int_0^t \|e\|^2 \, ds \le \int_0^t \|\rho\|^2 \, ds, \tag{2.29}$$

and the result claimed now follows by (2.23). □

The above type of error analysis of finite element methods for parabolic problems based on operators T_h generalizing the standard Galerkin solution operator of the elliptic problem was initiated in Bramble, Schatz, Thomée, and Wahlbin [37] for homogeneous parabolic equations and followed up in Thomée [228] for inhomogeneous equations. The method used here as a particular example was introduced in Nitsche [180]. Other examples include Babuška's method with Lagrangian multipliers [10], the method of interpolated boundary conditions by Berger, Scott, and Strang [24], [213], an alternative method of Nitsche [181] which uses the bilinear form (2.3) with $\gamma = 0$ under an additional assumption ensuring that the functions in S_h are small on $\partial\Omega$, and also a so called mixed method which we shall consider in Chapter 17.

Another way of dealing with the problem of a curved boundary was considered in Bramble, Dupont, and Thomée [32] and Dupont [83] where the finite element method is based on an approximating polygonal domain with a correction built into the boundary values.

Problems with vanishing initial data but with inhomogeneous and nonsmooth boundary data are considered in Lasiecka [151].

3. Nonsmooth Data Error Estimates

In this chapter we shall first discuss a smoothing property of the solution operator of a homogeneous parabolic equation which shows that the solution is regular for positive time even if the initial data are not. We shall then demonstrate that an analogous behavior for the finite element solution implies that optimal order convergence takes place for positive time even for nonsmooth initial data. We also show some other results which elucidate the relation between the convergence of the finite element solution and the regularity of the exact solution.

We begin by introducing some function spaces which are convenient in describing the regularity of the solution of the initial boundary value problem for a homogeneous parabolic equation. Consider thus

$$\begin{aligned} u_t &= \Delta u \quad \text{in } \Omega, \quad \text{for } t > 0, \\ u &= 0 \quad \text{on } \partial\Omega, \quad \text{for } t > 0, \quad \text{with } u(\cdot, 0) = v \quad \text{in } \Omega, \end{aligned} \tag{3.1}$$

where Ω is a bounded domain in $\mathbb{R}^d$ with smooth boundary $\partial\Omega$. We associate with it the eigenvalue problem

$$-\Delta\varphi = \lambda\varphi \quad \text{in } \Omega, \quad \text{with } \varphi = 0 \quad \text{on } \partial\Omega. \tag{3.2}$$

As is well-known, this eigenvalue problem admits a nondecreasing sequence $\{\lambda_m\}_{m=1}^{\infty}$ of positive eigenvalues, which tend to ∞ with m, and a corresponding sequence $\{\varphi_m\}_{m=1}^{\infty}$ of eigenfunctions which form an orthonormal basis in $L_2 = L_2(\Omega)$, so that each $v \in L_2$ admits the representation $v = \sum_{m=1}^{\infty}(v, \varphi_m)\varphi_m$, and Parseval's relation,

$$(v, w) = \sum_{m=1}^{\infty} (v, \varphi_m)(w, \varphi_m),$$

holds.

For $s \geq 0$, let $\dot{H}^s = \dot{H}^s(\Omega)$ be the subspace of L_2 defined by

$$|v|_s = \Big(\sum_{m=1}^{\infty} \lambda_m^s (v, \varphi_m)^2 \Big)^{1/2} < \infty.$$

If we formally introduce nonnegative powers of the operator $-\Delta$ by

$$(3.3) \qquad (-\Delta)^s v = \sum_{m=1}^{\infty} \lambda_m^s (v, \varphi_m)\varphi_m, \quad \text{for } s \geq 0,$$

we may alternatively express the definition of $|\cdot|_s$ as

$$|v|_s = \|(-\Delta)^{s/2} v\| = ((-\Delta)^s v, v)^{1/2}.$$

Note that by (3.2), (3.3) agrees with the standard definition of $(-\Delta)^s$ for s integer when v is a finite linear combination of eigenfunctions.

We have the following characterization which makes this latter definition precise for s integer. Recall that for Ω an appropriately regular domain in $\mathbb{R}^d$ we define $H^s = H^s(\Omega)$, for s a nonnegative integer, by the norm

$$\|v\|_s = \|v\|_{H^s} = \Big(\sum_{|\alpha| \leq s} \|D^\alpha v\|^2 \Big)^{1/2}, \quad \text{where } \|\cdot\| = \|\cdot\|_{L_2}.$$

Lemma 3.1 *For s a nonnegative integer we have*

$$\dot{H}^s = \{ v \in H^s;\ \Delta^j v = 0 \ \text{ on } \partial\Omega, \ \text{ for } j < s/2 \},$$

where the boundary conditions are interpreted in the sense of traces in $L_2(\partial\Omega)$, and the norms $|\cdot|_s = \|\cdot\|_{\dot{H}^s}$ and $\|\cdot\|_s = \|\cdot\|_{H^s}$ are equivalent in $\dot{H}^s$, with

$$|v|_s = \begin{cases} \|\Delta^p v\|, & \text{if } s = 2p, \\ \|\nabla(\Delta^p v)\|, & \text{if } s = 2p+1. \end{cases}$$

Proof. We first show that if $v \in H^1$, with $v = 0$ on $\partial\Omega$, then $v \in \dot{H}^1$ and $|v|_1 \leq \|v\|_1$. In fact, for $v \in C_0^\infty(\Omega)$ we have

$$\lambda_m (v, \varphi_m) = (v, \lambda_m \varphi_m) = -(v, \Delta\varphi_m) = -(\Delta v, \varphi_m),$$

and hence, using Parseval's relation,

$$\begin{aligned} |v|_1^2 &= \sum_{m=1}^{\infty} \lambda_m (v, \varphi_m)^2 = -\sum_{m=1}^{\infty} (v, \varphi_m)(\Delta v, \varphi_m) \\ &= -(v, \Delta v) = \|\nabla v\|^2 \leq \|v\|_1^2. \end{aligned}$$

Since $C_0^\infty(\Omega)$ is dense in $\{v \in H^1;\ v = 0 \ \text{ on } \partial\Omega\}$, this shows the result.

For $v \in H^{2p+1}$ with $\Delta^j v = 0$ on $\partial\Omega$ for $j \leq p$ we have hence

$$\begin{aligned} |v|_{2p+1}^2 &= \sum_{m=1}^{\infty} \lambda_m^{2p+1} (v, \varphi_m)^2 = \sum_{m=1}^{\infty} \lambda_m (v, \lambda_m^p \varphi_m)^2 \\ &= \sum_{m=1}^{\infty} \lambda_m ((-\Delta)^p v, \varphi_m)^2 = \|\nabla(\Delta^p v)\|^2 \leq C\|v\|_{2p+1}^2. \end{aligned}$$

Similarly, if $v \in H^{2p}$ with $\Delta^j v = 0$ on $\partial\Omega$ for $j < p$, we have

$$|v|_{2p}^2 = \sum_{m=1}^{\infty} \lambda_m^{2p} (v, \varphi_m)^2 = \sum_{m=1}^{\infty} (v, \lambda_m^p \varphi_m)^2$$
$$= \sum_{m=1}^{\infty} ((-\Delta)^p v, \varphi_m)^2 = \|\Delta^p v\|^2 \le C\|v\|_{2p}^2.$$

We have thus shown that if $v \in H^s$ and $\Delta^j v = 0$ on $\partial\Omega$ for $j < s/2$, then $v \in \dot{H}^s$ and $|v|_s \le C\|v\|_s$.

We now turn to the opposite inclusion. Let $s = 2p$ and let $\widetilde{v}$ be any linear combination of finitely many of the eigenfunctions φ_m. Then, by the above computation, $\|\Delta^p \widetilde{v}\| = |\widetilde{v}|_{2p}$. On the other hand, by the well-known regularity estimate for the elliptic operator Δ^p (cf., e.g., Lions and Magenes [156]), we have in view of the boundary conditions $\Delta^j \widetilde{v} = 0$ on $\partial\Omega$ for $j < p$, that $\|\widetilde{v}\|_{2p} \le C\|\Delta^p \widetilde{v}\| = C|\widetilde{v}|_{2p}$. Since the $\widetilde{v}$ are dense in $\dot{H}^{2p}$, we conclude that $v \in \dot{H}^{2p}$ implies $v \in H^{2p}$ and $\|v\|_{2p} \le C|v|_{2p}$ for $v \in \dot{H}^{2p}$. By a trace inequality (cf. [156]),

$$\|\Delta^j v\|_{L_2(\partial\Omega)} \le C\|\Delta^j v\|_1 \le C\|v\|_{2p}, \quad \text{for } j < p,$$

and since the $\Delta^j \widetilde{v}$ vanish on $\partial\Omega$, we conclude that this holds for $\Delta^j v$ as well.

The proof for s odd is similar; for $s = 1$ one uses Friedrichs' inequality (1.4) which shows $\|\widetilde{v}\|_1 \le C\|\nabla \widetilde{v}\| = C|\widetilde{v}|_1$. □

We emphasize that the boundary conditions in $\dot{H}^r$ are quite restrictive in applications. For instance, in the one-dimensional situation with $\Omega = (0,1)$, the very regular function $v(x) = x(1-x)$, which vanishes at $x = 0, 1$, belongs to $\dot{H}^2$, but not to $\dot{H}^3$, because $\Delta v = v'' = -2$ does not vanish at $x = 0, 1$.

The solution of our initial boundary value problem (3.1) may now be represented, with $E(t)$ the associated solution operator, as

$$u(x,t) = (E(t)v)(x) = \sum_{m=1}^{\infty} e^{-t\lambda_m} (v, \varphi_m)\varphi_m(x).$$

Setting $D_t = \partial/\partial t$, we note that a solution $u(t) = E(t)v$ of (3.1) which is in $\mathcal{C}^\infty(\bar{\Omega} \times [0,\infty))$ satisfies $\Delta^j u(t) = D_t^j u(t) = 0$ on $\partial\Omega$ for $t > 0$, and hence the initial data also satisfy $\Delta^j v = 0$ on $\partial\Omega$ for any $j \ge 0$, so that $v \in \dot{H}^s$ for any $s \ge 0$. Even when the initial function v is less regular it is still the case that $u(t) \in \dot{H}^s$ for any $t > 0$ and any $s \ge 0$, as follows from the following regularity result. We remark that this is related to the fact that $E(t)$ is an analytic semigroup on L_2, which is a topic we will discuss in more detail in Chapter 5 below.

Lemma 3.2 *If $v \in L_2$ then the solution $u(t) = E(t)v$ of (3.1) belongs to $\dot{H}^s$ for any $s \ge 0$, if $t > 0$. If $0 \le s \le q$ and $l \ge 0$, and if $v \in \dot{H}^s$, we have*

$$|D_t^l E(t)v|_q \le Ct^{-(q-s)/2-l}|v|_s, \quad \textit{for } t > 0.$$

Proof. We have with $C = \sup_{\tau>0}(\tau^{q-s+2l}e^{-2\tau})$

$$|D_t^l E(t)v|_q^2 = |(-\Delta)^l E(t)v|_q^2 = \sum_{m=1}^{\infty} \lambda_m^{q+2l} e^{-2\lambda_m t}(v,\varphi_m)^2$$

$$\leq Ct^{-(q-s)-2l} \sum_{m=1}^{\infty} \lambda_m^s (v,\varphi_m)^2 = Ct^{-(q-s)-2l}|v|_s^2. \qquad \square$$

We now return to the discussion of the spatially semidiscrete approximation of our initial value problem within the framework introduced in Chapter 2. We assume thus again that $\{S_h\}$ is a family of finite dimensional subspaces of L_2, and $\{T_h\}$ a family of operators $T_h : L_2 \to S_h$, approximating the exact solution operator T of the Dirichlet problem

$$-\Delta u = f \quad \text{in } \Omega, \qquad \text{with } u = 0 \quad \text{on } \partial\Omega,$$

such that

(i) *T_h is selfadjoint, positive semidefinite on L_2, and positive definite on S_h.*

(ii) *There is a positive integer $r \geq 2$ such that*

$$\|(T_h - T)f\| \leq Ch^s\|f\|_{s-2}, \quad \text{for } 2 \leq s \leq r, \quad f \in H^{s-2}.$$

The semidiscrete analogue of (3.1) is then defined as

$$T_h u_{h,t} + u_h = 0, \quad \text{for } t > 0, \quad \text{with } u_h(0) = v_h. \tag{3.4}$$

or, if we set $\Delta_h = -T_h^{-1} : S_h \to S_h$,

$$u_{h,t} - \Delta_h u_h = 0, \quad \text{for } t > 0, \quad \text{with } u_h(0) = v_h.$$

The error estimate proved earlier in Theorem 2.3 for the inhomogeneous equation shows for the homogeneous equation that if v_h is chosen so that $\|v_h - v\| \leq h^r\|v\|_r$, e.g., for $v_h = P_h v$ or $v_h = R_h v$, then, if $v \in \dot{H}^{r+\varepsilon}$ with $\varepsilon > 0$, we have, for t bounded,

$$\|u_h(t) - u(t)\| \leq C_\varepsilon h^r |v|_{r+\varepsilon}. \tag{3.5}$$

In fact, in this case, by Lemma 3.2,

$$\|u_t(s)\|_r = \|\Delta u(s)\|_r \leq C\|u(s)\|_{r+2} \leq C|u(s)|_{r+2} \leq Cs^{-(1-\varepsilon/2)}|v|_{r+\varepsilon},$$

so that

$$\int_0^t \|u_t\|_r\, ds \leq C|v|_{r+\varepsilon} \int_0^t s^{-(1-\varepsilon/2)}\, ds = 2C\varepsilon^{-1}t^{\varepsilon/2}|v|_{r+\varepsilon}.$$

Since $\|v\|_r \leq C|v|_{r+\varepsilon}$, Theorem 2.3 therefore shows (3.5).

We shall prove the following slightly sharper smooth data error estimate:

Theorem 3.1 *Assume that (i) and (ii) hold, that* $v \in \dot{H}^r$, *and that*

$$\|v_h - v\| \leq Ch^r |v|_r. \tag{3.6}$$

Then we have for the error in the semidiscrete parabolic problem (3.4)

$$\|u_h(t) - u(t)\| \leq Ch^r |v|_r, \quad \text{for } t \geq 0.$$

The proof of this result will depend on the following:

Lemma 3.3 *Assume that* T_h *is positive semidefinite on* L_2 *and that*

$$T_h e_t + e = \rho, \quad \text{for } t \geq 0, \quad \text{with } T_h e(0) = 0. \tag{3.7}$$

Then

$$\|e(t)\|^2 \leq C\Big(\|\rho(t)\|^2 + \frac{1}{t}\int_0^t (\|\rho\|^2 + s^2\|\rho_t\|^2)\, ds\Big), \quad \text{for } t > 0.$$

Proof. Taking inner products of (3.7) by $2e_t$ we find

$$2(T_h e_t, e_t) + \frac{d}{dt}\|e\|^2 = 2(\rho, e_t),$$

or, since T_h is positive semidefinite,

$$\frac{d}{dt}\|e\|^2 \leq 2(\rho, e_t) = 2\frac{d}{dt}(\rho, e) - 2(\rho_t, e).$$

Multiplying by t, we obtain

$$\frac{d}{dt}(t\|e\|^2) \leq 2\frac{d}{dt}(t(\rho, e)) - 2t(\rho_t, e) + \|e\|^2 - 2(\rho, e),$$

so that, after integration,

$$t\|e(t)\|^2 \leq 2t\|\rho(t)\|\,\|e(t)\| + \int_0^t (\|e\|^2 + 2\|\rho\|\,\|e\| + 2s\|\rho_t\|\,\|e\|)\, ds,$$

or

$$\|e(t)\|^2 \leq C\Big(\|\rho(t)\|^2 + \frac{1}{t}\int_0^t (\|e\|^2 + \|\rho\|^2 + s^2\|\rho_t\|^2)\, ds\Big). \tag{3.8}$$

We now recall from (2.29) that

$$\int_0^t \|e\|^2\, ds \leq \int_0^t \|\rho\|^2\, ds.$$

Together with (3.8), this completes the proof. □

As an immediate consequence we have:

Lemma 3.4 *Under the assumptions of Lemma 3.3, we have*

$$\|e(t)\| \le C \sup_{s\le t}(s\|\rho_t(s)\| + \|\rho(s)\|), \quad \textit{for } t \ge 0.$$

We also note for later use that the coefficient of the first term on the right can be made small:

Lemma 3.5 *Under the assumptions of Lemma 3.3 we have, for $\varepsilon > 0$ arbitrary,*

$$\|e(t)\| \le \varepsilon \sup_{s\le t}(s\|\rho_t(s)\|) + C_\varepsilon \sup_{s\le t}\|\rho(s)\|, \quad \textit{for } t \ge 0.$$

Proof. It follows by an obvious modification of the proof of Lemma 3.3 that

$$\|e(t)\|^2 \le \frac{\varepsilon^2}{t}\int_0^t s^2\|\rho_t\|^2\,ds + C_\varepsilon\big(\|\rho(t)\|^2 + \frac{1}{t}\int_0^t \|\rho\|^2\,ds\big),$$

and hence the result. □

Proof of Theorem 3.1. We first note that it is sufficient to consider the case $v_h = P_h v$; using (3.6) this follows at once from

$$\|E_h(t)(P_h v - v_h)\| \le \|P_h v - v_h\| \le \|P_h v - v\| + \|v - v_h\| \le Ch^r|v|_r,$$

where $E_h(t)$ denotes the solution operator of the semidiscrete problem.

Recall from (2.25) that the error $e = u_h - u$ satisfies the equation

$$T_h e_t + e = \rho, \quad \text{where } \rho = -(T_h - T)\Delta u = -(T_h - T)u_t, \tag{3.9}$$

and from (2.28) that $T_h e(0) = 0$ for $v_h = P_h v$. We are thus in a position to apply Lemma 3.4. Using (ii) and Lemma 3.4, we have, for $s \ge 0$,

$$\|\rho(s)\| \le Ch^r\|u_t(s)\|_{r-2} \le Ch^r\|u(s)\|_r \le Ch^r|v|_r,$$

and similarly,

$$s\|\rho_t(s)\| \le Ch^r s\|u_t(s)\|_r \le Ch^r s\|u(s)\|_{r+2} \le Ch^r|v|_r.$$

These estimates complete the proof of the theorem. □

Note that in the proof of this theorem, the estimate of assumption (ii) is used only for $f \in \dot H^{r-2}$ and not for general $f \in H^{r-2}$. Similar remarks apply to Theorems 3.2 and 3.3 below.

We now turn to the situation when v is not regular enough to belong to $\dot H^r$. We shall show that, nevertheless, we have optimal order convergence for t positive.

We shall first consider the case of the standard Galerkin method when $S_h \subset H_0^1$ and satisfies (1.10), and when T_h in (3.4) is defined by

$$(\nabla T_h f, \nabla\chi) = (f, \chi), \quad \forall \chi \in S_h. \tag{3.10}$$

Recall that in this case, with $R_h u \in S_h$ defined by (1.22) we have $R_h = -T_h\Delta$, and hence $\|R_h u - u\| \le Ch\|u\|_1$. We then have the following:

Theorem 3.2 *With T_h defined by the standard Galerkin method, i.e., by* (1.10) *and* (3.10), *and with $v_h = P_h v$, we have for the error in the semidiscrete parabolic problem* (3.4)

$$\|u_h(t) - u(t)\| \le Ch^r t^{-r/2}\|v\|, \quad \text{for } t > 0. \tag{3.11}$$

Proof. We first prove, as usual with $e = u_h - u$, that

$$\|e(t)\| \le Cht^{-1/2}\|v\|, \tag{3.12}$$

and then show the theorem from this by an iteration argument.

We shall apply Lemma 3.3. This time (3.9) holds with $\rho = R_h u - u$, and hence, using the estimate for the elliptic projection of Lemma 1.1 and the regularity estimate of Lemma 3.2 for the exact solution,

$$\|\rho(t)\| = \|R_h u(t) - u(t)\| \le Ch\|u(t)\|_1 \le Cht^{-1/2}\|v\|.$$

Recalling the definition of the norm in $\dot{H}^1$ we also have

$$\int_0^t \|\rho\|^2\,ds \le Ch^2 \int_0^t \|u\|_1^2\,ds \le Ch^2 \int_0^t \sum_{m=1}^{\infty} \lambda_m (u(s), \varphi_m)^2\,ds$$
$$\le Ch^2 \int_0^{\infty} \sum_{m=1}^{\infty} \lambda_m e^{-2\lambda_m s}(v, \varphi_m)^2\,ds \le Ch^2 \sum_{m=1}^{\infty} (v, \varphi_m)^2 = Ch^2\|v\|^2.$$

Finally, in the same way,

$$\int_0^t s^2\|\rho_t\|^2\,ds \le Ch^2 \int_0^t s^2\|u_t\|_1^2\,ds \le Ch^2 \int_0^t s^2\|u\|_3^2\,ds$$
$$\le Ch^2 \int_0^{\infty} \sum_{m=1}^{\infty} \lambda_m^3 (v, \varphi_m)^2 s^2 e^{-2\lambda_m s}\,ds \le Ch^2 \sum_{m=1}^{\infty} (v, \varphi_m)^2 = Ch^2\|v\|^2.$$

By Lemma 3.3 these estimates show (3.12).

Introducing now the error operator $F_h(t)$ by

$$e(t) = F_h(t)v = E_h(t)P_h v - E(t)v = u_h(t) - u(t),$$

with $E_h(t)$ the solution operator of (3.4), (3.11) may be stated as

$$\|F_h(t)v\| \le Ch^r t^{-r/2}\|v\|, \quad \text{for } t > 0. \tag{3.13}$$

Since clearly $F_h(t)$ is bounded in L_2, it is no restriction to assume $ht^{-1/2} \le 1$. We have the identity

$$F_h(t) = F_h(t/2)E(t/2) + E(t/2)F_h(t/2) + F_h(t/2)^2.$$

In fact, using our definitions and the semigroup property $E(t+s) = E(t)E(s)$, and similarly for $E_h(t)$, the right-hand side equals

$$(E_h(t/2)P_h - E(t/2))E(t/2) + E(t/2)(E_h(t/2)P_h - E(t/2)) \\ + (E_h(t/2)P_h - E(t/2))^2 = E_h(t/2)^2P_h - E(t/2)^2 = F_h(t).$$

We have, using Theorem 3.1 and Lemma 3.2,

$$\|F_h(t/2)E(t/2)v\| \le Ch^r|E(t/2)v|_r \le Ch^r t^{-r/2}\|v\|.$$

Noting that $F_h(t/2)$ and $E(t/2)$ are selfadjoint, we see that the product $E(t/2)F_h(t/2)$ is the adjoint of $F_h(t/2)E(t/2)$ and thus has the same norm, considered as an operator on L_2, so that

$$\|E(t/2)F_h(t/2)v\| \le Ch^r t^{-r/2}\|v\|.$$

Also, by (3.12),

$$\|F_h(t/2)^2v\| \le Cht^{-1/2}\|F_h(t/2)v\|,$$

so that, altogether,

$$\|F_h(t)v\| \le Ch^r t^{-r/2}\|v\| + Cht^{-1/2}\|F_h(t/2)v\|.$$

By repeated application we have, since $ht^{-1/2} \le 1$,

$$\|F_h(t)v\| \le Ch^r t^{-r/2}\|v\| + C(ht^{-1/2})^s\|F_h(t/2^s)v\|.$$

Choosing $s = r$ and noting that $\|F_h(t/2^r)v\| \le 2\|v\|$ completes the proof of (3.13), and thus of the theorem. □

We note in passing that since u_h and u are bounded, t may be replaced by $t + h^2$ in the bound in (3.11).

We shall now turn to the more general situation when we only know that (i) and (ii) hold. We have the following:

Theorem 3.3 *Assume that (i), (ii) hold, and that $v_h = P_h v$. Then we have for the error in the semidiscrete parabolic problem* (3.4)

$$\|u_h(t) - u(t)\| \le Ch^r t^{-r/2}\|v\|, \quad \text{for } t > 0.$$

Proof. We shall prove the result for $r = 2$. The same bootstrapping argument as in Theorem 3.2 may then be used to complete the proof.

Recalling the error equation (3.9) and setting $\widetilde{\rho}(t) = \int_0^t \rho(s)\,ds$, we shall prove for the error $e = u_h - u$

$$\|e(t)\| \le Ct^{-1}\sup_{s\le t}\{s^2\|\rho_t(s)\| + s\|\rho(s)\| + \|\widetilde{\rho}(s)\|\}. \tag{3.14}$$

Assuming that this has already been accomplished, we have by (ii) and Lemma 3.2,

$$s\|\rho(s)\| = s\|(T_h - T)u_t(s)\| \le Ch^2 s\|u_t(s)\| \le Ch^2\|v\|$$

and

$$s^2\|\rho_t(s)\| = s^2\|(T_h - T)u_{tt}(s)\| \le Ch^2 s^2\|u_{tt}(s)\| \le Ch^2\|v\|.$$

Since

$$\begin{aligned}\|\widetilde{\rho}(s)\| &= \|\int_0^s (T_h - T)u_t d\sigma\| = \|(T_h - T)(u(s) - v)\| \\ &\le Ch^2(\|u(s)\| + \|v\|) \le Ch^2\|v\|,\end{aligned}$$

we conclude $\|e(t)\| \le Ch^2 t^{-1}\|v\|$, which is the desired estimate for $r = 2$.

It remains to prove (3.14). For this we set $w = te$ and note that by (3.9) w satisfies

$$T_h w_t + w = \eta := t\rho + T_h e.$$

By Lemma 3.5 we therefore find

$$\|w(t)\| \le \varepsilon \sup_{s\le t}(s\|\eta_t(s)\|) + C_\varepsilon \sup_{s\le t}\|\eta(s)\|.$$

Here

$$\|\eta(s)\| \le s\|\rho(s)\| + \|T_h e(s)\|,$$

and, using (3.9),

$$\begin{aligned}s\|\eta_t(s)\| &\le s^2\|\rho_t(s)\| + s\|\rho(s)\| + s\|T_h e_t(s)\| \\ &\le s^2\|\rho_t(s)\| + 2s\|\rho(s)\| + s\|e(s)\| = s^2\|\rho_t(s)\| + 2s\|\rho(s)\| + \|w(s)\|.\end{aligned}$$

With $\varepsilon = 1/2$, say, we conclude, for all $t \ge 0$,

$$\|w(t)\| \le \tfrac{1}{2}\sup_{s\le t}\|w(s)\| + C\sup_{s\le t}(s^2\|\rho_t(s)\| + s\|\rho(s)\| + \|T_h e(s)\|).$$

Choosing $\tau = \tau(t)$ such that $\sup_{s\le t}\|w(s)\| = \|w(\tau)\|$, we have

$$\|w(t)\| \le \|w(\tau)\| \le C\sup_{s\le t}(s^2\|\rho_t(s)\| + s\|\rho(s)\| + \|T_h e(s)\|). \tag{3.15}$$

We now estimate $T_h e$. For this purpose we integrate the error equation (3.9) over $(0, t)$, keeping in mind that $T_h e(0) = 0$, to obtain

$$T_h e + \widetilde{e} = T_h \widetilde{e}_t + \widetilde{e} = \widetilde{\rho}, \quad \text{with } \widetilde{e}(t) = \int_0^t e\, ds.$$

It follows from Lemma 3.4 that

$$\|\widetilde{e}(t)\| \le C\sup_{s\le t}(s\|\widetilde{\rho}_t(s)\| + \|\widetilde{\rho}(s)\|) \le C\sup_{s\le t}(s\|\rho(s)\| + \|\widetilde{\rho}(s)\|),$$

and hence also

$$\|T_h e(t)\| \le \|\widetilde{e}(t)\| + \|\widetilde{\rho}(t)\| \le C \sup_{s\le t}(s\|\rho\| + \|\widetilde{\rho}\|).$$

Combining this with our previous estimate (3.15), we have

$$\|w(t)\| \le C \sup_{s\le t}(s^2\|\rho_t\| + s\|\rho\| + \|\widetilde{\rho}\|),$$

which is (3.14). The proof of the theorem is now complete. □

We shall now show some similar estimates for time derivatives of the error (or the error in the time derivatives). Recall the notation $D_t = \partial/\partial t$.

Theorem 3.4 *Assume that (i) and (ii) hold, and that* $v_h = P_h v$. *Then we have for the error in the semidiscrete parabolic problem* (3.4)

$$\|D_t^l(u_h(t) - u(t))\| \le Ch^r t^{-r/2-l}\|v\|, \quad \text{for } t > 0, \ l \ge 0.$$

Proof. The proof will be by induction over l, with the case $l = 0$ clear by Theorem 3.3. Assume thus the result already shown for $l - 1$ with $l \ge 1$. Setting $e^{(l)} = D_t^l e$, etc., we have, by differentiation of (3.9),

$$T_h e_t^{(l)} + e^{(l)} = \rho^{(l)}.$$

Multiplication by $t^{r/2+l}$ gives, since $T_h e_t^{(l-1)} = T_h e^{(l)}$,

$$\begin{aligned} T_h(t^{r/2+l}e^{(l)})_t + t^{r/2+l}e^{(l)} &= t^{r/2+l}\rho^{(l)} + (\tfrac{1}{2}r + l)t^{r/2+l-1}T_h e^{(l)} \\ &= t^{r/2+l}\rho^{(l)} + (\tfrac{1}{2}r + l)t^{r/2+l-1}(\rho^{(l-1)} + e^{(l-1)}), \end{aligned}$$

and application of Lemma 3.5 yields (note that $T_h(t^{r/2+l}e^{(l)}(t)) = 0$ for $t = 0$)

$$\begin{aligned} t^{r/2+l}\|e^{(l)}(t)\| \le{}& \varepsilon \sup_{s\le t}(s^{r/2+l}\|e^{(l)}(s)\|) \\ &+ C_\varepsilon \sup_{s\le t}\Big(\sum_{j=-1}^{1} s^{r/2+l+j}\|\rho^{(l+j)}(s)\| + s^{r/2+l-1}\|e^{(l-1)}(s)\|\Big). \end{aligned}$$

Now since $\rho^{(q)} = -(T_h - T)u^{(q+1)}$, we have by (ii) and Lemma 3.2, for any $q \ge 0$,

$$s^{r/2+q}\|\rho^{(q)}(s)\| \le Ch^r s^{r/2+q}\|u^{(q+1)}(s)\|_{r-2} \le Ch^r\|v\|,$$

and, using our induction assumption, $s^{r/2+l-1}\|e^{(l-1)}(s)\| \le Ch^r\|v\|$. Hence, choosing $\varepsilon < 1$ above, the result follows in the same way as in the proof of Theorem 3.3. □

Recall that the convergence rate for the solution u_h of (3.4) is of order $O(h^r)$, uniformly down to $t = 0$, if the initial v data belong to $\dot{H}^r$. If the order of regularity of v is lower, only a correspondingly weaker convergence estimate may be proved, uniformly for $t \geq 0$. We have also seen that even without any regularity assumption on initial data, the rate of convergence is $O(h^r)$ for t bounded away from 0, but the bound then depends in a singular way on t as t tends to 0. The order of this singularity depends on the smoothness of v. These remarks are put into quantitative form in the following theorem.

Theorem 3.5 *Assume that (i), (ii) hold, and that* $v_h = P_h v$. *If* $v \in \dot{H}^s$ *and* $0 \leq s \leq q \leq r$, *then we have for the error in the semidiscrete parabolic problem* (3.4)

$$\|u_h(t) - u(t)\| \leq C h^q t^{-(q-s)/2} |v|_s, \quad \textit{for } t > 0. \tag{3.16}$$

Proof. We first show (3.16) for $s = q$. We write

$$v = \sum_{h^2\lambda_j \leq 1} (v, \varphi_j)\varphi_j + \sum_{h^2\lambda_j > 1} (v, \varphi_j)\varphi_j = v_1 + v_2.$$

By Theorem 3.1 we have $\|F_h(t)v_1\| \leq C h^r |v_1|_r$, where

$$|v_1|_r^2 = \sum_{h^2\lambda_j \leq 1} \lambda_j^r (v, \varphi_j)^2 \leq h^{-2(r-q)} \sum_{j=1}^{\infty} \lambda_j^q (v, \varphi_j)^2 = h^{2(q-r)} |v|_q^2.$$

Further, by the stability of the discrete and continuous problems,

$$\begin{aligned} \|F_h(t)v_2\|^2 \leq C\|v_2\|^2 &= C \sum_{h^2\lambda_m \geq 1} (v, \varphi_m)^2 \\ &\leq C h^{2q} \sum_{m=1}^{\infty} \lambda_m^q (v, \varphi_m)^2 = C h^{2q} |v|_q^2. \end{aligned}$$

Thus

$$\|u_h(t) - u(t)\| = \|F_h(t)(v_1 + v_2)\| \leq C h^q |v|_q.$$

For a general s with $0 \leq s \leq q$, we now write

$$v = \sum_{t\lambda_m \leq 1} (v, \varphi_m)\varphi_m + \sum_{t\lambda_m > 1} (v, \varphi_m)\varphi_m = v_I + v_{II}.$$

Using the result for $s = q$ we have

$$\begin{aligned} \|F_h(t)v_I\|^2 \leq C h^{2q} |v_I|_q^2 &= C h^{2q} \sum_{t\lambda_m \leq 1} \lambda_m^q (v, \varphi_m)^2 \\ &= C h^{2q} t^{-(q-s)} \sum_{t\lambda_m \leq 1} (t\lambda_m)^{q-s} \lambda_m^s (v, \varphi_m)^2 \leq C h^{2q} t^{-(q-s)} |v|_s^2. \end{aligned} \tag{3.17}$$

We also note that

$$\|F_h(t)v_{II}\| \leq C h^q t^{-q/2} \|v_{II}\|.$$

This follows at once by stability for $h^2t^{-1} > 1$, and for $h^2t^{-1} \le 1$ by Theorem 3.2 since then $h^r t^{-r/2} \le h^q t^{-q/2}$. Since

$$\|v_{II}\|^2 = \sum_{t\lambda_m > 1} (v, \varphi_m)^2 \le t^s \sum_{m=1}^{\infty} \lambda_m^s (v, \varphi_m)^2 = t^s |v|_s^2,$$

we conclude

$$\|F_h(t) v_{II}\| \le C h^q t^{-(q-s)/2} |v|_s. \tag{3.18}$$

Together (3.17) and (3.18) show our claim. □

We shall now briefly describe an alternative way of deriving the above nonsmooth data error estimates in the case of the standard Galerkin method, in which the main technical device is the use of a dual backward inhomogeneous parabolic equation with vanishing final data, and which avoids the use of the operators T_h and T. We begin with an auxiliary error estimate for the initial boundary value problem

$$\begin{aligned} u_t - \Delta u &= f \quad \text{in } \Omega, \quad t > 0, \\ u &= 0 \quad \text{on } \partial\Omega, \ t > 0, \quad u(0) = 0 \quad \text{in } \Omega, \end{aligned} \tag{3.19}$$

and its semidiscrete analogue

$$(u_{h,t}, \chi) + (\nabla u_h, \nabla \chi) = (f, \chi), \quad \chi \in S_h, \ t > 0, \quad u_h(0) = 0. \tag{3.20}$$

Lemma 3.6 *Let $e = u_h - u$, where u_h and u are the solutions of* (3.20) *and* (3.19). *Then*

$$\int_0^t (\|e_t\|^2 + h^{-2}\|e\|_1^2)\, ds \le C \int_0^t \|f\|^2\, ds, \quad \textit{for } t \ge 0.$$

Proof. We have

$$(e_t, \chi) + (\nabla e, \nabla \chi) = 0 \quad \forall \chi \in S_h, \ t \ge 0, \tag{3.21}$$

and hence writing $e = \theta + \rho$ in the usual way, since $\theta \in S_h$,

$$\begin{aligned} (e_t, e) + (\nabla e, \nabla e) &= (e_t, \rho) + (\nabla e, \nabla \rho) \le \|e_t\|\,\|\rho\| + \|\nabla e\|\,\|\nabla \rho\| \\ &\le C(h^2\|e_t\|^2 + h^{-2}\|\rho\|^2 + \|\nabla \rho\|^2) + \tfrac{1}{2}\|\nabla e\|^2. \end{aligned}$$

By integration and using the standard estimates for ρ, and since $e(0) = 0$, it follows that

$$\int_0^t \|e\|_1^2\, ds \le C h^2 \int_0^t (\|e_t\|^2 + \|u\|_2^2)\, ds.$$

Further, since $e_t = u_{h,t} - u_t$,

$$\int_0^t \|e_t\|^2\,ds \le C \int_0^t (\|u_t\|^2 + \|u_{h,t}\|^2)\,ds.$$

By simple energy arguments we find

$$\int_0^t (\|u_t\|^2 + \|u_{h,t}\|^2 + \|u\|_2^2)\,ds \le C \int_0^t \|f\|^2\,ds.$$

Together these estimates complete the proof. □

Out next lemma concerns the homogeneous parabolic equation and its semidiscrete analogue

$$(3.22)\qquad (u_{h,t}, \chi) + (\nabla u_h, \nabla\chi) = 0, \quad \forall \chi \in S_h,\ t > 0, \quad u_h(0) = P_h v.$$

Lemma 3.7 *Let $e = u_h - u$, where u_h and u are the solutions of (3.22) and (3.1), with $v_h = P_h v$. Then*

$$\int_0^t \|e\|^2\,ds \le Ch^2\|v\|^2, \quad \textit{for } t > 0.$$

Proof. We shall show the estimate for a fixed $t = t_0$. For this purpose consider the backward problem

$$(3.23)\qquad \begin{aligned} -z_t - \Delta z &= e \quad \text{in } \Omega, \quad \text{for } t \le t_0,\\ z &= 0 \quad \text{on } \partial\Omega, \quad \text{for } t \le t_0, \quad z(t_0) = 0 \quad \text{in } \Omega, \end{aligned}$$

and let z_h be the solution of the corresponding semidiscrete problem

$$-(z_{h,t}, \chi) + (\nabla z_h, \nabla\chi) = (e, \chi), \quad \forall \chi \in S_h,\ t \le t_0, \quad z_h(t_0) = 0.$$

Noting that (3.21) holds also in the present case we use this with $\chi = z_h$ to obtain

$$\begin{aligned} \|e\|^2 &= -(e, z_t + \Delta z) = -\frac{d}{dt}(e, z) + (e_t, z) + (\nabla e, \nabla z)\\ &= -\frac{d}{dt}(e, z) - (e_t, z_h - z) - (\nabla e, \nabla(z_h - z))\\ &= -\frac{d}{dt}(e, z_h) + (e, z_{h,t} - z_t) - (\nabla e, \nabla(z_h - z)). \end{aligned}$$

The error $\delta = z_h - z$ satisfies

$$-(\chi, \delta_t) + (\nabla\chi, \nabla\delta) = 0, \quad \forall \chi \in S_h, \quad \text{for } t \le t_0,$$

and recalling that $e = \theta + \rho$, with $\theta \in S_h$, we find

$$\|e\|^2 = -\frac{d}{dt}(e, z_h) + (\rho, \delta_t) - (\nabla\rho, \nabla\delta).$$

By integration, noting that $z_h(t_0) = e(0) = 0$,

$$\int_0^{t_0} \|e\|^2 \, ds \le \varepsilon \int_0^{t_0} (\|\delta_t\|^2 + h^{-2}\|\delta\|_1^2) \, ds + C_\varepsilon \int_0^{t_0} (\|\rho\|^2 + h^2\|\rho\|_1^2) \, ds.$$

Using Lemma 3.6 for the backward problem (3.23), we have

$$\int_0^{t_0} (\|\delta_t\|^2 + h^{-2}\|\delta\|_1^2) \, ds \le \widetilde{C} \int_0^{t_0} \|e\|^2 \, ds,$$

and if $\widetilde{C}\varepsilon \le 1/2$ we may conclude

$$\int_0^{t_0} \|e\|^2 \, ds \le C \int_0^{t_0} (\|\rho\|^2 + h^2\|\rho\|_1^2) \, ds \le Ch^2 \int_0^{t_0} \|u\|_1^2 \, ds \le Ch^2\|v\|^2,$$

which completes the proof. Note that C is independent of t_0. □

Using this lemma we finally show the estimate (3.12) in the present case. The bootstrapping argument of the proof of Theorem 3.2 then implies (3.11).

Lemma 3.8 *Under the assumptions of Lemma 3.7, we have*

$$\|e(t)\| \le Cht^{-1/2}\|v\|, \quad \text{for } t > 0.$$

Proof. Since $\rho(t)$ is bounded as desired by $\|\rho(t)\| \le Ch\|u(t)\|_1 \le Cht^{-1/2}\|v\|$, it remains to consider $\theta = u_h - R_h u$, which satisfies

$$(\theta_t, \chi) + (\nabla\theta, \nabla\chi) = -(\rho_t, \chi), \quad \forall \chi \in S_h, \quad \text{for } t \ge 0.$$

Choosing $\chi = 2\theta$ we obtain, after multiplication by t,

$$\frac{d}{dt}(t\|\theta\|^2) + 2t\|\nabla\theta\|^2 = \|\theta\|^2 - 2t(\rho_t, \theta).$$

Integration yields

$$\begin{aligned} t\|\theta\|^2 &\le C \int_0^t \|\theta\|^2 \, ds + C \int_0^t s^2\|\rho_t\|^2 \, ds \\ &\le C \int_0^t \|e\|^2 \, ds + C \int_0^t (\|\rho\|^2 + s^2\|\rho_t\|^2) \, ds \le Ch^2\|v\|^2, \end{aligned}$$

where in the last step we have used Lemma 3.7 and the estimates for ρ and ρ_t of the proof of Theorem 3.2. This completes the proof. □

The above nonsmooth data results are related to the following smoothing property of $E_h(t)$ which is analogous to that of $E(t)$ shown in Lemma 3.2.

Lemma 3.9 *Assume that (i) holds. Then for each $l \ge 0$ there is a constant C_l, with $C_0 = 1$, such that, for the solution $u_h(t) = E_h(t)v_h$ of (3.4) with $f = 0$,*

$$\|D_t^l E_h(t)v_h\| \le C_l t^{-l}\|v_h\|, \quad \text{for} \quad t > 0.$$

Proof. Letting λ_j^h and $\varphi_j^h, j = 1, \dots, N_h$, be the eigenvalues and orthonormal eigenfunctions of the positive definite operator $-\Delta_h$ we may write

$$E_h(t)v_h = \sum_{j=1}^{N_h} e^{-\lambda_j^h t}(v_h, \varphi_j^h)\varphi_j^h,$$

from which we conclude

$$\begin{aligned}\|D_t^l E_h(t)v_h\|^2 &= \sum_{j=1}^{N_h} (\lambda_j^h)^{2l} e^{-2\lambda_j^h t}(v_h, \varphi_j^h)^2 \le C_l t^{-2l} \sum_{j=1}^{N_h} (v_h, \varphi_j^h)^2 \\ &= C_l t^{-2l}\|v_h\|^2, \quad \text{where } C_l = \sup_{s>0}(s^{2l}e^{-2s}). \qquad \square\end{aligned}$$

We note that as a consequence of this Lemma 3.9, the time derivatives of the error caused by choosing other initial data than $P_h v$ in Theorems 3.4 and 3.6 may be bounded by

$$\|D_t^l E_h(t)(v_h - P_h v)\| \le C t^{-l}\|v_h - P_h v\|.$$

We shall complete this discussion by using our error estimates for the homogeneous problem to show that in order to obtain optimal order error estimates for the inhomogeneous equation, with time bounded away from zero, stringent regularity assumptions only have to be imposed near the time at which the error estimate is sought. We consider thus

$$\begin{aligned} u_t - \Delta u = f \quad &\text{in } \Omega, \quad t > 0, \\ u = 0 \quad &\text{on } \partial\Omega, \quad t > 0, \quad u = v \quad \text{in } \Omega, \quad \text{for } t = 0,\end{aligned}$$

and the semidiscrete analogue of this problem

$$T_h u_{h,t} + u_h = T_h f, \quad \text{for } t > 0, \quad \text{with } u_h(0) = v_h. \tag{3.24}$$

We shall prove the following:

Theorem 3.6 *Assume that (i) and (ii) hold, and that $v_h = P_h v$. Then for any $l \ge 0$, $t \ge \delta > 0$, we have for the error in the semidiscrete parabolic problem (3.24), for $t \ge \delta$,*

$$\|D_t^l(u_h(t) - u(t))\| \le Ch^r\Big(\|v\| + \int_0^t \|f\|\,ds + \sum_{j\le l+1}\int_{t-\delta}^t \|D_t^j u(s)\|_r\,ds\Big).$$

Proof. We shall consider a fixed $t = t_0 > \delta$. Let $\varphi \in \mathcal{C}^\infty$ be such that $\varphi(t) = 1$ for $t \ge -3\delta/4$, $\varphi(t) = 0$ for $t \le -\delta$. Set $\varphi_1(t) = \varphi(t - t_0)$. We now write $u = u_1 + u_2 + u_3$, where $u_1 = u\,\varphi_1$ and u_2 is the solution of the homogeneous equation,

(3.25) $$u_{2,t} - \Delta u_2 = 0, \quad \text{for } t > 0, \quad \text{with } u_2(0) = v.$$

Since

(3.26) $$u_{1,t} - \Delta u_1 = f_1 := f\varphi_1 + u\varphi_1', \quad \text{for } t > 0, \quad \text{with } u_1(0) = 0,$$

it follows that u_3 satisfies

(3.27) $$u_{3,t} - \Delta u_3 = f_3 := f(1 - \varphi_1) - u\varphi_1', \quad \text{for } t > 0, \quad \text{with } u_3(0) = 0.$$

We notice that f_1 and f_3 vanish for $t \le t_0 - \delta$ and $t \ge t_0 - 3\delta/4$, respectively.

Let $u_{j,h}$, $j = 1,2,3$, be the semidiscrete approximations of problems (3.26), (3.25), and (3.27) with $u_{1,h}(0) = u_{3,h}(0) = 0$, $u_{2,h}(0) = P_h v$, and set $e_j = u_{j,h} - u_j$. Since, by linearity, $e = u_h - u = \sum_{j=1}^{3} e_j$, it suffices to estimate $e_j(t_0)$, $j = 1,2,3$, by the right-hand side of the estimate claimed.

Consider first the error in u_1. It follows by Theorem 2.3 that since $D_t^l u_1$ satisfies the equation resulting from (3.26) by differentiation and $D_t^l u_{1,h}$ its discrete counterpart, with both these functions vanishing for small t,

$$\|D_t^l e_1(t_0)\| \le Ch^r \int_0^{t_0} \|D_t^{l+1} u_1\|_r \, ds \le Ch^r \sum_{j \le l+1} \int_{t_0-\delta}^{t_0} \|D_t^j u\|_r \, ds.$$

For u_2, the solution of the homogeneous equation, we have by Theorem 3.4 above

$$\|D_t^l e_2(t_0)\| \le Ch^r t_0^{-r/2-l} \|v\| \le C(\delta) h^r \|v\|.$$

For the purpose of dealing with u_3, finally, we utilize again the error operator $F_h(t) = E_h(t)P_h - E(t)$, and recall that by above

$$\|D_t^l F_h(t) v\| \le Ch^r \|v\|, \quad \text{for } t \ge \delta/4.$$

We observe now that by superposition we may write, for $t > t_0 - \delta/2$,

$$e_3(t) = \int_0^t F_h(t-s) f_3(s) \, ds = \int_0^{t_0 - 3\delta/4} F_h(t-s) f_3(s) \, ds,$$

and hence

$$D_t^l e_3(t_0) = \int_0^{t_0 - 3\delta/4} D_t^l F_h(t_0 - s) f_3(s) \, ds.$$

Therefore, since $t_0 - s$ is bounded below,

$$\begin{aligned} \|D_t^l e_3(t_0)\| &\le Ch^r \int_0^{t_0 - 3\delta/4} \|f_3(s)\| \, ds \\ &\le Ch^r \int_0^{t_0} (\|f\| + \|u\|) \, ds \le Ch^r \big(\|v\| + \int_0^{t_0} \|f\| \, ds \big). \end{aligned}$$

Here the last step follows by the fact that

$$\frac{d}{dt}\|u\|^2 + 2\|\nabla u\|^2 = 2(f,u) \le 2\|f\|\,\|u\|,$$

and hence in the standard way

$$\|u(t)\| \le \|v\| + \int_0^t \|f\|\,ds, \quad \text{for } t \ge 0.$$

This completes the proof. □

We shall close this chapter by using Lemma 3.9 to show an almost optimal order error estimate for the inhomogeneous problem (3.4), in which the error bound does not contain any time derivative of the exact solution. In this we need to assume the inverse estimate

$$\|\Delta_h \chi\| \le C h^{-\beta}\|\chi\|, \quad \text{for } \chi \in S_h, \quad \text{with } \beta > 0. \tag{3.28}$$

For the standard Galerkin method, with the inverse assumption (1.12), valid for quasiuniform triangulations, the estimate (3.28) holds with $\beta = 2$, since, for $\chi \in S_h$,

$$\|\Delta_h\chi\|^2 = -(\nabla(\Delta_h\chi), \nabla\chi) \le \|\nabla(\Delta_h\chi)\|\,\|\nabla\chi\| \le Ch^{-2}\|\Delta_h\chi\|\,\|\chi\|,$$

and similarly one obtains for Nitsche's method, using Lemma 2.1,

$$\|\Delta_h\chi\|^2 = -N_\gamma(\Delta_h\chi,\chi) \le C|||\Delta_h\chi|||\,|||\chi||| \le Ch^{-2}\|\Delta_h\chi\|\,\|\chi\|.$$

The estimate (3.28) holds for more general families of triangulations than quasiuniform ones.

Theorem 3.7 *Assume that (i), (ii), and* (3.28) *hold, and let u_h and u be the solutions of* (2.20) *and* (2.14), *with $v_h = R_h v := -T_h\Delta v$. Then*

$$\|u_h(t) - u(t)\| \le Ch^r \max\big(1, \log\frac{t}{h^\beta}\big) \sup_{0\le s\le t}\|u(s)\|_r, \quad \textit{for } t \ge 0.$$

Proof. As on previous occasions we write

$$u_h - u = (u_h - R_h u) + (R_h u - u) = \theta + \rho.$$

From (2.23) we have

$$\|\rho(t)\| \le Ch^r\|u(t)\|_r,$$

and it remains to bound $\theta = u_h - R_h u$. We obtain by (3.24), since $R_h T = T_h$,

$$T_h\theta_t + \theta = T_h f - (T_h R_h u_t + R_h u) = -T_h\rho_t = -T_h P_h \rho_t,$$

where in the last step we have used (2.24). We therefore have

$$\theta_t - \Delta_h\theta = -P_h\rho_t, \quad \text{for } t > 0, \quad \text{with } \theta(0) = 0,$$

and hence by Duhamel's principle, as in (1.37) for the standard Galerkin method,

$$\theta(t) = -\int_0^t E_h(t-s)P_h\rho_t(s)\,ds.$$

By integration by parts we obtain

$$\theta(t) = E_h(t)P_h\rho(0) - P_h\rho(t) - \int_0^t E_h'(t-s)P_h\rho(s)\,ds,$$

which shows

$$\|\theta(t)\| \le \Big(\|E_h(t)\| + 1 + \int_0^t \|E_h'(s)\|\,ds\Big) \sup_{0\le s\le t}\|\rho(s)\|. \tag{3.29}$$

In order to estimate the integral, we may bound the integrand for small s by $Ch^{-\beta}$. In fact, applying the inverse assumption (3.28) and Lemma 3.9 we have

$$\|E_h'(t)v_h\| = \|\Delta_h E_h(t)v_h\| \le Ch^{-\beta}\|E_h(t)v_h\| \le Ch^{-\beta}\|v_h\|.$$

Thus,

$$\int_0^t \|E_h'(s)\|\,ds \le C. \quad \text{for } t \le h^\beta.$$

Since by Lemma 3.9 also $\|E_h'(t)v_h\| \le Ct^{-1}\|v_h\|$, we have

$$\int_{h^\beta}^t \|E_h'(s)\|\,ds \le C\Big|\int_{h^\beta}^t \frac{ds}{s}\Big| = C\log\frac{t}{h^\beta}, \quad \text{for } t \ge h^\beta.$$

Since $E_h(t)$ is bounded, we conclude from (3.29) and (2.23) that

$$\|\theta(t)\| \le Ch^r \max\big(1, \log\frac{t}{h^\beta}\big) \sup_{0\le s\le t}\|u(s)\|_r,$$

which completes the proof. □

The smooth data result of Theorem 3.1 is from Bramble, Schatz, Thomée, and Wahlbin [37]. Results for nonsmooth data for the homogeneous equation were first discussed by spectral representation in Blair [28], Thomée [225], Helfrich [117], Fujita and Mizutani [103], and Bramble, Schatz, Thomée, and Wahlbin [37], and later by the energy method in Luskin and Rannacher [166], Sammon [205], and Thomée [228]. The use of the backward parabolic problem in the nonsmooth data estimates was proposed in Luskin and Rannacher [167]. Theorem 3.6 is a special case of a result in [228].

4. More General Parabolic Equations

In this chapter we shall briefly discuss the generalization of our previous error analysis to initial-boundary value problems for more general parabolic equations, in which we allow the elliptic operator to have coefficients depending on both x and t, to contain lower order terms, and to be nonselfadjoint and nonpositive. In order not to have to account for possible exponential growth of stability constants and error bounds we restrict our considerations to a finite interval in time.

We consider thus the initial boundary value problem

$$(4.1)\qquad \begin{aligned} u_t + A(t)u &= f \quad \text{in } \Omega, \qquad \text{for } t \in J, \\ u &= 0 \quad \text{on } \partial\Omega, \quad \text{for } t \in J, \quad u(0) = v \quad \text{in } \Omega, \end{aligned}$$

where Ω is a domain in $\mathbb{R}^d$ with smooth boundary $\partial\Omega$, $J = (0, \bar{t}\,]$, and $A(t)$ denotes the elliptic operator

$$A(t)u := -\sum_{j,k=1}^{d} \frac{\partial}{\partial x_j}\Big(a_{jk}\frac{\partial u}{\partial x_k}\Big) + \sum_{j=1}^{d} a_j \frac{\partial u}{\partial x_j} + a_0 u,$$

where a_{jk}, a_j, and a_0 are $\mathcal{C}^\infty$ functions on $\bar{\Omega} \times \bar{J}$, $a_{jk} = a_{kj}$, and

$$\sum_{j,k=1}^{d} a_{jk}(x,t)\xi_j\xi_k \ge c_0|\xi|^2, \quad \text{with } c_0 > 0, \quad \text{for } (x,t) \in \bar{\Omega} \times \bar{J}.$$

Associating with $A(t)$ the bilinear form

$$A(t; v, w) = \int_\Omega \Big(\sum_{j,k=1}^{d} a_{jk}\frac{\partial v}{\partial x_k}\frac{\partial w}{\partial x_j} + \sum_{j=1}^{d} a_j \frac{\partial v}{\partial x_j} w + a_0 v w\Big)\, dx,$$

we may write the parabolic problem in variational form as

$$(4.2)\qquad \begin{aligned} (u_t, \varphi) + A(t; u, \varphi) &= (f, \varphi), \quad \forall \varphi \in H_0^1 = H_0^1(\Omega), \quad t \in J, \\ u(0) &= v. \end{aligned}$$

This time the bilinear form is not necessarily positive definite, but one easily shows Gårding's inequality

$$A(t;v,v) \geq c\|v\|_1^2 - \kappa\|v\|^2, \quad \forall v \in H_0^1, \quad \text{with } c>0,\ \kappa \in \mathbb{R}. \tag{4.3}$$

In fact, we have for $v \in H_0^1$

$$\begin{aligned}
A(t;v,v) &+ \kappa\|v\|^2 \\
&= \int_\Omega \Big(\sum_{j,k=1}^d a_{jk} \frac{\partial v}{\partial x_k}\frac{\partial v}{\partial x_j} + \sum_{j=1}^d a_j v \frac{\partial v}{\partial x_j} + (a_0+\kappa)v^2 \Big)\, dx \\
&= \int_\Omega \Big(\sum_{j,k=1}^d a_{jk} \frac{\partial v}{\partial x_k}\frac{\partial v}{\partial x_j} + (\kappa + a_0 - \tfrac12 \sum_{j=1}^d \frac{\partial a_j}{\partial x_j}) v^2 \Big)\, dx \\
&\geq c\|v\|_1^2, \quad \text{with } c>0, \quad \text{if } \kappa > \sup_{\Omega\times J} \big(\tfrac12 \sum_{j=1}^d \frac{\partial a_j}{\partial x_j} - a_0 \big);
\end{aligned}$$

we shall consider κ to be fixed in this manner in the sequel.

One may show, cf. [96], that problem (4.1) admits a unique solution which belongs to any space $H^s = H^s(\Omega)$, together with its time derivatives, for $t \in J$, and that the regularity estimate (1.20) holds, provided f and v are regular enough and satisfy the appropriate compatibility conditions on $\partial\Omega$ for $t=0$. Here we restrict ourselves to showing the following stability estimate in $L_2 = L_2(\Omega)$, namely

$$\|u(t)\| \leq C\big(\|v\| + \int_0^t \|f\|\, ds\big), \quad \text{for } t \in \bar{J}. \tag{4.4}$$

For this we choose $\varphi = u$ in (4.2) to obtain, in view of (4.3),

$$\tfrac12 \frac{d}{dt}\|u\|^2 + c\|u\|_1^2 \leq \|f\|\,\|u\| + \kappa\|u\|^2. \tag{4.5}$$

As in the proof of (1.29) this shows

$$\|u(t)\| \leq \|v\| + \int_0^t \|f\|\, ds + \kappa \int_0^t \|u\|\, ds,$$

from which (4.4) follows by Gronwall's lemma. Note that in contrast to the case treated in Chapters 1-3 where the operator $A(t) = A$ was assumed independent of t and selfadjoint, the method of eigenfunction expansion is not suitable here.

We now associate with the parabolic problem (4.1) the time-dependent Dirichlet problem

$$A_\kappa(t)u := A(t)u + \kappa u = f \ \text{ in } \Omega, \quad \text{with } u=0 \text{ on } \partial\Omega, \quad \text{for } t \in J,$$

or, in weak form,

$$A_\kappa(t;u,\varphi) := A(t;u,\varphi) + \kappa(u,\varphi) = (f,\varphi), \quad \forall \varphi \in H_0^1, \ t \in J.$$

We denote by $T(t) : L_2 \to H^2 \cap H_0^1$ the solution operator of this problem, so that

$$(4.6) \qquad A_\kappa(t;T(t)f,\varphi) = (f,\varphi), \quad \forall \varphi \in H_0^1, \quad \text{for } t \in \bar{J},$$

and recall the elliptic regularity estimate, cf. [96],

$$(4.7) \qquad \|T(t)f\|_s \le C\|f\|_{s-2}, \quad \text{for } s \ge 2, \ t \in \bar{J}.$$

Introducing $\check{u} = e^{-t\kappa}u$ as a new dependent variable in (4.1), we have

$$(4.8) \qquad \check{u}_t + A_\kappa(t)\check{u} = \check{f}, \quad \text{where } \check{f} = e^{-t\kappa}f,$$

or

$$(\check{u}_t,\varphi) + A_\kappa(t;\check{u},\varphi) = (\check{f},\varphi), \quad \forall \varphi \in H_0^1, \quad \text{for } t \in J,$$

or also

$$T(t)\check{u}_t + \check{u} = T(t)\check{f}, \quad \text{for } t \in J, \quad \text{with } \check{u}(0) = v.$$

For the purpose of defining approximate solutions of (4.1), let $\{S_h\}$ be a family of finite dimensional subspaces of L_2 and $T_h(t) : L_2 \to S_h$ approximations of $T(t)$ with certain properties to be stated below. Consider then as an approximate solution of (4.1) a function $u_h : J \to S_h$ such that

$$(4.9) \quad u_h(t) = e^{\kappa t}\check{u}_h(t), \quad \text{where } T_h\check{u}_{h,t} + \check{u}_h = T_h\check{f}, \ \text{for}\, t \in J, \ \ \check{u}_h(0) = v_h.$$

We note that boundedness for positive time of $\check{u}$ and $\check{u}_h$ in (4.8) and (4.9) correspond to exponential growth of u and u_h when $\kappa > 0$.

For brevity we shall often omit the variable t in the notation below and simply write A for $A(t)$, $A(v,w)$ for $A(t;v,w)$, T_h for $T_h(t)$, etc.

We now describe the conditions which will be placed upon the operators T_h for the function u_h defined by (4.9) to be a good approximation of the exact solution of (4.1). The first two conditions correspond to those for the model problem treated earlier with the second one modified to allow for the variation in time of the coefficients. The third condition will bound the degree of nonselfadjointness of T_h and is automatically satisfied in the selfadjoint case. We assume thus that for $t \in J$, with C independent of t, and with $'$ denoting differentiation with respect to t,

(i) $(f, T_h f) \ge 0$ *for* $f \in L_2$, *and* $(\chi, T_h\chi) > 0$ *for* $0 \ne \chi \in S_h$;

(ii) *for some integer* $r \ge 2$ *and for* $2 \le s \le r$,

$$\|(T_h - T)f\| + \|(T_h' - T')f\| \le Ch^s\|f\|_{s-2}, \quad \text{for } f \in H^{s-2};$$

(iii) $|(T_h f, g) - (f, T_h g)| \le C(f, T_h f)^{1/2} \|T_h g\|$, *for* $f, g \in L_2$.

As a first example, consider the case that $S_h \subset H_0^1$ and T_h is associated with the standard Galerkin method, so that

$$(4.10) \qquad A_\kappa(t; T_h(t) f, \chi) = (f, \chi), \quad \forall \chi \in S_h, \ t \in J.$$

Then the semidiscrete equation in (4.9) is equivalent to

$$(\check{u}_{h,t}, \chi) + A_\kappa(\check{u}_h, \chi) = (\check{f}, \chi), \quad \forall \chi \in S_h, \ t \in J,$$

which in turn, with $u_h(t) = e^{\kappa t} \check{u}_h(t)$, reduces to the standard weak formulation

$$(u_{h,t}, \chi) + A(u_h, \chi) = (f, \chi), \quad \forall \chi \in S_h, \ t \in J.$$

We shall prove that our conditions are valid for this choice of the T_h.

Lemma 4.1 *Let T_h be defined by (4.10), with S_h satisfying* (1.10). *Then (i), (ii), and (iii) hold.*

Proof. We have at once by (4.10) and (4.3)

$$(f, T_h f) = A_\kappa(T_h f, T_h f) \ge c\|T_h f\|_1^2 \ge 0,$$

which shows the first part of (i) and also that equality holds only if $T_h f = 0$. Assume now that $T_h f = 0$ and that $f = \chi \in S_h$. Using (4.10) once more we have $\|\chi\|^2 = A_\kappa(T_h \chi, \chi) = 0$, so that $\chi = 0$, showing the second part of (i).

We now turn to condition (ii). It is well known, and proved in essentially the same way as for the selfadjoint case, that

$$(4.11) \qquad \|(T_h - T)f\| + h\|(T_h - T)f\|_1 \le Ch^s \|f\|_{s-2}, \qquad \text{for } 2 \le s \le r,$$

and it thus remains to prove the corresponding result for the time derivative. For this purpose, set $w = Tf$, $w_h = T_h f$ and $e = w_h - w$, so that $(T_h' - T')f = e_t$. Differentiating the equation $A_\kappa(e, \chi) = 0$ we obtain, with $A'(\cdot,\cdot)$ the bilinear form obtained from $A(\cdot,\cdot)$ by differentiating the coefficients with respect to t, noting that $A_\kappa'(\cdot,\cdot) = A'(\cdot,\cdot)$,

$$(4.12) \qquad A_\kappa(e_t, \chi) + A'(e, \chi) = 0, \quad \forall \chi \in S_h, \ t \in J.$$

Hence, for any $\chi \in S_h$,

$$c\|e_t\|_1^2 \le A_\kappa(e_t, e_t) = A_\kappa(e_t, e_t + \chi) + A'(e, e_t + \chi) - A'(e, e_t).$$

From this we conclude

$$\|e_t\|_1^2 \le C(\|e_t\|_1 + \|e\|_1) \inf_{\chi \in S_h} \|w_t - \chi\|_1 + C\|e\|_1 \|e_t\|_1,$$

and hence easily, using (4.11),

$$\|e_t\|_1 \le C(\|e\|_1 + \inf_{\chi\in S_h} \|w_t - \chi\|_1) \le Ch^{s-1}(\|f\|_{s-2} + \|w_t\|_s), \quad 2 \le s \le r.$$

Here $\|w_t\|_s \le C\|w\|_s$, which follows since $w_t \in H_0^1$ is the solution of the Dirichlet problem

$$A_\kappa(w_t, \varphi) = -A'(w, \varphi), \quad \forall \varphi \in H_0^1,$$

and $\|w\|_s \le C\|f\|_{s-2}$ by (4.7) so that $\|e_t\|_1 \le h^{s-1}\|f\|_{s-2}$ for $2 \le s \le r$.

In order to show the L_2-norm bound stated for e_t, let A_κ^* be the adjoint of A_κ and ψ the solution of

$$A_\kappa^* \psi = \varphi \quad \text{in } \Omega, \quad \text{with } \psi = 0 \text{ on } \partial\Omega.$$

We then have, again by application of (4.12),

$$(e_t, \varphi) = A_\kappa(e_t, \psi) = A_\kappa(e_t, \psi - \chi) + A'(e, \psi - \chi) - A'(e, \psi), \quad \forall \chi \in S_h,$$

whence, using Green's formula in the last term,

$$\begin{aligned}|(e_t, \varphi)| &\le C(\|e_t\|_1 + \|e\|_1) \inf_{\chi\in S_h} \|\psi - \chi\|_1 + C\|e\|\,\|\psi\|_2 \\ &\le C\big(h(\|e_t\|_1 + \|e\|_1) + \|e\|\big)\|\psi\|_2.\end{aligned}$$

By the elliptic regularity estimate $\|\psi\|_2 \le C\|\varphi\|$ this, together with the error bounds already derived, shows

$$\|e_t\| \le C(h(\|e_t\|_1 + \|e\|_1) + \|e\|) \le Ch^s\|f\|_{s-2}, \quad \text{for } 2 \le s \le r,$$

which completes the proof of (ii).

By our definitions we have with $v_h = T_h f$, $w_h = T_h g$,

$$\begin{aligned}(T_h f, g) - (f, T_h g) &= A_\kappa(T_h g, T_h f) - A_\kappa(T_h f, T_h g) \\ &= \int_\Omega \sum_{j=1}^d a_j\Big(\frac{\partial w_h}{\partial x_j} v_h - \frac{\partial v_h}{\partial x_j} w_h\Big)\,dx \\ &= -\int_\Omega \sum_{j=1}^d \Big(2a_j \frac{\partial v_h}{\partial x_j} w_h + \frac{\partial a_j}{\partial x_j} v_h w_h\Big)\,dx,\end{aligned} \tag{4.13}$$

and hence

$$\begin{aligned}|(T_h f, g) - (f, T_h g)| &\le C\|v_h\|_1\|w_h\| = C\|T_h f\|_1\|T_h g\| \\ &\le C(f, T_h f)^{1/2}\|T_h g\|,\end{aligned}$$

which shows (iii). The proof of the lemma is now complete. □

Another example of a family of operators $T_h(t)$ satisfying our above conditions (i), (ii), (iii) is provided by the generalization to the present context of Nitsche's method described in Chapter 2 where the bilinear form used is now defined by

$$N_{\kappa,\gamma}(t;\varphi,\psi)$$
$$= A_\kappa(t;\varphi,\psi) - \langle \frac{\partial\varphi}{\partial\nu},\psi\rangle - \langle \varphi, \frac{\partial\psi}{\partial\nu} + \sum_{j=1}^{2} a_j n_j \psi\rangle + \gamma h^{-1}\langle\varphi,\psi\rangle,$$

with $\partial/\partial\nu = \Sigma_{jk} n_j a_{jk} \partial/\partial x_k$ the conormal derivative.

We return to the initial-boundary value problem (4.1), and begin our error analysis in L_2 with the following simple result for the inhomogeneous equation which generalizes Theorem 2.3. We note that condition (iii) does not enter in this result.

Theorem 4.1 *Assume that (i) and (ii) hold. Then we have for the error in the semidiscrete parabolic problem* (4.9)

$$\|u_h(t) - u(t)\| \le C\|v_h - v\| + Ch^r\Big(\|v\|_r + \int_0^t \|u_t\|_r\,ds\Big), \quad \text{for } t \in \bar{J}.$$

Proof. With the above notation, set $e = \check{u}_h - \check{u} = e^{-\kappa t}(u_h - u)$. We have then the error equation

$$T_h e_t + e = \rho := (T_h - T)A_\kappa \check{u}.$$

Recalling Lemma 2.4, we hence have

$$\|e(t)\| \le \|e(0)\| + C\Big(\|\rho(0)\| + \int_0^t \|\rho_t\|\,ds\Big).$$

Here, by (ii),

$$\|\rho(0)\| \le Ch^r\|A_\kappa v\|_{r-2} \le Ch^r\|v\|_r,$$

and

$$\begin{aligned}\|\rho_t\| &\le \|(T_h' - T')A_\kappa\check{u}\| + \|(T_h - T)(A'\check{u} + A_\kappa\check{u}_t)\| \\ &\le Ch^r(\|\check{u}\|_r + \|\check{u}_t\|_r) \le Ch^r(\|u\|_r + \|u_t\|_r),\end{aligned}$$

where $A' = A'(t)$ denotes the operator obtained from $A(t)$ by differentiation of its coefficients with respect to t. Hence, since J is bounded,

$$\int_0^t \|\rho_t\|\,ds \le Ch^r\Big(\|v\|_r + \int_0^t \|u_t\|_r\,ds\Big),$$

which completes the proof. □

We now turn to the homogeneous equation

$$u_t + A(t)u = 0 \quad \text{in } \Omega, \quad \text{for } t \in J, \tag{4.14}$$

again with the initial-boundary conditions of (4.1), and its semidiscrete counterpart, to find $u_h(t) : J \to S_h$ such that $\check{u}_h(t) = e^{-\kappa t} u_h(t)$ satisfies

$$T_h \check{u}_{h,t} + \check{u}_h = 0, \quad \text{for } t \in J, \quad \text{with } \check{u}_h(0) = v_h. \tag{4.15}$$

As an example of a nonsmooth data error estimate of continuous piecewise linear functions.

Theorem 4.2 *Assume that (i), (ii) with $r = 2$, and (iii) hold, and that $v_h = P_h v$. Then we have for the error in the semidiscrete homogeneous parabolic problem* (4.15)

$$\|u_h(t) - u(t)\| \le C h^2 t^{-1} \|v\|, \quad \text{for } t \in J.$$

Before we give the proof we shall derive some auxiliary technical results, and begin the proof by showing some bounds for T_h'.

Lemma 4.2 *Assume that (ii) holds with $r = 2$. Then, for $f \in L_2$,*

$$|(T_h' f, f)| \le C((T_h f, f) + h^2 \|f\|^2),$$

and

$$\|T_h' f\| \le C(\|T_h f\| + h^2 \|f\|).$$

Proof. We shall show the continuous counterparts of these estimates, namely

$$|(T' f, f)| \le C(Tf, f) \quad \text{and} \quad \|T' f\| \le C\|Tf\|.$$

The desired results then easily follow by (ii), as for instance, for the first inequality,

$$\begin{aligned} |(T_h' f, f)| &= |(T' f, f) + ((T_h' - T')f, f)| \le C((Tf, f) + h^2 \|f\|^2) \\ &\le C((T_h f, f) + h^2 \|f\|^2). \end{aligned}$$

For the continuous inequalities, recall definition (4.6) and note that we may identify the adjoint of T in L_2 with the operator $T^* : L_2 \to H^2 \cap H_0^1$ defined by $A_\kappa(\varphi, T^* g) = (\varphi, g)$, for $\varphi \in H_0^1$, that is, as the solution of the Dirichlet problem corresponding to the elliptic operator A_κ^*. For

$$(Tf, g) = A_\kappa(Tf, T^* g) = (f, T^* g), \quad \forall f, g \in L_2.$$

Differentiating (4.6) we have $A_\kappa(T' f, \varphi) + A'(Tf, \varphi) = 0$, and we find

$$\begin{aligned} |(T' f, f)| &= |A_\kappa(T' f, T^* f)| = |A'(Tf, T^* f)| \le C\|Tf\|_1 \|T^* f\|_1 \\ &\le C(f, Tf)^{1/2} (T^* f, f)^{1/2} = C(f, Tf), \end{aligned}$$

which is the first of the desired inequalities.

Further, with $\varphi \in L_2$, $(T'f,\varphi) = A_\kappa(T'f, T^*\varphi) = -A'(Tf, T^*\varphi)$, and using Green's formula to transfer all derivatives onto the second factor,

$$|(T'f,\varphi)| \le C\|Tf\|\,\|T^*\varphi\|_2 \le C\|Tf\|\,\|\varphi\|,$$

which shows the second estimate claimed. □

We next show the following analogue of Lemma 3.5.

Lemma 4.3 *Assume that (i), (ii) with $r = 2$, and (iii) hold and that*

$$T_h e_t + e = \rho, \quad \text{for } t \in J, \quad \text{with } T_h e(0) = 0. \tag{4.16}$$

Then for each $\varepsilon > 0$ there is a C_ε such that

$$\|e(t)\| \le \varepsilon \sup_{s\le t}(s\|\rho_t(s)\|) + C_\varepsilon \sup_{s\le t}\|\rho(s)\|, \quad \text{for } t \in J.$$

Proof. As in the proof of Lemma 3.5 we shall show

$$\|e(t)\|^2 \le \frac{\varepsilon^2}{t}\int_0^t s^2\|\rho_t\|^2\,ds + C_\varepsilon\Big(\|\rho(t)\|^2 + \frac{1}{t}\int_0^t \|\rho\|^2\,ds\Big),$$

which immediately implies the desired conclusion. For this purpose we multiply (4.16) by $2te_t$ and obtain after some manipulation,

$$2t(T_h e_t, e_t) + \frac{d}{dt}(t\|e\|^2) = 2\frac{d}{dt}(t(\rho,e)) - 2(\rho,e) - 2t(\rho_t,e) + \|e\|^2,$$

and hence, after integration and obvious estimates, and using (i),

$$t\|e(t)\|^2 \le \varepsilon^2\int_0^t s^2\|\rho_t\|^2\,ds + C_\varepsilon\big(t\|\rho(t)\|^2 + \int_0^t (\|\rho\|^2 + \|e\|^2)\,ds\big).$$

In order to complete the proof we now show

$$\int_0^t \|e\|^2\,ds \le C\int_0^t \|\rho\|^2\,ds. \tag{4.17}$$

For this we multiply (4.16) by $2e$ to obtain

$$\frac{d}{dt}(T_h e, e) + 2\|e\|^2 = 2(\rho,e) + (T_h' e, e) + \big((T_h e, e_t) - (T_h e_t, e)\big).$$

Here, using (iii) and (4.16), we have

$$\begin{aligned} |(T_h e, e_t) - (T_h e_t, e)| &\le C(T_h e, e)^{1/2}\|T_h e_t\| \le C(T_h e, e)^{1/2}(\|\rho\| + \|e\|) \\ &\le C(T_h e, e) + C\|\rho\|^2 + \tfrac12\|e\|^2, \end{aligned}$$

and by Lemma 4.2, for small h, $|(T_h' e, e)| \le C(T_h e, e) + \frac14\|e\|^2$. Hence,

$$\frac{d}{dt}(T_h e, e) + \|e\|^2 \le C(\|\rho\|^2 + (T_h e, e)).$$

By Gronwall's lemma and our assumptions that $T_h e(0) = 0$, the time interval J is bounded, and $(T_h e, e) \ge 0$ by (i), this yields (4.17) and thus completes the proof of the lemma. □

We shall also need the following regularity result.

Lemma 4.4 *For each $j \ge 0$ we have for the solution of* (4.14) *with $u(0) = v$*

$$\|D_t^j u(t)\| \le Ct^{-j}\|v\|, \quad \textit{for } t \in J. \tag{4.18}$$

Proof. We shall only give the proof for $j = 1$; for $j = 0$ it follows from the stability property (4.4) and for other values of j, see [217]. As for (4.4) we shall use an energy argument. We may assume $\kappa = 0$; otherwise we transform the equation as earlier by $\check{u}(t) = e^{-\kappa t}u(t)$.

Differentiating and choosing $\varphi = 2t^2 u_t$ in (4.2) (with $f = 0$) we obtain

$$\frac{d}{dt}(t^2\|u_t\|^2) + 2t^2(A(u_t, u_t) + A'(u, u_t)) = 2t\|u_t\|^2,$$

and hence by integration and obvious estimates, using (4.3) with $\kappa = 0$,

$$t^2\|u_t(t)\|^2 + \int_0^t s^2\|u_t\|_1^2\,ds \le C\int_0^t \|u\|_1^2\,ds + C\int_0^t s\|u_t\|^2\,ds.$$

Integrating (4.5) (with $f = 0$) we find that the first term on the right is bounded by $C\|v\|^2$. To bound the second term we note, cf. (4.13), that

$$|A(v, w) - A(w, v)| \le C\|v\|_1\|w\|,$$

and hence

$$\frac{d}{dt}A(u, u) = A(u, u_t) + A(u_t, u) + A'(u, u) \le 2A(u, u_t) + \|u_t\|^2 + C\|u\|_1^2.$$

With $\varphi = 2u_t$ in (4.2) (with $f = 0$) we therefore get

$$\|u_t\|^2 + \frac{d}{dt}A(u, u) \le C\|u\|_1^2,$$

and hence, after multiplication by t and integration

$$\int_0^t s\|u_t\|^2\,ds + t\|u(t)\|_1^2 \le C\int_0^t \|u\|_1^2\,ds \le C\|v\|^2.$$

Together these estimates show $t^2\|u_t(t)\|^2 \le C\|v\|^2$ which is (4.18) for $j = 1$. □

Proof of Theorem 4.2. By our definitions the error $e = \check{u}_h - \check{u} = e^{-\kappa t}(u_h - u)$ satisfies the equation

$$T_h e_t + e = \rho = -(T_h - T)\check{u}_t, \quad \text{for } t \in J. \tag{4.19}$$

Note also that with $v_h = P_h v$ we have $T_h e(0) = 0$. For since $e(0) = P_h v - v$ is orthogonal to S_h, we have by (iii), for any $\chi \in S_h$,

$$|(T_h e(0), \chi)| = |(T_h e(0), \chi) - (e(0), T_h \chi)| \le C(T_h e(0), e(0))^{1/2} \|T_h \chi\| = 0.$$

We shall prove now, using Lemma 4.3, that with $\widetilde{\rho}(t) = \int_0^t \rho\, ds$

$$\|e(t)\| \le C t^{-1} \sup_{s \le t} (s^2 \|\rho_t\| + s\|\rho\| + \|\widetilde{\rho}\| + h^2 \|e\|), \quad \text{for } t \in J. \tag{4.20}$$

Let us first complete the proof under the assumption that this inequality has already been proved. We have by (ii) and Lemma 4.4

$$s\|\rho(s)\| = s\|(T_h - T)\check{u}_t(s)\| \le C h^2 s \|\check{u}_t(s)\| \le C h^2 \|v\|,$$

and

$$\begin{aligned} s^2 \|\rho_t(s)\| &\le s^2 \|(T_h' - T')\check{u}_t\| + s^2 \|(T_h - T)\check{u}_{tt}\| \\ &\le C h^2 s^2 (\|\check{u}_t(s)\| + \|\check{u}_{tt}(s)\|) \le C h^2 \|v\|. \end{aligned}$$

Since

$$\widetilde{\rho}(t) = -\int_0^t (T_h - T)\check{u}_t\, ds = -\big[(T_h - T)\check{u}(s)\big]_0^t + \int_0^t (T_h' - T')\check{u}\, ds,$$

we also have

$$\|\widetilde{\rho}(s)\| \le C h^2 \sup_{y \le s} \|\check{u}(y)\| \le C h^2 \|v\|,$$

and the stability of the solution operators gives at once

$$\|e(s)\| \le \|\check{u}_h(s)\| + \|\check{u}(s)\| \le 2\|v\|.$$

Inserted into (4.20) these estimates show $\|e(t)\| \le C h^2 t^{-1} \|v\|$, which is the desired result.

In order to show (4.20), we set $w = te$. We shall demonstrate

$$\|w(t)\| \le C \sup_{s \le t} (s^2 \|\rho_t\| + s\|\rho\| + \|T_h e\|), \tag{4.21}$$

and thereafter

$$\|T_h e(t)\| \le C \sup_{s \le t} (s\|\rho\| + \|\widetilde{\rho}\| + h^2 \|e\|). \tag{4.22}$$

Together these estimates imply (4.20).

We begin with (4.21), and note that w satisfies

$$T_h w_t + w = \omega = t\rho + T_h e.$$

We observe, using (4.19), Lemma 4.2, and the boundedness of T_h',

$$\|\omega_t\| = \|t\rho_t + \rho + T_h e_t + T_h' e\| \le C(t\|\rho_t\| + \|\rho\| + \|e\|).$$

Hence by Lemma 4.3 we have with ε suitable, since $w(0) = 0$, that for $t \in J$,

$$\begin{aligned}\|w(t)\| &\le \varepsilon \sup_{s\le t}(s\|\omega_t(s)\|) + C_\varepsilon \sup_{s\le t}\|\omega(s)\| \\ &\le \tfrac{1}{2}\sup_{s\le t}\|w(s)\| + C\sup_{s\le t}(s^2\|\rho_t\| + s\|\rho\| + \|T_h e\|),\end{aligned}$$

which yields (4.21).

For (4.22) we integrate the error equation (4.16) and obtain for $\widetilde{e}(t) = \int_0^t e\,ds$, taking note of $T_h e(0) = 0$,

$$T_h e + \widetilde{e} \equiv T_h \widetilde{e}_t + \widetilde{e} = \widetilde{\rho} + \int_0^t T_h' e\,ds. \tag{4.23}$$

Since $\widetilde{e}(0) = 0$, we may again apply Lemma 4.3 and obtain

$$\|\widetilde{e}(t)\| \le \varepsilon \sup_{s\le t}(s\|\widetilde{\rho}_t\| + s\|T_h' e\|) + C_\varepsilon \sup_{s\le t}(\|\widetilde{\rho}\| + \|\int_0^s T_h' e\,dy\|),$$

and hence, using also Lemma 4.2 to estimate $T_h' e$,

$$\begin{aligned}\|\widetilde{e}(t)\| \le& \varepsilon \sup_{s\le t}(s\|\rho\| + Cs\|T_h e\| + Csh^2\|e\|) + C_\varepsilon \sup_{s\le t}\|\widetilde{\rho}\| \\ &+ C_\varepsilon \int_0^t (\|T_h e\| + h^2\|e\|)\,ds.\end{aligned}$$

It follows from (4.23) that

$$\begin{aligned}&\|T_h e(t)\| \le \|\widetilde{e}\| + \|\widetilde{\rho}\| + \|\int_0^t T_h' e\,ds\| \\ &\le \varepsilon C\bar{t}\sup_{s\le t}\|T_h e(s)\| + C_\varepsilon \sup_{s\le t}(s\|\rho\| + \|\widetilde{\rho}\| + h^2\|e\|) + C_\varepsilon \int_0^t \|T_h e\|\,ds.\end{aligned}$$

Choosing ε such that $\varepsilon C\bar{t} < 1$ this gives, for $t \in \bar{J}$,

$$\|T_h e(t)\| \le C\sup_{s\le t}(s\|\rho\| + \|\widetilde{\rho}\| + h^2\|e\|) + C\int_0^t \|T_h e\|\,ds.$$

The desired inequality (4.22) now follows by an application of Gronwall's lemma. This completes the proof of the theorem. □

The material in this chapter is taken from Huang and Thomée [125], where several examples of approximate solution operators of the elliptic problem satisfying (i), (ii), and (iii) are given, cf. also Sammon [205], Luskin and Rannacher [167], and Lasiecka [150]. For a thorough treatment of the semidiscrete problem in the present generality, see also Fujita and Suzuki [104], where both the theory of evolution operators due to Sobolevskii [217], Kato [134], Kato and Tanabe [135], and energy arguments such as those presented here are used.

The idea of reducing the regularity requirements on the initial data at the expense of a singularity in the error estimates at $t = 0$ has been used also for more complicated problems such as for the Navier-Stokes equations in Heywood and Rannacher [122], and for Biot's consolidation problem in Murad, Thomée, and Loula [173]. In both these cases optimal order error estimates may be derived for positive time without having to satisfy certain nonlocal conditions for the initial data, needed for the solution to be smooth for $t \geq 0$.

5. Negative Norm Estimates and Superconvergence

In this chapter we shall extend our earlier error estimates in L_2 and H^1 to estimates in norms of negative order. It will turn out that if the accuracy in L_2 of the family of approximating spaces is $O(h^r)$ with $r > 2$, then the error bounds in norms of negative order is of higher order than $O(h^r)$. In certain situations these higher order bounds may be applied to show error estimates for various quantities of these higher orders, so called *superconvergent* order estimates. We shall exemplify this by showing how certain integrals of the solution of the parabolic problem, and, in one space dimension, the values of the solution at certain points may be calculated with high accuracy using the semidiscrete solution.

We shall begin by considering the stationary problem. Let Ω be a domain in $\mathbb{R}^d$ with smooth boundary $\partial\Omega$ and consider the Dirichlet problem

$$Au = f \quad \text{in } \Omega, \quad \text{with } u = 0 \quad \text{on } \partial\Omega, \tag{5.1}$$

with A the elliptic operator defined by

$$Au = -\sum_{j,k=1}^{d} \frac{\partial}{\partial x_j}\Big(a_{jk}\frac{\partial u}{\partial x_k}\Big) + a_0 u, \tag{5.2}$$

where the coefficients are smooth functions of x and (a_{jk}) is uniformly positive definite and a_0 nonnegative in $\bar{\Omega}$. In variational form this problem may be stated as

$$A(u,\varphi) = (f,\varphi), \quad \forall \varphi \in H_0^1 = H_0^1(\Omega),$$

where now

$$A(u,v) = \int_\Omega \Big(\sum_{j,k=1}^{d} a_{jk}\frac{\partial u}{\partial x_k}\frac{\partial v}{\partial x_j} + a_0 uv\Big)\, dx. \tag{5.3}$$

Here we have chosen the operator A in the above general form rather than the Laplacian for the purpose of a subsequent application in one space dimension.

Let $\{S_h\}$ denote a family of finite dimensional subspaces of H_0^1 satisfying our standard approximation assumption (1.10) with $r \geq 2$. We may then pose the standard Galerkin finite element problem to find $u_h \in S_h$ such that

(5.4) $$A(u_h, \chi) = (f, \chi), \quad \forall \chi \in S_h.$$

(We have assumed $S_h \subset H_0^1$ for simplicity only; what follows also carries over to the general framework with solution operators T and T_h employed in Chapters 2-4.)

In the same way as for our earlier model problem it follows that the discrete elliptic problem (5.4) has a unique solution $u_h \in S_h$, and that, with u the solutions of (5.1),

(5.5) $$\|u_h - u\| + h\|u_h - u\|_1 \le Ch^q \|u\|_q, \quad \text{for } 1 \le q \le r.$$

We shall now see that for $r > 2$, the duality argument used to show the L_2 norm estimate above also yields an error estimate in a negative order norm. We introduce such negative norms by

(5.6) $$\|v\|_{-s} = \sup\Big\{\frac{(v, \varphi)}{\|\varphi\|_s}; \ \varphi \in H^s\Big\}, \quad \text{for } s \ge 0 \text{ integer.}$$

Although this may be used to define a space $H^{-s} = H^{-s}(\Omega) \supset L_2$, we shall only use the negative norm here as a quantitative measure for functions in L_2. Note that here we do not require boundary conditions on φ in the definition (5.6) of $\|v\|_{-s}$, as will be done sometimes in the rest of the book.

Theorem 5.1 *Let u_h and u be the solutions of* (5.4) *and* (5.1). *Then*

$$\|u_h - u\|_{-s} \le Ch^{q+s}\|u\|_q, \quad \textit{for } 0 \le s \le r-2, \ 1 \le q \le r.$$

Proof. We shall demonstrate that, for $e = u_h - u$,

(5.7) $$|(e, \varphi)| \le Ch^{q+s}\|u\|_q \|\varphi\|_s, \quad \forall \varphi \in H^s,$$

which immediately implies the desired estimate. For this purpose, we introduce the solution $\psi = T\varphi$ of $A\psi = \varphi$ in Ω, with $\psi = 0$ on $\partial\Omega$, and recall that $\|\psi\|_{s+2} \le C\|\varphi\|_s$, for any $s \ge 0$. By the orthogonality of the error to S_h with respect to $A(\cdot, \cdot)$ we obtain

$$(e, \varphi) = (e, A\psi) = A(e, \psi) = A(e, \psi - \chi), \quad \forall \chi \in S_h,$$

and hence, for $0 \le s \le r-2$,

$$|(e, \varphi)| \le C\|e\|_1 \inf_{\chi \in S_h} \|\psi - \chi\|_1 \le Ch^{s+1}\|e\|_1 \|\psi\|_{s+2} \le Ch^{s+1}\|e\|_1\|\varphi\|_s.$$

Here, by (5.5), $\|e\|_1 \le Ch^{q-1}\|u\|_q$ which shows (5.7). □

Note, in particular, the case $s = r-2$, $q = r$, in which the result of Theorem 5.1 reads

$$\|u_h - u\|_{-(r-2)} \le Ch^{2r-2}\|u\|_r.$$

Since $2r-2>r$ for $r>2$ the order of accuracy in this estimate is then higher than in the standard $O(h^r)$ error estimate in the L_2-norm.

We remark that the error estimate of Theorem 5.1 may also be expressed as

$$\|(R_h - I)u\|_{-s} \le Ch^{q+s}\|u\|_q, \quad \text{for } 0 \le s \le r-2,\ 1 \le q \le r,$$

where $R_h : H_0^1 \to S_h$ is the Ritz projection defined by

$$A(R_h u, \chi) = A(u, \chi), \quad \forall \chi \in S_h.$$

Further, with T the solution operator of (5.1) and defining the approximate solution operator $T_h : L_2 \to S_h$ of the elliptic problem by

$$A(T_h f, \chi) = (f, \chi), \quad \forall \chi \in S_h,$$

the properties (i) and (ii) used in Chapters 2 and 3 are valid also in the present case, with (ii) now extended to negative orders as

$$\|(T_h - T)f\|_{-s} \le Ch^{q+2+s}\|f\|_q, \quad \text{for } 0 \le s,\ q \le r-2.$$

The operators T_h and T will be used extensively in our analysis below.

We note in passing that for the L_2-projection of $v \in H^q \cap H_0^1$ onto S_h we have

$$\|(P_h - I)v\|_{-s} = \sup_{\varphi} \frac{((P_h - I)v, (I - P_h)\varphi)}{\|\varphi\|_s} \le Ch^{q+s}\|v\|_q, \quad 0 \le s, q \le r,$$

so that, in particular $\|(P_h - I)v\|_{-r} \le Ch^{2r}\|v\|_r$ if $v \in H^r \cap H_0^1$.

As a very simple application of a negative norm error estimate, assume we are interested in evaluating the integral

$$F(u) = \int_\Omega u\,\psi\,dx, \quad \text{where } \psi \in H^{r-2}, \tag{5.8}$$

and where u is the solution of (5.1). Then, for the obvious approximation

$$F(u_h) = \int_\Omega u_h \psi\,dx, \tag{5.9}$$

where u_h is the solution of (5.4) we find

$$\begin{aligned}|F(u_h) - F(u)| &= |(u_h - u, \psi)| \le \|u_h - u\|_{-(r-2)}\|\psi\|_{r-2}\\ &\le Ch^{2r-2}\|u\|_r\,\|\psi\|_{r-2},\end{aligned}$$

which is an error estimate of superconvergent order $O(h^{2r-2})$.

We shall consider one more example of these ideas, which concerns superconvergent nodal approximation in the two-point boundary value problem

(5.10) $$Au = -\frac{d}{dx}(a_{11}\frac{du}{dx}) + a_0 u = f \quad \text{in } (0,1), \quad u(0) = u(1) = 0.$$

We shall now work with the finite-dimensional space defined by the partition $0 = x_0 < x_1 < \cdots < x_M = 1$, with $x_{i+1} - x_i \le h$, and, with $I_i = (x_i, x_{i+1})$,

(5.11) $$S_h = \{\chi \in \mathcal{C}([0,1]);\ \chi|_{I_i} \in \Pi_{r-1},\ 0 \le i < M,\ \chi(0) = \chi(1) = 0\},$$

where Π_{r-1} denotes the set of polynomials of degree at most $r-1$. Clearly this family satisfies our approximation assumption (1.10).

Let $g = g^{\bar{x}}$ denote the Green's function of the two-point boundary value problem (5.10) with singularity at the partition point $\bar{x}$, which we now consider fixed, so that

(5.12) $$w(\bar{x}) = A(w, g), \quad \forall w \in H_0^1 = H_0^1((0,1)).$$

Applied to the error $e = u_h - u$ in the discrete solution u_h, and using the orthogonality of e to S_h we find

$$e(\bar{x}) = A(e, g) = A(e, g - \chi), \quad \forall \chi \in S_h,$$

so that

$$|e(\bar{x})| \le C\|e\|_1 \inf_{\chi \in S_h} \|g - \chi\|_1 \le Ch^{r-1}\|u\|_r \inf_{\chi \in S_h} \|g - \chi\|_1.$$

Note now that although $g^{\bar{x}}$ is not a smooth function at $\bar{x}$, it may still be approximated well by a function in S_h, since it is smooth except at $\bar{x}$, and the discontinuity of the derivative at $\bar{x}$ can be accommodated in S_h. In particular, we have

$$\inf_{\chi \in S_h} \|g - \chi\|_1 \le Ch^{r-1}(\|g\|_{H^r((0,\bar{x}))} + \|g\|_{H^r((\bar{x},1))}) \le Ch^{r-1}.$$

Thus $|e(\bar{x})| \le Ch^{2r-2}\|u\|_r$, that is, superconvergence occurs at the nodes of the partition.

This latter example is the reason why the more general form of A has been used in this chapter; for $A = -d^2/dx^2$, the Green's function $g^{\bar{x}}$ is linear outside $\bar{x}$ and so $g^{\bar{x}} \in S_h$. We may then conclude that $e(\bar{x}) = 0$, which is a degenerate case.

For our analysis of the parabolic problem it will be convenient to use instead of the negative norm introduced above, such a norm defined by

$$|v|_{-s} = \|T^{s/2}v\| = (T^s v, v)^{1/2}, \quad \text{for } s \ge 0,$$

where, as before, $T = A^{-1}$ denotes the exact solution operator of the elliptic problem. Again, this may be used to define a space $\dot{H}^{-s} = \dot{H}^{-s}(\Omega)$, but we think of it as a norm on L_2. With $\dot{H}^s = \{\psi \in H^s;\ A^j\psi = 0, \text{ on } \partial\Omega, \text{ for } j < s/2\}$ we have the following:

Lemma 5.1 *For s a nonnegative integer, the norm $|v|_{-s}$ is equivalent to* $\sup\{(v,\psi)/\|\psi\|_s;\ \psi\in\dot{H}^s\}$.

Proof. In fact, with $\{\lambda_j\}_{j=1}^\infty$ and $\{\varphi_j\}_{j=1}^\infty$ the eigenvalues and orthonormal eigenfunctions of A (with Dirichlet boundary conditions), an equivalent norm on $\dot{H}^s$ to our standard Sobolev norm $\|\psi\|_s$ for integer $s\ge 0$ is (cf. Lemma 3.1)

$$(5.13)\qquad |\psi|_s=(A^s\psi,\psi)^{1/2}=\Big(\sum_{j=1}^\infty \lambda_j^s(\psi,\varphi_j)^2\Big)^{1/2},$$

and we have at once, since the eigenvalues of the compact operator T are $\{\lambda_j^{-1}\}_{j=1}^\infty$, corresponding to the eigenvectors $\{\varphi_j\}_{j=1}^\infty$,

$$(5.14)\qquad |v|_{-s}=(T^s v,v)^{1/2}=\Big(\sum_{j=1}^\infty \lambda_j^{-s}(v,\varphi_j)^2\Big)^{1/2}.$$

Since $(v,\psi)=\sum_{j=1}^\infty(v,\varphi_j)(\psi,\varphi_j)$, (5.13) and (5.14) easily show

$$\sup\big\{(v,\psi)/|\psi|_s;\ \psi\in\dot{H}^s\big\}=|v|_{-s}.$$

By the equivalence of $|\psi|_s$ and $\|\psi\|_s$ this shows our claim. □

Since, for integer $s\ge 0$,

$$(v,\psi)\le\|v\|_{-s}\,\|\psi\|_s\le C\|v\|_{-s}|\psi|_s,\quad \forall\psi\in\dot{H}^s,$$

it follows from Lemma 5.1 that $|v|_{-s}\le C\|v\|_{-s}$, and Theorem 5.1 therefore immediately implies the following:

Lemma 5.2 *We have, for $0\le s,\ q\le r-2$, with s integer,*

$$|(R_h-I)v|_{-s}\le Ch^{s+q+2}\|v\|_{q+2},\quad \textit{for } v\in H^{q+2}\cap H_0^1.$$

Note, in particular, $|(R_h-I)v|_{-(r-2)}\le Ch^{2r-2}\|v\|_r$, and that in terms of T_h and T we have

$$(\text{ii}')\qquad |(T_h-T)f|_{-s}\le Ch^{s+q+2}\|f\|_q,\quad \text{for } 0\le s,\ q\le r-2,\ f\in H^q.$$

For the analysis of the parabolic problem we introduce also a *discrete negative seminorm* on L_2 by

$$|v|_{-s,h}=\|T_h^{s/2}v\|=(T_h^s v,v)^{1/2};$$

it corresponds to the discrete semi-inner product $(v,w)_{-s,h}=(T_h^s v,w)$. Since T_h is positive definite on S_h, $|v|_{-s,h}$ and $(v,w)_{-s,h}$ define a norm and an inner product there. The following lemma shows that this discrete negative seminorm is equivalent to the corresponding continuous negative norm, modulo a small error.

Lemma 5.3 *We have, for s a nonnegative integer with $0 \le s \le r$,*

$$|v|_{-s,h} \le C\big(|v|_{-s} + h^s\|v\|\big) \quad \text{and} \quad |v|_{-s} \le C\big(|v|_{-s,h} + h^s\|v\|\big).$$

Proof. We show the first inequality by induction over s. The result is trivial for $s = 0$ and also clear for $s = 1$, since

$$|v|_{-1,h}^2 = (T_h v, v) = (Tv, v) + ((T_h - T)v, v) \le |v|_{-1}^2 + Ch^2\|v\|^2,$$

by (ii). Now let $1 \le s \le r - 1$ and assume that it is proved up to s. We have

$$|v|_{-(s+1),h} = |T_h v|_{-(s-1),h} \le |Tv|_{-(s-1),h} + |(T_h - T)v|_{-(s-1),h}.$$

By the induction assumption

$$|Tv|_{-(s-1),h} \le C(|Tv|_{-(s-1)} + h^{s-1}\|Tv\|) = C(|v|_{-(s+1)} + h^{s-1}|v|_{-2}).$$

Using, for instance, our above spectral representations of the norms, we have easily $|v|_{-2} \le C(h^2\|v\| + h^{-(s-1)}|v|_{-(s+1)})$, so that we may conclude

$$|Tv|_{-(s-1),h} \le C(|v|_{-(s+1)} + h^{s+1}\|v\|).$$

Further, by the induction assumption and (ii′) with $q = 0$ (recall that $s-1 \le r - 2$),

$$|(T_h - T)v|_{-(s-1),h} \le C\big(|(T_h - T)v|_{-(s-1)} + h^{s-1}\|(T_h - T)v\|\big) \le Ch^{s+1}\|v\|,$$

which completes the proof. By interchanging the roles of T and T_h, the second inequality follows analogously. □

With A and $A(\cdot,\cdot)$ as in (5.2) and (5.3), we now direct our attention to the parabolic initial-boundary value problem

$$u_t + Au = f \quad \text{in } \Omega, \qquad \text{for } t > 0, \tag{5.15}$$
$$u = 0 \quad \text{on } \partial\Omega, \quad \text{for } t > 0, \quad u(\cdot, 0) = v \quad \text{in } \Omega,$$

and pose the corresponding semidiscrete problem

$$(u_{h,t}, \chi) + A(u_h, \chi) = (f, \chi), \ \forall \chi \in S_h, \ t > 0, \ u_h(0) = v_h \in S_h. \tag{5.16}$$

We shall show the following.

Theorem 5.2 *Let $0 \le s \le r - 2$ and assume $v_h \in S_h$ and v are such that*

$$|v_h - v|_{-s} + h^s\|v_h - v\| \le Ch^{s+r}\|v\|_r. \tag{5.17}$$

Then we have for the solutions of (5.16) *and* (5.15)

$$|u_h(t) - u(t)|_{-s} \le Ch^{s+r}\big(\|v\|_r + \int_0^t \|u_t(y)\|_r\, dy\big).$$

Proof. We recall from our earlier analysis that $e = u_h - u$ satisfies

$$T_h e_t + e = \rho, \quad \text{where } \rho = (R_h - I)u,$$

and also (cf. Lemma 2.4) that if T_h is nonnegative with respect to the semi-inner product $(\cdot,\cdot)$ then, for the corresponding seminorm $\|\cdot\|$,

$$\|e(t)\| \le \|e(0)\| + C\big(\|\rho(0)\| + \int_0^t \|\rho_t\|\,dy\big).$$

We now note that, since T_h^{s+1} is positive semidefinite in L_2, we have $(T_h v, v)_{-s,h} = (T_h^{s+1} v, v) \ge 0$ so that we may conclude that

$$|e(t)|_{-s,h} \le |e(0)|_{-s,h} + C\Big(|\rho(0)|_{-s,h} + \int_0^t |\rho_t|_{-s,h}\,dy\Big). \tag{5.18}$$

By our assumptions and Lemma 5.3, $|e(0)|_{-s,h} \le Ch^{s+r}\|v\|_r$. Further, by Lemmas 5.3 and 5.2, $|\rho|_{-s,h} \le Ch^{s+r}\|u\|_r$ so that, in particular, $|\rho(0)|_{-s,h} \le Ch^{s+r}\|v\|_r$ and similarly $|\rho_t|_{-s,h} \le Ch^{s+r}\|u_t\|_r$. Inserted into (5.18) these estimates show

$$|e(t)|_{-s,h} \le Ch^{s+r}\Big(\|v\|_r + \int_0^t \|u_t\|_r\,dy\Big),$$

and hence

$$|e(t)|_{-s} \le C(|e(t)|_{-s,h} + h^s\|e(t)\|) \le Ch^{s+r}\Big(\|v\|_r + \int_0^t \|u_t\|_r\,dy\Big),$$

which completes the proof. □

As a first simple application, consider the approximation of the integral in (5.8) by that in (5.9), where u and u_h are solutions of (5.15) and (5.16), and where we now assume $\psi \in \dot{H}^{r-2}$. Then, provided (5.17) holds, we have

$$\begin{aligned}|F(u_h) - F(u)| &= |(u_h - u, \psi)| \le |u_h - u|_{-(r-2)}|\psi|_{r-2}\\ &\le Ch^{2r-2}\big(\|v\|_r + \int_0^t \|u_t\|_r\,dy\big),\end{aligned}$$

again exhibiting a superconvergent order error bound.

The assumption above for the choice of initial values is satisfied as usual by, for instance, $v_h = P_h v$ and $v_h = R_h v$, if $v \in H^r \cap H_0^1$. In our application below we shall need also a negative norm estimate for a time derivative of the error at positive time, which we now state, for simplicity only for $v_h = P_h v$. Here and below we often write D_t for $\partial/\partial t$.

Theorem 5.3 *Let $j \geq 0$, $0 \leq s \leq r-2$, and $\delta > 0$, and let $v_h = P_h v$. Then we have, for u_h and u the solutions of (5.16) and (5.15), with $v_h = P_h v$,*

$$|D_t^j(u_h(t) - u(t))|_{-s} \leq C_\delta h^{r+s}\Big(\sum_{l=0}^{j} \|D_t^l u(t)\|_r + \int_{t-\delta}^{t} \|D_t^{j+1} u\|_r \, dy + \int_0^t \|u_t\|_{s+2} \, dy\Big), \quad \text{for } t \geq \delta > 0.$$

We shall not demonstrate this theorem in detail but only remark that the proof uses the ideas of the proof of Theorem 3.6. One thus multiplies the solution by a cut-off function permitting one to consider separately one problem with u vanishing in $(0, t-\delta/2)$ and another with u vanishing in $(t-\delta, t)$. For the first of these problems an estimate for the time derivatives may be obtained from Theorem 5.2 by differentiation. For the second problem one uses the fact, easily established by spectral representation, that for a solution of the homogeneous semidiscrete equation $T_h u_{h,t} + u_h = 0$ one has

$$|D_t^j u_h(t)|_{-s,h} \leq C_\delta |u_h(t-\delta/2)|_{-(r-2),h}, \quad \text{for } t \geq \delta,\ 0 \leq s \leq r-2.$$

Using either an inverse estimate or a standard energy argument, it is also possible to prove a similar estimate for the gradient of the error so that altogether we have, for any $j \geq 0$ and u appropriately smooth,

$$|D_t^j(u_h(t) - u(t))|_{-s} \leq C_\delta(u) h^{r+s}, \quad \text{for } t \geq \delta > 0,\ -1 \leq s \leq r-2.$$

We shall show now that if more care is exercised in the choice of discrete initial data, then the negative norm error estimates for the time derivatives can be made to hold uniformly down to $t = 0$. For this purpose note that $u_h^{(j)} = D_t^j u_h$ satisfies the semidiscrete equation in (5.16), where $j \geq 1$ is fixed. We first show that the initial data v_h may be chosen in such a way that

$$u_h^{(j)}(0) = P_h u^{(j)}(0). \tag{5.19}$$

In particular, we may then apply Theorem 5.2 to $u_h^{(j)}$. To accomplish this we introduce the discrete elliptic operator $A_h = T_h^{-1} : S_h \to S_h$, so that

$$u_{h,t} + A_h u_h = P_h f, \quad \text{for } t > 0,$$

and hence by differentiation

$$u_{h,t}^{(l)} + A_h u_h^{(l)} = P_h f^{(l)}, \quad \text{for } l \geq 0,\ t > 0. \tag{5.20}$$

By the equations satisfied by $u_h^{(j-1)}, \dots, u_h$ we have for the initial data

$$\begin{aligned} u_h^{(j)}(0) &= u_{h,t}^{(j-1)}(0) = -A_h u_h^{(j-1)}(0) + P_h f^{(j-1)}(0) \\ &= A_h^2 u_h^{(j-2)}(0) - A_h P_h f^{(j-2)}(0) + P_h f^{(j-1)}(0) = \cdots \\ &= (-A_h)^j v_h + \sum_{l=0}^{j-1} (-A_h)^{j-1-l} P_h f^{(l)}(0). \end{aligned}$$

After multiplication by T_h^j and use of the differential equation in (5.15) this is seen to be equivalent to

$$v_h = (-T_h)^j u_h^{(j)}(0) - (-T_h)^j \sum_{l=0}^{j-1} (-A_h)^{j-1-l} P_h (u^{(l+1)}(0) + Au^{(l)}(0)).$$

Recalling that $T_h P_h = T_h$ and $R_h = T_h A$ this in turn may be written

$$v_h = P_h v + \sum_{l=0}^{j-1} (-T_h)^l P_h (R_h - I) u^{(l)}(0) + (-T_h)^j (u_h^{(j)}(0) - P_h u^{(j)}(0)).$$

We therefore find that condition (5.19) is equivalent to the choice

$$v_h = P_h v + \sum_{l=0}^{j-1} (-T_h)^l P_h (R_h - I) u^{(l)}(0). \tag{5.21}$$

Note that the $u^{(l)}(0)$ may be calculated from the differential equation in (5.15).

Since another possible choice of v_h in Theorem 5.2 is $R_h v$, we may require instead of (5.19) the relation

$$u_h^{(j)}(0) = R_h u^{(j)}(0), \tag{5.22}$$

which leads to an additional term in the sum in (5.21), or

$$\begin{aligned} v_h &= P_h v + \sum_{l=0}^{j} (-T_h)^l P_h (R_h - I) u^{(l)}(0) \\ &= R_h v + \sum_{l=1}^{j} (-T_h)^l (R_h - I) u^{(l)}(0). \end{aligned} \tag{5.23}$$

The type of construction of discrete initial data used in (5.21) and (5.23) is referred to as quasi-projections in the analysis of Douglas, Dupont and Wheeler [81].

We may now show the following.

Theorem 5.4 *Let $j > 0$, $0 \le s \le r-2$, and assume that v_h is given by (5.21) or (5.23). Then we have for the solutions of (5.16) and (5.15)*

$$|D_t^i(u_h(t) - u(t))|_{-s} \le Ch^{r+s}\Big(\sum_{l=i}^{j} \|D_t^l u(0)\|_{\max(r-2(l-i),s+2)} + \int_0^t \|D_t^{i+1}u\|_r\,dy\Big), \quad \textit{for } 0 \le i \le j,\ t \ge 0.$$

Proof. For $i = j$ this follows at once by application of Theorem 5.2 to $D_t^j u_h = u_h^{(j)}$ and $D_t^j u = u^{(j)}$, and recalling that (5.19) or (5.22) holds. Let now $0 \le i < j$ and consider first the choice (5.21). Then we may write $u_h = \widetilde{u}_h + \widetilde{\widetilde{u}}_h$ where

$$\widetilde{u}_{h,t} + A_h \widetilde{u}_h = P_h f, \quad \text{for } t \ge 0,$$
$$\widetilde{u}_h(0) = P_h v + \sum_{l=0}^{i-1} (-T_h)^l P_h (R_h - I) u^{(l)}(0),$$

and

$$\widetilde{\widetilde{u}}_{h,t} + A_h \widetilde{\widetilde{u}}_h = 0, \quad \text{for } t \ge 0,$$
$$\widetilde{\widetilde{u}}_h(0) = \sum_{l=i}^{j-1} (-T_h)^l P_h (R_h - I) u^{(l)}(0). \tag{5.24}$$

Then $\widetilde{u}_h^{(i)}(0) = P_h u^{(i)}(0)$, by the above construction, and hence, by the result just proved for $i = j$

$$|\widetilde{u}_h^{(i)}(t) - u^{(i)}(t)|_{-s} \le Ch^{r+s}\Big(\|D_t^i u(0)\|_r + \int_0^t \|D_t^{i+1}u\|_r\,dy\Big). \tag{5.25}$$

Further, by Lemma 5.3 and the stability in $|\cdot|_{-s,h}$ and $\|\cdot\|$,

$$|\widetilde{\widetilde{u}}_h^{(i)}(t)|_{-s} \le C(|\widetilde{\widetilde{u}}_h^{(i)}(t)|_{-s,h} + h^s \|\widetilde{\widetilde{u}}_h^{(i)}(t)\|)$$
$$\le C(|\widetilde{\widetilde{u}}_h^{(i)}(0)|_{-s,h} + h^s \|\widetilde{\widetilde{u}}_h^{(i)}(0)\|) = C(|A_h^i \widetilde{\widetilde{u}}_h(0)|_{-s,h} + h^s \|A_h^i \widetilde{\widetilde{u}}_h(0)\|).$$

Now, for $i \le l \le j-1$ we have

$$A_h^i T_h^l P_h (R_h - I) u^{(l)}(0) = T_h^{l-i} P_h (R_h - I) u^{(l)}(0),$$

and we conclude (note that $|P_h v|_{-s,h} = |v|_{-s,h}$ for $s > 0$)

$$|\widetilde{\widetilde{u}}_h^{(i)}(t)|_{-s} \le C \sum_{l=i}^{j-1} \big(|(R_h - I)u^{(l)}(0)|_{-(s+2(l-i)),h} + h^s |(R_h - I)u^{(l)}(0)|_{-2(l-i),h}\big).$$

Since by Lemmas 5.2 and 5.3

$$|(R_h - I)v|_{-(s+q),h} \le Ch^{r+s}\|v\|_{r-q}, \quad \text{if } s+q \le r-2,$$

and

$$|(R_h - I)v|_{-(s+q),h} \le Ch^{r+s}\|v\|_{s+2}, \quad \text{if } s+q > r-2,$$

this yields

$$|\widetilde{\widetilde{u}}_h^{(i)}(t)|_{-s} \le Ch^{r+s}\sum_{l=i}^{j-1}\|u^{(l)}(0)\|_{\max(r-2(l-i),s+2)}. \tag{5.26}$$

Together, (5.25) and (5.26) show the result for this choice of v_h. For v_h chosen by (5.23), the summation in (5.24), and hence also in (5.26), will extend to j, and the proof proceeds as before. □

Also H^1 estimates which are uniform down to $t = 0$ may be derived using an inverse estimate or, for v_h chosen to satisfy (5.23), by the standard energy argument. For an application below, we consider briefly the latter case. For $j = 0$, see Theorem 1.3.

Theorem 5.5 *Let $j > 0$ and assume v_h given by (5.23). Then for $0 \le i \le j$ and $t \ge 0$, we have, for the solutions of* (5.16) *and* (5.15),

$$\|D_t^i(u_h(t) - u(t))\|_1 \le Ch^{r-1}\Big(\|D_t^i u(t)\|_r + \sum_{l=i+1}^{j}\|D_t^l u(0)\|_{\max(r-2(l-i),1)} + \Big(\int_0^t \|D_t^{i+1}u\|_{r-1}^2\,dy\Big)^{1/2}\Big).$$

Proof. We consider first $i = j$ and write $u_h - u = (u_h - R_h u) + (R_h u - u) = \theta + \rho$. We have

$$(\theta_t^{(j)}, \chi) + A(\theta^{(j)}, \chi) = -(\rho_t^{(j)}, \chi), \quad \forall \chi \in S_h, \quad \text{for } t > 0,$$

with $\theta^{(j)}(0) = 0$ by (5.22). The standard energy argument with $\chi = \theta_t^{(j)}$ shows therefore

$$\|\theta^{(j)}(t)\|_1 \le C\Big(\int_0^t \|\rho_t^{(j)}\|^2\,dy\Big)^{1/2} \le Ch^{r-1}\Big(\int_0^t \|D_t^{j+1}u\|_{r-1}^2\,dy\Big)^{1/2}.$$

Since $\|\rho^{(j)}(t)\|_1 \le Ch^{r-1}\|D_t^j u(t)\|_r$, we conclude

$$\|u_h^{(j)}(t) - u^{(j)}(t)\|_1 \le Ch^{r-1}\Big(\|D_t^j u(t)\|_r + \Big(\int_0^t \|D_t^{j+1}u\|_{r-1}^2\,dy\Big)^{1/2}\Big),$$

which is the desired result for $i = j$. For $0 \le i < j$ we may now write $\theta = (\widetilde{u}_h - R_h u) + \widetilde{\widetilde{u}}_h = \widetilde{\theta} + \widetilde{\widetilde{u}}_h$, where

$$(\widetilde{\widetilde{u}}_{h,t},\chi)+A(\widetilde{\widetilde{u}}_h,\chi)=0,\quad \forall\chi\in S_h,\quad \text{for } t\geq 0,$$
$$\widetilde{\widetilde{u}}_h(0)=\sum_{l=i+1}^{j}(-T_h)^l(R_h-I)u^{(l)}(0).$$

Since $\widetilde{u}_h^{(i)}(0)=R_h u^{(i)}(0)$, we obtain as above for $\widetilde{u}_h-u=\widetilde{\theta}+\rho$ that

$$\|\widetilde{u}_h^{(i)}(t)-u^{(i)}(t)\|_1\leq Ch^{r-1}\Big(\|D_t^i u(t)\|_r+\big(\int_0^t\|D_t^{i+1}u\|_{r-1}^2\,dy\big)^{1/2}\Big),$$

and for $\widetilde{\widetilde{u}}_h$ we have

$$\|\widetilde{\widetilde{u}}_h^{(i)}(t)\|_1\leq C\|\widetilde{\widetilde{u}}_h^{(i)}(0)\|_1\leq C\|A_h^i\widetilde{\widetilde{u}}_h(0)\|_1\leq C\sum_{l=i+1}^{j}\|T_h^{l-i}(R_h-I)u^{(l)}(0)\|_1.$$

Here

$$\|T_h^{l-i}w\|_1\leq CA(T_h^{l-i}w,T_h^{l-i}w)^{1/2}=C(T_h^{l-i-1}w,T_h^{l-i}w)^{1/2}=C|w|_{1-2(l-i),h},$$

and we conclude

$$\|\widetilde{\widetilde{u}}_h^{(i)}(t)\|_1\leq C\sum_{l=i+1}^{j}|(R_h-I)u^{(l)}(0)|_{1-2(l-i),h}$$
$$\leq Ch^{r-1}\sum_{l=i+1}^{j}\|u^{(l)}(0)\|_{\max(r-2(l-i),1)},$$

which completes the proof. □

We shall apply our above estimates to obtain a superconvergence result in the case of C^0 elements in one space dimension. Note that the initial conditions (5.21) and (5.23) depend on j, but that all time derivatives of the error of orders at most j are bounded in Theorems 5.4 and 5.5. This will be useful in application of the following result.

Consider thus the problem

$$(5.27)\qquad u_t+Au=f\quad \text{in } (0,1),\qquad \text{for } t>0,$$
$$u=0\quad \text{at } x=0,1,\quad \text{for } t>0,\quad u(\cdot,0)=v\quad \text{in } (0,1),$$

where A is defined in (5.10), and the semidiscrete analogue (5.16) in the piecewise polynomial space S_h defined in (5.11). We then have the following result.

Theorem 5.6 *Let u_h and u be the solutions of* (5.16) *and* (5.27), *and let $\bar{x}$ be one of the nodes of the partition. Then, for any $n\geq 0$, we have for $e=u_h-u$*

$$|e(\bar{x},t)|\leq C\Big(h^{r-1}\sum_{j=0}^{n}\|D_t^j e\|_1+h^r\|D_t^{n+1}e\|+|D_t^{n+1}e|_{-2n}\Big).$$

We remark at once that by Theorems 5.4 and 5.5 this shows that under the appropriate choice of discrete initial data and regularity assumptions we have $|u_h(\bar{x},t) - u(\bar{x},t)| = O(h^{2r-2})$, for any $t > 0$.

Proof of Theorem 5.6. Let again $g = g^{\bar{x}}$ be the Green's function of A with zero boundary conditions and singularity at $\bar{x}$ so that (5.12) holds. Setting $L(u,v) = (u_t, v) + A(u,v)$, we have now, using the definition of the exact solution operator T,

$$\begin{aligned} e(\bar{x},t) &= A(e,g) = L(e,g) - (e_t,g) = L(e,g) - A(e_t,Tg) \\ &= L(e,g) - L(e_t,Tg) + (e_{tt},Tg) \\ &= \sum_{j=0}^{n} (-1)^j L(D_t^j e, T^j g) + (-1)^{n+1}(D_t^{n+1} e, T^n g). \end{aligned}$$

Recalling our definitions we find

$$L(e,\chi) = \big((u_{h,t},\chi) + A(u_h,\chi)\big) - \big((u_t,\chi) + A(u,\chi)\big) = 0, \quad \forall \chi \in S_h,$$

and, by differentiation, $L(D_t^j e, \chi) = 0$ for $\chi \in S_h$ Hence

$$\begin{aligned} |L(D_t^j e, T^j g)| &= \inf_{\chi \in S_h} |L(D_t^j e, T^j g - \chi)| \\ &\le \inf_{\chi \in S_h} \Big(\|D_t^{j+1} e\| \, \|T^j g - \chi\| + C \|D_t^j e\|_1 \|T^j g - \chi\|_1 \Big) \\ &\le C\big(h^r \|D_t^{j+1} e\| + h^{r-1} \|D_t^j e\|_1\big), \end{aligned}$$

where in the last step we have used the fact that $T^j g$ is continuous and smooth except possibly at $\bar{x}$. We have finally

$$|(D_t^{n+1} e, T^n g)| = |(T^n D_t^{n+1} e, g)| \le C \|T^n D_t^{n+1} e\| = C |D_t^{n+1} e|_{-2n},$$

which completes the proof of the theorem. □

Another type of application of negative norms to obtain superconvergent order error bounds is associated with situations when the partition is uniform in some interior subdomain Ω_0 of Ω, in a way we shall refrain from describing in detail here. For the elliptic problem and with $D^\alpha u$ a given derivative of the solution, one may then show an inequality of the form (see Nitsche and Schatz [185], Bramble, Nitsche, and Schatz [33])

$$\sup_{x \in \Omega_0} |Q_h u_h(x) - D^\alpha u(x)| \le C\big(h^r \|u\|_{H^s(\Omega_1)} + \|u_h - u\|_{-p}\big).$$

Here Q_h is a finite difference operator approximating the operator D^α to order $O(h^r)$, s is a number greater than r, p is arbitrary, and Ω_0 is contained in a compact subset of $\Omega_1 \subset \Omega$. The conclusion is that $D^\alpha u$ is approximated by $Q_h u_h$ to order $O(h^r)$ in Ω_0 provided u is smooth in Ω_1 and an $O(h^r)$ bound is available for the error $u_h - u$ in some negative order norm. It may also be shown that if the discrete solution u_h is convolved with a specific function ψ_h, a scaled version of the B-spline of order $r-2$ in $\mathbb{R}^d$, then Q_h may be defined in such a way that $\sup_{\Omega_0} |\psi_h * Q_h u_h - D^\alpha u| = O(h^{2r-2})$. This uses the negative norm estimate of Theorem 5.1 with $s = r-2, q = r$; the local averaging by means of the function ψ_h is associated with the use of the K-operator of Bramble and Schatz [36], see also Thomée [226]. Similar results have been derived for the parabolic problem; we shall not present these in detail here but refer to Bramble, Schatz, Thomée, and Wahlbin [37], Thomée [227], [228], and Nitsche [184].

In the case of nonuniform partitions it is also possible to find superconvergent order approximations to $u(x_0, t)$ for $x_0 \in \Omega, t > 0$, by using a local Green's function, see Louis [159]. We sketch this application in the elliptic case: Letting $x_0 \in \Omega_0 \subset \Omega$ and denoting by $G = G^{x_0}$ the Green's function of (5.1) with respect to Ω_0, with singularity at x_0, we have, for any smooth w vanishing on $\partial\Omega_0$,

$$w(x_0) = \int_{\Omega_0} Aw(y)\, G(y)\, dy.$$

Letting $\varphi \in \mathcal{C}_0^\infty(\Omega_0)$ and $\varphi \equiv 1$ in a neighborhood of x_0 we thus have, with g a function in $\in \mathcal{C}_0^\infty(\Omega_0)$ which can easily be determined, with $g \equiv 0$ near x_0,

$$u(x_0) = (A(\varphi u), G) = (\varphi A u, G) + (u, g) = (\varphi f, G) + (u, g).$$

If we approximate $u(x_0)$ by $\widetilde{u}_h(x_0) = (\varphi f, G) + (u_h, g)$, it is clear that

$$|\widetilde{u}_h(x_0) - u(x_0)| = |(u_h - u, g)| \le C\|u_h - u\|_{-(r-s)} = O(h^{2r-2}).$$

For the parabolic case, see [228].

The theory presented here was developed in Bramble, Schatz, Thomée, and Wahlbin [37] and Thomée [228]. For related material, see also Douglas, Dupont, and Wheeler [81]. Additional work on superconvergence for parabolic equations, not necessarily related to negative norm estimates, includes Thomée [224] where the first nodal superconvergence result for Galerkin methods was derived in the case of the Cauchy problem for the heat equation and using smooth splines, and several papers concerning superconvergent $O(h^2)$ approximations of the gradient of the solution in the piecewise linear case on triangulations that are almost uniform, see Thomée, Xu, and Zhang [234] and references therein. General references on superconvergence are Křížek and Neittaanmäki [142] and Wahlbin [243].

6. Maximum-Norm Estimates and Analytic Semigroups

The main purpose in this chapter is to discuss stability and smoothness estimates for the semidiscrete solution of the homogeneous heat equation with respect to the maximum-norm, and some consequences of such estimates for error bounds for problems with smooth and nonsmooth initial data. The semidiscrete solution is sought in a piecewise linear finite element space belonging to a quasiuniform family.

The proofs of the stability estimates are considerably more complicated than for those in the L_2-norm of our earlier chapters. We shall begin by demonstrating some preliminary such results from Schatz, Thomée and Wahlbin [209], using a weighted norm technique. We then reformulate our problem in abstract form using the concept of an analytic semigroup in a Banach space, and demonstrate that the stability and a certain smoothing property for a parabolic problem may also be expressed in terms of a bound for the resolvent of the associated elliptic operator. In the latter part of the chapter we then prove such a resolvent estimate for the discrete Laplacian in the maximum-norm by Bakaev, Thomée and Wahlbin [20], and then apply it together with the abstract theory to derive stability, smoothness, and error estimates, which are somewhat sharper than the preliminary ones in that a logarithmic factor ℓ_h may be removed. For the error estimates we need to do some auxiliary work in L_p with p large.

We consider thus the initial-boundary value problem

$$(6.1)\qquad \begin{aligned} &u_t = \Delta u \quad \text{in } \Omega, \qquad t > 0,\\ &u = 0 \qquad \text{on } \partial\Omega, \quad t > 0, \quad \text{with } u(\cdot,0) = v \quad \text{in } \Omega, \end{aligned}$$

where now for simplicity Ω is a smooth convex domain in the plane. $E(t)$ the solution operator of this problem, so that $u(t) = E(t)v$, we note that by the maximum-principle for the heat equation we have, for $v \in \mathcal{C}_0(\Omega)$, the continuous functions in Ω which vanish on $\partial\Omega$,

$$(6.2)\qquad \|E(t)v\|_{L_\infty} \le \|v\|_{L_\infty}, \quad \text{for } t \ge 0.$$

Our first interest is to show a discrete analogue of this estimate.

As in Chapter 1, let $S_h \subset H_0^1 = H_0^1(\Omega)$ denote the piecewise linear functions on a triangulation $\mathcal{T}_h = \{\tau_j\}$ of Ω with its boundary vertices on

$\partial\Omega$, and which vanish outside the polygonal domain $\Omega_h \subset \Omega$ defined by $\cup_j \bar{\tau}_j$. We assume that the $\mathcal{T}_h$ constitute a quasiuniform family. We consider the corresponding standard Galerkin semidiscrete problem, to find $u_h : [0,\infty) \to S_h$ such that

$$(u_{h,t}, \chi) + (\nabla u_h, \nabla \chi) = 0, \quad \forall \chi \in S_h, \quad t > 0, \quad \text{with } u_h(0) = v_h.$$

Recall from Chapter 1 that, with Δ_h the discrete Laplacian defined by (1.33), the semidiscrete problem may be written

$$u_{h,t} = \Delta_h u_h, \quad \text{for } t > 0, \quad \text{with } u_h(0) = v_h. \tag{6.3}$$

Introducing the solution operator of this problem, defined by $u_h(t) = E_h(t)v_h$, we shall now show that the discrete analogue of the maximum-norm stability estimate (6.2), associated with the maximum-principle for the heat equations, does not hold. For this purpose, recall that (6.3) may be written in matrix form as

$$\mathcal{B}\alpha'(t) + \mathcal{A}\alpha(t) = 0, \quad \text{for } t > 0, \quad \text{with } \alpha(0) = \gamma, \tag{6.4}$$

where $\alpha(t)$ is the vector of nodal values of $u_h(t)$ and $\mathcal{B}$ and $\mathcal{A}$ are the mass and stiffness matrices. We may therefore represent the nodal values of the solution of (6.4) as $\alpha(t) = e^{-\mathcal{B}^{-1}\mathcal{A}t}\gamma$, where the components of γ are the nodal values of v_h. The discrete analogue of (6.2) would therefore be equivalent to

$$\|\mathcal{M}(t)\|_\infty := \max_i \sum_{j=1}^{N_h} |m_{ij}(t)| \le 1, \text{ with } \mathcal{M}(t) = (m_{ij}(t)) = e^{-\mathcal{B}^{-1}\mathcal{A}t}. \tag{6.5}$$

To see that this cannot hold in general, we consider the one-dimensional example with $\Omega = (0,1)$, uniformly subdivided into intervals of length $h = 1/N$. Then $\mathcal{B}$ and $\mathcal{A}$ are symmetric tridiagonal Toeplitz matrices such that $h^{-1}\mathcal{B}$ has diagonal elements $\frac{2}{3}$ and bidiagonal elements $\frac{1}{6}$, and $h\mathcal{A}$ has diagonal elements 2 and bidiagonal elements -1. Setting $\mathcal{G} = h^2\mathcal{B}^{-1}\mathcal{A} = (g_{ij})$, which is independent of h, we have $\mathcal{M}(th^2) = I - t\mathcal{G} + O(t^2)$ as $t \to 0$, and

$$\|\mathcal{M}(th^2)\|_\infty = \max_i \big(1 - tg_{ii} + t\sum_{j\neq i} |g_{ij}|\big) + O(t^2), \quad \text{as } t \to 0.$$

Hence to show that (6.5) does not hold, it suffices to show that

$$\sum_{i\neq j} |g_{ij}| > |g_{ii}|, \quad \text{for some } i,\ 1 \le i \le N-1. \tag{6.6}$$

But the characteristic trigonometric polynomials associated with the Toeplitz matrices $h^{-1}\mathcal{B}$ and $h\mathcal{A}$ are $\frac{2}{3} + \frac{1}{3}\cos\theta$ and $2 - 2\cos\theta$, respectively, and the ratio of these has the Fourier series

$$\frac{2 - 2\cos\theta}{\frac{2}{3} + \frac{1}{3}\cos\theta} = \sum_{n=-\infty}^{\infty} b_n e^{in\theta},$$

where

$$b_0 = 6(\sqrt{3} - 1), \quad b_n = 6\sqrt{3}(-1)^n(2 - \sqrt{3})^n = b_{-n}, \quad \text{for } n > 0.$$

But the g_{ij} corresponding to points in the interior of Ω behave asymptotically like b_{i-j} for large N and we have

$$\sum_{n \neq 0} |b_n| = 6(3 - \sqrt{3}) \approx 7.61 > b_0 = 4.39.$$

In fact, already $|b_1| + |b_{-1}| = 5.57 > b_0$.

For fixed N, (6.6) may easily be checked, e.g., by MATLAB. For $N = 6$, i.e., with only 5 interior points we have

$$\sum_{i \neq 3} |g_{i3}| = 6.92 > |g_{23}| + |g_{43}| = 5.54 > |g_{33}| = 4.38.$$

It is worth noting that the lack of a maximum-principle in this case comes from the presence of the mass matrix, and that $\|e^{-t\mathcal{A}}\|_\infty \le 1$ for $t > 0$. This follows from $e^{-t\mathcal{A}} = \lim_{n\to\infty}(I + tn^{-1}\mathcal{A})^{-n}$ and

$$\|(I + k\mathcal{A})^{-1}\gamma\|_\infty \le \|\gamma\|_\infty = \max_j |\gamma_j|, \quad \text{for } k > 0.$$

This in turn follows from the discrete maximum-principle for the backward Euler five-point finite difference method. We shall return to this discussion in the context of the lumped mass method in Chapter 15.

Since the discrete analogue of (6.2) does not hold, we will thus have to be content with a weaker discrete maximum-norm stability estimate. In the next theorem, we show such a result in which the stability bound contains a logarithmic factor $\ell_h = \max(1, \log(1/h))$. With a more refined argument we shall later be able to remove this factor.

In the rest of this chapter, we denote the norm in $L_p(\Omega_0)$ by $\|\cdot\|_{L_p(\Omega_0)}$ or simply $\|\cdot\|_{L_p}$, if $\Omega_0 = \Omega$, and write similarly for the norm in the Sobolev space $W_p^s = W_p^s(\Omega)$, with s a nonnegative integer,

$$\|v\|_{W_p^s} = \begin{cases} \left(\sum_{|\alpha|\le s} \|D^\alpha v\|^p\right)^{1/p}, & \text{for } 1 \le p < \infty, \\ \max_{|\alpha|\le s} \|D^\alpha v\|_{L_\infty}, & \text{for } p = \infty. \end{cases}$$

As before, $\|\cdot\|$ and $\|\cdot\|_s$ are the norms in $L_2 = L_2(\Omega)$ and $H^s = H^s(\Omega)$.

Theorem 6.1 *Let S_h be the piecewise linear finite element spaces based on quasiuniform triangulations described above and let $E_h(t)$ the solution operator of* (6.3). *We then have*

$$\|E_h(t)v_h\|_{L_\infty} \le C\ell_h\|v_h\|_{L_\infty}, \quad \text{for } v_h \in S_h,\ t \ge 0.$$

Proof. We want to show that

$$|(E_h(t)v_h)(x)| \le C\ell_h \|v_h\|_{L_\infty}, \quad \text{for } t \ge 0,\ x \in \Omega.$$

For this purpose we introduce the *discrete delta-function* $\delta_h^x \in S_h$ defined by

$$(\delta_h^x, \chi) = \chi(x), \quad \forall \chi \in S_h,\ x \in \Omega, \tag{6.7}$$

and the *discrete fundamental solution* $\Gamma_h^x = \Gamma_h^x(t) = E_h(t)\delta_h^x \in S_h$, thus satisfying

$$\Gamma_{h,t}^x = \Delta_h \Gamma_h^x, \quad \text{for } t > 0, \quad \text{with } \Gamma_h^x(0) = \delta_h^x.$$

We may then represent the discrete solution operator as

$$(E_h(t)v_h)(x) = (\Gamma_h^x(t), v_h), \quad \text{for } x \in \Omega,\ t > 0. \tag{6.8}$$

In fact, since $E_h(t)$ is selfadjoint,

$$(\Gamma_h^x(t), v_h) = (E_h(t)\delta_h^x, v_h) = (\delta_h^x, E_h(t)v_h) = (E_h(t)v_h)(x).$$

It follows that

$$|(E_h(t)v_h)(x)| \le \|\Gamma_h^x(t)\|_{L_1} \|v_h\|_{L_\infty}, \quad \text{for } x \in \Omega,$$

and hence, in order to prove our theorem, it suffices to show

$$\|\Gamma_h^x(t)\|_{L_1} \le C\ell_h, \quad \forall x \in \Omega,\ t > 0. \tag{6.9}$$

For this purpose we define the *modified distance* between x and y by

$$\omega(y) = \omega_h^x(y) = (|y-x|^2 + h^2)^{1/2}. \tag{6.10}$$

From the Cauchy-Schwarz inequality we then have

$$\|\Gamma_h^x(t)\|_{L_1} \le \|(\omega_h^x)^{-1}\|\, \|\omega_h^x \Gamma_h^x(t)\| \le C\ell_h^{1/2} \|\omega_h^x \Gamma_h^x(t)\|, \tag{6.11}$$

since, with d the diameter of Ω,

$$\|(\omega_h^x)^{-1}\|^2 = \int_\Omega \frac{dy}{|y-x|^2 + h^2} \le 2\pi \int_0^d \frac{r\,dr}{r^2 + h^2} \le C\ell_h. \tag{6.12}$$

It thus remains to show the L_2-norm estimate

$$\|\omega_h^x \Gamma_h^x(t)\| \le C\ell_h^{1/2}, \quad \text{for } t \ge 0. \tag{6.13}$$

This will be accomplished by the energy method.

We shall repeatedly use well-known inverse properties of $\{S_h\}$, such as (1.12), which are valid when the triangulations are quasiuniform, as assumed in this chapter. We shall also need the following lemma by Descloux [70] (cf. also [59]) concerning the maximum-norm stability of the L_2-projection $P_h : L_2 \to S_h$.

Lemma 6.1 *With S_h as above, there is a positive constant c such that, if τ_0 is any triangle of $\mathcal{T}_h$ and $\Omega_0 \subset \Omega$ is disjoint from τ_0, then*

$$\|P_h v\|_{L_2(\Omega_0)} \le e^{-c\,\mathrm{dist}(\Omega_0,\tau_0)/h}\|v\|_{L_2(\tau_0)}, \quad \textit{if} \ \operatorname{supp} v \subset \tau_0. \tag{6.14}$$

Further, there is a positive constant C such that

$$\|P_h v\|_{L_\infty} \le C\|v\|_{L_\infty}. \tag{6.15}$$

Proof. Starting with $R_0 = \tau_0$, we define a sequence of sets R_j, $j = 0, 1, \dots,$ recursively by taking for R_k the union of (closed) triangles in $\mathcal{T}_h$ which are not in $\cup_{l<k} R_l$ and which are neighbors of the triangles of this set. By the quasiuniformity of the triangulations the points of R_k then have a distance to τ_0 which is bounded above and below by constants times $(k-1)h$. Letting $D_k = \cup_{l>k} R_l$, we shall show that, for some $\kappa > 0$,

$$\|P_h v\|^2_{L_2(D_k)} \le \kappa \|P_h v\|^2_{L_2(R_k)}, \quad \text{for } k \ge 1. \tag{6.16}$$

Assuming this for a moment, we denote the left-hand side by q_k and find thus $q_k \le \kappa(q_{k-1} - q_k)$, for $k \ge 1$, whence, with $\gamma = \kappa/(1+\kappa)$,

$$q_k \le \gamma q_{k-1} \le \gamma^k q_0 \le \gamma^k \|P_h v\|^2_{L_2(\Omega)} \le \gamma^k \|v\|^2_{L_2(\tau_0)}.$$

Defining c by $\gamma = e^{-2c}$, and with k the largest integer such that $D_k \supset \Omega_0$, this shows

$$\|P_h v\|_{L_2(\Omega_0)} \le \|P_h v\|_{L_2(D_k)} \le e^{-ck}\|P_h v\|_{L_2(\tau_0)} \le e^c e^{-c\mathrm{dist}(\Omega_0,\tau_0)/h}\|v\|_{L_2(\tau_0)},$$

which is (6.14) (possibly with a different choice of c).

In order to show (6.16) we note that since $\operatorname{supp} v \subset \tau_0$, we have $(P_h v, \chi) = 0$ for any $\chi \in S_h$ with $\operatorname{supp}\chi \subset D_{k-1} = D_k \cup R_k$ for $k \ge 1$. In particular, we may choose $\widetilde{\chi} \in S_h$ such that $\widetilde{\chi} = P_h v$ in D_k, $\widetilde{\chi} = 0$ in $\Omega \setminus D_{k-1}$. For the triangles of R_k, $\widetilde{\chi}$ coincides with $P_h v$ at one or two vertices and vanishes at the remaining two or one vertices. Then

$$0 = (P_h v, \widetilde{\chi}) = \|P_h v\|^2_{L_2(D_k)} + \int_{R_k} P_h v\, \widetilde{\chi}\, dx,$$

and hence

$$\|P_h v\|^2_{L_2(D_k)} \le \|P_h v\|_{L_2(R_k)} \|\widetilde{\chi}\|_{L_2(R_k)}.$$

By the definition of $\widetilde{\chi}$ it is now easy to see that there exists a $\kappa > 0$ such that, for each triangle τ of R_k, $\|\widetilde{\chi}\|_{L_2(\tau)} \le \kappa\|P_h v\|_{L_2(\tau)}$. Hence the corresponding inequality is valid with τ replaced by R_k, and we may conclude that (6.16) holds.

We may now finish the proof of the lemma by showing the maximum-norm stability estimate (6.15). Let τ_0 be a triangle where $P_h v$ attains its maximum value and set $v_j = v$ on τ_j and 0 otherwise. We then have $v = \sum_j v_j$ and

$$\|P_h v\|_{L_\infty} = \|P_h v\|_{L_\infty(\tau_0)} \le \sum_j \|P_h v_j\|_{L_\infty(\tau_0)}.$$

Using the local inverse estimate

$$\|\chi\|_{L_\infty(\tau_0)} \le Ch^{-1}\|\chi\|_{L_2(\tau_0)}, \tag{6.17}$$

together with (6.14) we have

$$\begin{aligned}\|P_h v_j\|_{L_\infty(\tau_0)} &\le Ch^{-1}\|P_h v_j\|_{L_2(\tau_0)}\\ &\le Ch^{-1}e^{-c\,\mathrm{dist}(\tau_0,\tau_j)/h}\|v_j\|_{L_2(\tau_j)} \le Ce^{-c\,\mathrm{dist}(\tau_0,\tau_j)/h}\|v\|_{L_\infty}.\end{aligned}$$

With R_k as above, the number of triangles of R_k is bounded by the number of triangles of $\cup_{l\le k}R_l$, and by quasiuniformity this is bounded by $C(kh)^2h^{-2} = Ck^2$, and hence

$$\|P_h v\|_{L_\infty} \le C\Big(\sum_k \sum_{\tau_j\subset R_k} e^{-ck}\Big)\|v\|_{L_\infty} \le C\sum_k k^2e^{-ck}\|v\|_{L_\infty} \le C\|v\|_{L_\infty},$$

which completes the proof. □

We remark that Lemma 6.1 generalizes to certain nonquasiuniform families of triangulations, see [59]. We shall not go into the details of this.

We note that the stability of P_h shows that $P_h v$ is an optimal order approximation to v in maximum-norm. In fact, using the estimate (1.43) for the interpolant we have, for $v \in \dot W^2_\infty$ (we denote $\dot W^s_p = W^s_p \cap H^1_0$,

$$\|P_h v - v\|_{L\infty} = \|(P_h - I)(v - I_h v)\|_{L\infty} \le C\|I_h v - v\|_{L\infty} \le Ch^2\|v\|_{W^2_\infty}.$$

Before we can finish the proof of Theorem 6.1 we need three additional technical lemmas. The first of these is related to the so-called superapproximation property of Nitsche and Schatz (cf. [185]). Note that for an $O(h)$ error estimate for $\nabla P_h u$ we normally need $u \in \dot H^2 = H^2 \cap H^1_0$.

Lemma 6.2 *There is a C such that, with $\omega = \omega_h^x$ defined in* (6.10),

$$\|\nabla(\omega^2\chi - P_h(\omega^2\chi))\| \le Ch(\|\chi\| + \|\omega\nabla\chi\|), \quad \forall \chi \in S_h,\ x \in \Omega.$$

Proof. Let $\varphi = I_h(\omega^2\chi)$ be the interpolant of $\omega^2\chi$ in S_h. With τ an arbitrary triangle of $\mathcal{T}_h$ we have, using Leibniz' rule and the facts that $|\nabla(\omega^2)| \le C\omega$, $D^\alpha(\omega^2) = C$ and $D^\alpha\chi = 0$ in τ for $|\alpha| = 2$,

$$\begin{aligned}\|\omega^2\chi - \varphi\|_{L_2(\tau)} &+ h\|\nabla(\omega^2\chi - \varphi)\|_{L_2(\tau)}\\ &\le Ch^2\sum_{|\alpha|=2}\|D^\alpha(\omega^2\chi)\|_{L_2(\tau)} \le Ch^2(\|\chi\|_{L_2(\tau)} + \|\omega\nabla\chi\|_{L_2(\tau)}),\end{aligned}$$

and hence, by squaring and summing over the triangles of $\mathcal{T}_h$, the corresponding inequality for Ω. By the inverse estimate (1.12) we have, since $P_h\varphi = \varphi$,

$$\|\nabla(\varphi - P_h(\omega^2\chi))\| \leq Ch^{-1}\|P_h(\varphi - \omega^2\chi)\| \leq Ch^{-1}\|\varphi - \omega^2\chi\|.$$

We have already bounded this latter quantity and hence the desired estimate now follows by the triangle inequality. □

The next lemma shows that the modified distance function in some sense compensates for the "singularity" of the discrete delta function.

Lemma 6.3 *With ω_h^x and δ_h^x defined by (6.10) and (6.7), there is a constant C such that*

$$\|\omega_h^x\delta_h^x\| \leq C, \quad \textit{for } x \in \Omega.$$

Proof. Let Ω_h be the domain defined by the union of the triangles of $\mathcal{T}_h$. Fixing x, we write $\omega = \omega_h^x$ and $\delta = \delta_h^x$, and let $\Omega_j = \{y \in \Omega_h;\ 2^{j-1}h < |y - x| \leq 2^j h\}$ for $j \geq 1$ and $\Omega_0 = \{y \in \Omega_h;\ |y - x| \leq h\}$. Clearly, $\omega_h^x(y) \leq (2^j + 1)h$, for $y \in \Omega_j$, and hence

$$\|\omega\delta\| \leq C\sum_{j\geq 0}\|\omega\delta\|_{L_2(\Omega_j)} \leq C\sum_{j\geq 0}2^j h\|\delta\|_{L_2(\Omega_j)}. \tag{6.18}$$

In order to bound $\|\delta\|_{L_2(\Omega_j)}$, let supp $\varphi \subset \Omega_j$ and let τ be the triangle containing x. We then have, using a local inverse estimate, cf. (6.17),

$$(\delta, \varphi) = (\delta, P_h\varphi) = (P_h\varphi)(x) \leq \|P_h\varphi\|_{L_\infty(\tau)} \leq Ch^{-1}\|P_h\varphi\|_{L_2(\tau)}. \tag{6.19}$$

Now, let $\{\tau_l\}_{l=1}^{M_{j,h}}$ be the triangles of $\mathcal{T}_h$ which intersect Ω_j, and set $\varphi_l = \varphi$ on τ_l and $\varphi_l = 0$ outside τ_l. By Lemma 6.1 we then have

$$\|P_h\varphi_l\|_{L_2(\tau)} \leq Ce^{-c2^j}\|\varphi_l\|_{L_2(\tau_l)} \leq Ce^{-c2^j}\|\varphi\|_{L_2(\Omega_j)}.$$

Since the number $M_{j,h}$ of such triangles is bounded by $C2^{2j}$, we find

$$\|P_h\varphi\|_{L_2(\tau)} \leq \sum_{l=1}^{M_{j,h}}\|P_h\varphi_l\|_{L_2(\tau)} \leq C2^{2j}e^{-c2^j}\|\varphi\|_{L_2(\Omega_j)},$$

and hence

$$\|\delta\|_{L_2(\Omega_j)} = \sup_{\|\varphi\|=1}(\delta, \varphi) \leq Ch^{-1}2^{2j}e^{-c2^j}.$$

Inserting this into (6.18) shows

$$\|\omega\delta\| \leq C\sum_{j=0}^{\infty}2^{3j}\,e^{-c2^j} \leq C,$$

which completes the proof. □

Recall that for functions in plane domains Ω, Sobolev's embedding theorem (see, e.g., Adams and Fournier [1]) asserts that $\|v\|_{L_\infty} \le C_\varepsilon \|v\|_{1+\varepsilon}$ for any $\varepsilon > 0$, but this inequality does not hold for all $v \in H^1$ when $\varepsilon = 0$. For functions in S_h, however, we have the following substitute. For later application we show this for more general than quasiuniform triangulations.

Lemma 6.4 *Assume that the family of triangulations underlying the* $\{S_h\}$ *are such that* $h_{\min} \ge Ch^\gamma$ *for some* $\gamma > 0$. *Then there is a* C *such that*

$$\|\chi\|_{L_\infty} \le C\ell_h^{1/2}\|\nabla\chi\|, \quad \forall \chi \in S_h.$$

Proof. By Sobolev's embedding theorem

$$(6.20) \qquad \|\varphi\|_{L_p} \le C_p\|\nabla\varphi\|, \quad \text{where } C_p = Cp^{1/2}, \quad \forall \varphi \in H_0^1,\ 2 \le p < \infty.$$

In fact, by a common proof of this result (cf., e.g., Stein [219]), C_p may be chosen as $C\|\,|x|^{-1}\|_{L_s(B)}$ with $1/s = 1/p + 1/2$ and B the unit ball in $\mathbb{R}^2$. Since this norm is bounded by $C(2-s)^{-1/s} = C((2+p)/4)^{1/p+1/2} \le Cp^{1/2}$, for large p, (6.20) follows. Applying first an inverse estimate (see Brenner and Scott [42]), and then (6.20) to $\chi \in S_h$, we find

$$\|\chi\|_{L_\infty} \le Ch_{min}^{-2/p}\|\chi\|_{L_p} \le Ch^{-2\gamma/p}\|\chi\|_{L_p} \le Ch^{-2\gamma/p}p^{1/2}\|\nabla\chi\|, \quad \forall \chi \in S_h.$$

The result stated now follows by taking $p = \ell_h = \log(1/h)$ for small h. □

We now return to the proof of Theorem 6.1, which we have reduced above to showing (6.13). Writing $\Gamma = \Gamma_h^x(t)$ and $\omega = \omega_h^x$ we consider the expression

$$\tfrac{1}{2}\frac{d}{dt}\|\omega\Gamma\|^2 + \|\omega\nabla\Gamma\|^2 = (\Gamma_t, \omega^2\Gamma) + (\nabla\Gamma, \nabla(\omega^2\Gamma)) - 2(\nabla\Gamma, \omega\,\Gamma\,\nabla\omega).$$

By the definition of Γ we have

$$(6.21) \qquad (\Gamma_t, \psi) + (\nabla\Gamma, \nabla\psi) = 0, \quad \forall \psi \in S_h,$$

and hence

$$(6.22) \qquad \begin{aligned} \tfrac{1}{2}\frac{d}{dt}\|\omega\Gamma\|^2 + \|\omega\nabla\Gamma\|^2 &= (\Gamma_t, \omega^2\Gamma - \psi) + (\nabla\Gamma, \nabla(\omega^2\Gamma - \psi)) \\ &\quad - 2(\omega\nabla\Gamma, \Gamma\,\nabla\omega) = I_1 + I_2 + I_3. \end{aligned}$$

We now choose $\psi = P_h(\omega^2\Gamma)$. Then $I_1 = 0$, and by Lemma 6.2 and the inverse estimate (1.12) we find

$$\begin{aligned} |I_2| &\le \|\nabla\Gamma\|\,\|\nabla(\omega^2\Gamma - \psi)\| \le \big(Ch^{-1}\|\Gamma\|\big)\big(Ch(\|\Gamma\| + \|\omega\nabla\Gamma\|)\big) \\ &\le C(\|\Gamma\|^2 + \|\Gamma\|\,\|\omega\nabla\Gamma\|). \end{aligned}$$

Further, since $\nabla\omega$ is bounded, $|I_3| \le C\|\Gamma\|\,\|\omega\nabla\Gamma\|$. Altogether,

$$\tfrac{1}{2}\frac{d}{dt}\|\omega\Gamma\|^2 + \|\omega\nabla\Gamma\|^2 \le C(\|\Gamma\|^2 + \|\Gamma\|\,\|\omega\nabla\Gamma\|),$$

so that, after a kick-back of $\|\omega\nabla\Gamma\|$ and integration,

$$\|\omega\Gamma(t)\|^2 + \int_0^t \|\omega\nabla\Gamma\|^2\,ds \le \|\omega\delta_h^x\|^2 + C\int_0^t \|\Gamma\|^2\,ds. \tag{6.23}$$

In view of Lemma 6.3 it remains now to prove

$$\int_0^t \|\Gamma\|^2 ds \le C\ell_h. \tag{6.24}$$

For this purpose, we note that with $T_h = (-\Delta_h)^{-1}$, Γ satisfies

$$T_h\Gamma_t + \Gamma = 0, \quad \text{for } t > 0, \quad \text{with } \Gamma(0) = \delta_h^x.$$

By taking inner products by 2Γ and integrating in time, we obtain

$$(T_h\Gamma,\Gamma) + 2\int_0^t \|\Gamma\|^2\,ds = (T_h\delta_h^x, \delta_h^x) = (T_h\delta_h^x)(x).$$

Setting $G_h^x = T_h\delta_h^x$, it thus suffices to show, since $(T_h\Gamma,\Gamma) \ge 0$, that

$$G_h^x(x) \le C\ell_h. \tag{6.25}$$

The function G_h^x may be thought of as a discrete Green's function; we have

$$(\nabla G_h^x, \nabla\chi) = (\nabla T_h\delta_h^x, \nabla\chi) = (\delta_h^x, \chi) = \chi(x), \quad \forall \chi \in S_h.$$

In particular, $G_h^x(x) = \|\nabla G_h^x\|^2$. In view of Lemma 6.4 this shows

$$G_h^x(x) \le \|G_h^x\|_{L_\infty} \le C\ell_h^{1/2}\|\nabla G_h^x\| = C\ell_h^{1/2} G_h^x(x)^{1/2},$$

and hence (6.25). This completes the proof of (6.9) and hence of Theorem 6.1. □

We remark for later use that Lemma 6.3, (6.23), and (6.24) also show

$$\int_0^t \|\omega\nabla\Gamma\|^2\,ds \le C\ell_h, \quad \text{for } t \ge 0. \tag{6.26}$$

We shall now apply the above stability result to obtain an error estimate for the semidiscrete solution of the inhomogeneous parabolic problem

$$\begin{aligned} u_t - \Delta u &= f \quad \text{in } \Omega, \quad \text{for } t > 0,\\ u &= 0 \quad \text{on } \partial\Omega, \quad \text{for } t > 0, \quad \text{with } u(\cdot,0) = v \quad \text{in } \Omega. \end{aligned} \tag{6.27}$$

With the semidiscrete analogue of (6.27) formulated as

$$u_{h,t} - \Delta_h u_h = P_h f, \quad \text{for } t > 0, \quad \text{with } u_h(0) = v_h, \tag{6.28}$$

we show the following.

Theorem 6.2 *Under the assumptions of Theorem 6.1, with $v_h = R_h v$ and $v = 0$ on $\partial\Omega$, we have for the error in the semidiscrete parabolic problem* (6.28)

$$\|u_h(t) - u(t)\|_{L_\infty} \le Ch^2\ell_h^2\Big(\|v\|_{W_\infty^2} + \int_0^t \|u_t\|_{W_\infty^2}\,ds\Big), \quad \text{for } t \ge 0.$$

Proof. We write as before $u_h - u = (u_h - R_h u) + (R_h u - u) = \theta + \rho$, where R_h is the Ritz projection onto S_h, and obtain from (1.46)

$$\|\rho(t)\|_{L_\infty} \le Ch^2\ell_h\,\|u(t)\|_{W_\infty^2} \le Ch^2\ell_h\big(\|v\|_{W_\infty^2} + \int_0^t \|u_t\|_{W_\infty^2}\,ds\big). \tag{6.29}$$

Moreover, with $\theta(0) = 0$ we have from (1.37)

$$\theta(t) = -\int_0^t E_h(t-s)P_h\rho_t(s)\,ds. \tag{6.30}$$

Applying the stability estimates for $E_h(t)$ and P_h shown above, together with (1.46), we find

$$\|E_h(t-s)P_h\rho_t(s)\|_{L_\infty} \le C\ell_h\,\|(R_h - I)u_t(s)\|_{L_\infty} \le Ch^2\ell_h^2\|u_t(s)\|_{W_\infty^2},$$

which together with (6.30) completes the proof. □

We remark that, by the stability result of Theorem 6.1, the same error bound as in Theorem 6.2 holds for any v_h such that $\|v_h - R_h v\|_{L_\infty} \le Ch^2\ell_h\|v\|_{W_\infty^2}$.

We now show a smoothing property of $E_h(t)$ in L_∞, cf. Lemma 3.9 for the corresponding L_2-result.

Theorem 6.3 *Under the assumptions of Theorem 6.1 we have*

$$\|E_h'(t)v_h\|_{L_\infty} \le Ct^{-1}\ell_h\|v_h\|_{L_\infty}, \quad \text{for } v_h \in S_h,\ t > 0. \tag{6.31}$$

Proof. Using (6.8), the proof of this result is first reduced, in the same way as in Theorem 6.1, to showing that the discrete fundamental solution $\Gamma = \Gamma_h^x(t)$ satisfies $t\|\Gamma_t(t)\|_{L_1} \le C\ell_h$, which in turn follows from the L_2-norm estimate $t\|\omega\Gamma_t(t)\| \le C\ell_h^{1/2}$, where $\omega = \omega_h^x$ is the discrete distance function from (6.10). For the latter inequality we differentiate (6.21) to see that Γ may be replaced by Γ_t in (6.22). After multiplication by t^2 we obtain, for any $\psi \in S_h$, the identity

$$\begin{aligned}\tfrac12\frac{d}{dt}(t^2\|\omega\Gamma_t\|^2) + t^2\|\omega\nabla\Gamma_t\|^2 &= t^2\big((\Gamma_{tt}, \omega^2\Gamma_t - \psi)\\ &\quad + (\nabla\Gamma_t, \nabla(\omega^2\Gamma_t - \psi)) - 2(\omega\nabla\Gamma_t, \nabla\omega\,\Gamma_t)\big) + t\|\omega\Gamma_t\|^2.\end{aligned}$$

The proof of the theorem then follows the lines of the proof of Theorem 6.1. Since we shall later derive the corresponding result without the factor ℓ_h, we shall not carry out the details. □

We shall now look at the stability problem from another angle, namely by use of the theory of analytic semigroups.

We consider thus an abstract initial value problem of the form

$$u' + Au = 0 \quad \text{for } t > 0, \quad \text{with } u(0) = v, \tag{6.32}$$

in a complex Banach space $\mathcal{B}$ with norm $\|\cdot\|$. We now assume that A is a closed, densely defined linear operator, with a resolvent set $\rho(A)$ such that

$$\rho(A) \supset \Sigma_\delta = \{z \in \mathbb{C} : \delta \le |\arg z| \le \pi\}, \quad \text{with } \delta \in (0, \tfrac{1}{2}\pi), \tag{6.33}$$

and such that the resolvent, $R(z; A) = (zI - A)^{-1}$, satisfies,

$$\|R(z; A)\| \le M|z|^{-1}, \quad \text{for } z \in \Sigma_\delta, \quad \text{with } M \ge 1. \tag{6.34}$$

Here and below $\|\cdot\|$ is also used to denote the operator norm for bounded linear operators in $\mathcal{B}$. We remark that since $0 \in \rho(A)$ it follows easily that $z \in \rho(A)$ for $|z| < 1/M_0$ where $M_0 = \|A^{-1}\|$, and that $\|R(z; A)\| \le 2M_0$ for $|z| \le 1/(2M_0)$, say. Hence the bound in (6.34) could have been replaced by $M_1(1 + |z|)^{-1}$.

The theory of semigroups shows that under these assumptions $-A$ is the infinitesimal generator of a uniformly bounded strongly continuous semigroup $E(t) = e^{-tA}$, $t \ge 0$, which is the solution operator of (6.32). Moreover, the solution of (6.32) may be represented as

$$u(t) = E(t)v = \frac{1}{2\pi i} \int_\Gamma e^{-zt} R(z; A)v \, dz, \tag{6.35}$$

where $\Gamma = \{z : |\arg z| = \psi, \ \psi \in (\delta, \frac{1}{2}\pi)\}$, with $\operatorname{Im} z$ decreasing along Γ (as is natural since we want to think of Γ as going around the spectrum $\sigma(A)$ of A in a positive sense). Because the integrand in (6.35) is analytic in Σ_δ, the curve Γ may be deformed to any other homotopic curve in Σ_δ which asymptotically approaches $|\arg z| = \psi$ as $|z| \to \infty$. The semigroup $E(t)$ may be extended to a semigroup $E(\tau)$ which is analytic for τ in a sector containing the positive real axis, and is therefore referred to as an *analytic semigroup.* It also has a smoothing property which is important for our purposes. All these properties are summarized in the following theorem (cf. Pazy [194], Theorem 2.5.2).

Theorem 6.4 *Let $E(t)$ be a strongly continuous semigroup in $\mathcal{B}$ generated by the densely defined linear operator $-A$. Then the following conditions are equivalent:*

(i) There are constants $M \ge 1$ and $\delta \in (0, \frac{1}{2}\pi)$ such that

$$\|R(z; A)\| \le M|z|^{-1}, \quad \text{for } z \in \Sigma_\delta \setminus \{0\}.$$

(ii) There is an $\varepsilon > 0$ such that $E(t)$ may be extended to a uniformly bounded analytic semigroup $E(\tau)$ in the sector $\{\tau; |\arg \tau| < \varepsilon\}$.

(iii) There is a constant $K \geq 1$ *such that*

$$\|E(t)\| + t\|E'(t)\| \leq K, \quad \text{for } t > 0.$$

If $\mathcal{B}$ is a Hilbert space and A selfadjoint and positive definite, then the spectrum $\sigma(A) = \mathbb{C} \setminus \rho(A)$ of A lies on the positive real axis and, with δ arbitrary in $(0, \frac{1}{2}\pi)$,

$$\|R(z;A)\| = \sup_{\lambda \in \sigma(A)} |z - \lambda|^{-1} \leq (\sin \delta)^{-1} |z|^{-1}, \quad \text{for } z \in \Sigma_\delta.$$

Thus A generates an analytic semigroup in $\mathcal{B}$.

For another example, let $\mathcal{B}$ be the complex-valued Hilbert space $L_2 = L_2(\Omega)$, with Ω a bounded domain in $\mathbb{R}^d$ with smooth boundary, and let A be a second order, not necessarily selfadjoint, partial differential operator of the form

$$Au = - \sum_{j,k=1}^{d} \frac{\partial}{\partial x_j}(a_{jk}\frac{\partial u}{\partial x_k}) + \sum_{j=1}^{d} b_j \frac{\partial u}{\partial x_j} + c_0 u, \tag{6.36}$$

with smooth, possibly complex-valued, coefficients in $\bar{\Omega}$, and $\mathcal{D}(A) = H^2 \cap H_0^1$. Assume that A is strongly elliptic so that, for the associated bilinear form,

$$\operatorname{Re} A(u,u) \geq c\|u\|_1^2, \quad \forall\, u \in H_0^1, \quad \text{with } c > 0. \tag{6.37}$$

In this case (6.33) and (6.34) hold, now for $\delta \in (0, \frac{1}{2}\pi)$ sufficiently close to $\frac{1}{2}\pi$. In fact, with $f \in L_2$ and z a given complex number, the complex-valued function $u = u_z = R(z;A)f$ is the solution of the Dirichlet problem

$$(zI - A)u = f \quad \text{in } \Omega, \quad \text{with } u = 0 \text{ on } \partial\Omega, \tag{6.38}$$

and the resolvent estimate to be shown may be expressed as

$$\|u\| \leq M|z|^{-1}\|f\|, \quad \text{for } z \in \Sigma_\delta,\ z \neq 0. \tag{6.39}$$

But, multiplying (6.38) by $-\bar{u}$ and integrating over Ω, we have

$$A(u,u) - z\|u\|^2 = -(f,u).$$

Taking real parts and using (6.37) we find

$$c\|u\|_1^2 - \operatorname{Re} z\, \|u\|^2 \leq \|f\|\, \|u\|, \tag{6.40}$$

and, similarly, by taking imaginary parts and using (6.40),

$$|\operatorname{Im} z|\, \|u\|^2 \leq \|f\|\, \|u\| + C\|u\|_1^2 \leq (C_1\|f\| + C_2 \operatorname{Re} z\, \|u\|)\|u\|.$$

Hence

$$(|\operatorname{Im} z| - C_2 \operatorname{Re} z)\|u\| \le C_1\|f\|.$$

Considering separately the cases $\operatorname{Re} z \le 0$ and $0 < \operatorname{Re} z \le \varepsilon|\operatorname{Im} z|$ with ε small enough, this shows (6.39) for δ sufficiently small. Thus $-A$ generates an analytic semigroup in $L_2(\Omega)$.

We remark that if $A_h : S_h \to S_h$ is the discrete version of the operator A in (6.36) in a finite element space $S_h \subset H_0^1$, defined by

$$(A_h\psi, \chi) = A(\psi, \chi), \quad \forall\, \psi, \chi \in S_h,$$

then $-A_h$ also generates an analytic semigroup $E_h(t)$ on $S_h \subset L_2$, and the constants M and δ in the resolvent estimate in *(i)*, with A replaced by A_h, are independent of h. In fact, the above proof of (6.39) remains valid for $u_h = R(z; A_h)f$, when $f \in S_h$, with the same constants.

Of particular interest to us in this chapter is the case when $A = -\Delta$ and $\mathcal{B}$ is associated with the maximum-norm. We first recall that the Laplacian generates a strongly continuous and analytic contraction semigroup $E(t)$ in L_p, thus with

$$(6.41)\quad \|E(t)v\|_{L_p} + t\|E'(t)v\|_{L_p} \le C\|v\|_{L_p}, \quad \text{for } t > 0,\ v \in L_p,\ 2 \le p < \infty.$$

For $p = \infty$ the situation is more delicate. Instead of L_∞ it is then often natural to work in the subspace $\mathcal{C} = \mathcal{C}(\bar{\Omega})$ of continuous functions in $\bar{\Omega}$, again normed with $\|\cdot\|_{L_\infty}$, or sometimes in $\mathcal{C}_0$ which consists of the functions in $\mathcal{C}$ which vanish on $\partial\Omega$.

It is a special case of a result from Stewart [220] that, if $D(A) = \mathcal{C}^2(\bar{\Omega}) \cap \mathcal{C}_0$, then, for any $\delta \in (0, \frac{1}{2}\pi)$,

$$(6.42)\qquad \|R(z; A)v\|_{L_\infty} \le M(1 + |z|)^{-1}\|v\|_{L_\infty}, \quad \text{for } z \in \Sigma_\delta, \quad v \in \mathcal{C}.$$

If we consider A as an operator in $\mathcal{C}$ it is not densely defined, and the solution operator is not strongly continuous at $t = 0$ since $E(t)v = 0$ on $\partial\Omega$ for $t > 0$, which is normally assumed in semigroup theory. However, in [220], A is shown to generate an analytic contraction semigroup $E(t)$ in $\mathcal{B} = \mathcal{C}_0$, so that in particular

$$(6.43)\qquad \|E(t)v\|_{L_\infty} + t\|E'(t)v\|_{L_\infty} \le C\|v\|_{L_\infty}, \quad \text{for } t > 0,\ v \in \mathcal{C}_0.$$

Note that in this case $Av \in \mathcal{C}_0$ when $v \in D(A)$. Using the resolvent estimate (6.42) we may extend the definition of $E(t)$ by (6.35) also to $v \in \mathcal{C}$, without loosing the stability and smoothing properties of (6.43). In fact, the solution operator may be further extended to L_∞, with the stability and smoothness properties retained. To see this, for $v \in L_\infty$, since then also $v \in L_2$ we may consider $E(t)v$ to be the result of the solution operator in L_2, which is known to be smooth for $t > 0$. Then also (6.41) holds, and since this estimate is uniform in p, we may let $p \to \infty$ to see that (6.43) holds also for $v \in L_\infty$.

To interpret the above theory in our present finite element context, we note that Theorem 6.4 implies that our stability and smoothing estimates of Theorems 6.1 and 6.3 together are equivalent to a bound for the resolvent $R(z; A_h)$. To study this equivalence in more detail, we now investigate more precisely than above how the constants δ and M in condition (i) depend on the constant K in (iii).

Lemma 6.5 *Let $E(t) = e^{-At}$ be an analytic semigroup in a Banach space $\mathcal{B}$. Then there are constants C and c independent of K such that if condition (iii) of Theorem 6.4 is valid, then condition (i) holds with $M = CK^2$, $\delta = \frac{1}{2}\pi - c/K^2$.*

Proof. We first show that *(iii)* implies that there is a positive ν such that $E(t)$ may be extended to the sector $\{\tau; |\arg\tau| \le \gamma = \nu/K\} \subset \mathbb{C}$, with $\|E(\tau)\| \le 2K$. In fact, using $E^{(n)}(t) = E'(t/n)^n$ and $n^n \le n!e^n$, we find by *(iii)* that

$$\|E^{(n)}(t)\| \le \|E'(t/n)\|^n \le \Big(\frac{Kn}{t}\Big)^n \le \Big(\frac{Ke}{t}\Big)^n n!, \quad \text{for } n \ge 1,$$

so that we may define

$$E(\tau) = E(t) + \sum_{n=1}^{\infty} \frac{E^{(n)}(t)}{n!}(\tau - t)^n,$$

with uniform convergence in $\mathcal{B}$, for $|\tau - t| \le \mu t/(Ke)$, with any $\mu < 1$. Thus $E(\tau)$ is analytic for $|\arg\tau| \le \arcsin(\mu/(Ke))$ and, if $\mu \le K/(1+K)$, we have

$$\|E(\tau)\| \le K + \sum_{n=1}^{\infty} \mu^n = K + \frac{\mu}{1-\mu} \le 2K.$$

It follows that $E(\tau)$ is analytic for $|\arg\tau| \le \arcsin(1/((K+1)e))$ and hence for $|\arg\tau| \le \nu/K$, with $\nu = \inf_{K\ge1}(K\arcsin(1/((K+1)e)))$.

Recalling that

$$R(z; A) = -\int_0^\infty e^{zt}E(t)\,dt, \quad \text{for } \operatorname{Re} z < 0, \tag{6.44}$$

we begin by showing *(i)* for $\operatorname{Re} z \le 0$. Setting $z = x + iy$ we first note that (6.44) and *(iii)* immediately yield

$$\|R(z; A)\| \le \int_0^\infty e^{xt}\|E(t)\|\,dt \le \frac{K}{|x|} \le \frac{K^2}{|x|}, \quad \text{when } x < 0. \tag{6.45}$$

By the analyticity of $E(\tau)$, we may shift the path of integration from the positive t-axis to $\arg\tau = \pm\gamma$, with $\gamma = \nu/K$, and obtain, with $\arg\tau = \gamma$ for $y > 0$, still with $x \le 0$, since $\sin\gamma \ge c/K$,

$$\|R(z;A)\| \le 2K \int_0^\infty e^{\rho(x\cos\gamma - y\sin\gamma)}\,d\rho \le \frac{2K}{y\sin\gamma} \le \frac{CK^2}{|y|}. \tag{6.46}$$

Hence, by (6.45) and (6.46),

$$\|R(z;A)\| \le \frac{CK^2}{\max(|x|,|y|)} \le \frac{2CK^2}{|z|},$$

and the same bound is obtained for $y < 0$ by taking $\arg\tau = -\gamma$, so that (i) is shown for $\operatorname{Re} z \le 0$.

For $\delta \le |\arg z| \le \frac{1}{2}\pi$ we have by Taylor expansion around iy that

$$R(z;A) = \sum_{n=0}^{\infty} R(iy;A)^{n+1}(-x)^n.$$

This series converges uniformly for $|x|\,\|R(i\,y;A)\| \le \kappa < 1$, and so, by (6.46) applied to $z = iy$, and if $CK^2|x|/|y| \le \kappa < 1$, we have for these z

$$\|R(z;A)\| \le \sum_{n=0}^{\infty} \Big|\frac{CK^2}{y}\Big|^{n+1} |x|^n \le \frac{CK^2}{|y|}\,\frac{1}{1-\kappa}.$$

This means that $\rho(A) \supset \{z; |x| \le \kappa(CK^2)^{-1}|y|\}$ and, together with the result already obtained for $\operatorname{Re} z \le 0$, that $\rho(A) \supset \Sigma_\delta$ where $\delta = \frac{1}{2}\pi - \arctan(\kappa/(CK^2))$. This completes the proof. □

A direct application of Lemma 6.5 to the semidiscrete solution operator $E_h(t) = e^{-tA_h}$, using the stability and smoothing properties in Theorems 6.1 and 6.3 shows that

$$\|R(z;A_h)\|_{L_\infty} \le C\ell_h^2|z|^{-1}, \quad \text{for } z \in \Sigma_{\delta_h}, \quad \text{where } \delta_h = \tfrac{1}{2}\pi - c\ell_h^{-2}. \tag{6.47}$$

On the other hand, by Theorem 6.4, this estimate shows a stability and smoothing estimate. To make the dependence of the resulting estimates estimates on h more precise, we show the following lemma. Here

$$\ell(t) = \max(1, \log(1/t)). \tag{6.48}$$

Lemma 6.6 *Let $E(t)$ be an analytic semigroup in a Banach space $\mathcal{B}$. There is a constant C independent of M and δ such that if condition (i) of Theorem 6.4 holds, then*

$$\|E(t)\| \le CM\ell(\cos\delta) \quad \text{and} \quad t\,\|E'(t)\| \le CM/\cos\delta, \quad \text{for } t > 0.$$

Proof. Since $R(z;A)$ is analytic in Σ_δ we may shift the integral in (6.35) to $\widetilde{\Gamma}_t = \Gamma_t \cup \Gamma_t^+ \cup \Gamma_t^-$, where

$$\Gamma_t^0 = \{t^{-1}e^{i\varphi};\ \delta \le |\varphi| \le \pi\} \quad \text{and} \quad \Gamma_t^{\pm} = \{r\,e^{\pm i\delta};\ t^{-1} \le r < \infty\}.$$

Then

$$\|\frac{1}{2\pi i}\int_{\Gamma_t^{\pm}} e^{-zt}R(z;A)\,dz\| \le \frac{1}{2\pi}\int_{t^{-1}}^{\infty} e^{-rt\cos\delta} Mr^{-1}\,dr$$
$$= \frac{M}{2\pi}\int_{\cos\delta}^{\infty} e^{-s}\frac{ds}{s} \le CM\ell(\cos\delta).$$

Further

$$\|\frac{1}{2\pi i}\int_{\Gamma_t^{0}} e^{-zt}R(z;A)\,dz\| \le \frac{1}{\pi}\int_{\delta}^{\pi} e^{-\cos\varphi} M d\varphi \le eM,$$

which completes the proof of the first inequality in the lemma. For the second inequality we take $\Gamma = \partial\Sigma_\delta$ to obtain

$$t\,\|E'(t)\| = \Big\|\frac{t}{2\pi i}\int_{\partial\Sigma_\delta} z\,e^{-zt}R(z;A)\,dz\Big\| \le \frac{t}{\pi}\int_0^{\infty} Me^{-rt\cos\delta}\,dr \le \frac{CM}{\cos\delta}. \qquad \square$$

We note, in particular, that if *(i)* is valid with $\delta \in (0, \frac{1}{2}\pi)$ fixed, then K may be chosen proportional to M.

By Lemma 6.6 we now see that since $\cos\delta_h = \cos(\frac{1}{2}\pi - c\ell_h^{-2}) \ge c\ell_h^{-2}$, the resolvent estimate (6.47) implies the stability and smoothing estimates

$$\|E_h(t)\|_{L_\infty} \le C\ell_h^2 \log\ell_h \quad \text{and} \quad t\|E_h'(t)\| \le C\ell_h^4, \quad \text{for } t > 0,$$

which are weaker than the original estimates of Theorems 6.1 and 6.3.

Thus, by Theorem 6.4, instead of showing the stability and smoothing estimates directly, one may prove a resolvent estimate, and to illustrate this we now show the following resolvent estimate for the discrete analogue $A_h = -\Delta_h$ in S_h of $A = -\Delta$, equipped with the maximum-norm. The proof by energy arguments will use the same components as that of Theorem 6.1.

Theorem 6.5 *For any $\delta \in (0, \frac{1}{2}\pi)$ there exists a constant $C = C_\delta$ such that*

$$\|R(z;A_h)\|_{L_\infty} \le C\ell_h(1+|z|)^{-1}, \quad \textit{for } z \in \Sigma_\delta. \tag{6.49}$$

Proof. For $x \in \Omega_h$ fixed and $\delta \in (0, \frac{1}{2}\pi)$, we will use the adjoint discrete Green's function

$$G_h^x(y,\bar z) = (R(\bar z;A_h)\delta_h^x)(y), \quad \text{for } z \in \Sigma_\delta, \tag{6.50}$$

where δ_h^x is the discrete delta-function defined in (6.7). We then have

$$(R(z;A_h)\chi)(x) = \big(\chi, G_h^x(\cdot,\bar z)\big), \quad \forall\chi \in S_h, \quad z \in \Sigma_\delta. \tag{6.51}$$

Since the sector Σ_δ is symmetric around the real axis it therefore suffices to show that, for any $x \in \Omega_h$ and $\delta \in (0, \pi/2)$, we have

(6.52) $$\|G_h^x(\cdot, z)\|_{L_1} \le C\ell_h |z|^{-1}, \quad \text{for } z \in \Sigma_\delta.$$

For brevity we write G for G_h^x. As in the proof of Theorem 6.1 we use the weight function $\omega(y) = \omega_h^x(y)$ in (6.10). By analogy to (6.11) and (6.13) it now suffices to show

(6.53) $$\|\omega G\| \le C\ell_h^{1/2}(1+|z|)^{-1}, \quad \forall x \in \Omega, \ z \in \Sigma_\delta.$$

For this we consider the expression

$$-z\|\omega G\|^2 + \|\omega \nabla G\|^2 = -z(G, \omega^2 G) + (\nabla G, \nabla(\omega^2 G)) - 2(\nabla G, \omega \nabla \omega\, G),$$

and note that G satisfies

(6.54) $$-z(G, \chi) + (\nabla G, \nabla \chi) = -(\delta_h^x, \chi), \quad \forall \chi \in S_h.$$

Choosing $\chi = P_h(\omega^2 G)$ and subtracting the resulting expression from the preceding one we have

(6.55) $$-z\|\omega G\|^2 + \|\omega \nabla G\|^2 = F,$$

where

$$F = (\nabla G, \nabla(\omega^2 G - P_h(\omega^2 G))) + (\delta_h^x, \omega^2 G) - 2(\nabla G, \omega \nabla \omega\, G) = F_1 + F_2 + F_3.$$

Since $\delta \le |\arg z| \le \pi$, this equation is of the form

$$e^{i\alpha} a + b = f, \quad \text{with } a, b > 0, \ 0 \le |\alpha| \le \pi - \delta,$$

and by multiplying by $e^{-i\alpha/2}$ and taking real parts we then have

$$a + b \le (\cos(\alpha/2))^{-1}|f| \le (\sin(\delta/2))^{-1}|f| = C_\delta |f|.$$

From (6.55) we therefore conclude

(6.56) $$|z| \ \|\omega G\|^2 + \|\omega \nabla G\|^2 \le C_\delta |F|, \quad \text{for } z \in \Sigma_\delta.$$

By Lemma 6.2 and using the easily shown inverse inequality $h\|\omega \nabla \chi\| \le C\|\omega \chi\|$, and the fact that $h \le \omega$, we find for the first term in F

$$|F_1| \le Ch\|\nabla G\|(\|G\| + \|\omega \nabla G\|) \le C(\|G\|^2 + \|G\| \ \|\omega \nabla G\|).$$

Further, by Lemma 6.3,

$$|F_2| \le \|\omega \delta_h^x\| \ \|\omega G\| \le C\|\omega G\| \le \varepsilon |z| \|\omega G\|^2 + C_\varepsilon |z|^{-1},$$

and since $\nabla \omega$ is bounded

$$|F_3| \le C\|G\| \ \|\omega \nabla G\|.$$

Thus, from (6.56), after kicking back $|z|\|\omega G\|^2$ and $\|\omega\nabla G\|^2$, we have

$$|z|\ \|\omega G\|^2 + \|\omega\nabla G\|^2 \le C(\|G\|^2 + |z|^{-1}). \tag{6.57}$$

To bound $\|G\|^2$ we note that, by (6.54),

$$-z\|G\|^2 + \|\nabla G\|^2 = -(\delta_h^x, G) = \bar{G}(x), \tag{6.58}$$

and hence, as in (6.56) and using Lemma 6.4,

$$|z|\ \|G\|^2 + \|\nabla G\|^2 \le |\bar{G}(x)| \le C\ell_h^{1/2}\|\nabla G\| \le \tfrac{1}{2}\|\nabla G\|^2 + C\ell_h,$$

or

$$|z|\ \|G\|^2 + \|\nabla G\|^2 \le C\ell_h, \quad \text{for } z \in \Sigma_\delta,\ x \in \Omega. \tag{6.59}$$

Together with (6.57) this shows

$$\|\omega G\| \le C\ell_h^{1/2}|z|^{-1},$$

which completes the proof of (6.53) for $|z| \ge 1$. Since also, using (6.59) in the third step,

$$\|\omega G\| \le C\|G\| \le C\|\nabla G\| \le C\ell_h^{1/2} \le C\ell_h^{1/2}(1+|z|)^{-1}, \quad \text{for } z \in \Sigma_\delta,\ |z| \le 1,$$

the proof is complete. □

Our next purpose is to improve the above result by showing that the logarithmic factor ℓ_h in the resolvent estimate of Theorem 6.5 may be removed. The proof of this is more delicate than that of Theorem 6.5, and will require some refined estimates for the Ritz and L_2-projections, and some technical estimates for the resolvent of the Laplacian with precise regularity properties. We shall not give complete proofs of all the auxiliary results.

We first discuss some properties of the Ritz and L_2-projection onto S_h that we shall need. We recall from Chapter 1 the almost stability property in maximum-norm,

$$\|R_h v\|_{L_\infty} \le C\ell_h\|v\|_{L_\infty}, \quad \text{for } v \in L_\infty. \tag{6.60}$$

An essential component in our analysis below is the fact that R_h is stable in W_∞^1, without a logarithmic factor, or, more precisely,

$$\|R_h v\|_{W_\infty^1} \le C\|v\|_{W_\infty^1(\Omega_h)} + Ch\|v\|_{W_\infty^1(\Omega\setminus\Omega_h)} \le C\|v\|_{W_\infty^1}, \quad v \in \dot{W}_\infty^1. \tag{6.61}$$

where Ω_h is the polygonal domain defined by the triangulation $\mathcal{T}_h$ of Ω. This was shown in [201] in the case of a convex polygonal domain in the plane; for a proof for a domain with smooth boundary in $\mathbb{R}^d$, see [20].

Also the L_2–projection P_h is bounded in $\dot{W}^1_\infty$, and

$$\|P_h v\|_{W^1_\infty} \le C\|v\|_{W^1_\infty(\Omega_h)}, \quad \text{for } v \in \dot{W}^1_\infty = W^1_\infty \cap H^1_0. \tag{6.62}$$

In fact, using the maximum-norm stability of P_h, together with an inverse inequality, we have

$$\begin{aligned} \|P_h v\|_{W^1_\infty} &\le \|I_h v\|_{W^1_\infty} + \|P_h(v - I_h)v\|_{W^1_\infty} \\ &\le C\|v\|_{W^1_\infty} + Ch^{-1}\|I_h v - v\|_{L_\infty} \le C\|v\|_{W^1_\infty}. \end{aligned}$$

Applying (6.61) to $I_h v - v$ we find, with $0 < \alpha < 1$,

$$\begin{aligned} \|R_h v - v\|_{W^1_\infty(\Omega_h)} &\le \|R_h(v - I_h v)\|_{W^1_\infty(\Omega_h)} + \|I_h v - v\|_{W^1_\infty(\Omega_h)} \\ &\le C\|I_h v - v\|_{W^1_\infty(\Omega_h)} + Ch\|v\|_{W^1_\infty(\Omega\setminus\Omega_h)} \le Ch^\alpha \|v\|_{W^{1+\alpha}_\infty}, \end{aligned}$$

and using the fact that $P_h v - R_h v = P_h(v - R_h v)$ it follows from (6.62) that

$$\|(P_h - R_h)v\|_{W^1_\infty} \le Ch^\alpha \|v\|_{W^{1+\alpha}_\infty}, \quad \text{for } v \in \dot{W}^{1+\alpha}_\infty, \quad 0 < \alpha < 1. \tag{6.63}$$

We now quote some resolvent estimates for the operator A from [220] which we shall need below. The first estimate contains (6.42) for $j = 0$.

Lemma 6.7 *For any $\delta \in (0, \frac{1}{2}\pi)$, there exists a constant C such that, for $z \in \Sigma_\delta$,*

$$\|R(z;A)v\|_{W^j_\infty} \le C(1 + |z|)^{-1+j/2}\|v\|_{L_\infty}, \quad \textit{for } j = 0,1, \quad v \in \mathcal{C}, \tag{6.64}$$

$$\|AR(z;A)v\|_{L_\infty} \le C(1 + |z|)^{-1/2}\|v\|_{W^1_\infty}, \quad \textit{for } v \in \dot{W}^1_\infty. \tag{6.65}$$

and

$$\|R(z;A)v\|_{W^{1+\alpha}_\infty} \le C(1 + |z|)^{-1+\alpha/2}\|v\|_{W^1_\infty}, \quad v \in \dot{W}_{\infty^1}. \tag{6.66}$$

Proof. The basic result is (6.64) which essentially is shown in [220], and which contains the central result (6.42) for $j = 0$. The remaining estimate are technical consequence of this result; for details, see [20].

We can now state and prove our logarithm free resolvent estimate for the discrete Laplacian.

Theorem 6.6 *Let S_h satisfy the assumptions of Theorem 6.1. Then for any $\delta \in (0, \frac{1}{2}\pi)$ there exists a constant $C = C_\delta$ such that*

$$\|R(z;A_h)\|_{L_\infty} \le C(1 + |z|)^{-1}, \quad \textit{for } z \in \Sigma_\delta. \tag{6.67}$$

Proof. In the first part of the proof we show the desired bound for $|z| \le \kappa h^{-2}$, with κ small enough, by using resolvent estimates for the continuous problem together with projections onto S_h. In the second part of the proof we then use a modification of the proof of Theorem 6.5 to show (6.67) for $|z| \geq \kappa h^{-2}$.

Our starting point for the first part of the proof is the identity

$$(6.68)\quad R(z;A_h)\chi = P_h R(z;A)\chi + A_h R(z;A_h)(R_h - P_h)\, R(z;A)\chi,\ \ \chi \in S_h,$$

The first term on the right is bounded at once as desired by the stability of P_h and (6.42).

To bound the second term on the right in (6.68) we shall first show that there exists $\kappa > 0$ such that, for any $\delta \in (0, \frac{1}{2}\pi)$ and $\chi \in S_h$,

$$(6.69)\quad \|A_h R(z;A_h)\chi\|_{L_\infty} \leq C(1+|z|)^{-1/2}\|\chi\|_{W^1_\infty}, \quad \text{for } z \in \Sigma_\delta,\ |z| \leq \kappa h^{-2}.$$

If this has been done we may conclude, using also the stability in $\dot{W}^1_\infty$ of P_h and R_h, and (6.64) with $j = 1$, that the second term in bounded by

$$(6.70)\quad C(1+|z|)^{-1/2}\|R(z;A)\chi\|_{W^1_\infty} \leq C(1+|z|)^{-1}\|\chi\|_{L_\infty}, \quad \text{for } z \in \Sigma_\delta,$$

which completes the proof.

To show (6.69) we now write, for $z \in \Sigma_\delta$,

$$(6.71)\quad A_h R(z;A_h)\chi = P_h A R(z;A)\chi - z\, A_h R(z;A_h)(R_h - P_h)\, R(z;A)\chi.$$

Here, by the stability of P_h and by (6.65),

$$(6.72)\qquad \|P_h A R(z;A)\chi\|_{L_\infty} \leq C(1+|z|)^{-1/2}\, \|\chi\|_{W^1_\infty} \quad \text{for } z \in \Sigma_\delta.$$

Using the operator norm

$$|||B_h||| = \sup_{\chi \in S_h} (\|B_h\chi\|_{L_\infty}/\|\chi\|_{W^1_\infty}),$$

the last term in (6.71) is bounded by

$$(6.73)\qquad |z|\, |||A_h R(z;A_h)|||\ \|(R_h - P_h)R(z;A)\chi\|_{W^1_\infty}.$$

For the last factor we have by (6.63) and (6.66), with $\alpha = 1/2$,

$$\begin{aligned}\|(R_h - P_h)R(z;A)\chi\|_{W^1_\infty} &\leq Ch^{1/2}\|R(z;A)\chi\|_{W^{3/2}_\infty}\\ &\leq Ch^{1/2}(1+|z|)^{-3/4}\|\chi\|_{W^1_\infty}.\end{aligned}$$

We therefore infer from (6.71) and (6.72) that for $z \in \Sigma_\delta$,

$$|||A_h R(z;A_h)||| \leq C(1+|z|)^{-1/2} + C_1 h^{1/2}(1+|z|)^{1/4}|||A_h R(z;A_h)|||.$$

It follows that, if $C_1 h^{1/2}(1+|z|)^{1/4} \leq 1/2$, we have

$$|||A_h R(z;A_h)||| \leq C(1+|z|)^{-1/2}, \quad z \in \Sigma_\delta.$$

This bounds the expression in (6.73) by

$$C|z|(1+|z|)^{-1/2}h^{1/2}(1+|z|)^{-3/4}\|\chi\|_{W^1_\infty} \le C(1+|z|)^{-1/2}\|\chi\|_{W^1_\infty},$$

which thus shows (6.69) for $|z| \le \kappa h^{-2}$, for h and κ small enough.

We now turn to the case $z \in \Sigma_\delta$, $|z| \ge \kappa h^{-2}$, which we shall treat with the technique of the proof of Theorem 6.5. We shall show that with $G = G_h^x(\cdot, z)$ defined in (6.50), we have for each $\delta \in (0, \pi/2)$ and $\kappa > 0$,

$$\|\omega^2 G\| \le Ch|z|^{-1}, \quad \text{for } z \in \Sigma_\delta, \quad |z| \ge \kappa h^{-2}, \quad x \in \Omega. \tag{6.74}$$

Once this has been shown it follows from $\|\omega^{-2}\| \le Ch^{-1}$ that

$$\|G\|_{L_1} \le C\|\omega^{-2}\| \cdot \|\omega^2 G\| \le C(Ch^{-1})\,(Ch|z|^{-1}) = C|z|^{-1}.$$

Since $|z| \ge c(1+|z|)$ for $|z| \ge \kappa h^{-2}$ and h small, this implies (6.67) by (6.51).

This time the energy argument uses the weight ω^2 rather than ω and we now study the expression

$$-z\|\omega^2 G\|^2 + \|\omega^2 \nabla G\|^2 = -z(G, \omega^4 G) + (\nabla G, \nabla(\omega^4 G)) - 4(\nabla G, \omega^3 \nabla\omega\, G).$$

Subtracting (6.54) with $\chi = P_h(\omega^4 G)$ we obtain

$$\begin{aligned} -z\|\omega^2 G\|^2 + \|\omega^2 \nabla G\|^2 = F &:= (\nabla G, \nabla(\omega^4 G - P_h(\omega^4 G))) \\ &+ (\delta_h^x, \omega^4 G) - 4(\nabla G, \omega^3 \nabla\omega\, G) = F_1 + F_2 + F_3, \end{aligned} \tag{6.75}$$

and we conclude this time, in the same way as earlier in (6.56),

$$|z|\;\|\omega^2 G\|^2 + \|\omega^2 \nabla G\|^2 \le C_\delta |F|, \quad \text{for } z \in \Sigma_\delta. \tag{6.76}$$

In the same way as in Lemma 6.2, and using the easily shown inverse inequality $h\|\omega^2 \nabla\chi\| \le C\|\omega^2 \chi\|$, and the fact that $h \le \omega$, we have

$$\|\omega^{-1}\nabla(\omega^4 G - P_h(\omega^4 G))\| \le Ch\big(\|\omega G\| + \|\omega^2 \nabla G\|\big) \le C\|\omega^2 G\|,$$

and hence

$$|F_1| \le C\|\omega \nabla G\|\;\|\omega^2 G\|.$$

Further, as in Lemma 6.3, we may show $\|\omega^2 \delta_h^x\| \le Ch$, and hence

$$|F_2| \le \|\omega^2 \delta_h^x\|\;\|\omega^2 G\| \le Ch\;\|\omega^2 G\|,$$

and, since $\nabla\omega$ is bounded,

$$|F_3| \le C\|\omega \nabla G\|\;\|\omega^2 G\|.$$

Thus, from (6.76),

$$|z|\;\|\omega^2 G\| \le C(\|\omega \nabla G\| + h). \tag{6.77}$$

By an inverse estimate and obvious estimates we next find that

$$\begin{aligned}\|\omega\nabla G\| &\le Ch^{-1}\|\omega\, G\| \le Ch^{-1}\|\omega^2 G\|^{1/2}\|G\|^{1/2}\\ &\le \varepsilon|z|\ \|\omega^2 G\| + C_\varepsilon|z|^{-1}h^{-2}\|G\| \le \varepsilon|z|\ \|\omega^2 G\| + C_\varepsilon\|G\|,\end{aligned}$$

where in the last step we have used $|z| \ge \kappa h^{-2}$. From (6.58) we have, again as in (6.56),

$$|z|\ \|G\| \le C\|\delta_h^x\| \le Ch^{-1}.$$

where the latter inequality follows from (6.19). Hence, from (6.77),

$$|z|\ \|\omega^2 G\| \le C(\|G\| + h) \le C(h^{-1}|z|^{-1} + h) \le Ch.$$

This completes the proof of (6.74) and hence of the theorem. □

Since the bound for the resolvent in Theorem 6.6 does not contain any logarithmic factor, the following logarithm free stability and smoothing estimates follow from Theorem 6.4.

Theorem 6.7 *Under the assumptions of Theorem 6.1 we have*

$$\|E_h(t)v_h\|_{L_\infty} + t\|E_h'(t)v_h\|_{L_\infty} \le C\|v_h\|_{L_\infty}, \quad \textit{for } v_h \in S_h,\ t \ge 0.$$

Using this result instead of the estimates of Theorems 6.1 and 6.3 we may easily show the following improvement of Theorem 6.2.

Theorem 6.8 *Under the assumptions of Theorem 6.1, with $v_h = R_h v$, we have, for the error in the semidiscrete parabolic problem* (6.28),

$$\|u_h(t) - u(t)\|_{L_\infty} \le Ch^2\ell_h\Big(\|v\|_{W_\infty^2} + \int_0^t \|u_t\|_{W_\infty^2}\, ds\Big), \quad \textit{for } t \ge 0.$$

We shall next show some smooth and nonsmooth data maximum-norm error estimates for the semidiscrete solution of the homogeneous parabolic problem (6.1).

In our analysis we shall need the Agmon-Douglis-Nirenberg [2] regularity estimate

$$(6.78) \qquad \|u\|_{W_p^2} \le Cp\|\Delta u\|_{L_p}, \quad \text{for } 2 \le p < \infty, \quad u \in \dot W_p^2 = W_p^2 \cap H_0^1.$$

In [2] this result is stated without precise accounting of the dependence of the bound upon p, but this may be determined by tracing its dependence on $q = p/(p-1)$ in the proof of (6.78) in [2] to the Calderòn-Zygmund lemma [43], in which the required estimate is contained.

As a preparation we now show some bounds for the error in the Ritz projection R_h of the exact solution of (6.1). We recall the maximum-norm stability bound of (6.60) with a logarithmic factor ℓ_h.

Lemma 6.8 *With u the solution of* (6.1) *we have, for $\rho = R_h u - u$,*

$$\|\rho(t)\|_{L_\infty} + t\|\rho_t(t)\|_{L_\infty} \le Ch^2\ell_h^2\|v\|_{W_\infty^2}, \quad \text{for } v \in \dot{W}_\infty^2. \tag{6.79}$$

Proof. We note that, with I_h the standard interpolation operator into S_h,

$$\|I_h u - u\|_{L_\infty} \le Ch^{2-2/p}\|u\|_{W_p^2}, \quad \text{for } 2 \le p < \infty,\ u \in \dot{W}_p^2. \tag{6.80}$$

This follows from the corresponding inequality for an individual triangle τ of $\mathcal{T}_h$, which in turn may be obtained by transformation to a reference triangle and application of the Bramble-Hilbert lemma, together with an obvious estimate for u on $\Omega \setminus \Omega_h$. Since $\rho = (R_h - I)u = (R_h - I)(u - I_h u)$ we obtain, using (1.45),

$$\|\rho\|_{L_\infty} \le C\ell_h\|I_h u - u\|_{L_\infty} \le C\ell_h h^{2-2/p}\|u\|_{W_p^2}, \tag{6.81}$$

and consequently, using (6.78) and (6.41),

$$\begin{aligned} \|\rho(t)\|_{L_\infty} &\le Ch^{2-2/p}\ell_h p\|\Delta u(t)\|_{L_p} \\ &\le Ch^{2-2/p}\ell_h p\|\Delta v\|_{L_p} \le Ch^{2-2/p}\ell_h p\|\Delta v\|_{L_\infty}. \end{aligned} \tag{6.82}$$

Choosing $p = \ell_h$ this shows the estimate stated for $\rho(t)$. To bound $\rho_t(t)$ we use the analogue of (6.82) from $t/2$ to t and then (6.41) to obtain

$$\|\rho_t(t)\|_{L_\infty} \le Ch^{2-2/p}\ell_h p\|\Delta u_t(t/2)\|_{L_p} \le Ch^{2-2/p}\ell_h p t^{-1}\|\Delta v\|_{L_p},$$

and finally take $p = \ell_h$. □

We are now ready for a smooth data error estimate.

Theorem 6.9 *Under our present assumptions, we have for the solutions of* (6.3) *and* (6.1), *with $v \in \dot{W}_\infty^2$ and $v_h = R_h v$,*

$$\|u_h(t) - u(t)\|_{L_\infty} \le Ch^2\ell_h^2\|v\|_{W_\infty^2}, \quad \text{for } t \ge 0.$$

Proof. Proceeding as in the proof of Theorem 6.2, $\rho(t)$ is bounded by Lemma 6.8. To estimate θ we write (6.30) as

$$\theta(t) = -\Big(\int_0^{t/2} + \int_{t/2}^t\Big) E_h(t-s)P_h\rho_t(s)\,ds = I + II.$$

Here, by Theorem 6.7, Lemma 6.1 and (6.79),

$$\|II\|_{L_\infty} \le C\int_{t/2}^t \|\rho_t\|_{L_\infty}\,ds \le Ch^2\ell_h^2\|v\|_{W_\infty^2}.$$

For I we integrate by parts to obtain

$$I = -[E_h(t-s)\,P_h\rho(s)]_0^{t/2} - \int_0^{t/2} E_h'(t-s)P_h\rho(s)\,ds,$$

and both terms are bounded as desired as above. □

For our nonsmooth data error estimate we shall need the following error bounds for the L_2 and Ritz projections.

Lemma 6.9 *With u the solution of* (6.1) *we have, for $\eta = (P_h - I)u$,*

$$t\|\eta(t)\|_{L_\infty} \le Ch^2\ell_h\|v\|_{L_\infty}, \tag{6.83}$$

and, for $\rho = (R_h - I)u$ and $\widetilde{\rho}(t) = \int_0^t \rho(s)\,ds$,

$$\|\widetilde{\rho}(t)\|_{L_\infty} + t\|\rho(t)\|_{L_\infty} + t^2\|\rho_t(t)\|_{L_\infty} \le Ch^2\ell_h^2\|v\|_{L_\infty}. \tag{6.84}$$

Proof. Using (6.82) together with (6.41 shows, since $\Delta E(t) = E'(t)$,

$$\|\rho(t)\|_{L_\infty} \le Ch^{2-2/p}\ell_h p\|E'(t/2)v\|_{L_p} \le Ch^{2-2/p}\ell_h p t^{-1}\|v\|_{L_p},$$

which with $p = \ell_h$ gives the bound for $\rho(t)$ in (6.84). The proofs of (6.83) and the bound for $\rho_t(t)$ are analogous, with one factor ℓ_h less in (6.83) because P_h is bounded in L_∞. To bound $\widetilde{\rho}(t)$, finally, we first show

$$\|(T_h - T)f\|_{L_\infty} \le Ch^2\ell_h^2\|f\|_{L_\infty}, \quad \text{with } T = A^{-1},\ T_h = R_hT. \tag{6.85}$$

For this we use the stability of P_h in L_∞ and (6.82) to obtain

$$\|(T_h - P_hT)f\|_{L_\infty} = \|P_h(R_h - I)Tf\|_{L_\infty} \le Ch^{2-2/p}\ell_h p\,\|f\|_{L_\infty},$$

where we have used the fact that $\Delta Tf = -ATf = -f$. Further, by (6.80), and (6.78),

$$\|(P_h - I)Tf\|_{L_\infty} \le Ch^{2-2/p}\|Tf\|_{W_p^2} \le Ch^{2-2/p}p\,\|f\|_{L_\infty}.$$

With $p = \ell_h$ these estimates together show (6.85). We now note

$$\begin{aligned}\widetilde{\rho}(t) &= (T_h - T)A\int_0^t u(s)\,ds = -(T_h - T)\int_0^t u_t(s)\,ds\\ &= -(T_h - T)(u(t) - v),\end{aligned} \tag{6.86}$$

so that (6.85) shows

$$\|\widetilde{\rho}(t)\|_{L_\infty} \le Ch^2\ell_h^2\|u(t) - v\|_{L_\infty} \le Ch^2\ell_h^2\|v\|_{L_\infty}.$$ □

We are now ready for our nonsmooth data error bound.

Theorem 6.10 *Under the assumptions of Theorem 6.1, we have for the error in the solution of* (6.3), *with $v_h = P_hv$,*

$$\|u_h(t) - u(t)\|_{L_\infty} \le Ch^2\ell_h^2t^{-1}\|v\|_{L_\infty}, \quad \text{for } t > 0.$$

Proof. We write this time $u_h - u = (u_h - P_h u) + (P_h u - u)$. By (6.83) it only remains to bound $\zeta = u_h - P_h u \in S_h$. We find

$$\zeta_t + A_h \zeta = A_h(R_h - P_h)u = -A_h P_h \rho, \quad \text{for } t > 0.$$

Since $\zeta(0) = 0$, we obtain by integration

$$\zeta(t) = \int_0^t E_h(t-s) A_h P_h \rho(s)\, ds = -\int_0^t E_h'(t-s) P_h \rho(s)\, ds.$$

Thus, with the obvious definitions of $I_j, II_j, j = 1, 2$, since $t = s + (t - s)$,

$$\begin{aligned} t\zeta(t) = &-\Big(\int_0^{t/2} + \int_{t/2}^t\Big) E_h'(t-s)\, s\, P_h \rho(s)\, ds \\ &- \Big(\int_0^{t/2} + \int_{t/2}^t\Big)(t-s)\, E_h'(t-s) P_h \rho(s)\, ds = I_1 + II_1 + I_2 + II_2. \end{aligned}$$

Here, using Lemma 6.3 and (6.84),

$$\|I_1\|_{L_\infty} \le C \int_0^{t/2} (t-s)^{-1} s\, \|\rho(s)\|_{L_\infty}\, ds \le C h^2 \ell_h^2 \|v\|_{L_\infty}.$$

Further, after integration by parts we have

$$II_1 = [E_h(t-s)\, s\, P_h \rho(s)]_{t/2}^t - \int_{t/2}^t E_h(t-s) P_h (s\rho_t(s) + \rho(s))\, ds, \tag{6.87}$$

and thus $\|II_1\|_{L_\infty} \le C h^2 \ell_h^2 \|v\|_{L_\infty}$ by Theorem 6.1, (6.84) and (6.84). For I_2 we integrate by parts to obtain, with $\widetilde{\rho} = (R_h - I)\widetilde{u}$,

$$\begin{aligned} I_2 = &-(t/2) E_h'(t/2) P_h \widetilde{\rho}(t/2) \\ &- \int_0^{t/2} (E_h'(t-s) + (t-s) E_h''(t-s)) P_h \widetilde{\rho}(s)\, ds, \end{aligned} \tag{6.88}$$

from which we find, by Theorem 6.3, applied twice to bound $E_h''(t) = (E_h'(t/2))^2$, and (6.84), that $\|I_2\|_{L_\infty} \le C h^2 \ell_h^2 \|v\|_{L_\infty}$. For II_2, finally, we have, using Theorem 6.3 and (6.84),

$$\|II_2\|_{L_\infty} \le C \int_{t/2}^t \|\rho(s)\|_{L_\infty}\, ds \le C h^2 \ell_h^2 \|v\|_{L_\infty}.$$

Together our estimates complete the proof. □

As we shall now see, it is possible to reduce the regularity assumptions even further than in Theorem 6.10 and still have an essentially $O(h^2)$ error estimate in the semidiscrete homogeneous problem for positive time, requiring only that the initial data are in L_1. For this we need again some technical preparations. Note first that $P_h v \in S_h$ is defined even for $v \in L_1$ by $(P_h v, \chi) = (v, \chi)$, $\forall \chi \in S_h$, and that P_h is stable in L_1 by duality, since it is stable in the maximum-norm. Also, $E(t)$ may be extended to L_1, with $\|E(t)v\|_{L_1} \le \|v\|_{L_1}$, by duality. The next theorem is a discrete analogue of the inequality

$$\|E(t)v\|_{L_\infty} \le Ct^{-1}\|v\|_{L_1}, \quad \text{for } t > 0. \tag{6.89}$$

To show this inequality, one may use the Gagliardo-Nirenberg inequality

$$\|u\|_{L_\infty} \le C\|Au\|^{1/2}\|u\|^{1/2}, \quad \text{for } u \in \dot{H}^2, \tag{6.90}$$

to conclude

$$\|E(t)v\|_{L_\infty} \le C\|AE(t)v\|^{1/2}\|E(t)v\|^{1/2} \le Ct^{-1/2}\|v\|.$$

By duality this also yields

$$\|E(t)v\| \le Ct^{-1/2}\|v\|_{L_1}.$$

and hence, by application of both of these inequalities,

$$\|E(t)v\|_{L_\infty} \le Ct^{-1/2}\|E(t/2)v\| \le Ct^{-1}\|v\|_{L_1}.$$

Theorem 6.11 *We have for the solution operator of* (6.3)

$$\|E_h(t)v_h\|_{L_\infty} \le Ct^{-1}\|v_h\|_{L_1}, \quad \textit{for } t > 0.$$

Proof. This follows exactly as (6.89) from the discrete analogue of (6.90),

$$\|\chi\|_{L_\infty} \le C\|A_h\chi\|^{1/2}\|\chi\|^{1/2}, \quad \forall \chi \in S_h. \tag{6.91}$$

To show this, we first derive the equivalent inequality for $T_h = A_h^{-1}$,

$$\|T_h\chi\|_{L_\infty} \le C\|\chi\|^{1/2}\|T_h\chi\|^{1/2}, \quad \forall \chi \in S_h. \tag{6.92}$$

For this we write

$$T_h\chi = P_h T\chi + (T_h - P_h T)\chi.$$

Here, by application of (6.90) to $v = Tu$, since P_h is bounded in L_∞,

$$\|P_h T\chi\|_{L_\infty} \le C\|\chi\|^{1/2}\|T\chi\|^{1/2} \le C\|\chi\|^{1/2}(\|T_h\chi\|^{1/2} + \|(T - T_h)\chi\|^{1/2}),$$

where, by the inverse estimate $\|A_h\chi\| \le Ch^{-2}\|\chi\|$,

$$\|(T - T_h)\chi\| \le Ch^2\|\chi\| \le C\|T_h\chi\|.$$

This completes the proof of (6.92) and thus of the theorem. □

We also need the following analogue of Lemma 6.9.

Lemma 6.10 *With the notation of Lemma 6.9 we have*

$$t^2\|\eta(t)\|_{L_\infty} \le Ch^2\ell_h\|v\|_{L_1}, \tag{6.93}$$

$$t^2\|\rho(t)\|_{L_\infty} + t^3\|\rho_t(t)\|_{L_\infty} + \|\widetilde{\rho}(t)\|_{L_1} \le Ch^2\ell_h^2\|v\|_{L_1}. \tag{6.94}$$

Proof. The first three inequalities follow easily by using $t/2$ as an intermediate time level from the corresponding estimates in Lemma 6.9, combined with (6.89). By duality it follows that the analogue of (6.85) holds also in L_1 and therefore by (6.86), since $E(t)$ is stable in L_1,

$$\|\widetilde{\rho}(t)\|_{L_1} = \|(T_h - T)(u(t) - v)\|_{L_1} \le Ch^2\ell_h^2\|v\|_{L_1}. \qquad \square$$

We are now ready for our error bound when v is only in L_1.

Theorem 6.12 *Under the assumptions of Theorem 6.1, we have for the solutions of* (6.3) *and* (6.1), *with* $v_h = P_h v$,

$$\|u_h(t) - u(t)\|_{L_\infty} \le Ch^2\ell_h^2 t^{-2}\|v\|_{L_1} \quad \textit{for } t > 0.$$

Proof. As in the proof of Theorem 6.10, we split the error into $\zeta = u_h - P_h u \in S_h$ and $\eta = P_h u - u$ where the latter term now is bounded by (6.93). To bound $t^2\zeta(t)$ we now write, with obvious notation,

$$\begin{aligned} t^2\zeta(t) &= -\Big(\int_0^{t/2} + \int_{t/2}^t\Big)(s^2 + 2s(t-s) + (t-s)^2)E_h'(t-s)P_h\rho(s)\,ds \\ &= \sum_{j=1}^3 (I_j + II_j). \end{aligned}$$

Here $\|I_1\|_{L_\infty} \le Ch^2\ell_h^2\|v\|_{L_1}$ by straightforward application of Theorem 6.3 and (6.94). We have

$$I_2 + II_2 = 2\int_0^t (t-s)E_h'(t-s)sP_h\rho(s)\,ds.$$

Combining Theorems 6.3 and 6.11 we have

$$\|E_h'(t)v_h\|_{L_\infty} \le Ct^{-1}\|E_h(t/2)v_h\|_{L_\infty} \le Ct^{-2}\|v_h\|_{L_1},$$

and by interpolation between this result and that of Theorem 6.3 we have

$$\|E_h'(t-s)P_h v\|_{L_\infty} \le C(t-s)^{-3/2}\|v\|_{L_2}.$$

Further,

$$\|\rho(s)\|_{L_2} \le Ch^2\|u(s)\|_{H^2} \le Ch^2 s^{-1}\|u(s/2)\|_{L_2} \le Ch^2 s^{-3/2}\|v\|_{L_1},$$

where the last inequality follows using the standard fundamental solution for Cauchy's problem. Thus

$$\|I_2 + II_2\|_{L_\infty} \le Ch^2 \int_0^t (t-s)^{-1/2} s^{-1/2}\, ds\, \|v\|_{L_1} = Ch^2 \|v\|_{L_1}.$$

Similarly

$$\|II_3\|_{L_\infty} \le Ch^2 \int_{t/2}^t (t-s)^{1/2} s^{-3/2}\, ds\, \|v\|_{L_1} \le Ch^2 \|v\|_{L_1}.$$

For II_1 we have, with the obvious modification of (6.87),

$$\|II_1\|_{L_\infty} = \| \int_{t/2}^t E_h'(t-s)\, s^2\, P_h \rho(s)\, ds \|_{L_\infty} \le Ch^2 \ell_h^2 \|v\|_{L_1}.$$

For I_3, finally, we integrate by parts as in (6.88) to obtain

$$\begin{aligned} I_3 &= -(t^2/4) E_h'(t/2) P_h \widetilde{\rho}(t/2) \\ &\quad - \int_0^{t/2} \big(2(t-s)E_h'(t-s) + (t-s)^2 E_h''(t-s)\big) P_h \widetilde{\rho}(s)\, ds, \end{aligned}$$

whence $\|I_3\|_{L_\infty} \le Ch^2 \ell_h^2 \|v\|_{L_1}$, by Theorems 6.3 and 6.11, and (6.94). The proof is now complete. □

The above analysis in the case of piecewise linear finite elements in two space dimensions is from Schatz, Thomée and Wahlbin [209]. Using similar techniques in one space dimension and with piecewise polynomials of arbitrary degree, an analogue of (6.1) was shown in Thomée and Wahlbin [232] with the bound including an additional factor of ℓ_h^2.

For piecewise polynomials of degree at least 3 and in 1, 2, and 3 space dimensions, Nitsche and Wheeler [186] showed that $E_h(t)P_h v$ is an almost best approximation of $E(t)v$ in the maximum-norm in a space-time domain, which implies maximum-norm stability in these cases, without a logarithmic factor, cf. also Nitsche [183] where logarithm-free error estimates were derived for $r \ge 2$ and d arbitrary. For Neumann boundary conditions it was shown in Schatz, Thomée and Wahlbin [210] that the restrictions in dimension and degree in the Nitsche-Wheeler result are not needed in this case and also that the corresponding smoothing estimates hold, so that the relevant bounds are valid for $r \ge 2$, and $d \ge 2$, without logarithmic factors. These conclusions were carried over the Dirichlet boundary conditions in Thomée and Wahlbin [233].

The alternative approach taken in the latter part of this chapter, building on the resolvent estimate (6.67) was developed in one dimension in Crouzeix, Larsson and Thomée [61]; the idea of using resolvent estimates as a basis for stability analysis had been exploited earlier in one space dimension in

Wahlbin [241]. The result of Theorem 6.6 was shown in d dimensions in Bakaev, Thomée and Wahlbin [20].

The assumption that we have used consistently in the above analysis that the families of triangulations $\mathcal{T}_h$ are quasiuniform is not a very desirable feature, but is required in the proof we have given. As was mentioned in connection with Lemma 6.1, the stability of the L_2-projection P_h was shown in Crouzeix and Thomée [59] also for certain classes of nonquasiuniform triangulations. The techniques used in the proof of this generalization is being applied in Bakaev, Crouzeix and Thomée [19], in which a resolvent estimate for A_h is derived, with a logarithmic factor $\ell_h^{1/2}$, for similarly nonquasiuniform triangulations.

For further maximum-norm error analyses, cf. Dobrowolski [71], [72], where also nonlinear situations are treated, Rannacher [200], and H. Chen [50]. For completeness we also quote Wheeler [245], [239] and Bramble, Schatz, Thomée and Wahlbin [37] where maximum-norm error bounds are derived from L_2-estimates for the parabolic problem when maximum-norm estimates are known for the stationary problem, as exemplified in Chapter 1. We remark that in our nonsmooth data error estimates we have always assumed that $P_h v$ is computed exactly; for the effect of numerical quadrature, see Wahlbin [240]. Finally, we quote the maximum-norm estimates by Fujii [102] for the lumped mass method which will be discussed in Chapter 13.

General references to semigroups of operators in Banach space are Hille and Phillips [124], Dunford and Schwartz [82], Yosida [248], and, particularly with reference to partial differential equations, Pazy [194] and Arendt [5].

7. Single Step Fully Discrete Schemes for the Homogeneous Equation

In this chapter we consider single step fully discrete methods for the initial boundary value problem for the homogeneous heat equation, and show analogues of our previous stability and error estimates in the spatially semidiscrete case for both smooth and nonsmooth data. Our approach is to first study the discretization with respect to time of an abstract parabolic equation in a Hilbert space setting by using rational approximations of the exponential, which allows the standard Euler and Crank-Nicolson procedures as special cases, and then to apply the results obtained to the spatially discrete problem investigated in the preceding chapters. The analysis uses eigenfunction expansions related to the elliptic operator occurring in the parabolic equation, which we assume positive definite.

We consider thus the initial boundary value problem for the homogeneous heat equation,

$$(7.1)\qquad \begin{aligned} &u_t = \Delta u \quad \text{in } \Omega, \qquad \text{for } t > 0,\\ &u = 0 \qquad \text{on } \partial\Omega, \quad \text{for } t > 0, \quad u(\cdot, 0) = v \quad \text{in } \Omega, \end{aligned}$$

where Ω is a bounded domain in $\mathbb{R}^d$ with smooth boundary $\partial\Omega$. We assume as in Chapters 2 and 3 that we are given a family of subspaces S_h of $L_2 = L_2(\Omega)$ and a corresponding family of operators $T_h : L_2 \to S_h$, approximating $T = (-\Delta)^{-1}$, with the properties

(i) *T_h is selfadjoint, positive semidefinite on L_2, and positive definite on S_h.*

(ii) *There is a positive integer $r \geq 2$ such that*

$$\|(T_h - T)f\| \leq Ch^s \|f\|_{s-2}, \quad \text{for } 2 \leq s \leq r, \quad f \in H^{s-2}.$$

The spatially semidiscrete problem is then to find $u_h : [0, \infty) \to S_h$ such that

$$(7.2)\qquad u_{h,t} = \Delta_h u_h \quad \text{for } t > 0, \quad \text{with } u_h(0) = v_h,$$

where $\Delta_h = -T_h^{-1} : S_h \to S_h$, is the discrete Laplacian. As earlier this problem may be thought of as a homogeneous linear system of ordinary differential equations. To define a fully discrete method we now want to discretize this system with respect to the time variable. For this purpose we introduce a time step k and look for approximations U_h^n in S_h of $u^n = u(t_n)$ where $t_n = nk$. In this chapter we will consider single step methods, i.e., methods which define U_h^{n+1} in terms of U_h^n only.

In order to make the discussion of the time discretization more clear we shall first separate it from the spatial discretization, and consider an evolution problem in a Hilbert space setting. Let thus $\mathcal{H}$ be a separable Hilbert space with norm $\|\cdot\|$, and assume that A is a linear, selfadjoint, positive definite, not necessarily bounded operator with a compact inverse, defined in $\mathcal{D}(A) \subset \mathcal{H}$, and consider the initial value problem

$$u' + Au = 0 \quad \text{for } t > 0, \quad \text{with } u(0) = v. \tag{7.3}$$

Included as applications are then both the case $\mathcal{H} = L_2$, with $A = -\Delta$ (where $\mathcal{D}(A) = H^2 \cap H_0^1$), and $\mathcal{H} = S_h$ (equipped with the L_2 inner product), with $A = -\Delta_h, \mathcal{D}(A) = S_h$. We shall normally think of $\mathcal{H}$ as a real Hilbert space in this chapter, but extensions to the complex case are obvious.

Since A^{-1} is compact, A has eigenvalues $\{\lambda_j\}_{j=1}^N$ and a corresponding basis of orthonormal eigenfunctions $\{\varphi_j\}_{j=1}^N$ (with $N \le \infty$), and we may write the solution operator of (7.3) as

$$u(t) = E(t)v = \sum_{j=1}^N e^{-\lambda_j t}(v, \varphi_j)\varphi_j. \tag{7.4}$$

For an arbitrary function $g(\lambda)$, defined on the spectrum $\sigma(A) = \{\lambda_j\}_{j=1}^N$ of A, we set

$$g(A)v = \sum_{j=1}^N g(\lambda_j)(v, \varphi_j)\varphi_j, \quad \text{for } v \in \mathcal{H}, \tag{7.5}$$

which is consistent with the standard eigenfunction expansion of Av, say. Note that by Parseval's relation we have for the operator norm of $g(A)$

$$\|g(A)\| = \sup_j |g(\lambda_j)| = \sup_{\lambda \in \sigma(A)} |g(\lambda)|. \tag{7.6}$$

As we have indicated already in Chapter 1, we may view the solution operator $E(t)$ of (7.3) as represented in (7.4) as the exponential e^{-tA}, and it then becomes natural to define a single step discrete method by approximation of $u(t_{n+1}) = E(k)u(t_n)$, using a rational function $r(\lambda)$ approximating $e^{-\lambda}$, so that U^n is defined for $n \ge 0$ recursively by

$$U^{n+1} = E_k U^n \quad \text{for } n \ge 0, \quad \text{where } E_k = r(kA), \quad \text{with } U^0 = v, \tag{7.7}$$

where the rational function $r(\lambda)$ is defined on $\sigma(kA)$.

To define the accuracy of this method we consider the scalar problem

$$u' + au = 0 \quad \text{for } t > 0, \quad \text{with } u(0) = 1. \tag{7.8}$$

The corresponding discrete solution is then defined by $U^{n+1} = r(ka)U^n$, and we say that the scheme (7.7) is accurate of order q if the exact solution of (7.8) satisfies this relation with an error of order $O(k^{q+1})$. Since the exact solution is $u(t) = e^{-at}$, this may be expressed as $r(ka) = e^{-ka} + O(k^{q+1})$, or

$$r(\lambda) = e^{-\lambda} + O(\lambda^{q+1}), \quad \text{as } \lambda \to 0. \tag{7.9}$$

In addition to accuracy conditions, $r(\lambda)$ will be assumed to satisfy certain boundedness conditions on the positive real axis.

Using the spectral representation (7.5) it follows that

$$U^n = E_k^n v = \sum_{j=1}^{N} r(k\lambda_j)^n (v, \varphi_j)\varphi_j.$$

We say that the operator E_k defined in (7.7) is stable in $\mathcal{H}$ if

$$\|E_k^n\| \leq C, \quad \text{for } n \geq 1.$$

By (7.6) this is equivalent to $|r(k\lambda)^n| \leq C$ for $n \geq 1$ and $\lambda \in \sigma(A)$, and this in turn holds if and only if

$$\sup_{\lambda \in \sigma(kA)} |r(\lambda)| \leq 1; \tag{7.10}$$

in this case Parseval's relation immediately shows that

$$\|U^n\|^2 \leq \sum_{j=1}^{N} |(v, \varphi_j)|^2 = \|v\|^2.$$

Condition (7.10) will be satisfied by the schemes studied below.

We illustrate our definitions by two familiar examples of time stepping schemes for the approximate solution of (7.1). In these we apply methods of the form (7.7) to the semidiscrete problem (7.2), with $\mathcal{H} = S_h \subset H_0^1$ and $A = A_h = -\Delta_h$ defined by the standard Galerkin method, so that

$$-(\Delta_h \psi, \chi) = (\nabla\psi, \nabla\chi), \quad \forall \psi, \chi \in S_h. \tag{7.11}$$

Our two examples are then provided by the backward Euler scheme

$$(U_h^{n+1}, \chi) + k(\nabla U_h^{n+1}, \nabla\chi) = (U_h^n, \chi), \quad \forall \chi \in S_h, \ n \geq 0, \tag{7.12}$$

and the Crank-Nicolson scheme

$$(7.13)\qquad (U_h^{n+1},\chi)+\tfrac12 k(\nabla U_h^{n+1},\nabla\chi)=(U_h^n,\chi)-\tfrac12 k(\nabla U_h^n,\nabla\chi),$$

for $\chi\in S_h, n\geq 0$. Written in operator form, (7.12) may be expressed

$$(I-k\Delta_h)U_h^{n+1}=U_h^n \quad\text{or}\quad U_h^{n+1}=(I-k\Delta_h)^{-1}U_h^n,$$

and similarly, for (7.13),

$$U_h^{n+1}=(I-\tfrac12 k\Delta_h)^{-1}(I+\tfrac12 k\Delta_h)U_h^n.$$

With $A=-\Delta_h$ these are both of the form (7.7) with

$$r(\lambda)=\frac{1}{1+\lambda}\quad\text{and}\quad r(\lambda)=\frac{1-\frac12\lambda}{1+\frac12\lambda},$$

respectively. Since in both cases $|r(\lambda)|\leq 1$ when $\lambda\geq 0$, and thus, in particular, on $\sigma(kA)=\sigma(-k\Delta_h)$, they satisfy condition (7.10).

For the abstract problem (7.3) the corresponding schemes are

$$U^{n+1}=(I+kA)^{-1}U^n,\quad\text{and}\quad U^{n+1}=(I+\tfrac12 kA)^{-1}(I-\tfrac12 kA)\,U^n.$$

Note that the rational functions $r(kA)$ employed here may also be expressed using (7.5).

We begin our error analysis for our time stepping scheme (7.7) with an estimate in the Hilbert space norm in the case that the initial data v are smooth in the sense that $v\in\mathcal{D}(A^q)$. In our analysis we shall use the spaces $\dot H^s=\mathcal{D}(A^{s/2})$ defined by the norms

$$|v|_s=(A^s v,v)^{1/2}=\|A^{s/2}v\|=\Big(\sum_{j=1}^N \lambda_j^s(v,\varphi_j)^2\Big)^{1/2}.$$

They are generalizations to the present context of the spaces $\dot H^s(\Omega)$ introduced in Chapter 3, where $\mathcal{H}=L_2(\Omega)$ and $A=-\Delta$.

Theorem 7.1 *Assume that the discretization scheme is accurate of order q and stable in $\mathcal{H}$, so that* (7.9) *and* (7.10) *hold. Then we have for the solutions of* (7.7) *and* (7.3)

$$\|U^n-u(t_n)\|\leq Ck^q|v|_{2q},\quad \textit{for } t_n\geq 0.$$

Proof. Introducing the function $F_n(\lambda)=r(\lambda)^n-e^{-n\lambda}$ and recalling the definition (7.5), we may write

$$(7.14)\qquad U^n-u(t_n)=r(kA)^n v-e^{-nkA}v=F_n(kA)v.$$

The result of Theorem 7.1 may then be expressed as

$$\|F_n(kA)v\| \le Ck^q|v|_{2q}, \quad \text{for } t_n \ge 0,$$

or, in terms of the operator norm in $\mathcal{H}$,

$$\|A^{-q}F_n(kA)\| \le Ck^q, \quad \text{for } t_n \ge 0.$$

In view of (7.6) this may be written as $|\lambda^{-q}F_n(\lambda)| \le C$ for $\lambda \in \sigma(kA)$, and this in turn will thus follow from

$$|F_n(\lambda)| \le C\lambda^q, \quad \text{for } \lambda \in \sigma(kA), \tag{7.15}$$

which we will now prove.

By (7.9) we have for λ_0 small enough that

$$|r(\lambda) - e^{-\lambda}| \le C\lambda^{q+1}, \quad \text{for } 0 \le \lambda \le \lambda_0.$$

We also find from this that, with λ_0 possibly further restricted,

$$|r(\lambda)| \le e^{-c\lambda}, \quad \text{for } 0 \le \lambda \le \lambda_0, \quad \text{with } 0 < c < 1. \tag{7.16}$$

Hence, with this λ_0,

$$\begin{aligned} |F_n(\lambda)| &= \Big|(r(\lambda) - e^{-\lambda}) \sum_{j=0}^{n-1} r(\lambda)^{n-1-j} e^{-j\lambda}\Big| \\ &\le Cn\lambda^{q+1}e^{-c(n-1)\lambda} \le C\lambda^q, \quad \text{for } 0 \le \lambda \le \lambda_0. \end{aligned} \tag{7.17}$$

By stability we also have

$$|F_n(\lambda)| \le |r(\lambda)^n| + e^{-n\lambda} \le 2 \le C\lambda^q, \quad \text{for } \lambda \ge \lambda_0, \ \lambda \in \sigma(kA).$$

Together these estimates show (7.15), and thus complete the proof. □

We now turn to the case that the initial data v are nonsmooth in the sense that they are only known to belong to $\mathcal{H}$ and not to $\mathcal{D}(A^s)$ for any $s > 0$. Our error estimates will then require further properties of the rational function $r(\lambda)$, and we therefore introduce the following classification of the discretizations in time. First, the rational function $r(\lambda)$ approximating $e^{-\lambda}$ will be said to be of type I, II, III, or IV, respectively, if

I: $|r(\lambda)| < 1$, for $0 < \lambda < \alpha$, with $\alpha > 0$;
II: $|r(\lambda)| < 1$, for $\lambda > 0$;
III: $|r(\lambda)| < 1$, for $\lambda > 0$, and $|r(\infty)| < 1$;
IV: $|r(\lambda)| < 1$, for $\lambda > 0$, and $r(\infty) = 0$.

Note that these conditions are successively more restrictive.

For the purpose of application of our results to the case that the equation (7.3) represents the spatially discrete problem (7.2) and thus $A = A_h$ is a bounded operator depending on the parameter h, we classify schemes of types

I and II further by saying that the scheme is of type I′ or II′, respectively, if, with $\lambda_{\max}$ denoting the largest eigenvalue of A,

I′: $r(\lambda)$ is of type I, and $k\lambda_{\max} \le \alpha_0$, for some α_0 with $0 < \alpha_0 < \alpha$,
II′: $r(\lambda)$ is of type II, and $k\lambda_{\max} \le \alpha_1$, for some α_1 with $0 < \alpha_1 < \infty$.

A scheme of type II, III or IV will thus simply be one for which $r(\lambda)$ is of type II, III or IV, respectively, with no restrictions on the relation between k and $\lambda_{\max}$.

We note in connection with schemes of types I′ and II′ satisfying (7.9) for some $q \ge 1$ that, setting $\lambda_0 = \alpha_0$ and α_1, respectively, we have $|r(\lambda)| < 1$ for $0 < \lambda \le \lambda_0$, and hence (7.16) holds. In particular, since $k\lambda \le \lambda_0$ for $\lambda \in \sigma(A)$, we have $|r(\lambda)| \le e^{-c\lambda}$ for $\lambda \in \sigma(kA)$, with $0 < c < 1$. This fact will be used repeatedly in the proofs below.

We shall briefly present some examples of schemes associated with our four classes of rational functions I, II, III, and IV.

Examples of schemes based on rational functions of types I, II, and IV are provided by the above diagonal, diagonal, and below diagonal entries of the Padé table for $e^{-\lambda}$, respectively. In fact, the general entry in this Padé table is given by

$$(7.18) \qquad r_{\mu\nu}(\lambda) = \frac{n_{\mu\nu}(\lambda)}{d_{\mu\nu}(\lambda)}, \quad \text{where}$$

$$n_{\mu\nu}(\lambda) = \sum_{j=0}^{\nu} \frac{(\mu+\nu-j)!\nu!}{(\mu+\nu)!j!(\nu-j)!}(-\lambda)^j, \quad d_{\mu\nu}(\lambda) = \sum_{j=0}^{\mu} \frac{(\mu+\nu-j)!\mu!}{(\mu+\nu)!j!(\mu-j)!}\lambda^j.$$

By the definition of the Padé approximant of $e^{-\lambda}$ as the rational function for which as many as possible of the coefficients in the Taylor series around $\lambda = 0$ agree with those of $e^{-\lambda}$, we have

$$(7.19) \qquad r_{\mu\nu}(\lambda) = e^{-\lambda} + O(\lambda^{\mu+\nu+1}), \quad \text{as } \lambda \to 0,$$

so that $r_{\mu\nu}(\lambda)$ approximates $e^{-\lambda}$ to order $q = \mu + \nu$. It is well known, and obvious from (7.18), that $r_{\mu\nu}(\lambda)$ is of type II for $\mu = \nu$ and type IV for $\mu > \nu$, and clearly, by (7.19), $r_{\mu\nu}(\lambda)$ is of type I for $\mu < \nu$.

In particular, $r_{01}(\lambda) = 1 - \lambda$, which gives the forward Euler scheme $U^{n+1} = (I - kA)U^n$. When $A = -\Delta_h$ this may be written

$$(7.20) \qquad (U_h^{n+1}, \chi) = (U_h^n, \chi) - k(\nabla U_h^n, \nabla \chi), \quad \forall \chi \in S_h.$$

This rational function is of type I with $\alpha = 2$. If, for instance, the inverse assumption (1.12) holds, then

$$\lambda_{\max} = \sup_{\chi \in S_h} \frac{\|\nabla \chi\|^2}{\|\chi\|^2} \le \kappa_0 h^{-2},$$

and hence (7.20) defines a type I′ scheme under the condition $k/h^2 \le \alpha_0/\kappa_0$, with $\alpha_0 < 2$.

The subdiagonal and diagonal Padé approximants with linear denominators are

$$r_{10}(\lambda) = \frac{1}{1+\lambda} \quad \text{and} \quad r_{11}(\lambda) = \frac{1-\lambda/2}{1+\lambda/2}.$$

They correspond to the backward Euler and Crank-Nicolson schemes discussed earlier and are of types IV and II, respectively.

As an example of a scheme of type III with $r(\infty) \neq 0$, we consider the so called Calahan scheme defined by

$$(7.21) \qquad r(\lambda) = 1 - \frac{\lambda}{1+b\lambda} - \frac{\sqrt{3}}{6}\Big(\frac{\lambda}{1+b\lambda}\Big)^2, \quad \text{with } b = \frac{1}{2}\Big(1+\frac{\sqrt{3}}{3}\Big).$$

To see that this $r(\lambda)$ is of type III, we note that, since $r(\lambda)$ is a decreasing function on $(0,\infty)$, it suffices to show that $r(\infty) > -1$. But this holds because

$$r(\infty) = 1 - \frac{1}{b} - \frac{\sqrt{3}}{6}\frac{1}{b^2} = 1 - \sqrt{3} > -1.$$

A simple calculation shows that $r(\lambda) - e^{-\lambda} = O(\lambda^4)$ as $\lambda \to 0$, so that the scheme is accurate of order $q = 3$. One advantage with this scheme is that the denominator is the square of a linear function. In this case the equation which has to be solved at each time step is of the form $(I + bkA)^2 U = W$, and this may be done in two steps, each of the same form $(I + bkA)X = Y$. In the finite dimensional case, when A is positive definite this means that the two systems have the same real-valued positive definite matrix. This is in contrast to, e.g., the method defined by the Padé approximant $r_{22}(\lambda)$, for which the quadratic denominator has two complex conjugate zeros and thus requires complex arithmetic.

We are now ready for the following nonsmooth data error estimate:

Theorem 7.2 *Assume that the discretization scheme is accurate of order q and of type I′, II′, or III. Then we have, for the solutions of (7.7) and (7.3),*

$$\|U^n - u(t_n)\| \le Ck^q t_n^{-q}\|v\|, \quad \textit{for } t_n > 0.$$

In case III the constant C is independent of A, and in cases I′ and II′ it depends only on the parameters α_0 and α_1, respectively.

Proof. With the notation of the proof of Theorem 7.1 we need to show that, in operator norm, $\|F_n(kA)\| \le Ck^q t_n^{-q}$ for $t_n > 0$, i.e., that

$$(7.22) \qquad |F_n(\lambda)| \le Ck^q t_n^{-q} = Cn^{-q}, \quad \text{for } \lambda \in \sigma(kA),\ n \ge 1.$$

Recall that for schemes of type I′ and II′, (7.16) holds with $\lambda_0 = \alpha_0$ and α_1, respectively, and note that for schemes of type III, (7.16) is valid for any $\lambda_0 > 0$. Hence we have, using (7.17),

$$|F_n(\lambda)| \le Cn^{-q}(n\lambda)^{q+1}e^{-cn\lambda} \le Cn^{-q}, \quad \text{for } 0 \le \lambda \le \lambda_0.$$

In cases I′ and II′, this completes the proof of (7.22) since then $k\lambda_{\max} \le \lambda_0$. For type III schemes we also need to consider λ large. We have, for $\lambda \ge \lambda_0 = 1$, say (recall that (7.16) now holds with λ_0 an arbitrary positive number), $e^{-n\lambda} \le e^{-n} \le Cn^{-q}$. Further, since $|r(\infty)| < 1$ we have $\sup_{\lambda\ge 1}|r(\lambda)| = e^{-c}$, with $c > 0$, so that $\sup_{\lambda \ge 1}|r(\lambda)^n| \le e^{-cn} \le Cn^{-q}$, and hence $\sup_{\lambda\ge 1}|F_n(\lambda)| \le Cn^{-q}$. This completes the proof. □

In the same way as in the spatially semidiscrete case, cf. Theorem 3.5, one may formulate a general result which expresses the relation between the regularity of data, the order of convergence, and the singularity of the error bound, and which includes both the smooth data and the nonsmooth data error estimates of Theorems 7.1 and 7.2.

Theorem 7.3 *Under the assumptions of Theorem 7.2 we have*

$$\|U^n - u(t_n)\| \le Ck^l t_n^{-(l-s)}|v|_{2s}, \quad \text{for } v \in \dot{H}^{2s},\ 0 \le s \le l \le q.$$

Proof. We note that since $F_n(\lambda)$ is bounded on $\sigma(kA)$, (7.15) and (7.22) hold with q replaced by l. Hence

$$|F(\lambda)| \le C(\lambda^l)^{s/l}(n^{-l})^{1-s/l} = C\lambda^s n^{-(l-s)}, \quad \text{for } \lambda \in \sigma(kA),\ n \ge 1,$$

from which the result follows as above. □

Although Theorem 7.2 does not cover schemes of type II without restrictions on $\lambda_{\max}$, it was discovered by Luskin and Rannacher [166] that a way of securing the estimate of Theorem 7.2 in the case of the diagonal Padé schemes is to start with a few steps of a corresponding subdiagonal scheme. We shall demonstrate this for the Crank-Nicolson scheme, starting with two steps of the backward Euler scheme, thus defining U^n by

$$(7.23)\qquad \begin{aligned} U^{n+1} &= r_1(kA)U^n, \quad \text{with } r_1(\lambda) = \frac{1-\lambda/2}{1+\lambda/2}, \quad \text{for } n \ge 2,\\ U^{n+1} &= r_0(kA)U^n, \quad \text{with } r_0(\lambda) = \frac{1}{1+\lambda}, \quad n = 0, 1, \quad U^0 = v. \end{aligned}$$

We then have the following result:

Theorem 7.4 *We have, for the solutions of* (7.23) *and* (7.3),

$$\|U^n - u(t_n)\| \le Ck^2 t_n^{-2}\|v\|, \quad \text{for } t_n > 0.$$

Proof. In the same way as in the proof of Theorem 7.2 it suffices to show (for $n = 1$ the estimate stated is obvious)

$$|\widetilde{F}_n(\lambda)| = |r_0(\lambda)^2 r_1(\lambda)^{n-2} - e^{-n\lambda}| \le Cn^{-2}, \quad \text{for } \lambda > 0,\ n \ge 2,$$

and since both terms in $\widetilde{F}_2(\lambda)$ are bounded, we may consider $n > 2$.

For large λ, $\lambda \geq \lambda_0$, say, we have, with c suitable,

$$|r_1(\lambda)| = \frac{1 - 2/\lambda}{1 + 2/\lambda} \leq e^{-c/\lambda}.$$

Hence, for these λ,

$$\begin{aligned} |r_0(\lambda)^2 r_1(\lambda)^{n-2}| &\leq C\lambda^{-2} e^{-c(n-2)/\lambda} \\ &\leq C(n-2)^{-2}((n-2)/\lambda)^2 e^{-c(n-2)/\lambda} \leq Cn^{-2}. \end{aligned}$$

It follows that

$$|\widetilde{F}_n(\lambda)| \leq Cn^{-2} + e^{-\lambda_0 n} \leq Cn^{-2}, \quad \text{for } \lambda \geq \lambda_0.$$

To consider $\lambda \leq \lambda_0$, we write

$$\widetilde{F}_n(\lambda) = r_0(\lambda)^2 (r_1(\lambda)^{n-2} - e^{-(n-2)\lambda}) + (r_0(\lambda)^2 - e^{-2\lambda}) e^{-(n-2)\lambda}.$$

By the argument of the proof of Theorem 7.2 we have, for $\lambda \leq \lambda_0$,

$$|r_1(\lambda)^{n-2} - e^{-(n-2)\lambda}| \leq C(n-2)^{-2} \leq Cn^{-2},$$

and $|r_0(\lambda)^2 - e^{-2\lambda}| \leq C\lambda^2$, so that

$$|\widetilde{F}_n(\lambda)| \leq Cn^{-2} + C\lambda^2 e^{-n\lambda} \leq Cn^{-2}, \quad \text{for } \lambda \leq \lambda_0.$$

Together these estimates complete the proof. □

Since the error bound in Theorem 7.2 is large for small t it appears natural to try to obtain a more uniform error bound by taking smaller time steps in the beginning of the computation. We shall analyze such a procedure for the backward Euler method. The method was briefly discussed in Chapter 1, using the energy method.

Let thus $0 = t_0 < t_1 < \cdots < t_n < \cdots$ be a partition of the positive time axis and set $J_n = (t_{n-1}, t_n)$ and $k_n = t_n - t_{n-1}$. We shall consider the approximation U^n of the solution of (7.3) at $t = t_n$ defined by

(7.24) $$\bar{\partial}_n U^n + AU^n = 0 \quad \text{for } n \geq 1, \quad \text{with } U^0 = v,$$

where $\bar{\partial}_n U^n = (U^n - U^{n-1})/k_n$. We begin with the following error estimate:

Theorem 7.5 *We have, for the solutions of* (7.24) *and* (7.3),

$$\|U^n - u(t_n)\| \leq \sum_{j=1}^{n} k_j \int_{J_j} \|u_{tt}\|\, dt, \quad \text{for } t_n \geq 0.$$

Proof. The solution may be represented as

$$U^n = E_{k_n} U^{n-1}, \quad \text{for } n \geq 1, \text{ with } E_k = (I + kA)^{-1}, \quad U^0 = v,$$

or, in concise form,

$$U^n = E_{n,1} v, \quad \text{where } \; E_{n,j} = E_{k_n} E_{k_{n-1}} \cdots E_{k_j} \quad \text{for } j \leq n.$$

The error $\eta^n = U^n - u^n$ then satisfies

$$\bar{\partial}_n \eta^n + A\eta^n = \omega^n := -\bar{\partial}_n u^n - Au^n = u_t^n - \bar{\partial}_n u^n. \tag{7.25}$$

Hence, we have

$$\eta^n = E_{k_n} \eta^{n-1} + k_n E_{k_n} \omega^n, \tag{7.26}$$

or, by repeated application, since $\eta^0 = 0$,

$$\eta^n = \sum_{j=1}^{n} k_j E_{n,j} \omega^j, \quad \text{for } n \geq 1. \tag{7.27}$$

As before, $\|E_k\| \leq 1$, so that $\|E_{n,j}\| \leq 1$, and thus

$$\|\eta^n\| \leq \sum_{j=1}^{n} k_j \|\omega^j\|.$$

Since, cf. (1.52),

$$\|\omega^j\| = \|u_t(t_j) - \bar{\partial}_j u(t_j)\| \leq \int_{J_j} \|u_{tt}\| \, dt,$$

the proof is complete. □

We shall now present an alternative error bound to that in Theorem 7.5, in which the sum over j is replaced by a maximum and where only the first order derivative of u with respect to time enters. We shall return in Chapter 12 to error estimates of this type for fully discrete methods, obtained by discretization in time of the spatially discrete problem, and applicable also to the inhomogeneous equation.

Theorem 7.6 *We have, for the solutions of* (7.24) *and* (7.3),

$$\|U^n - u(t_n)\| \leq \left(1 + \log \frac{t_n}{k_n}\right) \max_{j \leq n} \int_{J_j} \|u_t\| \, dt, \quad \textit{for } t_n > 0.$$

Proof. We write (7.27) in the form

$$\eta^n = \sum_{j=1}^{n} k_j A E_{n,j}\, A^{-1}\omega^j.$$

Our result will follow from

$$\sum_{j=1}^{n} k_j \|AE_{n,j}\| \le 1 + \log\frac{t_n}{k_n} \tag{7.28}$$

and

$$\|A^{-1}\omega^j\| \le \int_{J_j} \|u_t\|\,dt. \tag{7.29}$$

To show (7.28), we note that, by spectral representation,

$$\|AE_{n,j}\| \le \max_{\lambda\ge 0}\frac{\lambda}{(1+k_n\lambda)\cdots(1+k_j\lambda)} \le \frac{1}{k_j+\cdots+k_n} = \frac{1}{t_n - t_{j-1}}.$$

Hence

$$\sum_{j=1}^{n} k_j\|AE_{n,j}\| \le \sum_{j=1}^{n}\frac{k_j}{t_n - t_{j-1}} \le 1 + \sum_{j=1}^{n-1}\int_{J_j}\frac{dt}{t_n - t} = 1 + \log\frac{t_n}{k_n},$$

which shows (7.28). We have from (7.25)

$$\begin{aligned}\omega^j &= -\frac{1}{k_j}\int_{J_j} u_t\,dt - Au^j = A\Big(\frac{1}{k_j}\int_{J_j} u\,dt - u^j\Big)\\ &= A\frac{1}{k_j}\int_{J_j}(u(t) - u(t_j))\,dt = A\frac{1}{k_j}\int_{J_j}\int_{t_j}^{t} u_t(s)\,ds\,dt,\end{aligned}$$

from which (7.29) follows at once. The proof is now complete. □

Since, for most practical choices of the time steps, the logarithmic factor is of moderate size, one may use the result of Theorem 7.6, provided the behavior of u_t is known, to bound the error essentially uniformly in time by choosing the k_j such that $\int_{J_j}\|u_t\|\,dt$ is kept uniformly small. This may be accomplished by choosing k_j such that $k_j \max_{J_j}\|u_t\|$ is kept uniformly small.

For example, assume that $v \in \mathcal{D}(A^{1/2})$. Then the standard spectral argument, cf. Lemma 3.2, shows

$$\|u_t(t)\| \le Ct^{-1/2}\|A^{1/2}v\| = C_0 t^{-1/2},$$

and hence

$$\int_{J_j} \|u_t\|\,dt \le \begin{cases} 2C_0 k_1^{1/2}, & \text{for } j = 1,\\ C_0 k_j t_{j-1}^{-1/2}, & \text{for } j > 1. \end{cases}$$

With δ a small positive number, this suggests choosing $k_1 = \delta^2/(2C_0)^2$ and $k_j = \delta t_{j-1}^{1/2}/C_0$ for $j > 1$, for then $\int_{J_j} \|u_t\|\,dt \le \delta$ for $j \ge 1$, and since we easily find $t_n/k_n \le 1 + C_0\delta^{-1}t_{n-1}^{1/2}$, the error will therefore then be bounded by $\delta(1 + \log(1 + C_0\delta^{-1}t^{1/2}))$ for $t_n \le t$.

We now return to the spatially semidiscrete problem (7.2), which we now write as

$$u_{h,t} + A_h u_h = 0, \quad \text{for } t > 0, \quad \text{with } u_h(0) = v_h,$$

with $A_h = -\Delta_h = T_h^{-1} : S_h \to S_h$, where T_h satisfies assumptions (i) and (ii). We consider fully discrete schemes defined by application of our above time stepping procedure (7.7) to this semidiscrete equation. This defines the fully discrete approximation $U^n \in S_h$ of $u(t_n)$ recursively by

$$U_h^{n+1} = E_{kh} U_h^n, \quad \text{for } n \ge 0, \quad \text{where } E_{kh} = r(kA_h), \quad U^0 = v_h. \tag{7.30}$$

Assuming that $r(\lambda) = \alpha_0 \prod_j(\lambda + \beta_j)/\prod_j(\lambda + \gamma_j)$, the recursion formula in (7.30) may be written

$$\prod_j (kA_h + \gamma_j)U^{n+1} = \alpha_0 \prod_j (kA_h + \beta_j)U^n.$$

Hence, in order to determine U_h^{n+1} from U_h^n one needs to solve a sequence of equations of the form

$$(\alpha - k\beta\Delta_h)W = (\gamma - k\delta\Delta_h)V, \tag{7.31}$$

for W, with V given. Note that even when the rational function $r(\lambda)$ has real coefficients, the β_j and γ_j, and hence also the $\alpha, \beta, \gamma, \delta$ and the V and W, may be complex-valued ($A_h\psi$ may be thought of as being defined for complex ψ by linearity).

For example, consider the standard Galerkin method, so that $S_h \subset H_0^1(\Omega)$ and T_h is defined by (3.10). In this case (7.31) can be expressed as

$$\alpha(W, \chi) + \beta k(\nabla W, \nabla\chi) = \gamma(V, \chi) + \delta k(\nabla V, \nabla\chi), \quad \forall \chi \in S_h.$$

If $\{\Phi_j\}_{j=1}^{N_h}$ is a basis for S_h and $\mathcal{B} = ((\Phi_j, \Phi_k))$ and $\mathcal{A} = ((\nabla\Phi_j, \nabla\Phi_k))$ are the corresponding mass and stiffness matrices, and if ξ and η denote the vectors of coefficients of V and W with respect to $\{\Phi_j\}_{j=1}^{N_h}$, then the latter equation may also be written in matrix form as $(\alpha\mathcal{B} + \beta\mathcal{A})\eta = (\gamma\mathcal{B} + \delta\mathcal{A})\xi$. The backward Euler method (7.12) and the Crank-Nicolson method (7.13) are both of this form.

In the case of Nitsche's method discussed in Chapter 2, A_h is defined on S_h, which is now a subset of $H^1(\Omega)$, by

$$(A_h\psi, \chi) = N_\gamma(\psi, \chi), \quad \forall \psi, \chi \in S_h,$$

and the backward Euler method, e.g., takes the form

$$(U_h^{n+1}, \chi) + kN_\gamma(U_h^{n+1}, \chi) = (U^n, \chi), \quad \forall \chi \in S_h,\ n \geq 0.$$

As our first error estimate in the fully discrete case we now show the following nonsmooth data result, where the norm is again that in L_2.

Theorem 7.7 *Let the time discretization scheme be accurate of order q and of type I', II', or III, and assume that (i) and (ii) hold, and that $v_h = P_h v$. Then we have for the error in the fully discrete scheme* (7.30)

$$\|U_h^n - u(t_n)\| \leq C(h^r t_n^{-r/2} + k^q t_n^{-q})\|v\|, \quad \textit{for } t_n = nk > 0.$$

Proof. By Theorem 7.2, applied to the semidiscrete equation (7.2), we have

$$\|U_h^n - u_h(t_n)\| \leq Ck^q t_n^{-q}\|P_h v\| \leq Ck^q t_n^{-q}\|v\|, \quad \text{for } t_n > 0.$$

Further, by Theorem 3.2,

$$\|u_h(t) - u(t)\| \leq Ch^r t^{-r/2}\|v\|, \quad \text{for } t > 0.$$

The result stated now follows by the triangle inequality. □

We shall now turn to error estimates which hold uniformly down to $t = 0$. In this case, in order to obtain optimal order results, smoothness has to be required from the initial data. To express this we shall again use the spaces $\dot{H}^s = \dot{H}^s(\Omega)$ with norms $|\cdot|_s$ introduced in Chapter 3, consisting of functions u in $H^s(\Omega)$ with $A^j u = 0$ on $\partial\Omega$ for $j < s/2$. We recall from Theorem 7.1 that, without spatial discretization and with $A = -\Delta$, the requirement when the scheme is accurate of order q and stable in L_2 for a $O(k^q)$ error bound is that the initial data v are in $\dot{H}^{2q}$.

Our result in the smooth data case is then the following:

Theorem 7.8 *Let the time discretization scheme be of type I' or II, and assume that (i) and (ii) hold, that $v \in \dot{H}^{\max(r,2q)}(\Omega)$, and that $\|v_h - v\| \leq Ch^r|v|_r$. Then we have for the error in the fully discrete scheme* (7.30)

$$\|U_h^n - u(t_n)\| \leq C(h^r|v|_r + k^q|v|_{2q}), \quad \textit{for } t_n \geq 0.$$

We recall that for the semidiscrete problem (7.2), the energy method was used to show in Theorem 3.1 the error estimate

$$\|u_h(t) - u(t)\| \leq Ch^r|v|_r, \quad \text{for } t \geq 0,\ v \in \dot{H}^r(\Omega). \tag{7.32}$$

A direct application of Theorem 7.1 here gives

$$\|U_h^n - u_h(t_n)\| \leq Ck^q\|A_h^q v_h\|, \quad \text{for } t_n > 0,$$

but the bound on the right hand side now depends on h. In order to show the estimate stated, we shall combine (7.32) with the technique used in the proof of Theorem 7.2 and with the following easily verified identity.

Lemma 7.1 *We have, with* $T_h^0 = I$,

$$v = \sum_{j=0}^{q-1} T_h^j (T - T_h) A^{j+1} v + T_h^q A^q v, \quad \textit{for } v \in \dot{H}^{2q}(\Omega). \tag{7.33}$$

Proof. Since $TA = I$, the sum is telescoping, which shows the result. □

The following lemma will also be needed, where we again use the notation $F_n(\lambda) = r(\lambda)^n - e^{-n\lambda}$.

Lemma 7.2 *Let the discretization be of type I′ or II. Then*

$$\|F_n(kA_h)P_h T_h^j\| = \|F_n(kA_h)P_h T_h^j\|_{S_h} \le Ck^j, \quad \textit{for } 0 \le j \le q,\ n \ge 0.$$

Proof. We have (note $P_h T_h^j = T_h^j = (-\Delta_h)^{-j}$ for $j > 0$, and $T_h = T_h P_h$)

$$\|F_n(kA_h)P_h T_h^j\| \le k^j \sup_{\lambda \in \sigma(kA_h)} |\lambda^{-j} F_n(\lambda)|$$

and hence it suffices to show that $|\lambda^{-j} F_n(\lambda)| \le C$ for $\lambda \in \sigma(kA_h)$. As in the proof of Theorem 7.2, let λ_0 be a positive number such that $|r(\lambda)| < 1$ for $0 < \lambda \le \lambda_0$. Then, by our assumptions, we have for such λ,

$$|r(\lambda) - e^{-\lambda}| \le C\lambda^{j+1}, \quad 0 \le j \le q, \quad \text{and } |r(\lambda)| \le e^{-c\lambda}, \text{ with } 0 < c < 1.$$

Hence, for $\lambda \le \lambda_0$,

$$|\lambda^{-j} F_n(\lambda)| = |\lambda^{-j}(r(\lambda) - e^{-\lambda}) \sum_{l=0}^{n-1} r(\lambda)^{n-1-l} e^{-l\lambda}| \le Cn\lambda e^{-cn\lambda} \le C.$$

For schemes of type I′ this completes the proof. For $r(\lambda)$ of type II, the desired inequality follows trivially for $\lambda > \lambda_0$. □

Proof of Theorem 7.8. We first note that by the stability of the completely discrete scheme, it is no restriction of generality to assume that $v_h = P_h v$. For, by our assumptions,

$$\|E_{kh}^n (v_h - P_h v)\| \le \|v_h - v\| + \|P_h v - v\| \le Ch^r |v|_r.$$

Assuming thus $v_h = P_h v$ we may write $U^n - u_h(t_n) = F_n(kA_h)P_h v$. We now note that if we set

$$v_k = \sum_{k\lambda_l \le 1} (v, \varphi_l)\varphi_l,$$

where φ_l and λ_l are the eigenfunctions and eigenvalues of the differential operator A, with vanishing boundary values, then $v_k \in \dot{H}^s(\Omega)$ for each $s \ge 0$. Further, by the definition of the norm in $\dot{H}^s(\Omega)$, we find easily

(7.34) $$\|v - v_k\| \le k^q |v|_{2q},$$

(7.35) $$|v_k|_{2q} \le |v|_{2q},$$

(7.36) $$|v_k|_{r+2j} \le k^{-j} |v|_r, \quad \text{for } j = 0, \dots, q-1.$$

Applying now the identity (7.33) to v_k and setting for brevity $F_n = F_n(kA_h)P_h$, we may write

$$F_n v_k = \sum_{j=0}^{q-1} F_n T_h^j (T - T_h) A^{j+1} v_k + F_n T_h^q A^q v_k.$$

Here, by Lemma 7.2 and (7.35),

$$\|F_n T_h^q A^q v_k\| \le Ck^q \|A^q v_k\| = Ck^q |v_k|_{2q} \le Ck^q |v|_{2q}.$$

Further, using also property (ii) of T_h and (7.36), we obtain

$$\begin{aligned}\|F_n T_h^j (T - T_h) A^{j+1} v_k\| &\le Ck^j \|(T - T_h) A^{j+1} v_k\| \\ &\le Ck^j h^r \|A^{j+1} v_k\|_{r-2} \le Ck^j h^r |v_k|_{r+2j} \le Ch^r |v|_r, \quad \text{for } 0 \le j \le q-1.\end{aligned}$$

Together these estimates imply

$$\|F_n v_k\| \le C(h^r |v|_r + k^q |v|_{2q}).$$

Since obviously, by stability and (7.34),

$$\|F_n(v - v_k)\| \le 2\|v - v_k\| \le Ck^q |v|_{2q},$$

we conclude that

$$\|U_h^n - u_h(t_n)\| = \|F_n v\| \le C(h^r |v|_r + k^q |v|_{2q}).$$

In view of the estimate (7.32) for the semidiscrete problem this completes the proof. □

So far we have never had reason to use the property of a scheme to be of type IV. We shall close this chapter by proving, for later use, a smoothing property of time discretization schemes, including such schemes of type IV, which can be thought of as a discrete analogue of the property defining an analytic semigroup (cf. Lemma 3.2). We formulate this result in the Hilbert space framework.

Lemma 7.3 *Let A be a positive definite operator in a Hilbert space $\mathcal{H}$ as in the beginning of this chapter, and let the discretization scheme (7.7) for the initial value problem (7.3) be accurate of order $q \ge 1$ and of type I', II', or IV. Then, for each $j \ge 0$,*

$$\|A^j E_k^n v\| \le C t_n^{-j} \|v\|, \quad \textit{for } t_n \ge t_j,\ v \in \mathcal{H}.$$

Proof. We have by (7.6) that $\|A^j E_k^n\| = \sup_{\lambda\in\sigma(A)} |\lambda^j r(k\lambda)^n|$. Considering first schemes of types I′ and II′ we recall from above that in these cases $|r(k\lambda)| \le e^{-ck\lambda}$, for $\lambda \in \sigma(A)$, with $0 < c < 1$. Therefore, since $t_n = nk$,

$$|\lambda^j r(k\lambda)^n| \le \lambda^j e^{-cnk\lambda} \le Ct_n^{-j}, \quad \text{for } \lambda \in \sigma(A),$$

which proves the desired estimate, in fact for $t_n > 0$.

For rational functions of type IV we shall show below that

$$|r(\lambda)| \le \frac{1}{1+c\lambda}, \quad \text{for } \lambda \ge 0, \quad \text{with } c > 0. \tag{7.37}$$

Assuming this we have now

$$\|A^j E_k^n\| \le k^{-j} \sup_{\lambda\ge 0} |\lambda^j r(\lambda)^n| \le k^{-j} \sup_{\lambda\ge 0} \frac{\lambda^j}{(1+c\lambda)^n}.$$

For $0 \le \lambda \le 1$, say, we have

$$\frac{\lambda^j}{(1+c\lambda)^n} \le \lambda^j e^{-c_1 n\lambda} \le Cn^{-j}, \quad \text{with } c_1 > 0,$$

whereas, for $\lambda \ge 1$ and $n \ge j$,

$$\frac{\lambda^j}{(1+c\lambda)^n} \le \Big(\frac{\lambda}{1+c\lambda}\Big)^j \frac{1}{(1+c)^{n-j}} \le Cn^{-j}.$$

Together these inequalities complete the proof.

It remains to show (7.37). For $\lambda \le \lambda_0$, with λ_0 sufficiently small, this is clear from (7.16). On the other hand, since $r(\infty) = 0$, the degree of the numerator of $r(\lambda)$ is less than that of its denominator. Hence for $c > 0$ sufficiently small, we have $\lim_{\lambda\to\infty} |(1+c\lambda)r(\lambda)| < 1$, so that, for some $\lambda_1 > 0$, we have $(1+c\lambda)|r(\lambda)| < 1$, for $\lambda > \lambda_1$. Finally, since $|r(\lambda)| < 1$ for $\lambda > 0$, we may choose $c > 0$ so small that $(1+c\lambda)|r(\lambda)| < 1$, for $\lambda_0 \le \lambda \le \lambda_1$. This completes the proof. □

The presentation in this chapter originates in Baker, Bramble, and Thomée [21] where fully discrete schemes were considered directly without first stydying the abstract time dependent differential equation. For more information about rational approximations of $e^{-\lambda}$ of the types discussed here and suitable for the solution of stiff ordinary differential equations, see, e.g., Hairer and Wanner [113].

The results above generalize directly to parabolic equations of the form $u_t + Au = 0$ where the elliptic operator A is selfadjoint, positive definite, and time independent. Nonselfadjoint operators have been analyzed in LeRoux [152], [153], where both smooth and nonsmooth data are considered, using the Dunford-Taylor spectral representation; in [152] the operator is allowed to depend on t. For such methods, see also Suzuki [222], Fujita and Suzuki [104] and references therein; our next chapter is devoted to this approach. The case of time-dependent operators has also been studied by energy arguments in, e.g., Huang and Thomée [127], Luskin and Rannacher [166], [167], Sammon [206] and Karakashian [133]. In the selfadjoint time dependent case a combination of spectral and energy arguments has been used in Bramble and Sammon [35].

Conditions II–IV for the rational function are related to the concept of $A(0)$-stability which we will return to in the next chapter as a special case of $A(\theta)$-stability.

8. Single Step Fully Discrete Schemes for the Inhomogeneous Equation

In this chapter we shall continue our study of single step fully discrete methods and turn now to the approximation of the inhomogeneous heat equation. Following the approach of Chapter 7 we shall first consider discretization in time of an ordinary differential equation in a Hilbert space setting, and then apply our results to the spatially discrete equation. In view of the work in Chapter 7 for the homogeneous equation with given initial data, we now restrict ourselves to the case that the initial data vanish.

We consider thus first the abstract initial value problem

$$u' + Au = f, \quad \text{for } t > 0, \quad \text{with } u(0) = 0, \tag{8.1}$$

in a Hilbert space $\mathcal{H}$, where A is a linear, selfadjoint, positive definite, not necessarily bounded operator with a compact inverse T, defined on $\mathcal{D}(A) \subset \mathcal{H}$. As before, we could have $\mathcal{H} = L_2(\Omega)$ and $A = -\Delta$, or $\mathcal{H} = S_h$ and $A = A_h = -\Delta_h$.

Generalizing from the case of the homogeneous equation, we consider now a time stepping scheme of the form

$$\begin{aligned} U^{n+1} &= E_k U^n + k(Q_k f)(t_n), \quad \text{for } n \ge 0, \quad \text{with } U^0 = 0, \\ &\text{where } E_k v = r(kA)v, \quad Q_k f(t) = \sum_{i=1}^{m} p_i(kA) f(t + \tau_i k). \end{aligned} \tag{8.2}$$

Here, with k the time step and $t_n = nk$, $r(\lambda)$ and $\{p_i(\lambda)\}_{i=1}^m$ are rational functions which are bounded on the spectrum of kA, uniformly in k, and $\{\tau_i\}_{i=1}^m$ are distinct real numbers, which for simplicity we assume in $[0,1]$.

We shall begin by discussing the accuracy of this discretization. For this purpose we consider the simple scalar ordinary differential equation problem

$$u' + au = f, \quad \text{for } t > 0, \quad \text{with } u(0) = 0, \tag{8.3}$$

where $a > 0$, and its discrete analogue which now reduces to

$$U^{n+1} = r(ka)U^n + k \sum_{i=1}^{m} p_i(ka) f(t_n + \tau_i k), \quad \text{for } n \ge 0, \quad U^0 = 0. \tag{8.4}$$

We shall say that the time discretization scheme (8.2) is accurate of order q if the solution of (8.3) satisfies (8.4) with an error which is $O(k^{q+1})$, as $k \to 0$, for any choice of a and f. We have the following:

Lemma 8.1 *The time discretization scheme* (8.2) *is accurate of order* q *if and only if*

$$\text{(j)} \qquad r(\lambda) = e^{-\lambda} + O(\lambda^{q+1}), \quad \text{as } \lambda \to 0,$$

and, for $0 \le l \le q$,

$$\text{(jj)} \qquad \sum_{i=1}^{m} \tau_i^l p_i(\lambda) = \frac{l!}{(-\lambda)^{l+1}} \Big(e^{-\lambda} - \sum_{j=0}^{l} \frac{(-\lambda)^j}{j!}\Big) + O(\lambda^{q-l}), \quad \text{as } \lambda \to 0,$$

or, equivalently,

$$\text{(jj}'\text{)} \qquad \sum_{i=1}^{m} \tau_i^l p_i(\lambda) = \int_0^1 s^l e^{-\lambda(1-s)} ds + O(\lambda^{q-l}), \quad \text{as } \lambda \to 0.$$

Proof. We begin by showing the necessity of (j) and (jj), (jj′). The exact solution of (8.3) satisfies

$$u(t_{n+1}) = e^{-ka} u(t_n) + k \int_0^1 e^{-ka(1-s)} f(t_n + sk)\, ds.$$

Choosing $f = 0$ we have, if the scheme is of order q,

$$u(t_{n+1}) = e^{-ka} u(t_n) = r(ka) u(t_n) + O(k^{q+1}), \quad \text{as } k \to 0,$$

or $r(ka) = e^{-ka} + O(k^{q+1})$ as $k \to 0$, for each $a > 0$, showing (j).

It remains to show that (jj) and (jj′) follow from

$$\int_0^1 e^{-ka(1-s)} f(t_n + sk)\, ds = \sum_{i=1}^{m} p_i(ka) f(t_n + \tau_i k) + O(k^q), \quad \text{as } k \to 0.$$

Developing $f(t_n + \tau_i k)$ in a Taylor series around t_n we find, since $f^{(l)}(t_n)$, $l = 0, \dots, q$, as well as ka, are arbitrary,

$$\int_0^1 s^l e^{-ka(1-s)} ds = \sum_{i=1}^{m} \tau_i^l p_i(ka) + O(k^{q-l}), \quad \text{as } k \to 0,$$

which yields (jj)′. Since an elementary calculation shows that

$$\frac{1}{l!} \int_0^1 s^l e^{-\lambda(1-s)} ds = \frac{1}{(-\lambda)^{l+1}} \sum_{j=l+1}^{\infty} \frac{(-\lambda)^j}{j!},$$

we find that (jj) and (jj′) are equivalent.

The sufficiency of the conditions follows by reversing the above arguments. □

From a computational point of view it would be convenient to choose the rational functions $p_i(\lambda)$ such that their denominators are all the same as that of $r(\lambda)$, for, if with $n(\lambda), n_i(\lambda)$, and $d(\lambda)$ polynomials, we have

$$r(\lambda) = \frac{n(\lambda)}{d(\lambda)}, \quad \text{and} \quad p_i(\lambda) = \frac{n_i(\lambda)}{d(\lambda)}, \quad \text{for } i = 1, \dots, m,$$

then the scheme (8.2) may be written simply as

$$d(kA)U^{n+1} = n(kA)U^n + k\sum_{i=1}^{m} n_i(kA)f(t_n + \tau_i k).$$

One way of achieving this, as well as the conditions of Lemma 8.1, is to first choose $r(\lambda)$ such that (j) holds, then to select $\{\tau_i\}_{i=1}^m$ as $m = q$ distinct real numbers in $[0, 1]$, and finally to solve the system

$$(8.5) \qquad \sum_{i=1}^{q} \tau_i^l p_i(\lambda) = \frac{l!}{(-\lambda)^{l+1}}\Big(r(\lambda) - \sum_{j=0}^{l} \frac{(-\lambda)^j}{j!}\Big), \quad l = 0, \dots, q-1,$$

for $\{p_i(\lambda)\}_{i=1}^q$. Since the matrix of the coefficients on the left is of Vandermonde's type, and thus nonsingular, this results in rational functions $p_i(\lambda)$ which are linear combinations of those on the right hand side of (8.5). In particular, the only singularities of the right hand sides of (8.5), and hence of the $p_i(\lambda)$, are those of $r(\lambda)$, and the $p_i(\lambda)$ thus have the same denominators as $r(\lambda)$. If $r(\lambda)$ is bounded for large λ, then the right hand sides of (8.5) are small for large λ, and hence the numerator of $p_i(\lambda)$ is of lower degree than its denominator. Note that the condition (j) together with (8.5) implies that (jj) holds. This is evident for $0 \le l \le q-1$, and for $l = q$ condition (jj) reads

$$(8.6) \qquad \sum_{i=1}^{m} \tau_i^q p_i(\lambda) = \frac{q!}{(-\lambda)^{q+1}} \sum_{j=q+1}^{\infty} \frac{(-\lambda)^j}{j!} + O(1) = O(1), \quad \text{as } \lambda \to 0.$$

Since by (j) each right hand side in (8.5) is bounded for small λ, this also holds for the $p_i(\lambda)$, which shows (8.6).

Choosing $q = m = 2$, $\tau_1 = 0$, $\tau_2 = 1$, $r(\lambda) = (1 - \frac{1}{2}\lambda)/(1 + \frac{1}{2}\lambda)$, this procedure gives the Crank-Nicolson type scheme

$$(8.7) \qquad (I + \tfrac{1}{2}kA)U^{n+1} = (I - \tfrac{1}{2}kA)U^n + \tfrac{1}{2}(f(t_{n+1}) + f(t_n)),$$

For certain schemes, the number m of quadrature points could be less than q. An example of this is provided by the Crank-Nicolson scheme

$$(8.8) \qquad (I + \tfrac{1}{2}kA)U^{n+1} = (I - \tfrac{1}{2}kA)U^n + kf(t_n + \tfrac{1}{2}k),$$

for which $q = 2$, $m = 1$, $\tau_1 = \frac{1}{2}$, $r(\lambda) = (1 - \frac{1}{2}\lambda)/(1 + \frac{1}{2}\lambda)$, The relations (j) and (jj) here reduce to

$$\frac{1-\frac{1}{2}\lambda}{1+\frac{1}{2}\lambda} = e^{-\lambda} + O(\lambda^3),$$

and

$$\frac{1}{1+\frac{1}{2}\lambda} = -\frac{1}{\lambda}(e^{-\lambda}-1) + O(\lambda^2),$$
$$\frac{1}{2}\frac{1}{1+\frac{1}{2}\lambda} = \frac{1}{\lambda^2}(e^{-\lambda}-1+\lambda) + O(\lambda),$$
$$\frac{1}{4}\frac{1}{1+\frac{1}{2}\lambda} = -\frac{2}{\lambda^3}(e^{-\lambda}-1+\lambda-\tfrac{1}{2}\lambda^2) + O(1),$$

respectively, as $\lambda \to 0$.

A frequently employed family of schemes which fits into our framework is the Runge-Kutta methods. For the linear equation (8.1) such a method takes the form

$$U^{n+1} = U^n + k\sum_{j=1}^m b_j(-AU_{nj} + f(t_n+\tau_j k)),$$

where the intermediate U_{nj} are determined from the linear system

$$U_{ni} = U^n + k\sum_{j=1}^m g_{ij}(-AU_{nj} + f(t_n+\tau_j k)), \quad i = 1,\dots,m.$$

Here the quadrature points τ_j are distinct numbers in $[0,1]$ and the coefficients g_{ij} and b_j are associated with the quadrature formulas

$$(8.9) \qquad \int_0^1 \varphi\,dt \approx \sum_{j=1}^m b_j\varphi(\tau_j), \quad \int_0^{\tau_i} \varphi\,dt \approx \sum_{j=1}^m g_{ij}\varphi(\tau_j),\ i = 1,\dots,m.$$

The method is implicit unless the matrix $\mathcal{G} = (g_{ij})$ is strictly lower triangular. We shall assume that $\mathcal{G}$ has no eigenvalues in $(-\infty, 0]$, so that, in particular the method is implicit and $\sigma(\lambda) = (I+\lambda\mathcal{G})^{-1}$ exists for $\lambda \geq 0$. After elimination of the $U_{ni}, i = 1,\dots,m$, these equations take the form (8.2) where

$$(8.10) \qquad (p_1(\lambda),\cdots,p_m(\lambda)) = (b_1,\cdots,b_m)\sigma(\lambda), \quad r(\lambda) = 1-\lambda\sum_{j=1}^m b_j p_j(\lambda).$$

It is known that such a method is accurate of order q if the quadrature formulas in (8.9) are exact for polynomials of degree $q-1$ and $q-2$, respectively.

We shall return to a discussion of the choice of the discretization scheme later in this chapter.

Our purpose is now to analyze the error in the fully discrete method (8.2) for the inhomogeneous abstract equation (8.1).

We shall assume that E_k is stable in $\mathcal{H}$, so that $|r(\lambda)| \le 1$ for $\lambda \in \sigma(kA)$ cf. (7.10); this condition is satisfied for all operators E_k (or discretization schemes for the corresponding homogeneous equation) of types I′ and II of our previous classification.

In our first result we shall prove that if the scheme is accurate of order q, then the error in the time discretization of (8.1) is $O(k^q)$, provided certain assumptions on the data are satisfied. We employ again the spaces $\dot{H}^s = \mathcal{D}(A^{s/2})$ introduced in Chapter 7 and the corresponding norm $|v|_s = (A^s v, v)^{1/2} = \|A^{s/2} v\|$. We shall often use the notation $f^{(l)}$ for $(d/dt)^l f$ in the sequel.

Theorem 8.1 *Assume that the time discretization scheme in* (8.2) *is accurate of order q and that E_k is of type I′ or II. Then, if $f^{(l)}(t) \in \dot{H}^{2p-2l}$ for $l < q$, when $t \ge 0$, we have for the solutions of* (8.2) *and* (8.1), *when $t_n \ge 0$,*

$$\|U^n - u(t_n)\| \le Ck^q \Big(t_n \sum_{l=0}^{q-1} \sup_{s \le t_n} |f^{(l)}(s)|_{2q-2l} + \int_0^{t_n} \|f^{(q)}\|\, ds \Big). \tag{8.11}$$

Proof. We have at once from (8.2) that

$$U^n = k \sum_{j=0}^{n-1} E_k^{n-1-j} Q_k f(t_j).$$

Setting as usual $E(t) = e^{-tA}$ we may write for the solution of (8.1)

$$u(t_n) = \int_0^{t_n} E(t_n - s) f(s)\, ds = k \sum_{j=0}^{n-1} E(t_{n-1-j}) I_k f(t_j),$$
$$\text{where } I_k g(t) = \int_0^1 E(k - sk) g(t + sk)\, ds.$$

With this notation, the error $e^n = U^n - u(t_n)$ may be represented as

$$\begin{aligned} e^n &= k \sum_{j=0}^{n-1} \big(E_k^{n-1-j} Q_k f(t_j) - E(t_{n-1-j}) I_k f(t_j) \big) \\ &= k \sum_{j=0}^{n-1} (E_k^{n-1-j} - E(t_{n-1-j})) I_k f(t_j) \\ &\quad + k \sum_{j=0}^{n-1} E_k^{n-1-j} (Q_k - I_k) f(t_j) = e_1^n + e_2^n. \end{aligned} \tag{8.12}$$

Using Theorem 7.3 to bound the error operator for the homogeneous equation we have, since $E(k - sk)$ commutes with $E_k^n - E(t_n)$,

$$
(8.13)\qquad
\begin{aligned}
\|e_1^n\| &\le k \sum_{j=0}^{n-1} \int_0^1 \|(E_k^{n-1-j} - E(t_{n-1-j}))f(t_j + sk)\|\, ds \\
&\le Ck^{q+1} \sum_{j=0}^{n-1} \int_0^1 |f(t_j + sk)|_{2q}\, ds = Ck^q \int_0^{t_n} |f|_{2q}\, ds,
\end{aligned}
$$

which is bounded by the right hand side of (8.11).

In order to estimate e_2^n, we write

$$
\begin{aligned}
I_k f(t_j) &= \int_0^1 E(k - sk) f(t_j + sk)\, ds \\
&= \sum_{l=0}^{q-1} \frac{k^l}{l!} \int_0^1 E(k - sk) s^l\, ds\, f^{(l)}(t_j) + R_{q,1} f(t_j),
\end{aligned}
$$

and

$$
\begin{aligned}
Q_k f(t_j) &= \sum_{i=1}^{m} p_i(kA) f(t_j + \tau_i k) \\
&= \sum_{l=0}^{q-1} \frac{k^l}{l!} \Big(\sum_{i=1}^{m} \tau_i^l p_i(kA)\Big) f^{(l)}(t_j) + R_{q,2} f(t_j),
\end{aligned}
$$

where

$$
\begin{aligned}
R_{q,1} f(t_j) &= \int_0^1 E(k - sk) \Big(\int_{t_j}^{t_j + sk} \frac{(t_j + sk - \tau)^{q-1}}{(q-1)!} f^{(q)}(\tau)\, d\tau \Big)\, ds, \\
R_{q,2} f(t_j) &= \sum_{i=1}^{m} p_i(kA) \int_{t_j}^{t_j + \tau_i k} \frac{(t_j + \tau_i k - s)^{q-1}}{(q-1)!} f^{(q)}(s)\, ds.
\end{aligned}
$$

We conclude thus that

$$
(8.14)\qquad (Q_k - I_k) f(t_j) = \sum_{l=0}^{q-1} \frac{k^l}{l!} b_l(kA) f^{(l)}(t_j) + R_q f(t_j),
$$

where

$$
b_l(\lambda) = \sum_{i=1}^{m} \tau_i^l p_i(\lambda) - \int_0^1 s^l e^{-(1-s)\lambda}\, ds,
$$

and where $R_q f = R_{q,1} f + R_{q,2} f$ satisfies

$$
(8.15)\qquad \|R_q f(t_j)\| \le Ck^{q-1} \int_{t_j}^{t_{j+1}} \|f^{(q)}\|\, ds.
$$

By (jj′) we have $b_l(\lambda) = O(\lambda^{q-l})$, as $\lambda \to 0$, and hence $|b_l(\lambda)| \le C\lambda^{q-l}$ on $\sigma(kA)$ so that

$$\|k^l b_l(kA)v\| \le k^q \sup_{\lambda\in\sigma(kA)} |\lambda^{l-q} b_l(\lambda)|\,\|A^{q-l}v\| \le Ck^q|v|_{2q-2l}. \tag{8.16}$$

Together with (8.14) and (8.15) this shows

$$\|(Q_k - I_k)f(t_j)\| \le Ck^q \sum_{l=0}^{q-1} |f^{(l)}(t_j)|_{2q-2l} + Ck^{q-1}\int_{t_j}^{t_{j+1}} \|f^{(q)}\|\,ds,$$

so that

$$\|e_2^n\| \le Ck^q\Big(t_n \sum_{l=0}^{q-1} \sup_{s\le t_n} |f^{(l)}(s)|_{2q-2l} + \int_0^{t_n} \|f^{(q)}\|\,ds\Big).$$

The proof of the theorem is now complete. □

We observe that in the above analysis, in order to obtain optimal order convergence, $f^{(l)}(t)$ was required to belong to $\dot{H}^{2q-2l}$ for $t \ge 0$. In the case $A = -\Delta$ this means, in particular, that in addition to smoothness, f and its derivatives with respect to time are required to satisfy certain boundary conditions on $\partial\Omega$ for $t \ge 0$. This is unsatisfactory in that, except at $t = 0$, such boundary conditions are not needed to ensure existence and smoothness of the exact solution of (8.1). In an attempt to reduce these assumptions we shall first note that if the operator $E_k = r(kA)$ has the stronger smoothing property of schemes of types I′, II′ and IV (cf. Lemma 7.3), then the above regularity requirements may be considerably weakened, except in a short interval preceding the point t at which the error estimate is sought.

Theorem 8.2 *Assume that the time discretization scheme in* (8.2) *is accurate of order q and that $E_k = r(kA)$ is of type I', II' or IV. Then there is a $C > 0$ such that, if $0 < \delta \le t_n \le \bar{t}$ and $f^{(l)}(t) \in \dot{H}^{2q-2l}$ for $l < q$ and $t_n - \delta \le t \le t_n$, we have*

$$\begin{aligned}\|U^n - u(t_n)\| \le{}& Ck^q\Big(\sum_{l=0}^{q-1} \big(\|f^{(l)}(0)\| \\ &+ \sup_{t_n-\delta\le s\le t_n} |f^{(l)}(s)|_{2q-2l}\big) + \int_0^{t_n} \|f^{(q)}\|\,ds\Big).\end{aligned} \tag{8.17}$$

Proof. In order to estimate $e^n = U^n - u(t_n)$, we choose $\varphi \in C^\infty(\mathbb{R})$ such that $\varphi(t) = 1$ for $t \ge -\delta/2$, $\varphi(t) = 0$ for $t \le -\delta$, and write, with t_n the point at which we want to estimate the error,

$$f(t) = f(t)\varphi(t - t_n) + f(t)(1 - \varphi(t - t_n)) = f_1(t) + f_2(t),$$

so that $f_1(t) = 0$ for $t \le t_n - \delta$ and $f_2(t) = 0$ for $t \ge t_n - \delta/2$. The solutions of (8.2) and (8.1) are then obtained by linearity from the solutions corresponding to f_1 and f_2. By the proof of Theorem 8.1 the contribution to the error from f_1 is bounded by the right hand side of (8.11), with $f(t)$ replaced by $f(t)\varphi(t-t_n)$, which is bounded by the right hand side of (8.17).

In order to bound the contribution from f_2 it suffices then to show (8.17) in the case that f vanishes for $t \geq t_n - \delta/2$. As in the proof of Theorem 8.1, we write $U^n - u(t_n) = e^n = e_1^n + e_2^n$, with e_1^n and e_2^n defined by (8.12). Using now the nonsmooth data estimate of Theorem 7.2 in (8.13) we obtain, since $t_{n-1-j} \geq \delta/2 > 0$, for all nonvanishing terms of e_1^n, that

$$\|e_1^n\| \leq Ck \sum_{j=0}^{n-1} \int_0^1 k^q \|f(t_j + sk)\|\, ds \leq Ck^q \int_0^{t_n} \|f\|\, ds.$$

For e_2^n we have with the above notation

$$\begin{aligned} \|e_2^n\| &\leq k \sum_{j=0}^{n-1} \|E_k^{n-1-j}(Q_k - I_k) f(t_j)\| \\ &\leq Ck \sum_{j=0}^{n-1} \sum_{l=0}^{q-1} k^l \|E_k^{n-1-j} b_l(kA) f^{(l)}(t_j)\| + Ck \sum_{j=0}^{n-1} \|R_q f(t_j)\|. \end{aligned}$$

Lemma 7.3 shows that for $t_{n-1-j} \geq c\delta > 0$ and any p

$$\|E_k^{n-1-j} v\| = \|E_k^{n-1-j} A^p T^p v\| \leq C \|T^p v\|.$$

Hence, since $b_l(\lambda) = O(\lambda^{q-l})$ for small λ,

$$k^l \|E_k^{n-1-j} b_l(kA) v\| \leq Ck^q \|(kA)^{-(q-l)} b_l(kA) v\| \leq Ck^q \|v\|,$$

so that, using also the above estimate (8.15) for $R_q f(t_j)$, since $t_n \leq \bar{t}$,

$$\begin{aligned} \|e_2^n\| &\leq Ck \sum_{j=0}^{n-1} \sum_{l=0}^{q-1} k^q \|f^{(l)}(t_j)\| + Ck^q \int_0^{t_n} \|f^{(q)}\|\, ds \\ &\leq Ck^q \Big(\sum_{l=0}^{q-1} \|f^{(l)}(0)\| + \int_0^{t_n} \|f^{(q)}\| ds \Big). \end{aligned}$$

This completes the proof. □

Our next purpose is to reduce our assumptions even further on the behavior of $f^{(l)}(t)$ on $\partial\Omega$, for $t > 0$, by a more careful analysis of the error and by imposing additional conditions on the time discretization in (8.2). We shall begin with a slight reformulation of the conditions for accuracy and set

$$\gamma_l(\lambda) = \frac{l!}{(-\lambda)^{l+1}} \Big(r(\lambda) - \sum_{j=0}^{l} \frac{(-\lambda)^j}{j!} \Big) - \sum_{i=1}^{m} \tau_i^l p_i(\lambda), \quad \text{for } l = 0, \ldots, q-1,$$

$$\gamma_q(\lambda) = \frac{q!}{(-\lambda)^{q+1}} \Big(r(\lambda) - \sum_{j=0}^{q} \frac{(-\lambda)^j}{j!} \Big).$$

With this notation it follows easily from Lemma 8.1 that (8.2) is accurate of order q if and only if

$$\gamma_l(\lambda) = O(\lambda^{q-l}), \quad \text{as } \lambda \to 0, \quad \text{for } l = 0, \dots, q. \tag{8.18}$$

Note that $\gamma_q(\lambda) = O(1)$ as $\lambda \to 0$ is equivalent with (j). We shall say that the time discretization scheme (8.2) is strictly accurate of order q_0, where $q_0 \le q$, if

$$\gamma_l(\lambda) = 0, \quad \text{for } l = 0, \dots, q_0 - 1. \tag{8.19}$$

The conditions (8.5) which were used above in the construction of particular schemes of order q may then be expressed by saying that these schemes are also strictly accurate of order q. In particular, the second order Crank-Nicolson schemes (8.7) and (8.8) are both strictly accurate of order 2.

In our next result we shall show an error estimate for schemes satisfying (8.19) and in which no artificial boundary conditions are imposed for $t > 0$. This time we shall prefer to express our result in terms of the solution rather than the data, and remark that it is appropriate to assume that u and its derivatives with respect to time are in $\dot H^2 = D(A)$ but not in $\dot H^s$ for $s \ge 3$; in the application to $A = -\Delta$ this corresponds to saying that u and its derivatives in time may be assumed to vanish on $\partial\Omega$, but that further boundary conditions are unnatural.

Theorem 8.3 *Assume that the scheme* (8.2) *is both accurate and strictly accurate of order q and that $E_k = r(kA)$ is stable in $\mathcal{H}$. Then we have under the appropriate regularity assumptions, for $t_n \ge 0$,*

$$\|U^n - u(t_n)\| \le Ck^q\Big(t_n \sup_{s\le t_n} |u^{(q)}(s)|_2 + \int_0^{t_n} \|u^{(q+1)}\|\,ds\Big).$$

Proof. The error $e^n = U^n - u(t_n)$ satisfies

$$e^{n+1} = E_k e^n + \varphi^n, \quad \text{for } n \ge 0, \quad \text{with } e^0 = 0, \tag{8.20}$$

where

$$\begin{aligned}\varphi^n &= -u(t_{n+1}) + E_k u(t_n) + kQ_k f(t_n)\\ &= -u(t_{n+1}) + r(kA)u(t_n) + k\sum_{i=1}^m p_i(kA)(u' + Au)(t_n + \tau_i k).\end{aligned}$$

Taylor expansions with respect to k give

$$\begin{aligned}\varphi^n = &-\sum_{l=0}^q \frac{k^l}{l!}u^{(l)}(t_n) + k\sum_{i=1}^m p_i(kA)\sum_{l=0}^{q-1}\frac{(\tau_i k)^l}{l!}(u^{(l+1)} + Au^{(l)})(t_n)\\ &+ r(kA)u(t_n) - \int_{t_n}^{t_{n+1}} \frac{(t_{n+1}-s)^q}{q!}u^{(q+1)}(s)\,ds\\ &+ k\sum_{i=1}^m p_i(kA)\int_{t_n}^{t_n+\tau_i k}\frac{(t_n+\tau_i k - s)^{q-1}}{(q-1)!}(u^{(q+1)} + Au^{(q)})(s)\,ds,\end{aligned}$$

or

$$\varphi^n = -\sum_{l=0}^{q} \frac{k^l}{l!} h_l(kA) u^{(l)}(t_n) + R_1^n + R_2^n,$$

where we have set

$$h_l(\lambda) = 1 - l \sum_{i=1}^{m} \tau_i^{l-1} p_i(\lambda) - \lambda \sum_{i=1}^{m} \tau_i^l p_i(\lambda), \quad \text{for } 1 \le l \le q-1,$$

$$h_0(\lambda) = 1 - r(\lambda) - \lambda \sum_{i=1}^{m} p_i(\lambda), \quad h_p(\lambda) = 1 + q \sum_{i=1}^{m} \tau_i^{q-1} p_i(\lambda).$$

We have at once

$$\|R_1^n\| + \|R_2^n\| \le Ck^q \int_{t_n}^{t_{n+1}} (\|u^{(q+1)}\| + \|Au^{(q)}\|)\, ds.$$

A simple calculation shows that, with $\gamma_{-1}(\lambda) = 0$,

$$h_l(\lambda) = l\gamma_{l-1}(\lambda) + \lambda\gamma_l(\lambda), \quad \text{for } l = 0, \dots, q, \tag{8.21}$$

and since the scheme is strictly accurate of order q, we have thus that $h_l(kA) = 0$ for $l < q$. In the expression for φ^n it remains only to estimate the term with $l = q$. We have $h_q(\lambda) = -\lambda\gamma_q(\lambda)$, and hence, since $\gamma_q(kA)$ is bounded,

$$\|k^q h_q(kA) u^{(q)}(t_n)\| \le k^{q+1} \|\gamma_q(kA) A u^{(q)}\| \le Ck^{q+1} |u^{(q)}(t_n)|_2.$$

Altogether, we have thus

$$\|\varphi^n\| \le Ck^{q+1} \sup_{t_n \le s \le t_{n+1}} |u^{(q)}(s)|_2 + Ck^q \int_{t_n}^{t_{n+1}} \|u^{(q+1)}\|\, ds,$$

and hence, using the stability of E_k in (8.20),

$$\|e^n\| \le \sum_{j=0}^{n-1} \|\varphi^j\| \le Ck^q \Big(t_n \sup_{s \le t_n} |u^{(q)}(s)|_2 + \int_0^{t_n} \|u^{(q+1)}\|\, ds \Big).$$

This completes the proof of the theorem. □

If the scheme is accurate of order q we have by (8.21) and (8.18) that $h_l(\lambda) = O(\lambda^{q-l+1})$ as $\lambda \to 0$, for $l = 0, \dots, q$. However, if it is not strictly accurate of order q so that $h_l(\lambda) \ne 0$ for some $l < q$, then this will bring an additional term to the truncation error φ^n of the form

$$-\frac{k^l}{l!} h_l(kA) u^{(l)}(t_n) = -\frac{k^{l+1}}{l!} \widetilde{h}_l(kA) A u^{(l)}(t_n), \quad \text{with } \widetilde{h}_l(\lambda) = h_l(\lambda)/\lambda.$$

Since $\widetilde{h}_l(\lambda) = O(\lambda^{q-l})$ for small λ, we conclude as in (8.16) that if $u^{(l)}$ belongs to the appropriate spaces $\dot{H}^s$, then

$$\|\frac{k^l}{l!}h_l(kA)u^{(l)}(t_n)\| \le Ck^{q+1}|u^{(l)}(t_n)|_{2q+2-2l}.$$

After summation the contribution to the total error will still be of the correct order $O(k^q)$ but, as in Theorem 8.1, undesirable boundary conditions will have been imposed. If these are not satisfied, a reduction of the order of convergence has to be expected.

In our next result we shall see, however, that if the scheme is strictly accurate of order $q-1$, and an additional condition is satisfied, then an optimal order error estimate holds without any assumption of artificial boundary conditions.

Theorem 8.4 *Assume that the scheme* (8.2) *is accurate of order q and strictly accurate of order $q-1$, that $E_k = r(kA)$ is stable in $\mathcal{H}$, and that $\kappa(\lambda) = h_{q-1}(\lambda)/(\lambda(1-r(\lambda)))$ is bounded on $\sigma(kA)$, uniformly in k. Then, under the appropriate regularity assumptions,*

$$\begin{aligned}\|U^n - u(t_n)\| \le & Ck^q\Big(\sup_{s\le t_n}|u^{(q-1)}(s)|_2 \\ & + t_n \sup_{s\le t_n}|u^{(q)}(s)|_2 + \int_0^{t_n}\|u^{(q+1)}\|\,ds\Big), \quad \textit{for } t_n\ge 0.\end{aligned}$$

Proof. It follows from the above that the contribution to the global error of the additional term is, with $\widetilde{h}_{q-1}(\lambda) = h_{q-1}(\lambda)/\lambda$,

$$S_n = -\sum_{j=0}^{n-1} E_k^{n-1-j}\frac{k^q}{(q-1)!}\widetilde{h}_{q-1}(kA)Au^{(q-1)}(t_j).$$

By the definition of $\kappa(\lambda)$ we have $\widetilde{h}_{q-1}(kA) = \kappa(kA)(I-E_k)$, and hence

$$\begin{aligned}-(q-1)!S_n &= k^q\kappa(kA)\sum_{j=0}^{n-1}E_k^{n-1-j}(I-E_k)Au^{(q-1)}(t_j) \\ &= k^q\kappa(kA)\Big(Au^{(q-1)}(t_{n-1}) - \sum_{j=1}^{n-1}E_k^{n-j}\int_{t_{j-1}}^{t_j}Au^{(q)}\,ds - E_k^nAu^{(q-1)}(0)\Big).\end{aligned}$$

We conclude

$$\|S_n\| \le Ck^q\Big(\sup_{s\le t_n}|u^{(q-1)}(s)|_2 + t_n\sup_{s\le t_n}|u^{(q)}(s)|_2\Big),$$

which, together with the estimate of Theorem 8.3, shows our claim. □

It is clear that using the technique of the proof of Theorem 8.2 above, the regularity assumptions imposed in the latter two theorems may be further reduced for $t \le t_n - \delta$, with $\delta > 0$, provided E_k has the appropriate smoothing properties. We shall not insist on the details.

We shall now return to the discussion of the accuracy conditions for the time discretization. Recall from Lemma 8.1 and the subsequent discussion that (8.2) is accurate of order q if and only if (j) holds together with

$$\gamma_l(\lambda) = \frac{l!}{(-\lambda)^{l+1}}\Big(r(\lambda) - \sum_{j=0}^{l} \frac{(-\lambda)^j}{j!}\Big) - \sum_{i=1}^{m} \tau_i^l p_i(\lambda) = O(\lambda^{q-l}), \tag{8.22}$$
$$\text{as } \lambda \to 0, \quad \text{for } l = 0, \dots, q-1.$$

For the case that the number m of quadrature points is less than q we shall give an alternative characterization of a scheme of order q which may be used to construct such schemes.

Lemma 8.2 *Let $m < q$. Then the time discretization scheme in* (8.2) *is accurate of order q if and only if* (j) *holds together with*

(jj‴) $\qquad \gamma_l(\lambda) = O(\lambda^{q-l}), \quad$ *as* $\lambda \to 0, \quad$ *for* $l = 0, \dots, m-1,$

and, with $\omega(\tau) = \prod_{i=1}^m (\tau - \tau_i)$,

(jjj) $\qquad \displaystyle\int_0^1 \omega(\tau)\tau^j \, d\tau = 0, \quad$ *for* $j = 0, \dots, q-m-1.$

Proof. We first note that (jjj) is equivalent to the existence of $b_1, \dots, b_m$ such that

(jjj′) $\qquad \displaystyle\int_0^1 \varphi(\tau)\, d\tau = \sum_{i=1}^m b_i \varphi(\tau_i), \quad \forall \varphi \in \Pi_{q-1}.$

In fact, it follows by (jjj) that the integrand in (jjj′) may be replaced by its Lagrange interpolation polynomial, which shows (jjj′). The converse is trivial.

To show the necessity of (jjj), it thus suffices to show (jjj′) for $\varphi = \tau^l$, $l = 0, \dots, q-1$. But using the definition of $\gamma_l(\lambda)$, we have by (j) and (jj″) that

$$\gamma_l(0) = \frac{1}{l+1} - \sum_{i=1}^m \tau_i^l p_i(0) = 0, \quad \text{for } l = 0, \dots, q-1,$$

so that with $b_i = p_i(0)$,

$$\int_0^1 \tau^l \, d\tau = \frac{1}{l+1} = \sum_{i=1}^m b_i \tau_i^l, \quad \text{for } l = 0, \dots, q-1.$$

We now turn to the sufficiency of the conditions, and it suffices then to show that (j), (jj‴) and (jjj) imply

$$\gamma_l(\lambda) = O(\lambda^{q-l}), \quad \text{as } \lambda \to 0, \quad \text{for } l = m, \dots, q-1. \tag{8.23}$$

We have by integration by parts and by (j),

$$\frac{(-\lambda)^{l+1}}{l!} \int_0^1 e^{-\lambda(1-\tau)} \tau^l \, d\tau = r(\lambda) - \sum_{j=0}^{l} \frac{(-\lambda)^j}{j!} + O(\lambda^{q+1}), \quad \text{as } \lambda \to 0,$$

and hence

$$\gamma_l(\lambda) = \int_0^1 e^{-\lambda(1-\tau)} \tau^l d\tau - \sum_{i=1}^{m} \tau_i^l p_i(\lambda) + O(\lambda^{q-l}), \quad \text{as } \lambda \to 0.$$

For $\omega(\tau)$ as above we write $\omega(\tau) = \sum_{j=0}^m \alpha_j \tau^j$. Then, since $\omega(\tau_i) = 0$, we obtain by expanding the integral and using (jjj), for $l = 0, \dots, q-m-1$, as $\lambda \to 0$,

$$\sum_{j=0}^{m} \alpha_j \gamma_{j+l}(\lambda) = \int_0^1 e^{-\lambda(1-\tau)} \tau^l \omega(\tau) d\tau + O(\lambda^{q-m-l}) = O(\lambda^{q-m-l}).$$

Since $\alpha_m = 1$, we may conclude the proof of (8.23) by successively setting $l = 0, 1, \dots, q-m-1$ in this formula, and using (jj‴). □

Applying the lemma we may now construct, for any given q and m with $q/2 \le m \le q$, a scheme which is accurate of order q and strictly accurate of order m: We start with a $r(\lambda)$ such that (j) holds (and with the desired stability properties), then select the distinct numbers $\{\tau_i\}_{i=1}^m \subset [0,1]$ so that (jjj) is satisfied, and finally determine the rational functions $\{p_i(\lambda)\}_{i=1}^m$ from

$$\sum_{i=1}^{m} \tau_i^l p_i(\lambda) = \frac{l!}{(-\lambda)^{l+1}} \Big(r(\lambda) - \sum_{j=0}^{l} \frac{(-\lambda)^j}{j!} \Big), \quad l = 0, \dots, m-1. \tag{8.24}$$

Note that the matrix of this system is nonsingular since the τ_i are distinct, and that the $p_i(\lambda)$ will have the same denominators as $r(\lambda)$. Note also that the condition $q \le 2m$ is necessary for the existence of $\{\tau_i\}_{i=1}^m$ so that (jjj) holds; for $q = 2m$ the points are uniquely determined as the Gaussian points of order m in $[0,1]$.

For example, let $r(\lambda)$ denote the fourth order diagonal Padé approximant of $e^{-\lambda}$,

$$r(\lambda) = \frac{1 - \frac{1}{2}\lambda + \frac{1}{12}\lambda^2}{1 + \frac{1}{2}\lambda + \frac{1}{12}\lambda^2} = e^{-\lambda} + O(\lambda^5), \quad \text{as } \lambda \to 0,$$

so that $q = 4$. Choose now $m = 2$ and $\tau_{1,2} = \frac{1}{2} \mp \frac{\sqrt{3}}{6}$, the Gaussian points of order 2. The system (8.24) then reduces to

$$p_1(\lambda) + p_2(\lambda) = -\tfrac{1}{\lambda}(r(\lambda) - 1),$$
$$(\tfrac{1}{2} - \tfrac{\sqrt{3}}{6})p_1(\lambda) + (\tfrac{1}{2} + \tfrac{\sqrt{3}}{6})p_2(\lambda) = \tfrac{1}{\lambda^2}(r(\lambda) - 1 + \lambda),$$

which results in the scheme

$$(I + \tfrac{1}{2}kA + \tfrac{1}{12}k^2A^2)U^{n+1} = (I - \tfrac{1}{2}kA + \tfrac{1}{12}k^2A^2)U^n$$
$$+ \tfrac{1}{2}k\Big((I - \tfrac{\sqrt{3}}{6}kA)f(t_n + (\tfrac{1}{2} - \tfrac{\sqrt{3}}{6})k) + (I + \tfrac{\sqrt{3}}{6}kA)f(t_n + (\tfrac{1}{2} + \tfrac{\sqrt{3}}{6})k)\Big).$$

We have here

$$\gamma_2(\lambda) = -\frac{2}{\lambda^3}\Big(\frac{1 - \frac{1}{2}\lambda + \frac{1}{12}\lambda^2}{1 + \frac{1}{2}\lambda + \frac{1}{12}\lambda^2} - 1 + \lambda - \tfrac{1}{2}\lambda^2\Big)$$
$$- \tfrac{1}{2}(\tfrac{1}{2} - \tfrac{\sqrt{3}}{6})^2 \frac{1 - \frac{\sqrt{3}}{6}\lambda}{1 + \frac{1}{2}\lambda + \frac{1}{12}\lambda^2} - \tfrac{1}{2}(\tfrac{1}{2} + \tfrac{\sqrt{3}}{6})^2 \frac{1 + \frac{\sqrt{3}}{6}\lambda}{1 + \frac{1}{2}\lambda + \frac{1}{12}\lambda^2} = 0,$$

so that the scheme is actually strictly accurate of order 3. Since

$$\kappa(\lambda) = \frac{h_3(\lambda)}{\lambda(1 - r(\lambda))} = \frac{h_3(\lambda)}{\lambda^2}(1 + \tfrac{1}{2}\lambda + \tfrac{1}{12}\lambda^2)$$
$$= \frac{\gamma_3(\lambda)}{\lambda}(1 + \tfrac{1}{2}\lambda + \tfrac{1}{12}\lambda^2) = O(1), \quad \text{as } \lambda \to 0,$$

this function is bounded for $\lambda \geq 0$, and thus Theorem 8.4 applies.

With the same $r(\lambda)$, we may prefer to choose instead the three quadrature points $\tau_1 = 0$, $\tau_2 = \frac{1}{2}$, $\tau_3 = 1$. We then have

$$\int_0^1 \omega(\tau)\, d\tau = \int_0^1 \tau(\tau - \tfrac{1}{2})(\tau - 1)\, d\tau = 0,$$

so that (jjj) holds since $q - m - 1 = 0$. We now solve the system

$$p_1(\lambda) + p_2(\lambda) + p_3(\lambda) = -\frac{1}{\lambda}(r(\lambda) - 1) = \frac{1}{1 + \frac{1}{2}\lambda + \frac{1}{12}\lambda^2},$$
$$\tfrac{1}{2}p_2(\lambda) + p_3(\lambda) = \frac{1}{\lambda^2}(r(\lambda) - 1 + \lambda) = \frac{\frac{1}{2} + \frac{1}{12}\lambda}{1 + \frac{1}{2}\lambda + \frac{1}{12}\lambda^2},$$
$$\tfrac{1}{4}p_2(\lambda) + p_3(\lambda) = -\frac{2}{\lambda^3}(r(\lambda) - 1 + \lambda - \tfrac{1}{2}\lambda^2) = \frac{\frac{1}{3} + \frac{1}{12}\lambda}{1 + \frac{1}{2}\lambda + \frac{1}{12}\lambda^2},$$

to obtain the scheme

$$(I + \tfrac{1}{2}kA + \tfrac{1}{12}k^2A^2)U^{n+1} = (I - \tfrac{1}{2}kA + \tfrac{1}{12}k^2A^2)U^n$$
$$+ k\big((\tfrac{1}{6} - \tfrac{1}{12}kA)f(t_n) + \tfrac{2}{3}f(t_n + \tfrac{1}{2}k) + (\tfrac{1}{6} + \tfrac{1}{12}kA)f(t_{n+1})\big),$$

which is then strictly accurate of order at least 3. Since a simple calculation shows that

$$\gamma_3(\lambda) = \frac{6}{\lambda^4}(r(\lambda) - 1 + \lambda - \tfrac{1}{2}\lambda^2 + \tfrac{1}{6}\lambda^3) - \tfrac{1}{8}p_2(\lambda) - p_3(\lambda) = 0,$$

the scheme is, in fact, strictly accurate of order 4, and Theorem 8.3 applies.

We shall now apply our results to the analysis if fully discrete approximations of the model parabolic partial differential equation, and consider thus, with Ω a bounded domain in $\mathbb{R}^d$ with smooth boundary, the problem

$$\begin{aligned} u_t - \Delta u &= f \quad \text{in } \Omega, \quad t > 0, \\ u &= 0 \quad \text{on } \partial\Omega,\ t > 0, \quad \text{with } u(\cdot, 0) = 0 \quad \text{in } \Omega. \end{aligned} \tag{8.25}$$

Assuming as earlier that we are given a pair of families $\{S_h\}$ and $\{T_h\}$, satisfying the properties (i) and (ii) of Chapter 2, and setting $A_h = -\Delta_h = T_h^{-1}$ on S_h, the fully discrete schemes will be obtained by applying our time stepping procedures analyzed above to the semidiscrete analogue of (8.25), i.e.,

$$u_{h,t} + A_h u_h = f_h := P_h f, \quad \text{for } t > 0, \quad \text{with } u_h(0) = 0, \tag{8.26}$$

where P_h is the L_2–projection onto S_h. Our fully discrete analogue is thus obtained by replacing A and f by $-\Delta_h$ and $P_h f$ in (8.2) so that

$$U_h^{n+1} = E_{kh} U_h^n + k(Q_{kh} P_h f)(t_n), \quad \text{for } n \geq 0, \quad \text{with } U^0 = 0, \tag{8.27}$$

where

$$E_{kh} v = r(kA_h)v \quad \text{and} \quad Q_{kh} f(t) = \sum_{i=1}^{m} p_i(kA_h) P_h f(t + \tau_i k).$$

Our purpose now is thus to derive error estimates for (8.27) in L_2 which extend to the present case those obtained above for the abstract problem (8.2). We begin with a fully discrete version of Theorem 8.1. The spaces $\dot{H}^s = \dot{H}^s(\Omega)$ are defined as earlier using $A = -\Delta$; we note that s may be negative.

Theorem 8.5 *Assume that the time discretization scheme* (8.2) *is accurate of order q and that E_{kh} is of type I' or II. Let U_h^n and u be the solutions of* (8.27) *and* (8.25), *respectively. Then, if $f^{(l)}(t) \in \dot{H}^{\max(r, 2q) - 2l}(\Omega)$ for $l < q$, when $t \geq 0$, we have, for $t_n \geq 0$,*

$$\begin{aligned} \|U_h^n - u(t_n)\| &\leq C h^r t_n \sum_{l=0}^{q-1} \sup_{s \leq t_n} |f^{(l)}(s)|_{r-2l} \\ &\quad + C k^q \Big(t_n \sum_{l=0}^{q-1} \sup_{s \leq t_n} |f^{(l)}(s)|_{2q-2l} + \int_0^{t_n} \|f^{(q)}\|\, ds \Big). \end{aligned} \tag{8.28}$$

Proof. We first bound the error in the spatially semidiscrete solution. Writing $F_h(t) = E_h(t)P_h - E(t)$ as in Chapter 3, we have by Theorem 3.1,

$$\|u_h(t_n) - u(t_n)\| \le \int_0^{t_n} \|F_h(t-s)f(s)\|\,ds \le Ch^r \int_0^{t_n} |f(s)|_r\,ds,$$

which is bounded by the right hand side of (8.28). In order to bound the remaining part or the error, $U^n - u_h(t_n)$, we proceed as in the proof of Theorem 8.1, with A replaced by $A_h = -\Delta_h$, adding subscripts h to indicate the dependence on h. In particular, the analogue of (8.12) holds for $U^n - u_h(t_n) = e_h^n = e_{1,h}^n + e_{2,h}^n$ and $f_h = P_h f$. Applying Theorem 7.8 to the terms in $e_{1,h}^n$ we obtain then

$$\|e_{1,h}^n\| \le C\Big(h^r \int_0^{t_n} |f|_r\,ds + k^q \int_0^{t_n} |f|_{2q}\,ds\Big),$$

which is bounded as claimed. For $e_{2,h}^n$, we note that, by (8.14),

$$(8.29)\qquad (Q_{kh} - I_{kh})f_h(t_j) = \sum_{l=0}^{q-1} \frac{k^l}{l!} b_l(kA_h) f_h^{(l)}(t_j) + R_q f_h(t_j),$$

where $R_q f_h(t_j)$ is bounded as in (8.15). For the purpose of dealing with the term involving $b_l(kA_h)$, we shall show that since $b_l(\lambda) = O(\lambda^{q-l})$ as $\lambda \to 0$, we have for $v \in \dot H^{\max(r,2q)-2l}(\Omega)$

$$(8.30)\qquad \|k^l b_l(kA_h)P_h v\| \le Ch^r|v|_{r-2l} + Ck^q|v|_{2q-2l}.$$

Assuming this we have

$$\|k^l b_l(kA_h) f_h^{(l)}(t_j)\| \le Ch^r|f^{(l)}(t_j)|_{r-2l} + Ck^q|f^{(l)}(t_j)|_{2q-2l}.$$

Hence by (8.29)

$$\begin{aligned}\|(Q_{kh} - I_{kh})P_h f(t_j)\| \le{}& Ch^r \sum_{l=0}^{q-1} |f^{(l)}(t_j)|_{r-2l} \\ &+ Ck^q \sum_{l=0}^{q-1} |f^{(l)}(t_j)|_{2q-2l} + Ck^{q-1}\int_{t_j}^{t_{j+1}} \|f^{(q)}\|\,ds,\end{aligned}$$

so that

$$\begin{aligned}\|e_2^n\| \le{}& Ck\sum_{j=0}^{n-1} \|(Q_{kh} - I_{kh})f_h(t_j)\| \le Ch^r t_n \sum_{l=0}^{q-1} \sup_{s\le t_n} |f^{(l)}(s)|_{r-2l} \\ &+ Ck^q\Big(t_n \sum_{l=0}^{q-1} \sup_{s\le t_n} |f^{(l)}(s)|_{2q-2l} + \int_0^{t_n} \|f^{(q)}\|\,ds\Big),\end{aligned}$$

which completes the proof.

It remains to show (8.30). Let $\{\varphi_j\}_{j=1}^{\infty}$ and $\{\lambda_j\}_{j=1}^{\infty}$ be the eigenfunctions and eigenvalues of $A = -\Delta$ and set $v_k = \sum_{k\lambda_j \le 1}(v, \varphi_j)\varphi_j$, so that (cf. the proof of Theorem 7.8)

$$\|v - v_k\| \le Ck^{q-l}|v|_{2q-2l}, \quad |v_k|_{2q-2l} \le C|v|_{2q-2l},$$
$$|v_k|_{r+2p} \le Ck^{-p-l}|v|_{r-2l}, \quad \text{for } 0 \le p \le q-l-1.$$

Recalling the identity (7.33), we have

$$v_k = \sum_{p=0}^{q-l-1} T_h^p(T - T_h)A^{p+1}v_k + T_h^{q-l}A^{q-l}v_k.$$

Setting now $B_{l,kh} = k^l b_l(kA_h)P_h : L_2(\Omega) \to S_h$, we have

$$\begin{aligned}\|B_{l,kh}T_h^p v\| &= k^l \|b_l(kA_h)A_h^{-p}P_h v\| \\ &\le k^{p+l} \sup_{\lambda \in \sigma(kA_h)} |\lambda^{-p}b_l(\lambda)|\, \|v\| \le Ck^{p+l}\|v\|, \quad \text{for } 0 \le p \le q-l.\end{aligned}$$

Hence, in particular,

$$\|B_{l,kh}T_h^{q-l}A^{q-l}v_k\| \le Ck^q\|A^{q-l}v_k\| \le Ck^q|v|_{2q-2l},$$

and, for $0 \le p \le q-l-1$,

$$\begin{aligned}\|B_{l,kh}T_h^p(T - T_h)A^{p+1}v_k\| &\le Ck^{p+l}\|(T - T_h)A^{p+1}v_k\| \\ &\le Ch^r k^{p+l}|v_k|_{r+2p} \le Ch^r|v|_{r-2l}.\end{aligned}$$

Finally,

$$\|B_{l,kh}(v - v_k)\| \le Ck^l\|v - v_k\| \le Ck^q|v|_{2q-2l}.$$

Together these estimates show (8.30). □

We proceed with a fully discrete variant of Theorem 8.3.

Theorem 8.6 *Assume that the scheme (8.27) is both accurate and strictly accurate of order q and that $|r(\lambda)| \le 1$ on $\sigma(kA_h)$ so that $E_{kh} = r(kA_h)$ is stable in L_2. Then, for the solutions of (8.27) and (8.25), we have, under the appropriate regularity assumptions, for $t_n \ge 0$,*

$$\begin{aligned}\|U_h^n - u(t_n)\| \le & Ch^r t_n \sup_{s \le t_n} \|u_t(s)\|_r \\ & + Ck^q\Big(t_n \sup_{s \le t_n} |u^{(q)}(s)|_2 + \int_0^{t_n} |u^{(q+1)}|_2\, ds\Big).\end{aligned}$$

Proof. With the standard decomposition of the error we have for $\rho^n = R_h u(t_n) - u(t_n)$, since $u(0) = 0$,

$$\|\rho^n\| \le Ch^r \|u(t_n)\|_r \le Ch^r t_n \sup_{s \le t_n} \|u_t(s)\|_r,$$

and it remains to consider $\theta^n = U^n - R_h u(t_n)$. We note that $w_h = R_h u$ satisfies the semidiscrete equation

$$w_{h,t} + A_h w_h = R_h u_t + P_h A u = P_h(f + \rho_t) =: g_h,$$

and introduce the solution of the corresponding fully discrete scheme

$$W^{n+1} = E_{kh} W^n + k(Q_{kh} g_h)(t_n), \quad \text{for } n \ge 0, \quad \text{with } W^0 = 0.$$

To estimate $W^n - w_h(t_n)$ we may now use Theorem 8.3 to obtain

$$\|W^n - w_h(t_n)\| \le Ck^q \Big(t_n \sup_{s \le t_n} \|A_h w_h^{(q)}(s)\| + \int_0^{t_n} \|w_h^{(q+1)}\| \, ds \Big).$$

Since $A_h w_h = A_h R_h u = P_h A u$, and $R_h = T_h A$ is bounded from $\dot{H}^2(\Omega)$ to L_2, this is bounded as desired.

It remains to consider $Z^n = U^n - W^n$, which satisfies

$$Z^{n+1} = E_{kh} Z^n + k(Q_{kh} P_h \rho_t)(t_n), \quad \text{for } n \ge 0, \quad \text{with } Z^0 = 0.$$

Using the stability of E_{kh} and the boundedness of Q_{kh}, we obtain

$$\|Z^n\| \le k \sum_{j=0}^{n-1} \|(Q_{kh} P_h \rho_t)(t_j)\| \le Ch^r t_n \sup_{s \le t_n} \|u_t(s)\|_r.$$

Together, our estimates show the theorem. □

We close with a fully discrete version of Theorem 8.4.

Theorem 8.7 *Assume that the scheme (8.27) is accurate of order q and strictly accurate or order $q-1$, that $|r(\lambda)| \le 1$ on $\sigma(kA_h)$ so that $E_{kh} = r(kA_h)$ is stable in L_2, and that, in addition, $\sigma(\lambda) = h_{q-1}(\lambda)/(\lambda(1 - r(\lambda)))$ is bounded on $\sigma(kA_h)$, uniformly in k and h. Then, for the solutions of (8.27) and (8.25), we have, under the appropriate regularity assumptions, for $t_n \ge 0$,*

$$\|U_h^n - u(t_n)\| \le Ch^r t_n \sup_{s \le t_n} \|u_t(s)\|_r$$
$$+ Ck^q \Big(|u^{(q-1)}(0)|_2 + t_n \sup_{s \le t_n} |u^{(q)}(0)|_2 + \int_0^{t_n} |u^{(q+1)}|_2 \, ds \Big).$$

Proof. With the notation of the proof of Theorem 8.6, we now use Theorem 8.4 instead of Theorem 8.3 to bound $W^n - w_h(t_n)$, which produces the additional term

$$Ck^q \|A_h w_h^{(q-1)}(0)\| = Ck^q \|P_h A u^{(q-1)}(0)\| \le Ck^q |u^{(q-1)}(0)|_2.$$ □

A large portion of this chapter is adapted from Brenner, Crouzeix and Thomée [40]. For work on Runge-Kutta type methods, see also Crouzeix [53], Lubich and Ostermann [161], [162] and Ostermann and Roche [189]; in the latter references particular attention is paid to the order of convergence in cases that the order of strict accuracy is lower than the order of accuracy, and it is shown that fractional order of convergence can then occur.

Error estimates that are optimal in $L_2(H_0^1) \cap H^{1/2}(L_2)$ space-time norms have been obtained for some simple time stepping methods by Baiocchi and F. Brezzi [15] in the case of vanishing initial data v and by Tomarelli [235] for nonvanishing v and the backward Euler method.

9. Single Step Methods and Rational Approximations of Semigroups

In this chapter we shall again study single step time stepping methods for a homogeneous parabolic equation in an abstract setting. This time we will use the semigroup approach and represent the time stepping operator as a Dunford-Taylor integral in the complex plane, which will allow us to treat more general elliptic operators than in the previous chapter. For the purpose of including also application to maximum-norm estimates with respect to a spatial variable, which will be given at the end of the chapter, the analysis will take place in a Banach space framework.

We consider thus, as earlier in Chapter 6, an initial value problem of the form

$$u' + Au = 0, \quad \text{for } t > 0, \quad \text{with } u(0) = v, \tag{9.1}$$

in a complex Banach space $\mathcal{B}$ with norm $\|\cdot\|$. We assume again that A is a closed, densely defined linear operator, such that, for its resolvent set $\rho(A)$,

$$\rho(A) \supset \Sigma_\delta = \{z \in \mathbb{C}; \delta \le |\arg z| \le \pi, z \ne 0\} \cup \{0\}, \quad \text{with } \delta \in (0, \tfrac{1}{2}\pi), \tag{9.2}$$

and such that its resolvent, $R(z;A) = (zI - A)^{-1}$, satisfies, in operator norm,

$$\|R(z;A)\| \le M|z|^{-1}, \quad \text{for } z \in \Sigma_\delta, \quad \text{with } M \ge 1. \tag{9.3}$$

We recall that $-A$ is then the infinitesimal generator of an analytic semigroup

$$E(t) = e^{-tA} = \frac{1}{2\pi i}\int_\Gamma e^{-zt} R(z;A)\,dz, \quad \text{for } t \ge 0, \tag{9.4}$$

which is the solution operator of (9.1), and where, e.g., $\Gamma = \{z; |\arg z| = \psi \in (\delta, \frac{1}{2}\pi)\}$, with $\operatorname{Im} z$ decreasing along Γ. It also has the stability and smoothing property

$$\|E(t)\| + t\|E'(t)\| \le K, \quad \text{for } t > 0,$$

and, conversely, these properties imply the resolvent estimate (9.3), cf. Theorem 6.4.

As in Chapter 7 we shall now discuss discretization in time of the initial value problem (9.1). Letting k denote the time step and $t_n = nk$, and letting $r(z)$ be a rational function defined on the spectrum $\sigma(kA)$ of kA, we define the approximation U^n of $u(t_n) = E(t_n)v$ by the recursion formula

$$U^{n+1} = E_k U^n, \quad \text{for } n \geq 0, \quad \text{where } E_k = r(kA), \quad \text{with } U^0 = v.$$

We may thus write $U^n = E_k^n v$.

We shall begin by discussing the stability of the operators E_k^n. We shall use the Dunford-Taylor spectral representation of a rational function of the operator A when this rational function is bounded in a sector in the right half-plane, as described in the following lemma.

Lemma 9.1 *Assume that* (9.2) *and* (9.3) *hold and let* $r(z)$ *be a rational function which is bounded for* $|\arg z| \leq \psi, |z| \geq \varepsilon > 0$, *where* $\psi \in (\delta, \frac{1}{2}\pi)$, *and for* $|z| \geq R$. *Then, if* $\varepsilon > 0$ *is so small that* $\{z; |z| \leq \varepsilon\} \subset \rho(A)$, *we have*

$$r(A) = r(\infty)I + \frac{1}{2\pi i} \int_{\gamma_\varepsilon \cup \Gamma_\varepsilon^R \cup \gamma^R} r(z) R(z; A)\, dz,$$

where $\Gamma_\varepsilon^R = \{z; |\arg z| = \psi, \varepsilon \leq |z| \leq R\}, \gamma_\varepsilon = \{z; |z| = \varepsilon, \ |\arg z| \leq \psi\}$, *and* $\gamma^R = \{z; |z| = R, \psi \leq |\arg z| \leq \pi\}$, *and with the closed path of integration oriented in the negative sense.*

Proof. See, e.g., [82], Theorem VII.9.4. In fact, this representation holds with $r(z)$ replaced by any function $f(z)$ which is analytic in a neighborhood of $\{z; |\arg z| \leq \psi, \ |z| \geq \varepsilon\}$, including at $z = \infty$. □

With $0 \leq \theta \leq \pi/2$ we say that the rational function $r(z)$ is $A(\theta)$-stable if

$$|r(z)| \leq 1, \quad \text{for} \quad |\arg z| \leq \theta. \tag{9.5}$$

In particular, if this holds with $\theta = \pi/2$, we say that $r(z)$ is A-stable. If $\theta = 0$, (9.5) reduces to $|r(\lambda)| \leq 1$ for $\lambda \geq 0$ and implies the stability condition (7.10) when A is a positive definite selfadjoint operator in a Hilbert space.

As earlier we say that $r(z)$ approximates e^{-z} to order $q \geq 1$ if

$$r(z) = e^{-z} + O(z^{q+1}), \quad \text{as } z \to 0; \tag{9.6}$$

if this is true for some $q \geq 1$ we say that $r(z)$ is consistent with e^{-z}.

We recall from Chapter 7 that the Padé approximants $r_{\mu\nu}(z)$ defined in (7.18) are accurate of order $\mu + \nu$. It is known, see, e.g., Hairer and Wanner [113], that $r_{\mu\nu}(z)$ is A-stable if and only if $0 \leq \mu - \nu \leq 2$, and further that the rational function associated with the Calahan scheme defined by (7.21) is A-stable. Other examples of $A(\theta)$-stable methods, with $\theta < \pi/2$, may be found in [113]. We note for later reference that $r_{\mu\nu}(\infty) = 0$ if $\mu > \nu$, and $|r_{\mu\mu}(\infty)| = 1$.

We now derive some useful bounds for $A(\theta)-$stable rational functions.

Lemma 9.2 *Assume that $r(z)$ is $A(\theta)$-stable with $\theta > 0$ and consistent with e^{-z}. Then for arbitrary $R > 0$ and $\psi \in (0, \theta)$ there are $c, C > 0$ and $\varepsilon \in (0, 1)$ such that*

$$(9.7) \qquad |r(z)| \le \begin{cases} e^{C|z|}, & \textit{for } |z| \le \varepsilon, \\ e^{-c|z|}, & \textit{for } |z| \le R, \ |\arg z| \le \psi. \end{cases}$$

Further, if $|r(\infty)| = 1$, there are $m, c, C > 0$, and $\omega \ge 1$, such that

$$(9.8) \qquad |r(z)| \le \begin{cases} e^{C|z|^{-m}}, & \textit{for } |z| \ge \omega, \\ e^{-c|z|^{-m}}, & \textit{for } |z| \ge R, \ |\arg z| \le \psi. \end{cases}$$

Proof. The first estimate follows at once since $r(z) = 1 + O(z)$ as $z \to 0$ by (9.6). This assumption also shows

$$|r(z)| \le e^{-\operatorname{Re} z} + C|z|^2 \le e^{-c|z|}, \quad \text{for } |z| \le \varepsilon, \ |\arg z| \le \psi,$$

since $\operatorname{Re} z \ge \cos\psi\,|z|$. By (9.5) and the maximum-principle we have $|r(z)| < 1$ for $|\arg z| < \theta, z \ne 0$. In particular, $|r(z)| < 1$ on the compact set $\{z;\ \varepsilon \le |z| \le R, |\arg z| \le \psi\}$, which implies the second estimate in (9.7) for c suitably chosen. If $|r(\infty)| = 1$ we may write, with $w = 1/z$,

$$(9.9) \qquad r(z) = a e^{bw^m + O(w^{m+1})}, \quad \text{with } |a| = 1, \ b \ne 0, \quad \text{as } w \to 0,$$

which immediately shows the first estimate of (9.8). By $A(\theta)$-stability we have $\operatorname{Re}(bw^m) \le 0$ for $|\arg w| \le \theta$ and hence $\operatorname{Re}(bw^m) \le -c|w|^m$ for $|\arg w| \le \psi$, which implies the second bound in (9.8). □

Note that if $r(z)$ is A-stable we must have $b < 0$ and $m = 1$ in (9.9). As an example, for the Crank-Nicolson method we have, with $w = 1/z$,

$$r(z) = \frac{1 - \frac{1}{2}z}{1 + \frac{1}{2}z} = -\frac{1 - 2w}{1 + 2w} = -r(4w) = -e^{-4w + O(w^2)}, \quad \text{as } w \to 0.$$

We are now ready to state the following stability result.

Theorem 9.1 *Let $r(z)$ be consistent with e^{-z} and $A(\theta)$-stable for some $\theta \in [\delta, \frac{1}{2}\pi]$. Assume that A satisfies (9.2) and (9.3). Then there is a $C = C_\delta$ such that*

$$\|E_k^n v\| \le CM\|v\|, \quad \textit{for } t_n \ge 0, \ v \in \mathcal{B}, \quad \textit{where } E_k = r(kA).$$

Proof. We shall show that, in operator norm,

$$(9.10) \qquad \|r(A)^n\| \le CM, \quad \text{for } n \ge 0.$$

We note that with A also kA satisfies (9.2) as well as (9.3) since, for $z \in \Sigma_\delta$,

$$\|R(z;kA)\| = \|k^{-1}(zk^{-1}I - A)^{-1}\| \le k^{-1}M|zk^{-1}|^{-1} = M|z|^{-1}.$$

Hence (9.10) applied to kA yields the desired bound $\|E_k^n\| = \|r(kA)^n\| \le CM$ for $n \ge 0$. We remark that if the bound in (9.3) had been replaced by $M_1(1+|z|)^{-1}$ this argument would have failed.

To show (9.10) we use Lemma 9.1 with $\psi \in (\delta, \theta)$. Since $r(z)$ is analytic at $z = 0$ we may replace the circular arc γ_ε by the complementary arc $\gamma^\varepsilon = \{z; |z| = \varepsilon; \psi \le |\arg z| \le \pi\} \subset \Sigma_\delta$. Using ε/n instead of ε we may write

$$r(A)^n = \kappa^n I + \frac{1}{2\pi i}\int_{\gamma^{\varepsilon/n}\cup\Gamma^R_{\varepsilon/n}\cup\gamma^R} r(z)^n R(z;A)\,dz, \quad \text{where } \kappa = r(\infty).$$

Clearly $\|\kappa^n I\| \le 1 \le M$. To bound the integrals over the three components of the path of integration, we first assume that $|\kappa| < 1$. We may then fix $R \ge 1$ large enough so that $|r(z)| \le 1$ for $|z| \ge R$, and hence

$$\Big\|\frac{1}{2\pi i}\int_{\gamma^R} r(z)^n R(z;A)\,dz\Big\| \le \frac{M}{2\pi}\int_{\gamma^R}\frac{|dz|}{|z|} \le M.$$

Further, with ε fixed as in Lemma 9.2,

$$\Big\|\frac{1}{2\pi i}\int_{\gamma^{\varepsilon/n}} r(z)^n R(z;A)\,dz\Big\| \le \frac{M}{2\pi}\int_{\gamma^{\varepsilon/n}} e^{Cn|z|}\frac{|dz|}{|z|} \le CM$$

and

$$\Big\|\frac{1}{2\pi i}\int_{\Gamma^R_{\varepsilon/n}} r(z)^n R(z;A)\,dz\Big\| \le \frac{M}{\pi}\int_{\varepsilon/n}^{\infty} e^{-cn\rho}\frac{d\rho}{\rho} \le CM, \tag{9.11}$$

which completes the proof in this case.

In the case $|\kappa| = 1$ we choose $R = \omega$ in (9.8) and use the analyticity of the integrand to exchange the arc $\gamma^R = \gamma^\omega$ in the above representation for $r(A)^n$ by $\Gamma_\omega^{\omega_n} \cup \gamma^{\omega_n}$, where $\omega_n = n^{1/m}\omega$. Here, by (9.8),

$$\Big\|\frac{1}{2\pi i}\int_{\gamma^{\omega_n}} r(z)^n R(z;A)\,dz\Big\| \le \frac{M}{2\pi}\int_{\gamma^{\omega_n}} e^{C\omega^{-m}}\frac{|dz|}{|z|} \le CM$$

and

$$\begin{aligned}\Big\|\frac{1}{2\pi i}\int_{\Gamma_\omega^{\omega_n}} r(z)^n R(z;A)\,dz\Big\| &\le \frac{M}{\pi}\int_{\omega}^{\omega_n} e^{-cn\rho^{-m}}\frac{d\rho}{\rho}\\ &= \frac{M}{\pi m}\int_{\omega^{-m}}^{n\omega^{-m}} e^{-c\rho}\frac{d\rho}{\rho} \le CM.\end{aligned}$$

Together these estimates complete the proof. □

For our error estimates we shall apply the following spectral representation of the semigroup.

Lemma 9.3 *Assume that* (9.2) *and* (9.3) *hold, let* $\psi \in (\delta, \frac{1}{2}\pi)$, *and let* j *be any integer. Then we have for* $\varepsilon > 0$ *sufficiently small*

$$A^j E(t) = \frac{1}{2\pi i} \int_{\gamma_\varepsilon \cup \Gamma_\varepsilon} e^{-zt} z^j R(z; A)\, dz,$$

where $\gamma_\varepsilon = \{z; |z| = \varepsilon, |\arg z| \le \psi\}$ *and* $\Gamma_\varepsilon = \{z; |\arg z| = \psi, |z| \ge \varepsilon\}$, *and where* $\operatorname{Im} z$ *is decreasing along* $\gamma_\varepsilon \cup \Gamma_\varepsilon$. *When* $j \ge 0$, *we may take* $\varepsilon = 0$.

Proof. For $j = 0$ this follows at once from (9.4). Since

$$AR(z; A) = zR(z; A) - I, \quad A^{-1}R(z; A) = z^{-1}R(z; A) + z^{-1}A^{-1},$$

and $\int_{\gamma_\varepsilon \cup \Gamma_\varepsilon} e^{-zt} z^j\, dz = 0$ for any j, the result for positive and negative j then easily follows by induction. When $j \ge 0$ the integrand is continuous at $z = 0$ so that we may let ε tend to 0. Note that since e^{-zt} has an essential singularity at $z = \infty$, the Dunford-Taylor representation of Lemma 9.1, with $r(z)$ replaced by e^{-zt}, does not apply. □

We now show a simple consequence of (9.5) and (9.6).

Lemma 9.4 *Assume that* $r(z)$ *is* $A(\theta)$*-stable and approximates* e^{-z} *to order* q. *Then for any* $\psi \in (0, \theta)$ *and* $R > 0$ *there are positive numbers* C *and* c *such that for* $|z| \le R$, $|\arg z| \le \psi$, *and* $n \ge 1$,

$$|r(z)^n - e^{-nz}| \le Cn|z|^{q+1} e^{-cn|z|}. \tag{9.12}$$

Proof. We first note that

$$|r(z) - e^{-z}| \le C|z|^{q+1}, \quad \text{for } |z| \le R, \ |\arg z| \le \psi.$$

By (9.6) this holds when $|z|$ is small and therefore, in view of (9.5), for $|z| \le R$. We next observe that, if $c \le \cos\psi$, then

$$|e^{-z}| = e^{-\operatorname{Re} z} \le e^{-c|z|}, \quad \text{for } |\arg z| \le \psi.$$

Using also Lemma 9.2 we hence obtain, for the z under consideration,

$$|r(z)^n - e^{-nz}| = |(r(z) - e^{-z}) \sum_{j=0}^{n-1} r(z)^j e^{-(n-1-j)z}| \le C|z|^{q+1} n e^{-(n-1)c|z|},$$

which proves the lemma. □

We begin our error estimates with the following estimate for smooth data.

Theorem 9.2 *Assume that* A *satisfies* (9.2) *and* (9.3), *and that* $r(z)$ *is accurate of order* q *and* $A(\theta)$*-stable with* $\theta \in (\delta, \frac{1}{2}\pi]$. *Then there is a constant* C *such that*

$$\|U^n - u(t_n)\| \le CMk^q \|A^q v\|, \quad \text{for } t_n \ge 0, \quad \text{if } v \in \mathcal{D}(A^q).$$

Proof. With $F_n(z) = r(z)^n - e^{-nz}$ we show that

$$\|F_n(A)A^{-q}\| = \|(r(A)^n - e^{-nA})A^{-q}\| \le CM.$$

As in the proof of Theorem 9.1 this then also holds with A replaced by kA, and thus shows the result stated. By Lemma 9.1 we have

$$r(A)^n A^{-q} = \frac{1}{2\pi i}\int_{\gamma_\varepsilon\cup\Gamma_\varepsilon^R\cup\gamma^R} r(z)^n z^{-q} R(z;A)\,dz,$$

and here, since the integrand is of order $O(|z|^{-q-1})$ for large z, we may let R tend to ∞. Using also Lemma 9.3 we therefore have

$$F_n(A)A^{-q} = \frac{1}{2\pi i}\int_{\gamma_\varepsilon\cup\Gamma_\varepsilon} F_n(z) z^{-q} R(z;A)\,dz.$$

By Lemma 9.4 we see that $F_n(z)z^{-q} = O(z)$ as $z \to 0$, and thus the integrand is bounded, so that we may let $\varepsilon \to 0$. It follows that

$$\|F_n(A)A^{-q}\| \le CM\int_0^\infty \big(|F_n(\rho e^{i\psi})| + |F_n(\rho e^{-i\psi})|\big)\,\frac{d\rho}{\rho^{q+1}}.$$

Since $r(z)^n$ and e^{-tz} are bounded on Γ_0 we find

$$\int_1^\infty |F_n(\rho e^{\pm i\psi})|\,\frac{d\rho}{\rho^{q+1}} \le C\int_1^\infty \frac{d\rho}{\rho^{q+1}} \le C,$$

and, using (9.12),

$$\int_0^1 |F_n(\rho e^{\pm i\psi})|\,\frac{d\rho}{\rho^{q+1}} \le Cn\int_0^1 e^{-cn\rho}\,d\rho \le C. \tag{9.13}$$

Together these estimates complete the proof. □

We now turn to an estimate for initial data which do not satisfy any regularity assumption in addition to $v \in \mathcal{B}$. We then need to require additionally that $|r(\infty)| < 1$, which will secure a certain smoothing property for the discrete solution operator E_k^n (cf. Theorem 7.2, schemes of type III). We will begin with the following lemma about the behavior of $r(z)^n$ for large z.

Lemma 9.5 *Assume that the rational function $r(z)$ is $A(\theta)$-stable with $\theta \le \frac{1}{2}\pi$, and that $|r(\infty)| < 1$. Then for any $\psi \in (0,\theta)$ and $R > 0$ there are positive C and c such that, with $\kappa = r(\infty)$,*

$$|r(z)^n - \kappa^n| \le C|z|^{-1}e^{-cn}, \quad \textit{for } |z| \ge R,\ |\arg z| \le \psi,\ n \ge 1.$$

Proof. Since $r(z) - \kappa$ vanishes at ∞ and $|r(z)| < 1$ for $|\arg z| \le \psi$, $z \ne 0$,

$$|r(z) - \kappa| \le C|z|^{-1} \text{ and } |r(z)| \le e^{-c}, \quad \text{for } |z| \ge R,\ |\arg z| \le \psi.$$

Hence, for these z,

$$|r(z)^n - \kappa^n| = |(r(z) - \kappa)\sum_{j=0}^{n-1} r(z)^j \kappa^{n-1-j}| \le C|z|^{-1} n e^{-cn} \le C|z|^{-1} e^{-cn},$$

which shows our claim. □

Theorem 9.3 *In addition to the assumptions of Theorem 9.2, let* $|r(\infty)| < 1$. *Then there is a constant* C *such that*

$$\|U^n - u(t_n)\| \le CMk^q t_n^{-q} \|v\|, \qquad \text{for } t_n > 0, \ v \in \mathcal{B}.$$

Proof. With $F_n(z)$ as above we now need to show

$$\|F_n(A)\| \le CMn^{-q}.$$

With $\kappa = r(\infty)$, we set $\widetilde{F}_n(z) = F_n(z) - \kappa^n z/(1+z)$. Since $|\kappa| < 1$ and $\|A(I+A)^{-1}\| \le 2M$, we have

$$\|\kappa^n A(I+A)^{-1}\| \le 2M|\kappa|^n \le CMn^{-q},$$

and it remains to show the same bound for the operator norm of $\widetilde{F}_n(A)$. Since $r(z)^n - \kappa^n z/(1+z)$ vanishes at $z = \infty$, we may use Lemmas 9.1 and 9.3 to see that with $\Gamma = \{z; |\arg z| = \psi\}$, $\psi \in (\delta, \theta)$,

$$\widetilde{F}_n(A) = \frac{1}{2\pi i}\int_\Gamma \widetilde{F}_n(z) R(z;A)\, dz.$$

Since $\widetilde{F}_n(z) = (r(z)^n - \kappa^n) + \kappa^n/(1+z) - e^{-nz}$, Lemma 9.5 shows

$$\int_1^\infty |\widetilde{F}_n(\rho e^{\pm i\psi})| \frac{d\rho}{\rho} \le \int_1^\infty \big((Ce^{-cn} + |\kappa|^n)\rho^{-2} + e^{-cn\rho}\rho^{-1}\big)\, d\rho \le Cn^{-q}.$$

Using also Lemma 9.4 and $|z/(1+z)| \le 1$ for $\operatorname{Re} z \ge 0$ we have

$$\begin{aligned} \int_0^1 |\widetilde{F}_n(\rho e^{\pm i\psi})| \frac{d\rho}{\rho} &\le \int_0^1 |F_n(\rho e^{\pm i\psi})| \frac{d\rho}{\rho} + |\kappa|^n \\ &\le Cn \int_0^\infty \rho^q e^{-cn\rho}\, d\rho + |\kappa|^n \le Cn^{-q}. \end{aligned} \tag{9.14}$$

Together these estimates complete the proof. □

The above approach may also be used to study single step methods for the inhomogeneous equation

$$u' + Au = f, \quad \text{for } t > 0, \quad \text{with } u(0) = v, \tag{9.15}$$

in our Banach space framework. We illustrate this by considering the Crank-Nicolson scheme

$$\bar\partial U^n + A\widehat{U}^n = f(t_{n-1/2}), \quad \text{for } n \ge 1, \quad \text{with } U^0 = v, \tag{9.16}$$

where $\widehat{U}^n = \frac12(U^n + U^{n-1}$, or

$$U^n = E_k U^{n-1} + Q_k f(t_{n-1/2}), \quad \text{with } E_k = r(kA), \quad Q_k = p_1(kA),$$

where

$$r(z) = (1-\tfrac12 z)/(1+\tfrac12 z), \quad p_1(z) = 1/(1+\tfrac12 z). \tag{9.17}$$

Since

$$U^n = E_k^n v + k\sum_{j=0}^{n-1} E_k^{n-j-1} Q_k f(t_{j+1/2}),$$

we find at once from Theorem 9.1 the stability estimate

$$\|U^n\| \le CM\Big(\|v\| + k\sum_{j=0}^{n-1}\|f(t_{j+1/2})\|\Big). \tag{9.18}$$

We now show the following error estimate.

Theorem 9.4 *Let U^n and u be the solutions of* (9.16) *and* (9.15). *Then*

$$\|U^n - u(t_n)\| \le CMk^2\int_0^{t_n}(\|u_{ttt}\| + \|Au_{tt}\|)\,d\tau, \quad \text{for } n \ge 0.$$

Proof. Setting $e^n = U^n - u(t_n)$ we have

$$\bar\partial e^n + A\widehat{e}^n = -\omega^n, \quad \text{for } n \ge 1, \quad \text{with } e^0 = 0,$$

where

$$\omega^n = (\bar\partial u(t_n) - u_t(t_{n-1/2})) - A(u(t_{n-1/2}) - \tfrac12(u(t_n) + u(t_{n-1})),$$

and hence,

$$e^n = E_k e^{n-1} - Q_k\omega^n, \quad \text{for } n \ge 1, \quad \text{with } e^0 = 0.$$

By iteration, using the stability estimate (9.18), and treating the terms in ω^n as in the proof of Theorem 1.6, leading to (1.57), we conclude

$$\|e^n\| \le CMk\sum_{j=1}^{n}\|\omega^j\| \le CMk^2\int_0^{t_n}(\|u_{ttt}\| + \|Au_{tt}\|)\,d\tau,$$

which completes the proof. □

The above results concerning time discretization of the abstract differential equation (9.1) may be applied to analyze fully discrete schemes for parabolic partial differential equations. We shall exemplify this by deriving maximum-norm error estimates for fully discrete methods for the homogeneous heat equation in two spatial variables, using piecewise linear approximation functions in space on quasiuniform triangulations of the spatial domain.

The problem we consider is thus, with $A = -\Delta$,

$$\begin{aligned} u_t + Au &= 0 \quad \text{in } \Omega, \quad \text{for } t > 0, \\ u &= 0 \quad \text{on } \partial\Omega, \quad \text{for } t > 0, \quad \text{with } u(\cdot,0) = v \quad \text{in } \Omega, \end{aligned} \tag{9.19}$$

where Ω is a convex domain in $\mathbb{R}^2$ with smooth boundary $\partial\Omega$, with the spatially discrete analogue defined by

$$u_{h,t} + A_h u_h = 0, \quad \text{for } t > 0, \quad \text{with } u_h(0) = v_h, \tag{9.20}$$

where $A_h = -\Delta_h$, with Δ_h the discrete Laplacian from (1.33).

We recall from Theorem 6.6 that for any $\delta \in (0, \frac{1}{2}\pi)$ we have, uniformly in h,

$$\|R(z; A_h)\|_{L_\infty} \le C|z|^{-1}, \quad \text{for } z \in \Sigma_\delta, \ z \neq 0, \tag{9.21}$$

and that hence the solution operator $E_h(t) = e^{-A_h t}$ of (9.19), the analytic semigroup on S_h generated by $-A_h$, satisfies, uniformly in h,

$$\|E_h(t)\|_{L_\infty} + t\,\|E_h'(t)\|_{L_\infty} \le C, \quad \text{for } t > 0. \tag{9.22}$$

Assume now that $r(z)$ is a rational function consistent with e^{-z}, that is $A(\theta)$-stable for some $\theta \in (0, \frac{1}{2}\pi]$. The fully discrete method obtained by discretization of (9.20) in time is then defined by

$$U_h^n = E_{kh}^n v_h, \quad \text{where } E_{kh} = r(kA_h). \tag{9.23}$$

We begin with a maximum-norm stability result.

Theorem 9.5 *Assume that $r(z)$ is consistent with e^{-z} and $A(\theta)$-stable for some $\theta \in (0, \frac{1}{2}\pi]$, and let U_h^n be defined by (9.23). Then we have, uniformly in h,*

$$\|U_h^n\|_{L_\infty} \le C\|v_h\|_{L_\infty}, \quad \text{for } n \ge 0. \tag{9.24}$$

Proof. This follows using the resolvent estimate of (9.21) for a $\delta \in (0, \theta)$, together with the stability result of Theorem 9.1. □

In the same way as in Chapter 7 we begin our error analysis with a nonsmooth data error estimate.

Theorem 9.6 *Let U_h^n and u be defined by (9.23) and (9.19), with $v_h = P_h v$, and assume that $r(z)$ is accurate of order q and $A(\theta)$-stable with $\theta \in (0, \frac{1}{2}\pi]$, and that $|r(\infty)| < 1$. Then we have, with C independent of h and k, for $v \in L_\infty$,*

$$\|U_h^n - u(t_n)\|_{L_\infty} \le C\big(h^2\ell_h^2 t_n^{-1} + k^q t_n^{-q}\big)\|v\|_{L_\infty}, \quad \textit{for } t_n > 0.$$

Proof. Let $u_h(t) = E_h(t)P_h v$ be the solution of (9.20) with $v_h = P_h v$. By (9.21) and our above argument we may apply Theorem 9.3 to obtain

$$\|U_h^n - u_h(t_n)\|_{L_\infty} = \|(E_{kh}^n - E_h(t_n))P_h v\|_{L_\infty} \le Ck^q t_n^{-q}\|P_h v\|_{L_\infty}.$$

Using the stability of P_h in L_∞ (Lemma 6.1) together with the estimate for $u_h - u$ of Theorem 6.10, this completes the proof. □

We now turn to a smooth data error estimate. Here $|r(\infty)| < 1$ is not needed. Note that for $v \in \mathcal{D}(A^q)$ we require $A^j v = 0$ on $\partial\Omega$ for $0 \le j < q$.

Theorem 9.7 *Let U_h^n and u be defined by (9.23) and (9.19), and assume that $r(z)$ is accurate of order q and $A(\theta)$-stable with $\theta \in (0, \frac{1}{2}\pi]$. Then if $v \in \mathcal{D}(A^q)$ and if $\|v_h - v\|_{L_\infty} \le Ch^2\ell_h^2\|v\|_{W_\infty^2}$, we have*

$$\|U_h^n - u(t_n)\|_{L_\infty} \le C\big(h^2\ell_h^2\|v\|_{W_\infty^2} + k^q\|A^q v\|_{L_\infty}\big), \quad \textit{for } n \ge 0.$$

Proof. In view of the stability property of Theorem 9.5 it is no loss of generality to assume $v_h = P_h v$. By Theorem 6.9 we have

$$\|u_h(t) - u(t)\|_{L_\infty} \le Ch^2\ell_h^2\|v\|_{W_\infty^2}, \quad \text{for } t \ge 0.$$

Hence, with $F_n(z) = r(z)^n - e^{-nz}$ and $F_n = F_n(kA_h)$, it remains to bound $U_h^n - u_h(t_n) = F_n P_h v$. For this purpose, we use Lemma 7.1 to write, with $T = A^{-1}$ and $T_h = A_h^{-1}P_h$,

$$v = \sum_{j=0}^{q-1} T_h^j(T - T_h)A^{j+1}\widetilde{v}_k + T_h^q A^q \widetilde{v}_k + (v - \widetilde{v}_k), \tag{9.25}$$

with $\widetilde{v}_k$ suitably chosen. We shall see below that this may be done so that, with C independent of p,

$$\begin{aligned} k^j\|A^j\widetilde{v}_k\|_{W_p^2} &\le Cp\|v\|_{W_p^2}, \quad \text{for } 2 \le p < \infty,\\ \|A^q\widetilde{v}_k\|_{L_\infty} &\le C\|A^q v\|_{L_\infty},\\ \|\widetilde{v}_k - v\|_{L_\infty} &\le Ck^q\|A^q v\|_{L_\infty}. \end{aligned} \tag{9.26}$$

Assuming this for a moment, we first note that by the stability properties of $E_h(t), E_{kh}^n$ and P_h, and by the last bound of (9.26),

$$\|F_n P_h(v - \widetilde{v}_k)\|_{L_\infty} \le C\|\widetilde{v}_k - v\|_{L_\infty} \le Ck^q\|A^q v\|_{L_\infty}.$$

For the remaining terms we apply Theorem 9.2 to A_h to obtain

$$\|F_n P_h w\|_{L_\infty} \le Ck^j \|A_h^j P_h w\|_{L_\infty}, \quad \text{for } 0 \le j \le q. \tag{9.27}$$

Note that if $r(z)$ is accurate of order q it is also accurate of order j with $1 \le j \le q$, which shows (9.27) for these j. The case $j = 0$ follows again directly by the stability properties of $E_h(t)$ and E_{kh}^n.

We recall from (6.81) that

$$\|(R_h - I)v\|_{L_\infty} \le Ch^{2-2/p}\ell_h \|v\|_{W_p^2}, \quad \text{for } 2 \le p < \infty.$$

Setting $w = T_h^j(T - T_h)A^{j+1}\widetilde{v}_k = T_h^j(I - R_h)A^j\widetilde{v}_k$ in (9.27) and choosing $p = \ell_h$ we therefore obtain, for $0 \le j \le q-1$,

$$\begin{aligned}\|F_n P_h T_h^j(T - T_h)A^{j+1}\widetilde{v}_k\|_{L_\infty} &\le Ck^j \|A_h^j T_h^j P_h (I - R_h)A^j\widetilde{v}_k\|_{L_\infty} \\ &\le Ch^{2-2/p}\ell_h k^j \|A^j \widetilde{v}_k\|_{W_p^2} \le Cph^{2-2/p}\ell_h \|v\|_{W_p^2} \le Ch^2\ell_h^2 \|v\|_{W_\infty^2}.\end{aligned}$$

Similarly,

$$\|F_n P_h T_h^q A^q \widetilde{v}_k\|_{L_\infty} \le Ck^q \|A^q \widetilde{v}_k\|_{L_\infty} \le Ck^q \|A^q v\|_{L_\infty}.$$

We have thus estimated all the terms in $F_n P_h v$ corresponding to the representation (9.25) in the way stated.

It remains to show that $\widetilde{v}_k$ may be chosen to satisfy (9.26). In Chapter 7 a corresponding construction was based on eigenfunction expansion of v and used Parseval's relation, but this is not appropriate here and we take instead

$$\widetilde{v}_k = s(kA)E(k)v, \text{ with } s(z) = \sum_{n=0}^{q} \frac{z^n}{n!} = e^z + O(z^{q+1}), \quad \text{as } z \to 0.$$

Note that $\widetilde{v}_k \in \dot{H}^s$ for any $s \ge 0$ when $k > 0$. Since $(-A)^l E(k) = E^{(l)}(k) = (E'(k/l))^l$ we have, using the smoothing property of $E(t)$ in (6.41), and the regularity estimate (6.78),

$$\begin{aligned}k^j \|A^j \widetilde{v}_k\|_{W_p^2} &\le Cp\|(kA)^j s(kA)E(k)Av\|_{L_p} \\ &\le Cp \sum_{l=j}^{q+j} k^l \|E^{(l)}(k)Av\|_{L_p} \le Cp\|v\|_{W_p^2}.\end{aligned}$$

Similarly we find

$$\|A^q \widetilde{v}_k\|_{L_\infty} = \|A^q s(kA)E(k)v\|_{L_\infty} \le \sum_{l=0}^{q} \frac{k^l}{l!} \|E^{(l)}(k)A^q v\|_{L_\infty} \le C\|A^q v\|_{L_\infty}.$$

To bound $\widetilde{v}_k - v$, finally, we recall that, for any $\delta \in (0, \frac{1}{2}\pi)$, we have

$$\|R(z;A)\|_{L_\infty} \le C|z|^{-1}, \quad \text{for } z \in \Sigma_\delta,\ z \ne 0. \tag{9.28}$$

Using Lemma 9.3 we may therefore write

$$\begin{aligned}\widetilde{v}_k - v &= T^q(s(kA)E(k) - 1)A^q v\\ &= \frac{1}{2\pi i}\int_\Gamma z^{-q}(s(kz)e^{-kz} - 1)R(z;A)\,dz\,A^q v,\end{aligned}$$

where $\Gamma = \{z; |\arg z| = \psi \in (\delta, \frac{1}{2}\pi)\}$. Hence

$$\begin{aligned}\|\widetilde{v}_k - v\|_{L_\infty} &\le Ck^q \int_0^\infty \rho^{-q-1}|s(\rho e^{i\psi})e^{-\rho e^{i\psi}} - 1|\,d\rho\,\|A^q v\|_{L_\infty}\\ &\le Ck^q\|A^q v\|_{L_\infty},\end{aligned}$$

where we have used the fact that the integrand is bounded on $[0,1]$ and bounded by $C(\rho^{-1}e^{-c\rho} + \rho^{-q-1})$ on $(1,\infty)$. The estimates of (9.26) are now shown, and the proof of the theorem is thus complete. □

We close this chapter with a maximum-norm error estimate for the fully discrete Crank-Nicolson method for the inhomogeneous heat equation, or (9.19) with a forcing term f on the right, with the spatially discrete analogue

$$u_{h,t} + A_h u_h = P_h f, \quad \text{for } t > 0, \quad \text{with } u_h(0) = v_h = R_h v, \tag{9.29}$$

where $A_h = -\Delta_h$, with Δ_h the discrete Laplacian from (1.33). The fully discrete method under consideration is then to find $U^n \in S_h$ for $n \ge 0$ such that, with $r(z)$ and $p_1(z)$ as in (9.17), and $E_{kh} = r(kA_h)$, $Q_{kh} = p_1(kA_h)$,

$$U_h^n = E_{kh}U_h^{n-1} + Q_{kh}P_h f(t_{n-1/2}), \quad \text{for } n \ge 1, \quad U_h^0 = R_h v. \tag{9.30}$$

We have the following error estimate.

Theorem 9.8 *Let U_h^n be defined by* (9.30) *and u be the solutions of the inhomogeneous version of* (9.19). *Then*

$$\begin{aligned}\|U_h^n - u(t_n)\|_{L_\infty} &\le Ch^2\ell_h\Big(\|v\|_{W_\infty^2} + \int_0^{t_n}\|u_t\|_{W_\infty^2}\,d\tau\Big)\\ &\quad + Ck^2\int_0^{t_n}(\|u_{ttt}\|_{L_\infty} + \|u_{tt}\|_{W_\infty^2})\,d\tau.\end{aligned}$$

Proof. Writing as usual $U_h^n - u(t_n) = \rho^n + \theta^n$, the first term $\rho^n = \rho(t_n)$ is bounded as desired by (6.29). For θ^n we have from (1.55),

$$\bar\partial\theta^n + A_h\widehat{\theta}^n = -P_h\omega^n, \quad \text{for } n \ge 1, \quad \text{with } \theta^0 = 0,$$

where

$$\begin{aligned}\omega^n =&\bar\partial\rho^n + (\bar\partial u(t_n) - u_t(t_{n-1/2}))\\ &- A\big(u(t_{n-1/2}) - \tfrac12(u(t_n)+u(t_{n-1}))\big) = \omega_1^n+\omega_2^n+\omega_3^n.\end{aligned}$$

In the same way as in the proof of Theorem 9.4 we find

$$\|\theta^n\|_{L_\infty} \le Ck\sum_{j=1}^n \|\omega^j\|_{L_\infty} \le Ck\sum_{j=1}^n (\|\omega_1^j\|_{L_\infty} + \|\omega_2^j\|_{L_\infty} + \|\omega_3^j\|_{L_\infty}).$$

Here, using (1.51) and again (6.29), we obtain

$$k\sum_{j=1}^n \|\omega_1^j\|_{L_\infty} \le \sum_{j=1}^n \int_{t_{j-1}}^{t_j} Ch^2\ell_h\|u_t\|_{W_\infty^2}\,d\tau = Ch^2\ell_h\int_0^{t_n}\|u_t\|_{W_\infty^2}\,d\tau,$$

and the remaining sum is bounded as in the Theorem 9.4 by

$$Ck\sum_{j=1}^n(\|\omega_2^j\|_{L_\infty} + \|\omega_3^j\|_{L_\infty}) \le Ck^2\int_0^{t_n}(\|u_{ttt}\|_{L_\infty} + \|u_{tt}\|_{W_\infty^2})\,d\tau.$$

Together these estimates complete the proof. □

Error estimates in Banach space of the type we have discussed above can be found in Brenner and Thomée [41], Piskarev [195], LeRoux [152], [153], Larsson, Thomée and Wahlbin [147], Crouzeix, Larsson, Piskarev and Thomée [63], Ashyralyev and Sobolevskii [6], Palencia [190], [191], Bakaev [16], [18], Fujita and Suzuki [104]; the ideas in the above proof of the general stability result for A-stable rational functions are from [190], [191]. For application to maximum-norm estimates, see Schatz, Thomée and Wahlbin [209], Palencia [192] and Crouzeix, Larsson and Thomée [61].

10. Multistep Backward Difference Methods

In this chapter we shall first consider approximations at equidistant time levels of parabolic equations in which the time derivate is replaced by a multistep backward difference quotient of maximum order consistent with the number of time steps involved. We show that when this order is at most 6, then the method is stable and has a smoothing property analogous to that of single step methods of type IV. We shall use these properties to derive both smooth and nonsmooth data error estimates. In the end of the chapter we shall also discuss the use of two-step backward difference operators with variable time steps.

We start by studying our parabolic problem in the Hilbert space framework used in earlier chapters, and consider thus the initial value problem for the abstract parabolic equation in a Hilbert space $\mathcal{H}$ given by

$$u' + Au = f(t), \quad \text{for } t > 0, \quad \text{with } u(0) = v, \tag{10.1}$$

where A is a selfadjoint positive definite operator with dense domain $\mathcal{D}(A)$ in $\mathcal{H}$ and with a compact inverse, and where f is a function of t with values in $\mathcal{H}$.

We shall study the numerical approximation of (10.1) by a q-step backward difference method: With k the time step and $t_n = nk$, we introduce the backward difference operator $\bar{\partial}_q$ by

$$\bar{\partial}_q U^n = \sum_{j=1}^{q} \frac{k^{j-1}}{j} \bar{\partial}^j U^n, \quad \text{where } \bar{\partial} U^n = (U^n - U^{n-1})/k,$$

and define our approximate solution U^n by

$$\begin{aligned} &\bar{\partial}_q U^n + AU^n = f^n, \quad \text{for } n \geq q, \quad \text{where } f^n = f(t_n), \\ &\qquad\qquad\qquad\qquad\quad \text{with } U^0, \dots, U^{q-1} \text{ given.} \end{aligned} \tag{10.2}$$

Note that we may also write, with coefficients α_j independent of k,

$$\bar{\partial}_q U^n = k^{-1} \sum_{j=0}^{q} \alpha_j U^{n-j}. \tag{10.3}$$

We observe that $\bar{\partial}_q$ is an approximation of d/dt which is accurate of order q. In fact, by Newton's backward difference formula we have for u smooth

$$u(t) = u^n + \sum_{j=1}^{q} \frac{(t-t_n)\cdots(t-t_{n-j+1})}{j!} \bar{\partial}^j u^n + R_q(u;t),$$

where

$$R_q(u;t) = \frac{(t-t_n)\cdots(t-t_{n-q})}{(q+1)!} u^{(q+1)}(\tau), \quad \text{with } \tau \in [t_{n-q}, t_n].$$

After differentiation and setting $t = t_n$ this shows

$$(u')^n = u'(t_n) = \bar{\partial}_q u^n + \frac{k^q}{q+1} u^{(q+1)}(\tau), \quad \text{with } \tau \in [t_{n-q}, t_n]. \tag{10.4}$$

Introducing the polynomial $\widetilde{\alpha}(\zeta) = \sum_{j=0}^{q} \alpha_j \zeta^j$, where the α_j are the coefficients in (10.3), and the translation operator $T_{-k}u(t) = u(t-k)$, this relation shows

$$(u')^n = k^{-1}\widetilde{\alpha}(T_{-k})u^n + O(k^q), \quad \text{as } k \to 0. \tag{10.5}$$

Applying this to $u(t) = e^t$ and replacing k by λ, we see that

$$\widetilde{\alpha}(e^{-\lambda}) = \lambda + O(\lambda^{q+1}), \quad \text{as } \lambda \to 0, \tag{10.6}$$

and one may also easily show that (10.6) implies (10.5).

For $q = 1$, (10.2) reduces to the backward Euler method

$$(U^n - U^{n-1})/k + AU^n = f^n, \quad \text{for } n \geq 1,$$

and the only starting value needed is $U^0 = v$. For $q = 2$, we have

$$(\tfrac{3}{2}U^n - 2U^{n-1} + \tfrac{1}{2}U^{n-2})/k + AU^n = f^n, \quad \text{for } n \geq 2.$$

Here both U^0 and U^1 are needed to start the procedure. In this case, it is natural to take $U^0 = v$ and to determine U^1 from one step of the backward Euler method, i.e., $\bar{\partial}U^1 + AU^1 = f^1$. Although this equation is only first order accurate, this will suffice to show a second order error estimate since it is only used once. For $q > 2$, starting values of accuracy $O(k^q)$ can be generated, e.g., by using the partial sums of the Taylor expansion of $u(t_j)$, i.e., with $u^{(l)} = (d/dt)^l u$,

$$U^j = \sum_{l=0}^{q-1} \frac{(jk)^l}{l!} u^{(l)}(0), \quad \text{for } j = 0, \dots, q-1. \tag{10.7}$$

Here the functions $u^{(l)}(0)$ can be computed from the differential equation in terms of data, so that, e.g., $u'(0) = f(0) - Av, u''(0) = f'(0) - A(f(0) - Av)$, etc. Note that some of the functions occurring are required to be in $\mathcal{D}(A)$. This choice is only appropriate when data are smooth. Starting values suitable for the nonsmooth data case will be discussed later.

It is known from the theory of numerical solution of stiff ordinary differential equation (cf., e.g., Hairer and Wanner [113]) that this method is $A(\theta)$-stable for some $\theta = \theta_q > 0$ when $q \le 6$. Our analysis here begins with the following stability result with respect to the norm $\|\cdot\|$ in $\mathcal{H}$.

Lemma 10.1 *Let $q \le 6$. Then there is a constant C, independent of the positive definite operator A, such that for the solution of* (10.2)

$$\|U^n\| \le C \sum_{j=0}^{q-1} \|U^j\| + Ck \sum_{j=q}^{n} \|f^j\|, \quad \textit{for } n \ge q. \tag{10.8}$$

The proof of this lemma is rather long and technical. Using eigenfunction expansions of U^n and f^n, it will be reduced to considering the scalar case in which $\mathcal{H} = \mathbb{R}$ and the operator A corresponds to multiplication by a positive scalar μ. With $\lambda = k\mu$, the solution $U^n = U^n(\lambda)$ then satisfies

$$(\alpha_0 + \lambda)U^n + \alpha_1 U^{n-1} + \cdots + \alpha_q U^{n-q} = g^n := kf^n, \quad \text{for } n \ge q, \tag{10.9}$$

with $U^0, \ldots, U^{q-1}$ given. The technical work is contained in the following two lemmas, the first of which shows $A(0)$-stability for $q \le 6$.

Lemma 10.2 *Let $q \le 6$ and let $P(\zeta; \lambda)$ be the characteristic polynomial of the difference equation* (10.9), *i.e.,*

$$P(\zeta; \lambda) = (\alpha_0 + \lambda)\zeta^q + \alpha_1 \zeta^{q-1} + \cdots + \alpha_q = \zeta^q(\widetilde{\alpha}(1/\zeta) + \lambda).$$

Then $P(\zeta; 0)$ has a simple zero at $\zeta = 1$ and the remaining zeros are in the interior of the unit disk. Further, for any $\lambda > 0$, the zeros of $P(\zeta; \lambda)$ are in the interior of the unit disk, and tend to 0 as λ tends to ∞.

Proof. It is obvious from (10.6) that $P(1; 0) = 0, P'_\zeta(1; 0) = 1$, and hence that $P(\zeta; 0)$ has a simple zero at $\zeta = 1$. It is further clear that all zeros tend to 0 as $\lambda \to \infty$. Since the roots depend continuously on λ, it therefore suffices to show that, except for the zero at 1 for $\lambda = 0$, there is no zero on the unit circle for $\lambda \ge 0$, or that $\widetilde{\alpha}(e^{i\theta}) + \lambda \ne 0$ when $\lambda \ge 0$, except when $\lambda = 0$ and $\theta \equiv 0 (\operatorname{mod} 2\pi)$. This can also be expressed by saying that $\widetilde{\alpha}(e^{i\theta})$ is never negative and vanishes in $[0, 2\pi)$ only at $\theta = 0$. We may write

$$\widetilde{\alpha}(e^{i\theta}) = \sum_{j=0}^{q} \alpha_j \cos j\theta + i \sum_{j=1}^{q} \alpha_j \sin j\theta = \xi(\theta) + i\eta(\theta),$$

and we thus want to show that $\eta(\theta) = 0$ implies $\xi(\theta) \ge 0$, with $\xi(\theta) = 0$ in $[0, 2\pi)$ only for $\theta = 0$. For each q there are polynomials $\widetilde{\xi}$ and $\widetilde{\eta}$ of degree q and $q - 1$, respectively, such that

$$\xi(\theta) = \widetilde{\xi}(\cos\theta), \quad \eta(\theta) = \sin\theta\,\widetilde{\eta}(\cos\theta).$$

To show our claim we only have to find the finite number of $\bar{\theta} \in [0, 2\pi)$ for which $\eta(\bar{\theta}) = 0$ and check that then $\xi(\bar{\theta}) > 0$ if $\bar{\theta} \neq 0$. For $q = 2$ this follows easily from

$$\begin{aligned}\xi(\theta) &= \tfrac{3}{2} - 2\cos\theta + \tfrac{1}{2}\cos 2\theta = (1 - \cos\theta)^2,\\ \eta(\theta) &= -2\sin\theta + \tfrac{1}{2}\sin 2\theta = \sin\theta(\cos\theta - 2).\end{aligned}$$

(In this case the quadratic equation $P(\zeta; \lambda) = 0$ could also be solved directly to find $\zeta_{1,2} = (2 \pm (1 - 2\lambda)^{1/2})^{-1}$ which are located as claimed.) For $q = 3, 4, 5, 6$ the claim is easily checked, e.g., using MATLAB. □

The proof can easily be extended to permit λ to be in a sector including the positive real axis, thus showing $A(\theta)$-stability with $\theta > 0$. We shall not pursue this here.

Lemma 10.3 *The solution of* (10.9) *may be written*

$$U^n = \sum_{j=q}^{n} \beta_{n-j}(\lambda) g^j + \sum_{s=0}^{q-1} \beta_{ns}(\lambda) U^s, \quad \textit{for } n \geq q, \tag{10.10}$$

where the $\beta_j(\lambda)$ and $\beta_{ns}(\lambda)$ are defined by

$$\widetilde{\beta}(\zeta) = \sum_{j=0}^{\infty} \beta_j(\lambda)\zeta^j := (\widetilde{\alpha}(\zeta) + \lambda)^{-1}, \quad \beta_{ns}(\lambda) = -\sum_{j=q-s}^{q} \beta_{n-s-j}(\lambda)\alpha_j.$$

If $q \leq 6$ there are positive constants c, C, and λ_0 such that

$$|\beta_j(\lambda)| \leq \begin{cases} Ce^{-cj\lambda}, & \textit{for } 0 < \lambda \leq \lambda_0,\\ C\lambda^{-1}e^{-cj}, & \textit{for } \lambda \geq \lambda_0.\end{cases} \tag{10.11}$$

Proof. For brevity we shall write β_j for $\beta_j(\lambda)$ and similarly for β_{ns}. Since the difference operator in (10.9) has constant coefficients it is clear that U^n may be represented in the form

$$U^n = \sum_{j=q}^{n} \gamma_{n-j} g^j + \sum_{s=0}^{q-1} \gamma_{ns} U^s, \quad \text{for } n \geq q,$$

and we want to identify the coefficients in this representation with those stated in the lemma. We begin by showing that $\gamma_j = \beta_j$ for $j \geq 0$. For this we choose $g^j = 1$ for $j = q$, $g^j = 0$ for $j > q$, and set $U^s = 0$ for $s \leq q - 1$, which gives $U^n = \gamma_{n-q}$ for $n \geq q$. Multiplying (10.9) by ζ^n and summing over $n \geq q$ we obtain

$$(\widetilde{\alpha}(\zeta) + \lambda)\widetilde{U}(\zeta) = \zeta^q, \quad \text{where } \widetilde{U}(\zeta) = \widetilde{U}(\zeta, \lambda) = \sum_{j=0}^{\infty} U^j\zeta^j,$$

and thus $\widetilde{U}(\zeta) = \zeta^q \widetilde{\beta}(\zeta)$. Since $U^n = \gamma_{n-q}$ for $n \geq q$, we also have

$$\widetilde{U}(\zeta) = \sum_{n=q}^{\infty} \gamma_{n-q} \zeta^n = \zeta^q \sum_{j=0}^{\infty} \gamma_j \zeta^j = \zeta^q \widetilde{\gamma}(\zeta).$$

Hence $\widetilde{\beta}(\zeta) = \widetilde{\gamma}(\zeta)$, which shows $\gamma_j = \beta_j$ for $j \geq 0$. Note that since U^n solves a homogeneous difference equation with constant coefficients for $n > q$ we have $|U^n| \leq C\kappa^n$ for some $\kappa > 0$ and hence the series defining $\widetilde{U}(\zeta)$ converges for ζ small.

For the γ_{ns} we assume $g^j = 0$ for $j \geq q$ and $U^j = 1$ for $j = s$, $U^j = 0$ for $0 \leq j \leq q-1, j \neq s$. This time multiplication of (10.9) by ζ^n for $n \geq q$ and summation gives

$$(\alpha_0 + \lambda) \sum_{n=q}^{\infty} U^n \zeta^n + \alpha_1 \zeta \sum_{n=q-1}^{\infty} U^n \zeta^n + \cdots + \alpha_q \zeta^q \sum_{n=0}^{\infty} U^n \zeta^n = 0.$$

Since now

$$\sum_{n=q-j}^{\infty} U^n \zeta^n = \begin{cases} \widetilde{U}(\zeta) - \zeta^s, & \text{if } q-j > s, \\ \widetilde{U}(\zeta), & \text{if } q-j \leq s, \end{cases}$$

we obtain

$$(\widetilde{\alpha}(\zeta) + \lambda)\widetilde{U}(\zeta) = \zeta^s \big(\widetilde{\alpha}(\zeta) + \lambda - \alpha_{q-s}\zeta^{q-s} - \cdots - \alpha_q \zeta^q\big)$$

or

$$\widetilde{U}(\zeta) = \zeta^s \big(1 - (\alpha_{q-s}\zeta^{q-s} + \cdots + \alpha_q \zeta^q)\widetilde{\beta}(\zeta)\big).$$

This time $U^n = \gamma_{ns}$ for $n \geq q$, and we have for ε small, since $n - s \neq 0$,

$$\begin{aligned} \gamma_{ns} &= \frac{1}{2\pi i} \int_{|\zeta|=\varepsilon} \zeta^{-n-1} \widetilde{U}(\zeta)\, d\zeta \\ &= -\frac{1}{2\pi i} \int_{|\zeta|=\varepsilon} \zeta^{-n-1+s} (\alpha_{q-s}\zeta^{q-s} + \cdots + \alpha_q \zeta^q) \widetilde{\beta}(\zeta)\, d\zeta \\ &= -\alpha_{q-s}\beta_{n-q} - \cdots - \alpha_q \beta_{n-s-q} = -\sum_{j=q-s}^{q} \beta_{n-s-j}\alpha_j, \end{aligned}$$

which completes the proof of the representation (10.10).

We now turn to the estimates (10.11). We first note that with Γ a closed curve in the complex plane which winds once around each zero of $P(\zeta; \lambda)$ in the positive sense, we have

$$\beta_j = \beta_j(\lambda) = \frac{1}{2\pi i} \int_\Gamma \frac{\zeta^{j+q-1}}{P(\zeta;\lambda)} d\zeta. \tag{10.12}$$

In fact, it follows by the definition of $\widetilde{\beta}(\zeta)$ that, for $\varepsilon > 0$ small,

$$\beta_j = \frac{1}{2\pi i} \int_{|\zeta|=\varepsilon} \zeta^{-j-1} \widetilde{\beta}(\zeta)\, d\zeta = \frac{1}{2\pi i} \int_{|\zeta|=\varepsilon} \frac{d\zeta}{\zeta^{j+1}(\widetilde{\alpha}(\zeta) + \lambda)},$$

and hence (10.12) is derived by introducing $1/\zeta$ as a new variable and then deforming the resulting contour $|\zeta| = 1/\varepsilon$.

By Lemma 10.2, the zeros of $P(\zeta;\lambda) = (\alpha_0 + \lambda) \prod_{l=1}^{q} (\zeta - \zeta_l(\lambda))$ are in the interior of the unit disk, and tend to zero as λ tends to infinity. We order these zeros so that $\zeta_l(\lambda)$ is continuous in λ for each l, and $\zeta_1(0) = 1$. Since $P(\zeta_1(\lambda);\lambda) = 0$ we find

$$\zeta_1(\lambda) = 1 - \frac{P'_\lambda(1;0)}{P'_\zeta(1;0)}\, \lambda + O(\lambda^2) = 1 - \lambda + O(\lambda^2), \quad \text{as } \lambda \to 0,$$

because $P'_\lambda(1;0) = P'_\zeta(1;0) = 1$ where the latter facts follow since $P(\zeta;0) = \zeta^q \widetilde{\alpha}(1/\zeta)$ and $\widetilde{\alpha}'(1) = -1$ by (10.5). As a result, there is a $\lambda_0 > 0$ such that $|\zeta_1(\lambda)| \le 1 - \lambda/2$ for $0 \le \lambda \le \lambda_0$, and such that $\zeta_1(\lambda)$ is a simple root of $P(\zeta;\lambda) = 0$ for $0 \le \lambda \le \lambda_0$. The remaining roots are bounded in absolute value by $1 - \delta$ for some positive constant δ, independently of $\lambda \ge 0$, and we may assume that λ_0 is so small that $|\zeta_1(\lambda)| > 1 - \delta/2$ for $0 \le \lambda \le \lambda_0$.

With the factorization $P(\zeta;\lambda) = (\zeta - \zeta_1(\lambda))\, Q(\zeta,\lambda)$ we have

$$\frac{1}{P(\zeta;\lambda)} = \frac{1}{(\zeta - \zeta_1(\lambda))\, Q(\zeta_1(\lambda),\lambda)} + \frac{R(\zeta,\lambda)}{Q(\zeta,\lambda)},$$

where

$$R(\zeta,\lambda) = \frac{Q(\zeta_1(\lambda),\lambda) - Q(\zeta,\lambda)}{(\zeta - \zeta_1(\lambda)) Q(\zeta_1(\lambda),\lambda)}.$$

Hence we obtain from (10.12)

$$\beta_j(\lambda) = \frac{\zeta_1(\lambda)^{j+q-1}}{Q(\zeta_1(\lambda),\lambda)} + \frac{1}{2\pi i} \int_\Gamma \zeta^{j+q-1} \frac{R(\zeta,\lambda)}{Q(\zeta,\lambda)}\, d\zeta, \tag{10.13}$$

where Γ may be taken to be the circle centered at the origin of radius $1-\delta/2$. In view of the above discussion, it is easily seen that the first term is bounded by $Ce^{-cj\lambda}$ for $\lambda \in (0, \lambda_0]$. For the second term we note that for each λ, $R(\zeta,\lambda)$ is a polynomial in ζ whose zeros depend continuously on λ, and therefore bounded independent of λ in $[0,\lambda_0]$. Hence $|R(\zeta,\lambda)| \le C$ for $0 \le \lambda \le \lambda_0$ and $|\zeta| \le 1$, which implies that the second term in (10.13) can be bounded by $Ce^{-\delta j/2}$. This verifies (10.11) for $0 < \lambda \le \lambda_0$.

For $\lambda \ge \lambda_0$ all zeros of $P(\cdot;\lambda)$ are bounded in modulus by $1 - \delta$ for some positive δ independent of λ, and (10.11) therefore easily follows in this case, taking Γ in (10.12) to be the circle $|z| = 1 - \delta/2$. □

We are now ready for the proof of our stability result.

Proof of Lemma 10.1. By superposition it suffices to show the result when all terms but one on the right in (10.8) vanish. The proof of the result in each of these situations is then reduced by eigenfunction expansion to the scalar case (10.9) with $\lambda = k\mu, \mu > 0$. We may then apply the representation (10.10) with only one term on the right present, and use the boundedness of the $|\beta_j(\lambda)|$, uniformly in λ, which follows from (10.11). For instance, in the case $U^j = 0$ for $j \le q-1$, and $f^j = 0$ for $j \ne s, q \le s \le n$, we have, with $\{\mu_l\}_{l=1}^\infty$ and $\{\varphi_l\}_{l=1}^\infty$ the eigenvalues and eigenfunctions of A,

$$U^n = k\beta_{n-s}(kA)f^s = k\sum_{l=1}^{\infty} \beta_{n-s}(k\mu_l)(f^s, \varphi_l)\varphi_l,$$

so that

$$\|U^n\| \le k \sup_{\lambda>0} |\beta_{n-s}(\lambda)|\, \|f^s\| \le Ck\|f^s\|.$$

The contributions from the discrete initial values are treated analogously. □

We now apply our stability lemma to derive a smooth data error estimate.

Theorem 10.1 *Let $q \le 6$. Then there is a constant C, independent of the positive definite operator A such that if U^n and u^n are solutions of* (10.2) *and* (10.1), *respectively, with u sufficiently smooth, we have*

$$\|U^n - u^n\| \le C\sum_{j=0}^{q-1} \|U^j - u^j\| + Ck^q \int_0^{t_n} \|u^{(q+1)}\|\,ds.$$

Proof. With $e^n = U^n - u^n$ we have

$$\bar\partial_q e^n + Ae^n = -\tau^n, \quad \text{for } n \ge q, \quad \text{where } \tau^n = \bar\partial_q u^n - (u')^n. \tag{10.14}$$

Application of Lemma 10.1 to e^n shows

$$\|e^n\| \le C\sum_{j=0}^{q-1} \|e^j\| + Ck\sum_{j=q}^{n} \|\tau^j\|, \quad \text{for } n \ge q. \tag{10.15}$$

By Taylor expansion around t_{j-q} we write, with $Q \in \Pi_q$,

$$u(t) = Q(t) + R(t), \quad \text{where } R(t) = \frac{1}{q!}\int_{t_{j-q}}^{t} (t-s)^q u^{(q+1)}(s)\,ds,$$

and since $\bar\partial_q Q - Q' = 0$ by (10.4) we have $\tau^j = \bar\partial_q R^j - (R')^j$. It follows that

$$k\|\tau^j\| \le C\sum_{l=j-q}^{j} \|R^l\| + Ck\|(R')^j\| \le Ck^q \int_{t_{j-q}}^{t_j} \|u^{(q+1)}\|\,ds, \tag{10.16}$$

and inserted into (10.15) this shows the theorem. □

Next we shall see how our stability result can be used to bound the error in the fully discrete solution of a parabolic partial differential equation. Our backward difference procedure will then be applied to an equation which has first been discretized in the spatial variables.

We consider the initial boundary value problem

$$(10.17)\qquad \begin{aligned} u_t - \Delta u &= f \quad \text{in } \Omega, \qquad \text{for } t>0,\\ u &= 0 \quad \text{on } \partial\Omega, \quad \text{for } t>0, \quad u(\cdot,0)=v \quad \text{in } \Omega, \end{aligned}$$

where Ω is a bounded domain in $\mathbb{R}^d$ with smooth boundary. We shall seek an approximate solution of (10.17) in a standard finite element space $S_h \subset H_0^1 = H_0^1(\Omega)$ with the $O(h^r)$ approximation property (1.10). With Δ_h the discrete Laplacian defined in (1.33) and P_h the L_2-projection onto S_h, the spatially semidiscrete problem is as earlier

$$(10.18)\qquad u_{h,t} - \Delta_h u_h = P_h f, \quad \text{for } t>0, \quad \text{with } u_h(0)=v_h.$$

This problem is of the same form as (10.1), and hence we can apply our multistep time discretization method to define a fully discrete approximation to the solution of (10.17). Here $\|\cdot\|$ and $\|\cdot\|_r$ are the norms in $L_2(\Omega)$ and $H^r = H^r(\Omega)$, respectively.

Theorem 10.2 *Let $q \le 6$, and let $U^n \in S_h$ and u be the solutions of* (10.2) *with $A=-\Delta_h$ and $P_h f$ instead of f, and of* (10.17)*, respectively. Then, for u sufficiently smooth,*

$$\|U^n - u^n\| \le C\sum_{j=0}^{q-1}\|U^j - u^j\| + Ch^r\Big(\|v\|_r + \int_0^{t_n}\|u_t\|_r\,ds\Big) + Ck^q\int_0^{t_n}\|u^{(q+1)}\|\,ds, \quad \text{for } n\ge 0.$$

Proof. With $R_h: H_0^1 \to S_h$ the Ritz projection, we write, as often before, $e^n = U^n - u^n = (U^n - R_h u^n) + (R_h u^n - u^n) = \theta^n + \rho^n$. In the standard way ρ^n is bounded as desired, and it remains to consider $\theta^n \in S_h$. We have

$$\bar\partial_q \theta^n - \Delta_h \theta^n = P_h \omega^n, \quad \text{for } n \ge q,$$

where

$$\omega^n = (R_h - I)\bar\partial_q u^n - \big(\bar\partial_q u^n - u_t^n\big) = \sigma^n + \tau^n.$$

Application of Lemma 10.1 to the present context therefore shows

$$\|\theta^n\| \le C\sum_{j=0}^{q-1}\|\theta^j\| + Ck\sum_{j=q}^{n}\|\sigma^j\| + Ck\sum_{j=q}^{n}\|\tau^j\|, \quad \text{for } n\ge q.$$

Here, since $\sum_{j=0}^q \alpha_j = 0$ by (10.6) we have

$$k\|\sigma^n\| \le Ch^r \|\sum_{j=0}^{q} \alpha_j (u^n - u^{n-j})\|_r \le Ch^r \int_{t_{n-q}}^{t_n} \|u_t\|_r \, ds,$$

and τ^n is bounded in (10.16). Together with $\|\theta^j\| \le \|U^j - u^j\| + \|\rho^j\|$ for $j \le q-1$, with the obvious bounds for the $\|\rho^j\|$, this completes the proof. □

For the error bound of Theorem 10.2 to be $O(h^r + k^p)$ we need to prescribe the starting values in an appropriate way. This could be done, e.g., by taking projections onto S_h, such as P_h or R_h, of the starting values in (10.7).

We now turn to the smoothing property of the backward difference method and begin again with the abstract Hilbert space problem (10.1). We have the following stability result. As earlier $|v|_s = \|A^{s/2}v\|$.

Lemma 10.4 *Let $q \le 6$ and $p \ge 0$, and let U^n be the solution of* (10.2). *Then we have, with C independent of the positive definite operator A,*

(10.19)
$$\begin{aligned} t_n^p \|U^n\|^2 + k \sum_{j=q}^{n} t_j^p |U^j|_1^2 &\le C \sum_{j=0}^{q-1} (|U^j|_{-p}^2 + k^p \|U^j\|^2) \\ &+ Ck \sum_{j=q}^{n} (|f^j|_{-p-1}^2 + t_j^p |f^j|_{-1}^2), \quad \text{for } n \ge q. \end{aligned}$$

Proof. We shall show that for the solution U^n of (10.9) we have

(10.20)
$$\begin{aligned} n^p (U^n)^2 + \lambda \sum_{j=q}^{n} j^p (U^j)^2 &\le C \sum_{j=0}^{q-1} (\lambda^{-p} + 1)(U^j)^2 \\ &+ C \sum_{j=q}^{n} (\lambda^{-p-1} + j^p \lambda^{-1})(g^j)^2. \end{aligned}$$

Setting $\lambda = k\mu, g^j = kf^j$, and multiplying by k^p, we obtain the result of the lemma for $\mathcal{H} = \mathbb{R}$ and $A = \mu$, from which the general result follows as earlier by eigenfunction expansion.

By linearity it suffices to consider separately the case when $U^j = 0$ for $j \le q-1$, and then the case when $g^j = 0$ for $j \ge q$.

We shall appeal to Lemma 10.3 and first note that as a result of that lemma, for $\beta_j = \beta_j(\lambda)$ as defined there,

(10.21)
$$n^p |\beta_n| + \lambda \sum_{j=0}^{\infty} j^p |\beta_j| \le C(1 + \lambda^{-p}), \quad \text{for } n \ge 0.$$

In fact, for $\lambda \le \lambda_0$, we have by (10.11), $n^p|\beta_n| \le Cn^p e^{-c\lambda n} \le C\lambda^{-p}$ and

$$\lambda \sum_{j=0}^{\infty} j^p |\beta_j| \le C\lambda \sum_{j=0}^{\infty} j^p e^{-c\lambda j} \le C\lambda^{-p},$$

and for $\lambda \geq \lambda_0$, the left-hand side of (10.21) is less than $Cn^pe^{-cn} + C\sum_{j=0}^{\infty} j^pe^{-cj}$, which is bounded.

For the case $U^j = 0$ for $j \leq q-1$ we have $U^n = \sum_{j=0}^{n-q} \beta_j g^{n-j}$, for $n \geq q$, so that using the Schwarz inequality and (10.21) with $p = 0$ we obtain

$$(U^n)^2 \leq \sum_{j=0}^{n-q} |\beta_j| \sum_{j=0}^{n-q} |\beta_j|(g^{n-j})^2 \leq C\lambda^{-1} \sum_{j=0}^{n-q} |\beta_j|(g^{n-j})^2.$$

Hence, since $n^p \leq C(j^p + (n-j)^p)$, we find using (10.21)

$$\begin{aligned} n^p(U^n)^2 &\leq C\lambda^{-1} \sum_{j=0}^{n-q} \big(j^p|\beta_j|(g^{n-j})^2 + (n-j)^p|\beta_j|(g^{n-j})^2\big) \\ &\leq C\lambda^{-1} \sum_{j=0}^{n-q} (\lambda^{-p} + (n-j)^p)(g^{n-j})^2, \end{aligned} \tag{10.22}$$

which is the desired estimate for the first term in (10.20). For the second term in (10.20) we obtain by summation of (10.22)

$$\begin{aligned} \lambda \sum_{n=q}^{N} n^p(U^n)^2 &\leq C \sum_{n=q}^{N} \sum_{j=0}^{n-q} \big(j^p|\beta_j|(g^{n-j})^2 + |\beta_j|(n-j)^p(g^{n-j})^2\big) \\ &\leq C \sum_{n=q}^{N} (g^n)^2 \sum_{j=0}^{N-q} j^p|\beta_j| + C \sum_{n=q}^{N} n^p(g^n)^2 \sum_{j=0}^{N-q} |\beta_j| \\ &\leq C\lambda^{-1} \sum_{n=q}^{N} (\lambda^{-p} + n^p)(g^n)^2, \end{aligned}$$

which completes the proof in the present case.

We now consider the case that $g^j = 0$, $j \geq q$, and assume first that $U^1 = \cdots = U^{q-1} = 0, U^0 = 1$. Then $U^n = -\tilde{\beta}_{n-q}\alpha_q$, and hence

$$n^p(U^n)^2 \leq Cn^p\beta_{n-q}^2 \leq C(1 + (n-q)^p)\beta_{n-q}^2 \leq C(1 + \lambda^{-p}).$$

From this we also obtain

$$\lambda \sum_{n=q}^{N} n^p(U^n)^2 \leq C\lambda \sum_{n=q}^{N} (1 + (n-q)^p)|\beta_{n-q}| \leq C(1 + \lambda^{-p}),$$

which shows (10.19). The arguments in the remaining cases that $U^j = \delta_{ij}$ for $j = 0, \ldots, q-1$ with $i = 1, \ldots, q-1$ are analogous, and with this the proof of the lemma is complete. □

We are now ready for the following nonsmooth data error estimate.

Theorem 10.3 *Let $q \le 6$, and let U^n and u be the solutions of* (10.2) *and* (10.1), *respectively, with $f \equiv 0$, and with the discrete initial values satisfying*

$$|U^j - u^j|_{-2q} + k^q \|U^j - u^j\| \le Ck^q \|v\|, \quad \text{for } j = 0, \dots, q-1. \tag{10.23}$$

Then, with C independent of the positive definite operator A,

$$\|U^n - u^n\| \le Ck^q t_n^{-q} \|v\|, \quad \text{for } n \ge 1. \tag{10.24}$$

Proof. Since the error $e^n = U^n - u^n$ satisfies (10.14), Lemma 10.4 shows

$$t_n^{2q} \|e^n\|^2 \le Ck^{2q} \|v\|^2 + Ck \sum_{j=q}^{n} (t_j^{2q} |\tau^j|_{-1}^2 + |\tau^j|_{-2q-1}^2), \quad \text{for } n \ge q.$$

In the same way as in (10.16), we have, for $l = 1, 2q+1$,

$$|\tau^j|_{-l}^2 \le Ck^{2q-1} \int_{t_{j-q}}^{t_j} |u^{(q+1)}(t)|_{-l}^2 \, dt, \quad \text{for } l = 1,\ 2q+1,\ j \ge q.$$

Except in the case $j = q, l = 1$, it follows that

$$kt_j^{2q+1-l} |\tau^j|_{-l}^2 \le Ck^{2q} \int_{t_{j-q}}^{t_j} t^{2q+1-l} |u^{(q+1)}(t)|_{-l}^2 \, dt.$$

Here $|u^{(q+1)}(t)|_{-l} \le C|u(t)|_{2q+2-l}$, and hence

$$k \sum_{j=q}^{n} (t_{j+1}^{2q} |\tau^{j+1}|_{-1}^2 + |\tau^j|_{-2q-1}^2) \le Ck^{2q} \int_0^{t_n} (t^{2q} |u(t)|_{2q+1}^2 + |u(t)|_1^2) \, dt.$$

Letting $\{\lambda_j\}_{j=1}^\infty$ and $\{\varphi_j\}_{j=1}^\infty$ denote the eigensystem of A we have

$$\int_0^{t_n} (t^{2q} |u(t)|_{2q+1}^2 + |u(t)|_1^2) \, dt$$
$$\le \int_0^\infty \sum_{j=1}^\infty (t^{2q} \lambda_j^{2q+1} + \lambda_j) e^{-2\lambda_j t} (v, \varphi_j)^2 \, dt \le C \sum_{j=1}^\infty (v, \varphi_j)^2 = C\|v\|^2.$$

For the remaining term we have, since $\sum_{j=0}^q \alpha_j = 0$,

$$\tau^q = k^{-1} \sum_{j=0}^q \alpha_j u(t_{q-j}) - u'(t_q) = k^{-1} \sum_{j=0}^q \alpha_j \int_0^{t_{q-j}} u' \, dt - u'(t_q),$$

and hence

$$kt_q^{2q} |\tau^q|_{-1}^2 \le Ck^{2q} \Big(\int_0^{t_q} |u'|_{-1}^2 \, dt + k|u'(t_q)|_{-1}^2 \Big) \le Ck^{2q} \|v\|^2.$$

Together these estimates show the error bound in (10.24) for $n \ge q$. For $n = 1, \dots, q-1$, it follows from (10.23) that $\|U^n - u^n\|$ is bounded, which shows (10.24) in this case. The proof is now complete. □

To satisfy (10.23) we may, e.g., choose the starting values

$$U^j = r(kA)^j v, \quad \text{for } j = 0, \dots, q-1, \tag{10.25}$$

where $r(\lambda)$ is a rational function of type IV which is accurate of order $q-1$ (cf. Chapter 7). In fact, by spectral representation, we then have

$$\begin{aligned} |U^j - u^j|_{-2s} &= \|A^{-s}(r(kA)^j - e^{-jkA})v\| \\ &\le k^s \sup_{\lambda>0} |\lambda^{-s}(r(\lambda)^j - e^{-j\lambda})| \, \|v\| \le Ck^s\|v\|, \quad \text{for } s = 0, q. \end{aligned}$$

Note that this choice of $U^1, \dots, U^{q-1}$ corresponds to applying a single step operator for the first $q-1$ steps, and that its accuracy only needs to be $O(k^{q-1})$ since it is only used a fixed number of times. For instance, if $q=2$, U^1 may be computed by the first-order backward Euler method. We remark that the bound for the second term in (10.23) is equivalent to $\|U^j\| \le C\|v\|$.

In order to show an L_2-norm error estimate for a fully discrete method, we now apply our above nonsmooth data error estimate to the solution of the semidiscrete equation (10.18). We recall that, with $v_h = P_h v$, the solution of (10.18) satisfies

$$\|u_h(t) - u(t)\| \le Ch^r t^{-r/2}\|v\|, \quad \text{for } t \ge 0. \tag{10.26}$$

We use $|v|_{-2q,h} = \|(-\Delta_h)^q v\|$ for a discrete analogue of $|v|_{-2q}$.

Theorem 10.4 *Assume $q \le 6$, and let U^n and u_h be the solutions of* (10.2) *and* (10.18), *with $v_h = P_h v$. Assume that the starting procedure is such that, for $j = 0, \dots, q-1$,*

$$|U^j - u_h(t_j)|_{-2q,h} + k^q\|U^j - u_h(t_j)\| \le Ck^q\|v\|.$$

Then

$$\|U^n - u(t_n)\| \le C(h^r t_n^{-r/2} + k^q t_n^{-q})\|v\|, \quad \textit{for } n \ge q, \ t_n \in J = (0, \bar{t}].$$

Proof. Using the triangle inequality this follows at once by Theorem 10.3, applied to (10.18), together with (10.26). □

Initial values satisfying the assumptions of the theorem may now be chosen in the form (10.25) with $A = -\Delta_h$ and with $r(\lambda)$ as discussed there.

We close this chapter with a discussion of the second order backward difference method with variable time steps for the abstract initial value problem (10.1) in our Hilbert space framework, with norms $\|v\|$ and $|v|_s = \|A^{s/2}v\|$.

Let $0 = t_0 < t_1 < \cdots < t_n < \cdots$ be a partition of the time axis and $k_n = t_n - t_{n-1}$ the variable step-sizes. We now introduce the variable step two step backward difference operator

$$\bar\partial_2 U^n = \frac{k_n + k_{n-1}}{k_{n-1}} \frac{U^n - U^{n-1}}{k_n} - \frac{k_n}{k_{n-1}} \frac{U^n - U^{n-2}}{k_n + k_{n-1}}. \tag{10.27}$$

This is a second order approximation of the time derivative in the sense that it is exact for polynomials of degree 2 and, as is easily checked, we have for smooth u

$$\bar\partial_2 u(t_n) = u'(t_n) + O(k_n^2 + k_{n-1}^2), \quad \text{as } k_n, k_{n-1} \to 0.$$

The approximation U^n of the solution $u(t_n)$ of (10.1) is now defined by

$$\begin{aligned} \bar\partial_2 U^n + AU^n &= f^n, \quad \text{for } n \geq 2, \\ \bar\partial U^1 + AU^1 &= f^1, \quad \text{and } U^0 = v; \end{aligned} \tag{10.28}$$

as before, since $\bar\partial_2$ involves three time levels, two starting values are needed.

We shall first show a stability result for (10.28), which generalizes the result of Lemma 10.1 when $q = 2$ and the time steps are constant. As we shall see, our analysis will also require that the stepsize ratio $\gamma_n = k_n/k_{n-1}$ is bounded by the number $\gamma^* = (2 + \sqrt{13})/3 \approx 1.86$. The stability result will contain the quantity

$$\Gamma_n = \sum_{j=2}^{n-2} [\gamma_j - \gamma_{j+2}]_+, \quad \text{where } [x]_+ = \max(x, 0). \tag{10.29}$$

We note that $\Gamma_n = 0$ if γ_n is nondecreasing (in particular when the k_j are constant), and $\Gamma_n = \gamma_2 + \gamma_3 - \gamma_{n-1} - \gamma_n \leq 2\gamma^*$ if γ_n is decreasing. For example, if $t_j = (j/N)^\alpha$, with $\alpha > 1$, then $k_j = (j^\alpha - (j-1)^\alpha)N^{-\alpha}$ and $\gamma_j = ((j^\alpha - (j-1)^\alpha)/((j-1)^\alpha - (j-2)^\alpha)$, and one easily finds that k_j increases and γ_j decreases to 1 as $j \to \infty$ (in particular $\gamma_j \leq \gamma^*$ except for a finite number of j). More generally, Γ_n is bounded if the number of changes in monotonicity in γ_n is bounded.

The rather technical proof of the stability lemma will use the following discrete form of Gronwall's lemma.

Lemma 10.5 *Assume that w_n, $n \geq 0$, satisfy*

$$w_n \leq \alpha_n + \sum_{k=0}^{n-1} \beta_k w_k, \quad \textit{for } n \geq 0,$$

where α_n is nondecreasing and $\beta_n \geq 0$. Then $w_n \leq \alpha_n \exp\big(\sum_{k=0}^{n-1} \beta_k\big)$.

Proof. Setting $u_m = \alpha_n + \sum_{k=0}^{m-1} \beta_k w_k$ we have, for $m \leq n$,

$$u_m = u_{m-1} + \beta_{m-1} w_{m-1} \leq (1 + \beta_{m-1}) u_{m-1} \leq e^{\beta_{m-1}} u_{m-1}.$$

Since $u_0 = \alpha_n$ the result follows. □

Lemma 10.6 *Let U^n be the solution of* (10.28), *and assume $\gamma_n \le \gamma^*$. Then we have, with C independent of A,*

$$\|U^n\| \le Ce^{C\Gamma_n}\Big(\|v\| + \sum_{j=1}^{n} k_j\|f^j\|\Big), \quad \textit{for } t_n \ge 0.$$

Proof. With $\gamma_n = k_n/k_{n-1}$ we set $\omega_n = k_{n-1}/(k_n+k_{n-1}) = 1/(1+\gamma_n)$ and $\psi_n = \psi(\gamma_n) = \gamma_n^2/(1+\gamma_n)^2 = (k_n/(k_n+k_{n-1}))^2$. We write (10.27) as

$$\bar\partial_2 U^n = \frac{1}{\omega_n k_n}(\delta_1 U^n - \psi_n \delta_2 U^n), \text{ where } \delta_l U^n = U^n - U^{n-l}. \tag{10.30}$$

For the purpose of using energy arguments we take inner products by $v = 2\omega_n k_n(U^n + \nu\delta_1 U^n)$ in the first equation of (10.28), where $\nu > 0$ is a parameter to be chosen below, to obtain, with $A(v,w) = (Av, w)$,

$$\begin{aligned} 2\omega_n k_n(\bar\partial_2 U^n, U^n + \nu\delta_1 U^n) &+ 2\omega_n k_n A(U^n, U^n + \nu\delta_1 U^n) \\ &= 2\omega_n k_n(f^n, U^n + \nu\delta_1 U^n), \quad \text{for } n \ge 2. \end{aligned} \tag{10.31}$$

We shall now carry out several technical manipulations with the terms of this equation to finally arrive at the stability estimate claimed.

Expanding the first term on the left-hand side of (10.31) we have

$$\begin{aligned} 2\omega_n k_n(\bar\partial_2 U^n, U^n + \nu\delta_1 U^n) &= 2(\delta_1 U^n, U^n) - 2\psi_n(\delta_2 U^n, U^n) \\ &+ 2\nu\|\delta_1 U^n\|^2 - 2\nu\psi_n(\delta_2 U^n, \delta_1 U^n) = I_1^n + I_2^n + I_3^n + I_4^n. \end{aligned} \tag{10.32}$$

Using the identity

$$2(\delta_l U^n, U^n) = \delta_l\|U^n\|^2 + \|\delta_l U^n\|^2, \quad \text{for } l = 1, 2,$$

we find

$$I_1^n = \delta_1\|U^n\|^2 + \|\delta_1 U^n\|^2,$$

and

$$I_2^n = -\psi_n\delta_2\|U^n\|^2 - \psi_n\|\delta_2 U^n\|^2.$$

Since $\delta_2 U^n = \delta_1 U^n + \delta_1 U^{n-1}$ we have

$$\|\delta_2 U^n\|^2 \le 2\|\delta_1 U^n\|^2 + 2\|\delta_1 U^{n-1}\|^2,$$

and hence

$$I_2^n \ge -\psi_n\delta_2\|U^n\|^2 - 2\psi_n\|\delta_1 U^n\|^2 - 2\psi_n\|\delta_1 U^{n-1}\|^2.$$

In the same way, since

$$2(\delta_1 U^n, \delta_1 U^{n-1}) \le \|\delta_1 U^n\|^2 + \|\delta_1 U^{n-1}\|^2,$$

we find

$$\begin{aligned} I_4^n &= -2\nu\psi_n\|\delta_1 U^n\|^2 - 2\nu\psi_n(\delta_1 U^{n-1}, \delta_1 U^n) \\ &\ge -3\nu\psi_n\|\delta_1 U^n\|^2 - \nu\psi_n\|\delta_1 U^{n-1}\|^2. \end{aligned}$$

Collecting terms we therefore obtain from (10.32)

$$\begin{aligned} &2\omega_n k_n(\bar\partial_2 U^n, U^n + \nu\delta_1 U^n) \\ &\ge \delta_1\|U^n\|^2 - \psi_n\delta_2\|U^n\|^2 + a_n\|\delta_1 U^n\|^2 - b_n\|\delta_1 U^{n-1}\|^2, \\ &\quad \text{where } a_n = 1 + 2\nu - (2+3\nu)\psi_n,\ b_n = (2+\nu)\psi_n. \end{aligned} \tag{10.33}$$

We proceed with the second term in (10.31). With $|v|_1 = A(v,v)^{1/2}$ we have, without the factor $\omega_n k_n$,

$$2A(U^n, U^n + \nu\delta_1 U^n) = 2|U^n|_1^2 + 2\nu A(U^n, \delta_1 U^n).$$

Since

$$2A(U^n, \delta_1 U^n) = \delta_1|U^n|_1^2 + |\delta_1 U^n|_1^2 \ge |U^n|_1^2 - |U^{n-1}|_1^2,$$

we find

$$\begin{aligned} 2\omega_n k_n A(U^n, U^n + \nu\delta_1 U^n) &\ge c_n k_n|U^n|_1^2 - d_n k_{n-1}|U^{n-1}|_1^2, \\ &\quad \text{where } c_n = (2+\nu)\omega_n,\ d_n = \nu\omega_n\gamma_n. \end{aligned} \tag{10.34}$$

Hence, using (10.31), (10.33) and (10.34) we thus obtain

$$\begin{aligned} &(\delta_1\|U^n\|^2 - \psi_n\delta_2\|U^n\|^2) + (a_n\|\delta_1 U^n\|^2 - b_n\|\delta_1 U^{n-1}\|^2) \\ &+ (c_n k_n|U^n|_1^2 - d_n k_{n-1}|U^{n-1}|_1^2) \le C k_n\|f^n\|(\|U^n\| + \|U^{n-1}\|), \end{aligned} \tag{10.35}$$

or, with obvious notation,

$$J_1^n + J_2^n + J_3^n \le \bar J^n.$$

We now sum this inequality from $n = 2$ to N. Beginning with the left hand side we have

$$\begin{aligned} \sum_{n=2}^N J_1^n &= \|U^N\|^2 - \|U^1\|^2 - \sum_{n=2}^N \psi_n\|U^n\|^2 + \sum_{n=0}^{N-2}\psi_{n+2}\|U^n\|^2 \\ &= (1-\psi_N)\|U^N\|^2 - \psi_{N-1}\|U^{N-1}\|^2 - (1-\psi_3)\|U^1\|^2 + \psi_2\|U^0\|^2 \\ &\quad - \sum_{n=2}^{N-2}(\psi_n - \psi_{n+2})\|U^n\|^2. \end{aligned}$$

Hence, noting that $\psi_n < 1$ and replacing negative terms in the last sum by 0,

$$\sum_{n=2}^{N} J_1^n \geq (1-\psi_N)\|U^N\|^2 - \psi_{N-1}\|U^{N-1}\|^2 - C\|U^1\|^2 - \sum_{n=2}^{N-2} [\psi_n - \psi_{n+2}]_+ \|U^n\|^2. \tag{10.36}$$

Moreover

$$\sum_{n=2}^{N} J_2^n = \sum_{n=2}^{N-1} (a_n - b_{n+1})\|\delta_1 U^n\|^2 + a_N\|\delta_1 U^N\|^2 - b_2\|\delta_1 U^1\|^2,$$

and

$$\sum_{n=2}^{N} J_3^n = \sum_{n=2}^{N-1} (c_n - d_{n+1})k_n |U^n|_1^2 + c_N k_N |U^N|_1^2 - d_2 k_1 |U^1|_1^2.$$

We shall show that if $\gamma_n \leq \gamma^*$ for all n, then $a_n - b_{n+1} \geq 0$ and $c_n - d_{n+1} \geq 0$, which implies

$$\sum_{n=2}^{N} (J_2^n + J_3^n) \geq -C(\|U^1\|^2 + \|U^0\|^2) - Ck_1|U^1|_1^2. \tag{10.37}$$

For the proof, assume first only that $\gamma_n \leq \gamma$ for all n. Since a_n and b_n are decreasing and increasing functions of ψ_n, and thus also of γ_n, we then have

$$\begin{aligned} a_n - b_{n+1} &\geq 1 + 2\nu - (2+3\nu)(\frac{\gamma}{1+\gamma})^2 - (2+\nu)(\frac{\gamma}{1+\gamma})^2 \\ &= 1 + 2\nu - 4(1+\nu)(\frac{\gamma}{1+\gamma})^2 \geq 0, \quad \text{if } \gamma \leq \frac{\sqrt{1+2\nu}}{2\sqrt{1+\nu} - \sqrt{1+2\nu}}, \end{aligned}$$

and, for similar reasons,

$$c_n - d_{n+1} \geq \frac{2+\nu}{1+\gamma} - \frac{\nu\gamma}{1+\gamma} \geq 0, \quad \text{if } \gamma \leq 1 + \frac{2}{\nu}.$$

Replacing these inequalities by equalities gives $\gamma = \gamma^* = (2+\sqrt{13})/3$ and $\nu = \nu^* = (1+\sqrt{13})/2$, so that choosing ν in this way, (10.37) holds for $\gamma_n \leq \gamma^*$.

From (10.35), (10.36) and (10.37) we now obtain, for $N \geq 2$,

$$\begin{aligned} &(1-\psi_N)\|U^N\|^2 \\ &\leq \psi_{N-1}\|U^{N-1}\|^2 + C(\|U^0\|^2 + \|U^1\|^2 + k_1|U^1|_1^2) \\ &+ C\sum_{n=2}^{N} k_n \|f^n\|(\|U^n\| + \|U^{n-1}\|) + \sum_{n=2}^{N-2} [\psi_n - \psi_{n+2}]_+ \|U^n\|^2. \end{aligned} \tag{10.38}$$

Here $[\psi_n - \psi_{n+2}]_+ \leq 2[\gamma_n - \gamma_{n+2}]_+$ since $\psi_n = \psi(\gamma_n)$ with ψ increasing and $\psi' \leq 2$. Further, for the terms in U^1 it follows after multiplication of the equation for U^1 in (10.28) by U^1 that

$$\|U^1\|^2 + ck_1|U^1|_1^2 \leq \|U^0\|^2 + 2k_1\|f^1\|\,\|U^1\|. \tag{10.39}$$

Since $\psi_N \leq \psi(\gamma^*) \leq 4/9 < 1$ and $\psi_{N-1}/(1-\psi_N) \leq (4/9)/(1-4/9) = 4/5 < 1$, we may divide (10.38) by $1 - \psi_N$ and apply (10.39) to obtain, for $N \geq 1$,

$$\begin{aligned} \|U^N\|^2 &\leq \tfrac{4}{5}\|U^{N-1}\|^2 + C\|U^0\|^2 \\ &+ C\sum_{n=1}^{N} k_n\|f^n\|(\|U^n\| + \|U^{n-1}\|) + C\sum_{n=2}^{N-2}[\gamma_n - \gamma_{n+2}]_+\|U^n\|^2. \end{aligned} \tag{10.40}$$

Now let M be such that $\|U^M\| = \max_{0\leq n\leq N}\|U^n\|$, $0 \leq M \leq N$. Then (10.40) holds with N replaced by M, and bounding one factor in each term on the right by $\|U^M\|$ and then canceling one such factor on both sides shows

$$\|U^M\| \leq C\|U^0\| + C\sum_{n=1}^{M} k_n\|f^n\| + C\sum_{n=2}^{M-2}[\gamma_n - \gamma_{n+2}]_+\|U^n\|.$$

Since $\|U^N\| \leq \|U^M\|$ and $M \leq N$, the same inequality is valid for $M = N$, and an application of Lemma 10.5 now completes the proof. □

We are now in a position to show the following error estimate.

Theorem 10.5 *Assume that $\gamma_* \leq \gamma \leq \gamma^*$, with $\gamma_* > 0$, and let U^n and u be the solutions of (10.28) and (10.1). Then, for $n \geq 0$,*

$$\|U^n - u^n\| \leq Ce^{C\Gamma_n}\Big(k_1\int_0^{t_1}\|u''\|\,ds + \sum_{j=1}^{n} k_j^2\int_{t_{j-1}}^{t_j}\|u'''\|\,ds\Big).$$

Proof. Let $e^n = U^n - u^n$. Then we have

$$\begin{aligned} (\bar\partial_2 e^n, v) + A(e^n, v) &= -(\bar\partial_2 u^n - (u')^n, v) \equiv -(\tau^n, v), \quad \text{for } n \geq 2, \\ (\bar\partial e^1, v) + A(e^1, v) &= -(\bar\partial u(t_1) - u'(t_1), v) \equiv -(\tau^1, v). \end{aligned}$$

By Lemma 10.6 we have, since $e^0 = 0$,

$$\|e^n\| \leq Ce^{C\Gamma_n}\sum_{j=1}^{n} k_j\|\tau^j\|.$$

Using Taylor's formula and (10.30) we find, for $j \geq 2$,

$$2\omega_j k_j \tau^j = \int_{t_{j-1}}^{t_j}(s - t_{j-1})^2 u'''(s)\,ds - \psi_j\int_{t_{j-2}}^{t_j}(s - t_{j-2})^2 u'''(s)\,ds,$$

with ω_j and ψ_j bounded away from 0 and ∞, and for τ^1 we have

$$k_1\tau^1 = -\int_0^{t_1} su''(s)\,ds.$$

Taking norms and using obvious estimates completes the proof. □

We remark that for constant time steps Theorem 10.5 shows

$$\|U^n - u(t_n)\| \le C\big(k\int_0^{t_1} \|u''\|\,ds + k^2\int_0^{t_n} \|u'''\|\,ds\big),$$

which is of order $O(k^2)$ if u is smooth.

With the example given in the discussion of Γ_n after (10.29) in mind, we note that by writing the equation in (10.28) in the form

$$U^n + \alpha_{n0}k_nAU^n = \alpha_{n1}U^{n-1} + \alpha_{n2}U^{n-2} + \alpha_{n3}k_nf^n, \quad \text{for } n \ge 2,$$

it is easy to see that the conclusions of Lemma 10.6 and Theorem 10.5 remain valid if the condition $\gamma_n \le \gamma^*$ is violated for at most a fixed finite number of n, but γ_n is bounded.

Multistep methods may also be considered in a Banach space framework, allowing the derivation of maximum-norm estimates for the concrete heat equation. We illustrate this with an analysis of the two-step backward difference method with constant time steps.

We consider thus the initial value problem for the homogeneous equation,

$$u' + Au = 0, \quad \text{for } t > 0, \quad \text{with } u(0) = v, \tag{10.41}$$

in a complex Banach space $\mathcal{B}$ with norm $\|\cdot\|$, where A is a closed densely defined linear operator such that, with the notation of Chapter 6, $\rho(A) \supset \Sigma_\delta$ for some $\delta \in (0, \frac{1}{2}\pi)$, and such that the resolvent estimate

$$\|R(z;A)\| \le M|z|^{-1}, \quad \text{for } z \in \Sigma_\delta$$

holds. We shall consider the two-step backward difference equation

$$\begin{aligned} &(\tfrac{3}{2}U^n - 2U^{n-1} + \tfrac{1}{2}U^{n-2})/k + AU^n = 0, \quad \text{for } n \ge 2,\\ &U^0 = v, \quad \bar\partial U^1 + AU^1 = 0. \end{aligned} \tag{10.42}$$

Solving step by step for U^n this shows that

$$U^n = r_n(kA)v, \tag{10.43}$$

where $r_n(z)$ is a rational function with poles at $z = -\frac{3}{2}, -1$. In fact, we may write (10.42) as a difference equation for the vector $(U^n, U^{n-1})^T$, viz.,

$$\begin{pmatrix} U^n \\ U^{n-1} \end{pmatrix} = R(kA) \begin{pmatrix} U^{n-1} \\ U^{n-2} \end{pmatrix} = R(kA)^{n-1} \begin{pmatrix} U^1 \\ U^0 \end{pmatrix} = R(kA)^{n-1} S(kA)v,$$

where

$$R(z) = \begin{pmatrix} \dfrac{2}{\frac{3}{2}+z} & -\dfrac{\frac{1}{2}}{\frac{3}{2}+z} \\ 1 & 0 \end{pmatrix}, \quad S(z) = \begin{pmatrix} \dfrac{1}{1+z} \\ 1 \end{pmatrix}.$$

It follows that, with $e_1 = (1,\ 0)$ the first unit vector,

$$r_n(z) = e_1\, R(z)^{n-1}\, S(z), \quad \text{for } n \geq 1, \quad \text{with } r_0(z) = 1. \tag{10.44}$$

Note that $r_n(\infty) = 0$ even though $R(\infty) \neq 0$. With the notation from the beginning of this chapter and (10.6) we have for the characteristic polynomial of the second order difference equation, with $\zeta = e^z$, for small z,

$$P(e^z; z) = (\tfrac{3}{2} + z)e^{2z} - 2e^z + \tfrac{1}{2} = e^{2z}(\widetilde{\alpha}(e^{-z}) + z) = O(z^3). \tag{10.45}$$

We note that this method is A–stable: The eigenvalues of $R(z)$ are $\zeta_{1,2}(z) = (2 \pm (1-2z)^{1/2})/(3+2z)$, and satisfy $|\zeta_{1,2}(z)| < 1$ for all z with $\operatorname{Re} z \geq 0$, exept at $z = 0$ where $\zeta_1(0) = 1$, $\zeta_2(0) = \frac{1}{3}$. For $z = \frac{1}{2}$ the eigenvalue is double, with $\zeta_{1,2}(\frac{1}{2}) = \frac{1}{2}$, and for other z they are simple. In fact, let D be the component of the set $\{z; |\zeta_{1,2}(z)| < 1\}$ containing $z = \frac{1}{2}$. Then for z on the boundary ∂D, there is an eigenvalue of the form $\zeta = e^{i\theta}$, and hence

$$P(e^{i\theta}; z) = (\tfrac{3}{2} + z)e^{2i\theta} - 2e^{i\theta} + \tfrac{1}{2} = 0.$$

For $\theta = 0$ we must have $z = 0$, and for $\theta \in (0, \pi]$ (after multiplication by $e^{-2i\theta}$),

$$\operatorname{Re} z = -(\tfrac{3}{2} - 2\cos\theta + \tfrac{1}{2}\cos 2\theta) = -(\cos\theta - 1)^2 < 0.$$

Thus $\partial D \subset \{z : \ \operatorname{Re} z < 0\} \cup \{0\}$, which shows the A–stability. We note that with $\zeta_2(z)$ the eigenvalue with smallest modulus, we have $|\zeta_2(z)| \leq \frac{1}{2}$, because $1/\zeta_{1,2}(z) = 2 \pm \sqrt{1-2z}$ and both these values cannot have modulus ≤ 2.

We begin with the following stability result.

Theorem 10.6 *There is a constant C such that for U^n defined by* (10.43)

$$\|U^n\| \leq CM\|v\|, \quad \textit{for } n \geq 0.$$

Proof. Since $r_n(\infty) = 0$ we have by Lemma 9.1, with Γ suitable,

$$U^n = r_n(kA)v = \frac{1}{2\pi i}\int_\Gamma r_n(kz)R(z;A)v\,dz, \tag{10.46}$$

and the result therefore follows as in the proof of Theorem 9.1 from the following lemma. □

Lemma 10.7 *For arbitrary $R > 0$ and $\psi \in (0, \frac{1}{2}\pi)$ there are ε, c, and $C > 0$ such that*

$$|r_n(z)| \le \begin{cases} C\, e^{Cn|z|}, & \text{for } |z| \le \varepsilon, \\ Ce^{-cn|z|}, & \text{for } |\arg z| \le \psi,\ |z| \le R. \end{cases}$$

Proof. There exists a Hermitian matrix $H = H(z)$ such that

$$R(z) = H(z) \begin{pmatrix} \zeta_1(z) & c(z) \\ 0 & \zeta_2(z) \end{pmatrix} H^*(z),$$

where $c(z)$ is bounded, and thus

$$R^n = R(z)^n = H \begin{pmatrix} \zeta_1^n & c_n \\ 0 & \zeta_2^n \end{pmatrix} H^*. \quad \text{where } c_n = c \sum_{j=0}^{n-1} \zeta_1^j \zeta_2^{n-1-j}.$$

Hence, using $\zeta_1(z) = 1 - z + O(z^2)$ for z small, we have, for $|z| \le \varepsilon$,

$$|\zeta_1(z)|^n \le e^{Cn|z|},\ |\zeta_2(z)|^n \le 2^{-n}, \quad \text{and } |c_n(z)| \le C \sum_{j=0}^{n-1} e^{cj|z|} 2^{-j} \le Ce^{Cn|z|}.$$

which shows $|R^n(z)| \le Ce^{Cn|z|}$ for $|z| \le \varepsilon$. Since $S(z)$ is bounded this yields the first bound of the lemma.

For $|z| \le \varepsilon$ and $|\arg z| \le \psi$ we have similarly

$$|\zeta_1(z)|^n \le e^{-n(\operatorname{Re} z + O(|z|^2))} \le e^{-cn|z|}, \quad |\zeta_2(z)|^n \le 2^{-n},$$

and, since $|\zeta_2(z)| \le \frac{3}{4}|\zeta_1(z)|$ for small $|z|$,

$$|c_n(z)| \le C|\zeta_1(z)|^{n-1} \le C\, e^{-cn|z|}.$$

On the compact set $\{z;\ \varepsilon \le |z| \le R,\ |\arg z| \le \psi\}$ we have $|\zeta_{1,2}(z)| \le \rho < 1$ and hence, with $\rho < \rho_1 < 1$, $|R(z)^n| \le C\rho_1^n \le Ce^{-cn|z|}$. Together these estimates show the second bound of the lemma. □

We now show the following error estimate which covers both smooth and nonsmooth initial data.

Theorem 10.7 *We have for the solutions of* (10.41) *and* (10.43)

$$\|U^n - u(t_n)\| \le CMk^j t_n^{-l} \|A^{j-l} v\|, \quad \text{for } 0 \le l \le j \le 2.$$

Proof. We note that

$$U^n - u(t_n) = r_n(kA)v - E(t_n)v = F_n(kA)v, \text{ with } F_n(z) = r_n(z) - e^{-nz}.$$

Using (10.46) and Lemma 9.3, we have, with the notation of that lemma,

$$U^n - u(t_n) = \frac{1}{2\pi i} \int_{\gamma_\varepsilon \cup \Gamma_\varepsilon} (r_n(kz) - e^{-nz})\, R(z; A)v\, dz.$$

The result therefore easily follows by the following lemma. □

Lemma 10.8 *We have for any* $\psi \in (0, \frac{1}{2}\pi)$, $n \geq 2$,

$$|r_n(z) - e^{-nz}| \leq \begin{cases} C|z|^2 e^{Cn|z|}, & \textit{for } |z| \leq \varepsilon, \\ C|z|^2 e^{-cn|z|}, & \textit{for } |\arg z| \leq \psi, \ |z| \leq R, \end{cases}$$

Proof. We have, with $Y(z) = (e^{-z}, 1)^T$,

$$\begin{aligned} r_n(z) - e^{-nz} &= e_1 \big(R(z)^{n-1} S(z) - e^{-(n-1)z} Y(z)\big) \\ &= e_1 \big(R(z)^{n-1} - e^{-(n-1)z} I\big) Y(z) + e_1 R(z)^{n-1} \big(S(z) - Y(z)\big) \\ &= e_1 \sum_{j=0}^{n-2} R(z)^j e^{-(n-2-j)z} \big(R(z) - e^{-z} I\big) Y(z) + e_1 R(z)^{n-1} \big(S(z) - Y(z)\big). \end{aligned}$$

Using (10.45) we find easily

$$R(z)Y(z) - e^{-z} Y(z) = O(z^3) \quad \text{and} \quad S(z) - Y(z) = O(z^2), \qquad \text{as } z \to 0.$$

Hence

$$|r_n(z) - e^{-nz}| \leq C|z|^3 \sum_{j=0}^{n-2} |R(z)|^j |e^{-(n-2-j)z}| + C|z|^2 |R(z)|^{n-1}.$$

The result stated now easily follows from the estimates for $|R(z)|^j$ of the proof of Lemma 10.7. □

We finally apply the above results to derive maximum-norm error estimates for the second order backward difference method for the model homogeneous heat equation in two spatial variables, using piecewise linear approximation functions in space on quasiuniform triangulations of the spatial domain. The problem we consider is thus, with $A = -\Delta$,

$$(10.47) \qquad \begin{aligned} u_t + Au &= 0 \quad \text{in } \Omega, \quad \text{for } t > 0, \\ u &= 0 \quad \text{on } \partial\Omega, \quad \text{for } t > 0, \quad \text{with } u(\cdot, 0) = v \quad \text{in } \Omega, \end{aligned}$$

where Ω is a convex domain in $\mathbb{R}^2$ with smooth boundary $\partial\Omega$, with the spatially discrete analogue defined by

$$(10.48) \qquad u_{h,t} + A_h u_h = 0, \quad \text{for } t > 0, \quad \text{with } u_h(0) = v_h,$$

where $A_h = -\Delta_h$, with Δ_h the discrete Laplacian defined in (1.33).

We recall that the resolvent estimate for A_h of Theorem 6.6 holds, so that our above abstract theory applies. With $r_n(z)$ defined in (10.44), the fully discrete method using the second order backward difference method for (10.48) in time then yields the solution

$$(10.49) \qquad U_h^n = r_n(kA_h) v_h, \quad \text{for } n \geq 1, \quad \text{with } U_h^0 = v_h$$

As an immediate consequence of Theorem 10.6 we then have the following maximum-norm stability result.

Theorem 10.8 *We have for the fully discrete solution of* (10.47), *defined by* (10.49),

$$\|U_h^n\|_{L_\infty} \le C\|v_h\|_{L_\infty}, \quad \text{for } n \ge 0.$$

As in the proof of Theorem 9.6 a combination of Theorems 10.7 and 6.10 show the following nonsmooth data error estimate.

Theorem 10.9 *Let U_h^n and u be defined by* (10.49), *with $v_h = P_h v$, and* (10.47), *respectively. Then, if $v \in L_\infty$, we have*

$$\|U_h^n - u(t_n)\|_{L_\infty} \le C\big(h^2\ell_h^2 t_n^{-1} + k^2 t_n^{-2}\big)\|v\|_{L_\infty}, \quad \text{for } t_n > 0.$$

We close with a smooth data error estimate.

Theorem 10.10 *Let U_h^n and u be defined by* (10.49) *and* (10.47). *Then, if $v \in \mathcal{D}(A^2)$ and if $\|v_h - v\|_{L_\infty} \le Ch^2\ell_h^2\|v\|_{W_\infty^2}$, we have*

$$\|U_h^n - u(t_n)\|_{L_\infty} \le C\big(h^2\ell_h^2\|v\|_{W_\infty^2} + k^2\|A^2 v\|_{L_\infty}\big), \quad \text{for } n \ge 0.$$

Proof. The proof is analogous to that of Theorem 9.7, with $q = 2$. The only modification needed is that we use the error operator $F_n = F_n(kA_h)$ with $F_n(z) = r_n(z) - e^{-nz}$ instead of $F_n(z) = r^n(z) - e^{-nz}$. □

Our presentation of multistep methods with constant time steps is adapted from Bramble, Pasciak, Sammon and Thomée [34]. For other work using spectral techniques, see Zlàmal [251], Crouzeix and Raviart [57], Crouzeix [55], LeRoux [152], [153] and Savaré [207]; in the latter references the results obtained for time-independent operators A are generalized to variable $A = A(t)$ by perturbation arguments.

The analysis in the last part of the chapter on the second order backward difference method with variable time steps is extracted from Becker [23], where more general time dependent and nonselfadjoint operators A are treated. The underlying energy argument for the two step method with constant time steps was given in McLean and Thomée [170] in the case of an integro-differential equation with a positive type memory term. Other multistep methods have been analyzed in, e.g., Crouzeix [54] and Dupont, Fairweather and Johnson [84].

Using spectral methods LeRoux [154] and Palencia and Garcia-Archilla [193] study certain higher order variable stepsize multistep methods with time independent elliptic operator and show stability results similar to those of [23] with a more restrictive Γ_N. For multistep methods with variable steps for ordinary differential equations, see, e.g., Crouzeix and Mignot [56] and Grigorieff [109].

11. Incomplete Iterative Solution of the Algebraic Systems at the Time Levels

In the fully discrete methods for the solution of parabolic equations which we have studied so far, a finite diminensional system of linear algebraic equations has to be solved at each time level of the time stepping procedure, and our analysis has always assumed that these systems are solved exactly. Because in applications these systems are of high dimension, direct methods are most often not appropriate, and iterative methods have to be used. Since the linear system to be solved at an individual time level is a discretization of an elliptic partial differential equation (with the step size occurring as a small parameter), methods normally used for elliptic problems are natural to apply here. In practice, only a moderate finite number of iterations can be carried out at each time level, and it thus becomes interesting to determine how many steps of the iterative algorithm are needed to guarantee that no loss occurs in the order of accuracy compared to the basic procedure in which the systems are solved exactly. For a successful iterative strategy it is also important to make a proper choice of the starting approximation at each time step.

Our purpose in this chapter is to study these questions for a simple model problem, under the appropriate assumptions on the iterative procedure. As an example of an iterative method satisfying our assumptions we shall discuss also in some detail the application of a multigrid algorithm.

As our model problem we shall take the standard backward Euler Galerkin piecewise linear finite element method for the approximate solution of the initial-boundary value problem

$$(11.1)\qquad \begin{aligned} u_t - \Delta u &= f \quad \text{in } \Omega, \qquad \text{for } t > 0,\\ u &= 0 \quad \text{on } \partial\Omega, \quad \text{for } t > 0, \quad \text{with } u(0) = v, \end{aligned}$$

where Ω is a bounded domain in $\mathbb{R}^2$, which we assume for simplicity in the latter part of this chapter to be convex and polygonal. The approximate solution is thus sought in a finite element space $S_h \subset H_0^1 = H_0^1(\Omega)$ satisfying our standard assumption (1.10) with $r = 2$.

Letting k denote the time step, $t_n = nk$, and using the bilinear form $A(v, w) = (\nabla v, \nabla w)$, our approximation scheme is to find $U_h^n \in S_h$, for $n \geq 0$, such that $U^0 = v_h$ and

$$(11.2)\qquad (U_h^n, \chi) + k(\nabla U_h^n, \nabla\chi) = (U_h^{n-1} + kf(t_n), \chi), \quad \forall \chi \in S_h, \ n \geq 1.$$

With Δ_h the discrete analogue of Δ defined by (1.33), and with $A_h = -\Delta_h$, which is positive definite on S_h, the equation at t_n may also be written

$$(I + kA_h)U_h^n = b^n := U_h^{n-1} + kP_h f^n, \tag{11.3}$$

where P_h denotes the L_2-projection onto S_h. This is thus the finite dimensional equation to which we want to apply an iterative solution procedure. As we know from Theorem 1.5, we have for the exact solution of (11.2)

$$\|U_h^n - u(t_n)\| \le C(u)(h^2 + k),$$

provided u is smooth enough and the discrete initial data $U_h^0 = v_h$ have been appropriately chosen, for instance as the elliptic projection $R_h v$ of v onto S_h.

The incomplete iteration backward Euler algorithm is now defined as follows: For $n \ge 2$ and U_h^{n-1} given, instead of the exact solution of (11.3), which we now denote $\bar{U}_h^n$, we define $U_h^n = U_h^{n,M_n}$ by taking a specified number M_n of steps of the iterative process

$$\begin{aligned} U_h^{n,m} = U_h^{n,m-1} - B_{kh}\big((I + kA_h)U_h^{n,m-1} - b^n\big), \quad \text{for } m \ge 1, \\ \text{with } U_h^{n,0} = 2U_h^{n-1} - U_h^{n-2}, \end{aligned} \tag{11.4}$$

where the operators B_{kh} will be chosen so that $U_h^{n,m}$ converges to $\bar{U}_h^n$ as $m \to \infty$. We note that the starting value $U_h^{n,0}$ is chosen as a second order accurate extrapolatory approximation to $\bar{U}_h^n$. We also note that

$$\begin{aligned} U_h^{n,m} - \bar{U}_h^n = D_{kh}^m(U_h^{n,0} - \bar{U}_h^n), \quad \text{for } m \ge 1, \\ \text{where } D_{kh} = I - B_{kh}(I + kA_h), \end{aligned}$$

and that convergence of $U_h^{n,m}$ to $\bar{U}_h^n$ is equivalent to $D_{kh}^m \to 0$ as $m \to \infty$. Since U_h^n is defined in this way only for $n \ge 2$ we assume, for simplicity, that $U_h^1 = \bar{U}_h^1$, i.e., that (11.3) is solved exactly for $n = 1$.

In the study of the stability and convergence properties of the method thus defined we shall use the norm

$$|\chi| = \big(\|\chi\|^2 + kA(\chi,\chi)\big)^{1/2} = \|(I + kA_h)^{1/2}\chi\|, \quad \text{for } \chi \in S_h, \tag{11.5}$$

and we shall make the assumption that the B_{kh} are chosen so that the iterative process (11.4) has the property that, with $c_0 > 0$ and $\kappa < 1$,

$$|U_h^{n,m} - \bar{U}_h^n| \le c_0\kappa^m |U_h^{n,0} - \bar{U}_h^n|, \quad \text{for } m \ge 1. \tag{11.6}$$

Expressed in terms of the operator norm corresponding to the norm $|\chi|$ for $\chi \in S_h$, this may be written

$$|D_{kh}^m| \le c_0\kappa^m, \quad \text{for } m \ge 1. \tag{11.7}$$

Such estimates are typical for preconditioned conjugate gradient iterative methods, with κ related to the condition number of the preconditioned system. As an example we shall discuss a case of the multigrid algorithm below, in which κ may be chosen independent of h.

We begin with the following error estimate for our fully discrete scheme with incomplete iteration in the case that the exact solution of (11.1) is sufficiently smooth. In this result we shall assume that the same number of iterations are taken at each time step, i.e., that M_n is independent of n. Later we shall discuss nonsmooth data error estimates for the homogeneous equation; in this case M_n will depend on n.

Theorem 11.1 *Assume that* (11.6) *holds with* $\kappa < 1$. *Let* $U_h^n = U_h^{n,M}$ *be the approximate solution of* (11.3) *defined by (11.4), with* M *independent of* n, *and let* $U_h^0 = v_h = R_h v$ *and* $U_h^1 = \bar{U}_h^1$. *Then there is a* $\delta > 0$ *such that, if* u *is a sufficiently smooth solution of* (11.1), *we have*

$$\|U_h^n - u(t_n)\| \le C_{\bar{t}}(u)(h^2 + k), \quad \text{for } t_n \le \bar{t}, \quad \text{if } c_0\kappa^M \le \delta.$$

Proof. With $u^n = u(t_n)$ we write in the standard manner

$$e^n = U_h^n - u^n = (U_h^n - R_h u^n) + (R_h u^n - u^n) = \theta^n + \rho^n,$$

and recall that $\|\rho^n\| \le C(u)h^2$. To estimate θ^n we note that, for $n \ge 2$,

$$\begin{aligned} \bar{\partial}\theta^n + A_h\theta^n = \sigma^n &:= (\bar{U}_h^n - U_h^{n-1})/k + A_h\bar{U}_h^n \\ &+ (U_h^n - \bar{U}_h^n)/k + A_h(U_h^n - \bar{U}_h^n) - \bar{\partial}R_h u^n - A_h R_h u^n, \end{aligned} \tag{11.8}$$

with the obvious simplification for $n = 1$ where $\bar{U}_h^1 = U_h^1$. Since $\bar{U}_h^n$ is the exact solution of (11.3) and $A_h R_h = -P_h\Delta$, we find, with $\omega^n = (U_h^n - \bar{U}_h^n)/k$ and $\tau^n = \bar{\partial}u^n - u_t^n$,

$$\begin{aligned} \sigma^n &= P_h f^n + (I + kA_h)\omega^n - \bar{\partial}R_h u^n + P_h\Delta u^n \\ &= P_h u_t^n - \bar{\partial}R_h u^n + (I + kA_h)\omega^n \\ &= (P_h - R_h)u_t^n - R_h\tau^n + (I + kA_h)\omega^n, \quad \text{for } n \ge 1, \end{aligned} \tag{11.9}$$

with $\omega^1 = 0$. A standard energy argument applied to (11.8), cf. Lemma 10.4 with $q = 1, p = 0$, shows, since $\theta_0 = 0$,

$$\|\theta^n\|^2 \le Ck\sum_{j=1}^n |\sigma^j|_{-1,h}^2, \quad \text{where } |\chi|_{-1,h} = \|A_h^{-1/2}\chi\|. \tag{11.10}$$

Here

$$\begin{aligned} |(P_h - R_h)u_t^n - R_h\tau^n|_{-1,h} &\le C\big(\|(P_h - R_h)u_t^n\| + \|R_h\tau^n\|\big) \\ &\le C_{\bar{t}}(u)(h^2 + k), \quad \text{for } t_n \le \bar{t}, \end{aligned} \tag{11.11}$$

and, since $(1+k\lambda)/\lambda$ is bounded for k bounded and λ bounded below,

$$|(I+kA_h)\omega^n|_{-1,h} \le C\|(I+kA_h)^{1/2}\omega^n\| = C|\omega^n|,$$

so that

$$|\sigma^n|_{-1,h} \le C_{\bar t}(u)(h^2+k) + C|\omega^n|.$$

Since $\omega^1 = 0$, we therefore infer from (11.10) that

$$\|\theta^n\|^2 \le C_{\bar t}(u)(h^2+k)^2 + Ck\sum_{j=2}^n |\omega^j|^2, \quad \text{for } t_n \le \bar t. \tag{11.12}$$

In order to estimate the term in ω^j we note that, by (11.6) and the triangle inequality, we have

$$|U_h^n - \bar U_h^n| \le \delta\big(|U_h^{n,0} - U_h^n| + |U_h^n - \bar U_h^n|\big), \quad \text{if } c_0\kappa^M \le \delta.$$

If $\delta/(1-\delta) \le \varepsilon$, with ε to be specified below, we then have

$$|\omega^n| = k^{-1}|U_h^n - \bar U_h^n| \le \varepsilon k^{-1}|U_h^{n,0} - U_h^n|.$$

Noting that $U_h^n - U_h^{n,0} = k^2\bar\partial^2 U_h^n$ by (11.4), we conclude that

$$|\omega^n| \le \varepsilon k|\bar\partial^2 U_h^n| \le \varepsilon k\big(|\bar\partial^2\theta^n| + |R_h\bar\partial^2 u^n|\big), \quad \text{for } n \ge 2. \tag{11.13}$$

Since $|v| \le C\|v\|_1$ for $k \le 1$, we have $|R_h\bar\partial^2 u^n| \le C\|\bar\partial^2 u^n\|_1 \le C_{\bar t}(u)$, so that

$$|\omega^n| \le C\varepsilon\big(|\bar\partial\theta^n| + |\bar\partial\theta^{n-1}|\big) + C_{\bar t}(u)k, \quad \text{for } n \ge 2,\ t_n \le \bar t. \tag{11.14}$$

Hence, by (11.12),

$$\|\theta^n\|^2 \le C_{\bar t}(u)(h^2+k)^2 + C\varepsilon^2 k\sum_{j=1}^n |\bar\partial\theta^j|^2, \quad \text{for } t_n \le \bar t. \tag{11.15}$$

We now need an estimate for the last term in (11.15). For this purpose we show the following lemma, which we express in the Hilbert space framework used earlier, for the backward Euler method

$$(I+kA)U^n = U^{n-1} + kf^n, \quad \text{for } n \ge 1, \quad \text{with } U^0 = v, \tag{11.16}$$

where A is a positive definite selfadjoint operator in the Hilbert space $\mathcal{H}$. In analogy with (11.5) we define $|v| = \|(I+kA)^{1/2}v\|$ where $\|\cdot\|$ is the norm in $\mathcal{H}$, and we also introduce the corresponding dual norm and the associated s-norms defined by

$$|v|_* = \|(I+kA)^{-1/2}v\| \quad \text{and} \quad |v|_{*,s} = |A^{s/2}v|_*. \tag{11.17}$$

For the purpose of later application the lemma is stated in a more general form than needed here.

Lemma 11.1 *Let U^n be the solution of* (11.16) *Then, for $p \geq 0$,*

$$k \sum_{j=1}^{n} t_j^p |\bar{\partial} U^j|^2 \leq C\big(k^{p-1}|v|^2 + |v|_{*,-p+1}^2\big) + Ck \sum_{j=1}^{n} (t_j^p |f^j|_*^2 + |f^j|_{*,-p}^2).$$

Assuming this for a moment, we can now complete the proof of Theorem 11.1. We apply Lemma 11.1 with $\mathcal{H} = S_h, A = A_h = -\Delta_h$, and $p = 0$, to (11.8), and obtain, since $\theta^0 = 0$,

$$k \sum_{j=1}^{n} |\bar{\partial}\theta^j|^2 \leq Ck \sum_{j=1}^{n} |\sigma^j|_*^2.$$

Using (11.9) and (11.11) we find, for $0 < t_j \leq \bar{t}$,

$$|\sigma^j|_* \leq C_{\bar{t}}(u)(h^2 + k) + C|(I + kA)\omega^j|_* = C_{\bar{t}}(u)(h^2 + k) + C|\omega^j|,$$

and thus, since $\omega^1 = 0$, using (11.14)

$$k \sum_{j=1}^{n} |\bar{\partial}\theta^j|^2 \leq C_{\bar{t}}(u)(h^2 + k)^2 + C\varepsilon^2 k \sum_{j=1}^{n} |\bar{\partial}\theta^j|^2.$$

Choosing ε sufficiently small yields

$$k \sum_{j=2}^{n} |\bar{\partial}\theta^j|^2 \leq C_{\bar{t}}(u)(h^2 + k)^2,$$

and hence, by (11.15),

$$\|\theta^n\| \leq C_{\bar{t}}(u)(h^2 + k) \quad \text{for } t_n \leq \bar{t},$$

which completes the proof of the theorem. □

Proof of Lemma 11.1. By eigenfunction expansion if suffices to consider the scalar case, with $A = \mu > 0$, in which case the statement reduces to

$$\begin{aligned}(1 + k\mu)k \sum_{j=1}^{n} t_j^p (\bar{\partial} U^j)^2 \leq & C\big(k^{p-1}(1 + k\mu) + (1 + k\mu)^{-1}\mu^{-p+1}\big)v^2 \\ & + C(1 + k\mu)^{-1} k \sum_{j=1}^{n} (t_j^p + \mu^{-p})(f^j)^2,\end{aligned}$$

or, replacing $k\mu$ by λ and kf^j by g^j,

$$\begin{aligned}\sum_{j=1}^{n} j^p (U^j - U^{j-1})^2 \leq & C(1 + (1 + \lambda)^{-2}\lambda^{-p+1})v^2 \\ & + C(1 + \lambda)^{-2} \sum_{j=1}^{n} (j^p + \lambda^{-p})(g^j)^2.\end{aligned}$$

We shall first show this for $g^j = 0$ for $j \geq 1$, and $v = 1$, and then for $v = 0$. The complete result then follows by linearity.

In the first case we have by the defining equation

$$U^n = (1+\lambda)^{-1}U^{n-1} = \cdots = (1+\lambda)^{-n}, \quad \text{for } n \geq 0,$$

and thus

$$U^j - U^{j-1} = -\lambda U^j = -\lambda(1+\lambda)^{-j}, \quad \text{for } j \geq 1. \tag{11.18}$$

Since

$$\sum_{j=1}^{\infty} j^p x^j \leq Cx(1-x)^{-p-1}, \quad \text{for } 0 \leq x < 1, \tag{11.19}$$

it follows that, for $\lambda > 0$,

$$\begin{aligned}\sum_{j=1}^{n} j^p (U^j - U^{j-1})^2 &= \lambda^2 \sum_{j=1}^{n} j^p (1+\lambda)^{-2j} \\ &\leq C\lambda^2(1+\lambda)^{-2}(1-(1+\lambda)^{-2})^{-p-1} \\ &= C\lambda^2(1+\lambda)^{2p}(2\lambda+\lambda^2)^{-p-1} \leq C\big(1+(1+\lambda)^{-2}\lambda^{-p+1}\big),\end{aligned}$$

which completes the proof in this case. (In the last step we only need to check the order of the functions for λ large and for λ small.)

In the other case we have, since $v = 0$,

$$U^j = \sum_{l=1}^{j} (1+\lambda)^{-(j+1-l)} g^l, \quad \text{for } j \geq 1,$$

and hence

$$U^j - U^{j-1} = (1+\lambda)^{-1} g^j - \lambda \sum_{l=1}^{j-1} (1+\lambda)^{-(j+1-l)} g^l, \quad \text{for } j \geq 2,$$

and, using Schwarz' inequality, for $j \geq 2$,

$$\begin{aligned}(U^j - U^{j-1})^2 &\leq 2(1+\lambda)^{-2}(g^j)^2 + 2\lambda^2 \Big(\sum_{l=1}^{j-1} (1+\lambda)^{-(j+1-l)} g^l\Big)^2 \\ &\leq 2(1+\lambda)^{-2}(g^j)^2 + 2\lambda(1+\lambda)^{-1} \sum_{l=1}^{j-1} (1+\lambda)^{-(j+1-l)} (g^l)^2.\end{aligned}$$

After multiplication by j^p, summation, and a change of the order of summation in the second term, we find, using again (11.19) and checking orders for λ small and large,

$$\sum_{j=1}^{n} j^p(U^j - U^{j-1})^2 \le C(1+\lambda)^{-2}\sum_{j=1}^{n} j^p(g^j)^2$$
$$+ C\lambda(1+\lambda)^{-1}\sum_{l=1}^{n-1}\sum_{j=l+1}^{n}(l^p + (j-l)^p)(1+\lambda)^{-(j+1-l)}(g^l)^2$$
$$\le C(1+\lambda)^{-2}\sum_{j=1}^{n}(j^p + \lambda^{-p})(g^j)^2.$$

This completes the proof of the lemma. □

We shall now study incomplete iteration in the case of the homogeneous equation with nonsmooth initial data. We recall that the exact solution of the backward Euler scheme (11.2) satisfies

$$\|U_h^n - u^n\| \le C(h^2 + k)t_n^{-1}\|v\|, \quad \text{for } t_n > 0,$$

and our ambition is to show that the incomplete iteration scheme can be designed so that this error estimate remains valid.

We shall begin by studying the time discretization in the Hilbert space setting so that the exact backward Euler method is (11.16), and the iterative scheme satisfies, for $m \ge 1$,

$$(11.20)\qquad \begin{aligned} U^{n,m} - \bar{U}^n &= D_k^m(U^{n,0} - \bar{U}^n), \ \text{where}\, D_k = I - B_k(I + kA),\\ U^{n,0} &= 2U^{n-1} - U^{n-2}, \end{aligned}$$

as in (11.4). After this we shall give the corresponding result for the fully discrete case.

We thus first demonstrate the following theorem, which shows that the desired nonsmooth data result holds provided the number or iterative steps is chosen appropriately larger at the earlier time levels, where the solution is less smooth.

Theorem 11.2 *Assume that* (11.6) *holds with* $\kappa < 1$. *For* $n \ge 3$, *let* $U^n = U^{n,M_n}$ *be the solution of the incomplete iterative scheme* (11.20) *for* (11.16) *with* $f \equiv 0$, *using* M_n *iterations at time level* t_n, *and let* $U^j = \bar{U}^j$ *for* $j = 1, 2$. *Then there is a* $\delta > 0$ *such that*

$$(11.21)\qquad \|U^n - u^n\| \le Ckt_n^{-1}\|v\|, \quad t_n > 0, \quad \text{if } \kappa^{M_n} \le \delta\min(t_n^{3/2}, 1).$$

Proof. With $\omega^n = (U^n - \bar{U}^n)/k$ and $\tau^n = \bar{\partial}u^n - u_t^n$, the error $e^n = U^n - u^n$ satisfies, cf. (11.8), (11.9),

$$\bar{\partial}e^n + Ae^n = \sigma^n := -\tau^n + (I + kA)\omega^n = -\tau^n + \widetilde{\omega}^n.$$

Therefore, application of Lemma 10.4 with $q = 1, p = 2$, gives, since $e^0 = 0$,

$$t_n^2\|e^n\|^2 \le Ck\sum_{j=1}^{n}(t_j^2|\sigma^j|_{-1}^2 + |\sigma^j|_{-3}^2),$$

where, as earlier $|v|_{-j} = \|A^{-j/2}v\|$. Now, for t_n bounded, and since A is positive definite,

$$\begin{aligned}t_j^2|\widetilde{\omega}^j|_{-1}^2 + |\widetilde{\omega}^j|_{-3}^2 \le C|\widetilde{\omega}^j|_{-1}^2 &= C\|A^{-1/2}(I+kA)\omega^j\|\\ &\le C\|(I+kA)^{1/2}\omega^j\| = C|\omega^j|, \quad \text{for } j\ge 1,\end{aligned}$$

and thus, since $\omega^1 = \omega^2 = 0$ by assumption,

$$t_n^2\|e^n\|^2 \le Ck\sum_{j=1}^{n}(t_j^2|\tau^j|_{-1}^2 + |\tau^j|_{-3}^2) + Ck\sum_{j=3}^{n}|\omega^j|^2. \tag{11.22}$$

We next show that

$$k\sum_{j=1}^{n}(t_j^2|\tau^j|_{-1}^2 + |\tau^j|_{-3}^2) \le Ck^2\|v\|^2. \tag{11.23}$$

Let $s=1$ or 3. Then, by the definition of τ^j,

$$|\tau^j|_{-s}^2 \le Ck\int_{t_{j-1}}^{t_j}|u_{tt}(y)|_{-s}^2\,dy,$$

and hence, for $j>1$ when $s=1$ and for $j\ge 1$ when $s=3$,

$$kt_j^{3-s}|\tau^j|_{-s}^2 \le Ck^2\int_{t_{j-1}}^{t_j}y^{3-s}|u_{tt}(y)|_{-s}^2\,dy. \tag{11.24}$$

To bound the sum in (11.23) we note that, by eigenfunction expansion,

$$\begin{aligned}\int_0^\infty y^{3-s}|u_{tt}(y)|_{-s}^2\,dy &\le \int_0^\infty y^{3-s}\sum_{l=1}^{\infty}\lambda_l^{4-s}e^{-2\lambda_l y}(v,\varphi_l)^2\,dy\\ &\le C\sum_{l=1}^{\infty}(v,\varphi_l)^2 = C\|v\|^2.\end{aligned} \tag{11.25}$$

Together these estimates show (11.23) except for the term corresponding to $j=1, s=1$. But for this term we have

$$\begin{aligned}kt_1^2|\tau^1|_{-1}^2 = k^3|\bar\partial u^1 - u_t^1|_{-1}^2 &\le Ck^3(|\bar\partial u^1|_{-1}^2 + |u_t^1|_{-1}^2)\\ &\le Ck^2\int_0^k|u_t|_{-1}^2\,dt + Ck^3|u(k)|_1^2 \le Ck^2\|v\|^2.\end{aligned}$$

The proof of (11.23) is now complete.

Combination of (11.22) and (11.23) gives

$$t_n^2\|e^n\|^2 \le Ck^2\|v\|^2 + Ck\sum_{j=3}^{n}|\omega^j|^2. \tag{11.26}$$

To bound the latter term we note that using assumption (11.6) we have this time

$$|U^n - \bar U^n| \le c_0\kappa^{M_n}(|U^{n,0} - U^n| + |U^n - \bar U^n|).$$

Letting ε be a positive number which is to be specified later, we may take $\delta = \delta(\varepsilon)$ in (11.21) small enough that $c_0\delta/(1-c_0\delta) \le \varepsilon$ so that

$$c_0\kappa^{M_n}/(1 - c_0\kappa^{M_n}) \le c_0\delta t_n^{3/2}/(1-c_0\delta) \le \varepsilon t_n^{3/2}.$$

Using the argument preceding (11.13), we then obtain

$$|U^n - \bar U^n| \le \varepsilon t_n^{3/2}|U^{n,0} - U^n| = \varepsilon t_n^{3/2}k^2|\bar\partial^2 U^n|,$$

and hence, for $j \ge 3$,

$$|\omega^j| \le \varepsilon t_j^{3/2}(k|\bar\partial^2 u^j| + k|\bar\partial^2 e^j|) \le \varepsilon t_j^{3/2}(k|\bar\partial^2 u^j| + |\bar\partial e^j| + |\bar\partial e^{j-1}|).$$

Thus

$$k\sum_{j=3}^{n}|\omega^j|^2 \le Ck^3\sum_{j=3}^{n}t_j^3|\bar\partial^2 u^j|^2 + C\varepsilon^2 k\sum_{j=2}^{n}t_j^3|\bar\partial e^j|^2. \tag{11.27}$$

We now estimate the first term on the right. Using the fact that $\bar\partial^2$ annihilates linear functions, Taylor's formula shows, for $j \ge 3$,

$$\begin{aligned} t_j^3\|\bar\partial^2 u^j\|^2 &\le Ct_j^3\Big\|\Big(\bar\partial^2\int_{t_{j-2}}^{t}(t-s)u_{tt}(s)\,ds\Big)_{t=t_j}\Big\|^2 \\ &\le Ct_j^3\Big(k^{-1}\int_{t_{j-2}}^{t_j}\|u_{tt}(s)\|\,ds\Big)^2 \le Ck^{-1}\int_{t_{j-2}}^{t_j}s^3\|u_{tt}(s)\|^2\,ds. \end{aligned}$$

Hence, using (11.25) with $s=0$,

$$k^3\sum_{j=3}^{n}t_j^3\|\bar\partial^2 u^j\|^2 \le Ck^2\|v\|^2.$$

Similarly, using one less term in the Taylor expansion, we have

$$k^3\sum_{j=3}^{n}t_j^3|\bar\partial^2 u^j|_2^2 \le C\|v\|^2.$$

Since $|v| = (\|v\|^2 + k(Av,v))^{1/2} \le (\|v\| + k|v|_2)^{1/2}\|v\|^{1/2} \le \|v\| + k|v|_2$ it follows that

$$k^3 \sum_{j=3}^{n} t_j^3 |\bar{\partial}^2 u^j|^2 \le Ck^2 \|v\|^2. \tag{11.28}$$

To estimate the last term of (11.27), we apply Lemma 11.1 to e^n to obtain

$$k \sum_{j=1}^{n} t_j^3 |\bar{\partial} e^j|^2 \le Ck \sum_{j=1}^{n} \big(t_j^3 |\tau^j|_*^2 + |\tau^j|_{*,-3}^2\big) + Ck \sum_{j=3}^{n} |\omega^j|^2, \tag{11.29}$$

where we have used $t_j^3 |\omega^j|_*^2 + |\omega^j|_{*,-3}^2 \le C|\omega^j|_*^2 \le C|\omega^j|^2$. Here, by (11.24) and (11.25),

$$k \sum_{j=2}^{n} t_j^3 |\tau^j|_*^2 \le k \sum_{j=2}^{n} t_j^3 \|\tau^j\|^2 \le Ck^2 \int_0^\infty y^3 \|u_{tt}(y)\|^2 \, dy \le Ck^2 \|v\|^2,$$

and we further have

$$k t_1^3 |\tau^1|_*^2 \le k^4 \|\tau^1\|^2 \le 2k^2 \|u^1 - u^0\|^2 + 2k^4 \|u_t^1\|^2 \le Ck^2 \|v\|^2.$$

By (11.23) we already know that

$$k \sum_{j=1}^{n} |\tau^j|_{*,-3}^2 \le k \sum_{j=1}^{n} |\tau^j|_{-3}^2 \le Ck^2 \|v\|^2.$$

Combining the above estimates with (11.29) we find

$$k \sum_{j=1}^{n} t_j^3 |\bar{\partial} e^j|^2 \le Ck^2 \|v\|^2 + k \sum_{j=3}^{n} |\omega^j|^2.$$

Together with (11.27) and (11.28) this shows

$$k \sum_{j=3}^{n} |\omega^j|^2 \le Ck^2 \|v\|^2 + C\varepsilon^2 k \sum_{j=3}^{n} |\omega^j|^2.$$

Choosing ε small enough, and combining the result with (11.26) completes the proof of the theorem. □

We shall now apply our above nonsmooth data error estimate to the fully discrete method for the homogeneous parabolic differential equation, i.e., (11.1) with $f = 0$, so that our time discretization procedure is applied to the spatially semidiscrete problem

$$u_{h,t} + A_h u_h = 0, \quad \text{for } t > 0, \quad \text{with } u_h(0) = v_h, \tag{11.30}$$

where $A_h = -\Delta_h$ with Δ_h the discrete Laplacian. Recall that if $v_h = P_h v$, then the solution of (11.30) satisfies

$$\|u_h(t) - u(t)\| \le Ch^2 t^{-1} \|v\|, \quad \text{for } t > 0. \tag{11.31}$$

We have the following:

Theorem 11.3 *Consider the fully discrete method* (11.2) *with* $f = 0$. *Let* $U_h^j = \bar{U}_h^j$ *for* $j = 1, 2$, *and for* $n \geq 3$, *let* U^n *be the solution of the incomplete iterative scheme* (11.4), *using* M_n *iterations at time level* t_n, *with* $f = 0$ *and* $v_h = P_h v$. *Assume that* (11.6) *holds. Then there is a* $\delta > 0$ *such that*

$$\|U_h^n - u^n\| \leq C(h^2 + k)t_n^{-1}\|v\|, \quad \textit{for } t_n > 0, \quad \textit{if } \kappa^{M_n} \leq \delta \min(t_n^{3/2}, 1).$$

Proof. This follows at once by Theorem 11.2, applied to (11.30), together with the estimate (11.31). □

To illustrate the above, we shall now present an example of a linear iteration method of the form (11.4) for the solution of the linear system (11.3), which has the convergence property (11.6) used in our analysis. The method will be expressed in abstract form, but is based on the V-cycle multigrid algorithm for the solution of the Dirichlet problem for Poisson's equation in a two-dimensional convex polygonal domain, in the way presented in Bramble [30].

Let S be a finite dimensional linear space with inner product $(\cdot, \cdot)$ and norm $\|\cdot\| = (\cdot, \cdot)^{1/2}$, and with a structure to be made precise presently, and let $M(\cdot, \cdot)$ be a symmetric, positive definite bilinear form on S. With the positive definite linear operator $M : S \to S$ defined by

$$(MU, V) = M(U, V), \quad \forall\, U, V \in S, \tag{11.32}$$

our concern is to solve the equation (corresponding to (11.3))

$$MU = b, \quad \text{for } b \in S, \tag{11.33}$$

by means of a linear iteration method of the form (cf. (11.4))

$$U^l = U^{l-1} - B(MU^{l-1} - b) = DU^{l-1} + Bb, \quad l = 1, 2, \ldots. \tag{11.34}$$

Here the operator $B : S \to S$ will be defined by a multilevel algorithm which we will now describe.

We assume that S is such that there is a nested sequence of subspaces $S_1 \subset S_2 \subset \cdots \subset S_J = S$, and define the positive definite local version $M_j : S_j \to S_j$ of M by

$$(M_j U, V) = M(U, V), \quad \forall\, U, V \in S_j. \tag{11.35}$$

In a way to be made precise below, we define approximations $B_j : S_j \to S_j$ of M_j^{-1} recursively, for $j = 1, \ldots, J$, starting with $B_1 = M_1^{-1}$, and finally set $B = B_J$. The algorithm is designed so that determining $B_j v$ for v given is less costly than to find $M_j^{-1}v$.

The goal is thus to calculate $Bx = B_J x$ in (11.34), with $x = MU^{l-1} - b$. To do so we shall first express the action of B_J in terms of that of B_{J-1}. Since at this point B_{J-1} is not known, we express it in terms of B_{J-2}, etc., till we get down to B_1. Since B_1 acts on the space S_1, which we assume to have a much lower dimension than S, we may take $B_1 = M_1^{-1}$, i.e., we may solve the equation $M_1 a = x_1$ exactly. Since we now know B_1, we can go back and calculate the action of B_2, which was expressed in terms of B_1. We proceed with B_3, and so on, till we arrive at B_J, which thus defines the action of B. Since at each iteration step the procedure makes us first go down in the scale of spaces S_j, from S_J to S_1, and then up again to S_J, it is referred to as the V-cycle algorithm.

A typical example of a situation such as the one just described is as follows: Assume that Ω is a polygonal domain, and that we want to solve the problem

$$-\Delta u = f \quad \text{in } \Omega, \quad \text{with } u = 0 \quad \text{on } \partial\Omega. \tag{11.36}$$

Let $\mathcal{T}_1$ be a coarse triangulation of Ω with maximal side length h_1. We may then construct a sequence of triangulations $\mathcal{T}_j$ of Ω, where $\mathcal{T}_{j+1}$ is obtained by subdividing each triangle of $\mathcal{T}_j$ into four by connecting the midpoints of its edges. With h_j the maximal side in $\mathcal{T}_j$ we then have $h_{j+1} = h_j/2$, or $h_j = 2^{-(j-1)}h_1$. We now define S_j to be the continuous piecewise linear functions on $\mathcal{T}_j$ which vanish on $\partial\Omega$. Clearly then $S_j \subset S_{j+1}$. The standard inner product in $L_2(\Omega)$ then induces an inner product in $S = S_J$ and we may define

$$M(U, V) = \int_\Omega \nabla U \cdot \nabla V \, dx, \quad \text{for } U, V \in S.$$

The discrete variational form of (11.36) is now

$$M(U, V) = (f_S, V) = \int_\Omega f\, V \, dx, \quad \forall V \in S,$$

where f_S is the L_2-projection of f onto S. The operator M_j defined in (11.35) is $-\Delta_j$ where $\Delta_j = \Delta_{h_j}$ denotes the discrete Laplacian in $S_j = S_{h_j}$, and the B_j are approximations of the $(-\Delta_j)^{-1}$. In each iteration step the only equation of the form $-\Delta_j W = g$ that has to be solved exactly is that associated with $-\Delta_1$, which is based on the coarsest triangulation.

We now specify how the action of the operator $B_j : S_j \to S_j$ is expressed in terms of that of $B_{j-1} : S_{j-1} \to S_{j-1}$. Letting thus $g \in S_j$, we define $B_j g \in S_j$ as the approximation of the solution $w \in S_j$ of $M_j w = g$ obtained in three steps, referred to as pre-smoothing, correction, and post-smoothing. The basic ingredient in the first and third steps is a preliminary approximation $Q_j : S_j \to S_j$ of M_j^{-1}, referred to as a smoothing operator, which has the property that the corresponding error operator $I - Q_j M_j$ particularly well reduces nonsmooth error components (or error components with high frequencies). We assume for simplicity of presentation that Q_j is symmetric. In the middle step the lower frequencies are reduced by projecting the residual onto S_{j-1}

and applying B_{j-1}. Letting $P_j : S \to S_j$ denote the orthogonal projection onto S_j, the three steps are then the following.

(i) *Set* $p = Q_j g$.

(ii) *Set* $q = B_{j-1}P_{j-1}y$, *where* $y = M_j p - g$.

(iii) *Set* $B_j g = v - Q_j(M_j v - g)$, *where* $v = p - q$.

Thus, in the first step, p is an approximation of $w = M_j^{-1}g$ with a relatively smooth error. The residual in this approximation is $y = M_j p - g = M_j(p-w)$. To get a better approximation we would therefore like to subtract from p a good approximation of the solution of $M_j z = y$. Since y has small nonsmooth components, it may be well represented in S_{j-1}, and we therefore now project onto S_{j-1} and use one step of the iterative method in S_{j-1} to find an approximation q of z such that $v = p - q$ is an improvement over p. Finally, this approximation is improved once more in S_j using the smoothing iteration.

The reduction of error in each step of the iteration (11.34) is determined by the operator $D = D_J = I - B_J M_J$, and the purpose of the convergence analysis is thus to estimate $|D|$ where $|\cdot|$ is a conveniently chosen norm. In order to do so we note that since $M(\cdot,\cdot)$ is symmetric and positive definite, it defines an inner product $[v,w] = M(v,w)$ and we now take $|\cdot|$ to be the corresponding norm, $|v| = [v,v]^{1/2}$. We shall also use the orthogonal projection $R_j : S \to S_j$ with respect to $[\cdot,\cdot]$. In our above application, $|\cdot|$ is the norm in $H_0^1(\Omega)$ and R_j is the Ritz projection onto S_j.

We define the error reduction operators $K_j, D_j : S_j \to S_j$ by

$$K_j = I_j - Q_j M_j \quad \text{and} \quad D_j = I_j - B_j M_j, \tag{11.37}$$

where I_j denotes the identity on S_j, we note that D_j satisfies the recursion

$$D_j = K_j(I_j - R_{j-1} + D_{j-1}R_{j-1})K_j, \quad \text{for } j = 2, \dots, J. \tag{11.38}$$

In fact, to calculate $D_j w$ with $w \in S_j$ given, we set $g = M_j w$ and define p, q, v as above. Then, we have by (iii) that $B_j M_j w = B_j g = K_j v + Q_j M_j w$, and hence

$$D_j w = w - B_j M_j w = w - K_j v - Q_j M_j w = K_j(w - v).$$

Further, since $P_{j-1}M_j = M_{j-1}R_{j-1}$ we find, by (2),

$$\begin{aligned} w - v &= w - p + q = w - p + B_{j-1}P_{j-1}M_j(p - w) \\ &= (I_j - B_{j-1}M_{j-1}R_{j-1})(w - p) = (I_j - R_{j-1} + D_{j-1}R_{j-1})(w - p). \end{aligned}$$

Finally, by (1), $w - p = w - Q_j g = K_j w$, which shows (11.38).

In order to analyze the algorithm thus defined, we now have to make some assumptions concerning the sequence of spaces S_j and the smoothing operators Q_j. Our analysis will be based on the following three hypotheses: With λ_j the maximal eigenvalue of M_j there are positive constants C_1, C_2, C_3 such that

$$(H_1) \qquad \|(I - R_j)v\| \leq C_1 \lambda_j^{-1/2} |(I - R_j)v|, \qquad \text{for } v \in S, \quad j = 1, \dots, J,$$

$$(H_2) \qquad \lambda_j / \lambda_{j-1} \leq C_2, \qquad \text{for} \qquad j = 2, \dots, J,$$

$$(H_3) \qquad |K_j v|^2 \leq |v|^2 - C_3 \lambda_j^{-1} \|M_j v\|^2, \quad \text{for } v \in S_j,\ j = 2, \dots, J.$$

The assumption (H_1) is an error estimate. In typical finite element applications it is proved by means of the Aubin-Nitsche duality argument and expresses the fact that the error in the Ritz projection R_j is smaller in the L_2-norm than in the energy norm. Assumption (H_2) means that the transition from S_{j-1} to S_j is not too rapid. Assumption (H_3) expresses the smoothing action of $K_j = I_j - Q_j M_j$: If v is an eigenvector of M_j with eigenvalue λ, then (H_3) implies $|K_j v|^2 \leq (1 - C_3 \lambda / \lambda_j)|v|^2$. High frequency eigenmodes are thus reduced in size by K_j more than low frequency modes. Note that for (H_3) to hold it is necessary that $C_3 \leq 1$ because otherwise the right hand side would be negative for the eigenvector associated with λ_j. Further, $C_3 = 1$ both for the "perfect smoother" $Q_j = M_j^{-1}$, and for $Q_j = \lambda_j^{-1} I_j$. In fact, in the first case $K_j = 0$ and

$$|v|^2 - \lambda_j^{-1} \|M_j v\|^2 \geq |v|^2 - \|M_j^{1/2} v\|^2 = 0,$$

and the second case is a special case of the following lemma.

Lemma 11.2 *Let* $Q_j = \mu I_j$ *with* $|\mu \lambda_j - 1| \leq \varepsilon < 1$. *Then* (H_3) holds with $C_3 = 1 - \varepsilon^2$.

Proof. We have $K_j = I_j - \mu M_j$ and hence for $v \in S_j$

$$|v|^2 - |K_j v|^2 = 2\mu [M_j v, v] - \mu^2 |M_j v|^2.$$

Here $[M_j v, v] = \|M_j v\|^2$ and $|M_j v|^2 \leq \lambda_j \|M_j v\|^2$, and hence

$$\begin{aligned} |v|^2 - |K_j v|^2 &\geq (2\mu - \mu^2 \lambda_j) \|M_j v\|^2 \\ &= (1 - (1 - \mu \lambda_j)^2) \lambda_j^{-1} \|M_j v\|^2 \geq (1 - \varepsilon^2) \lambda_j^{-1} \|M_j v\|^2, \end{aligned}$$

which proves the lemma. □

Assumption (H_3) is satisfied in finite element applications by other smoothers of practical interest, for example, the point and block Jacobi and Gauss-Seidel iterations, cf. [30].

Condition (H_3) implies that $|K_j| \leq (1 - C_3/\kappa_j)^{1/2}$, where κ_j is the condition number $\lambda_{\max}(M_j)/\lambda_{\min}(M_j)$ of M_j. In typical applications $\kappa_j \to \infty$ and hence $|K_j| \to 1$ as $j \to \infty$, which indicates a deterioration of the convergence rate of the smoothing iteration as j grows large. In contrast we shall show that $|D_j| \le \kappa < 1$, with κ independent of j. Hence the multigrid iteration defined above has a convergence rate which is uniform in the number of levels involved.

The assumptions (H_1), (H_2), and (H_3) enter our analysis combined into an inequality which we shall now state.

Lemma 11.3 *Assume that* (H_1), (H_2), and (H_3) hold. Then

$$|(I - R_{j-1})v|^2 \le C_0(|v|^2 - |K_j v|^2), \quad \forall v \in S_j, \quad \text{where } C_0 = C_1^2 C_2/C_3.$$

Proof. To prove this inequality we first use (H_1) to get

$$\begin{aligned}|(I-R_{j-1})v|^2 &= [(I - R_{j-1})v, v] = ((I - R_{j-1})v, M_j v)\\ &\le \|(I - R_{j-1})v\| \, \|M_j v\| \le C_1 \lambda_{j-1}^{-1/2} |(I - R_{j-1})v| \, \|M_j v\|,\end{aligned}$$

and hence

$$|(I - R_{j-1})v|^2 \le C_1^2 \lambda_{j-1}^{-1} \|M_j v\|^2.$$

In view of (H_2) and (H_3), this shows the lemma. □

We are now ready to state and prove a convergence result for the V-cycle algorithm. The main point to note in this result is that the bound is smaller than 1, independently of the dimension of S. This shows that condition (11.7) and thus (11.6) holds for this iteration method.

Theorem 11.4 *Assume that* (H_1), (H_2), and (H_3) hold. Then

$$|D| \le \kappa = 1 - 1/C_0, \quad \text{where } C_0 = C_1^2 C_2/C_3, \quad \text{with } D = I - B_J M_J.$$

Proof. Recalling the definitions of $K_j, D_j : S_j \to S_j$ in (11.37), we extend the scope of (11.38) to all of S by setting

$$\begin{aligned}\widetilde{D}_j &= I - R_j + D_j R_j = I - B_j M_j R_j,\\ \widetilde{K}_j &= I - R_j + K_j R_j = I - Q_j M_j R_j,\end{aligned}$$

and find that $\widetilde{D}_j = \widetilde{K}_j \widetilde{D}_{j-1} \widetilde{K}_j$. In fact, restricted to S_j this is the same as (11.38), and on the orthogonal complement of S_j, with respect to $[\cdot,\cdot]$, both sides reduce to the identity operator. Since $\widetilde{D}_1 = I - R_1$ we hence have

$$\widetilde{D}_J = \widetilde{K}_J \cdots \widetilde{K}_2 (I - R_1)^2 \widetilde{K}_2 \cdots \widetilde{K}_J,$$

and setting $E_1 = I - R_1$, and $E_j = \widetilde{K}_j E_{j-1}$, for $j = 2, \dots, J$, this yields

$$D = I - B_J M_J = \widetilde{D}_J = E_J E_J^*.$$

Note that E_J^* and E_J are the error reduction operators of the non-symmetric algorithms using only presmoothing and only postsmoothing, respectively.

Since $|D| = |E_J E_J^*| = |E_J^*|^2$ and $|E_J^*| = |E_J|$, it therefore suffices to estimate the latter norm. In order to do so we take $v \in S$ and consider the expressions

$$|E_{j-1}v|^2 - |E_j v|^2 = |E_{j-1}v|^2 - |\widetilde{K}_j E_{j-1}v|^2, \quad \text{for } j = 2, \ldots, J.$$

From the definition of $\widetilde{K}_j$ and Lemma 11.3, it follows that

$$\begin{aligned}
|\widetilde{K}_j w|^2 &= |(I - R_j)w|^2 + |K_j R_j w|^2 \\
&\le |(I - R_j)w|^2 + |R_j w|^2 - C_0^{-1}|(I - R_{j-1})R_j w|^2 \\
&= |w|^2 - C_0^{-1}|(R_j - R_{j-1})w|^2, \quad \text{for } w \in S,\ j = 2, \ldots, J.
\end{aligned}$$

Hence, setting $w = E_{j-1}v$,

$$C_0(|E_{j-1}v|^2 - |E_j v|^2) \ge |(R_j - R_{j-1})E_{j-1}v|^2 = |(R_j - R_{j-1})v|^2,$$

where in the last step we used the fact that $(I - E_{j-1})v \in S_{j-1}$ so that $(R_j - R_{j-1})(I - E_{j-1})v = 0$. This follows by induction and the recursion relation $I - E_j = I - E_{j-1} + Q_j M_j R_j E_{j-1}$ for $j = 2, \ldots, J$, with $I - E_1 = R_1$. By summation we therefore have

$$\begin{aligned}
C_0(|E_1 v|^2 - |E_J v|^2) &\ge \sum_{j=2}^{J} (|R_j v - R_{j-1}v|^2) \\
&= \sum_{j=2}^{J} (|R_j v|^2 - |R_{j-1}v|^2) = |v|^2 - |R_1 v|^2 = |(I - R_1)v|^2,
\end{aligned}$$

which, since $E_1 v = (I - R_1)v$, yields

$$|E_J v|^2 \le \kappa |(I - R_1)v|^2 \le \kappa |v|^2, \quad \text{with } \kappa = 1 - 1/C_0.$$

This implies the desired result. □

We end by illustrating how the abstract result of Theorem 11.4 can be applied to the backward Euler discretization (11.3) of the heat equation. We make the same assumptions for Ω and $S_h = S = S_J$ as in our above discussion of the elliptic problem (11.36). With $(\cdot,\cdot)$ and $\|\cdot\|$ the inner product and norm in $L_2 = L_2(\Omega)$, we use

$$[v, w] = M(v, w) = (v, w) + k(\nabla v, \nabla w) \quad \text{and} \quad |v| = [v, v]^{1/2},$$

where the former defines the operators $M_j : S_j \to S_j$. Letting ν_j denote the largest eigenvalue of $-\Delta_j$, the discrete analogue of $-\Delta$ on $S_j = S_{h_j}$, the largest eigenvalue of $M_j = I_j - k\Delta_j$ is then $\lambda_j = 1 + k\nu_j$. For our smoothing operator we choose $Q_j = \mu_j I_j$ with $|\mu_j \lambda_j - 1| \le \varepsilon < 1$. Our aim is now to check that the assumptions of Theorem 11.4 are satisfied, with constants that are independent of j and k.

We note first that (H_3) is an immediate consequence of Lemma 11.2. For (H_2), we recall that ν_j is bounded above and below by positive multiples of h_j^{-2}. Since $h_{j-1} = 2h_j$ we therefore have

$$\frac{\lambda_j}{\lambda_{j-1}} \le \frac{1 + c_2 k h_j^{-2}}{1 + c_1 k h_{j-1}^{-2}} = \frac{1 + c_2 k h_j^{-2}}{1 + \frac{1}{4} c_1 k h_j^{-2}} \le C.$$

We finally consider (H_1). For $k \le h_j^2$, we have $\lambda_j \le 1 + c_2 k h_j^{-2} \le C$ and hence

$$\|(I - R_j)v\| \le |(I - R_j)v| \le C\lambda_j^{-1/2}|(I - R_j)v|.$$

For $k \ge h_j^2$, we use an adaptation of the standard duality argument. Let $\psi \in L_2$ and let $w \in H_0^1 = H_0^1(\Omega)$ be the solution of

$$-k\Delta w + w = \psi \quad \text{in } \Omega, \quad \text{with } w = 0 \quad \text{on } \partial\Omega,$$

or

$$M(\phi, w) = (\phi, \psi), \quad \forall \phi \in H_0^1. \tag{11.39}$$

Then, for any $\chi \in S_j$,

$$((I - R_j)v, \psi) = M((I - R_j)v, w - \chi) \le |(I - R_j)v|\,|w - \chi|. \tag{11.40}$$

Here, with χ suitably chosen,

$$|w - \chi|^2 = \|w - \chi\|^2 + k\|\nabla(w - \chi)\|^2 \le C(h_j^4 + kh_j^2)\|w\|_2^2 \le Ckh_j^2\|w\|_2^2.$$

We now show that $k\|w\|_2 \le C\|\psi\|$, uniformly in k. In fact, since $-\Delta w = k^{-1}(\psi - w)$, the standard regularity estimate for elliptic problems shows $k\|w\|_2 \le C(\|w\| + \|\psi\|)$. But choosing $\phi = w$ in (11.39) we have $\|w\|^2 \le |w|^2 = M(w, w) = (w, \psi) \le \|w\|\,\|\psi\|$, so that $\|w\| \le \|\psi\|$, which completes the proof. Hence for this $\chi \in S_j$ we have

$$|w - \chi|^2 \le Ck^{-1}h_j^2\|\psi\|^2 \le C\lambda_j^{-1}\|\psi\|^2,$$

since $\lambda_j \le Ckh_j^{-2}$. Together with (11.40) this shows (H_1).

The assumptions of Theorem 11.4 are thus satisfied, and hence the multigrid algorithm studied produces a linear iterative method for the solution of the backward Euler Galerkin method at each time level, for which our results on incomplete iteration apply. More general multigrid methods may also be used, with more than one presmoothing, postsmoothing, and inner iteration step, and allowing more general smoothing iterations such as methods of Jacobi and Gauss-Seidel type, see [30]. We shall not pursue this further here.

The idea of using incomplete iteration was first analyzed for parabolic problems in Douglas, Dupont, and Ewing [78] and Bramble and Sammon [35] (cf. also Bramble [29], Keeling [136], Karakashian [133]) under the assumption that the exact solution is smooth. The above presentation is taken from Bramble, Pasciak, Sammon, and Thomée [34], where both smooth and nonsmooth solutions are considered for more general multistep backward difference schemes of the type considered in Chapter 10 above.

The use of multigrid methods for parabolic problems has been considered in, e.g., Bank and Dupont [22], Hackbusch [112], and Lubich and Ostermann [160]; the presentation here is extracted from Larsson, Thomée, and Zhou [149]. For the underlying basic material on multigrid methods for the elliptic problem, we refer to Bramble [30] and Bramble and Zhang [31].

12. The Discontinuous Galerkin Time Stepping Method

In the previous chapters we have considered fully discrete schemes for the heat equation which were derived by first discretizing in the space variables by means of a Galerkin finite element method, which results in a system of ordinary differential equations with respect to time, and then applying a finite difference type time stepping method to this system to define a fully discrete solution. In this chapter, we shall apply the Galerkin method also in the time variable and thus define and analyze a method which treats the time and space variables similarly. The approximate solution will be sought as a piecewise polynomial function in t of degree at most $q-1$, which is not necessarily continuous at the nodes of the defining partition.

As earlier, in order to avoid cumbersome notation we shall concentrate first on the discretization in time only. Let thus $\mathcal{H}$ be a Hilbert space and assume that A is a selfadjoint, positive definite, not necessarily bounded operator with compact inverse, defined in $\mathcal{D}(A) \subset \mathcal{H}$. Allowing thus as usual both spatially continuous and discrete operators, we consider the initial value problem

$$u' + Au = f, \quad \text{for } t > 0, \quad \text{with } u(0) = v. \tag{12.1}$$

In order to discretize this abstract ordinary differential equation we partition the $t-$axis in a not necessarily uniform fashion by $0 = t_0 < t_1 < \cdots < t_n < \cdots$ and set $J_n = (t_{n-1}, t_n]$, $k_n = t_n - t_{n-1}$, and $k = \max_n k_n$. With q a given positive integer we shall then look for an approximate solution of (12.1) which reduces to a polynomial of degree at most $q-1$ in t on each subinterval J_n, with coefficients in $\mathcal{H}$, or, equivalently, which belongs to the space

$$\mathcal{S}_k = \{X : [0,\infty) \to \mathcal{H};\ X\big|_{J_n} = \sum_{j=0}^{q-1} \psi_j t^j,\ \psi_j \in \mathcal{H}\}.$$

Note that these functions are allowed to be discontinuous at the nodal points, but are taken to be continuous to the left there. Note also that $X(0)$ has to be specified separately for $X \in \mathcal{S}_k$ since $0 \notin J_1$. For $X = X_k \in \mathcal{S}_k$ we denote by X^n and X_+^n the value of X and its limit from above at t_n, respectively, and write $\mathcal{S}_k^n$ for the restrictions to J_n of the functions in $\mathcal{S}_k$.

To introduce our discretization method, consider a fixed interval $[0, t_N]$, and note that the exact solution of (12.1) satisfies, for w smooth,

$$\int_0^{t_N} \big((u',w)+A(u,w)\big)\,dt = \int_0^{t_N} (f,w)\,dt,$$

and hence, after integration by parts in the first term, now if $w(t_N)=0$,

$$\int_0^{t_N} \big(-(u,w')+A(u,w)\big)\,dt = (v,w(0)) + \int_0^{t_N} (f,w)\,dt. \tag{12.2}$$

Here $A(u,w)$ is the bilinear form defined by $A(u,w)=(Au,w)=(u,Aw)$ for $u,w\in\mathcal{D}(A)$; it may be extended in a natural way by

$$A(u,w)=\sum_{j=1}^{\infty}\lambda_j(u,\varphi_j)(w,\varphi_j), \quad \text{for } u,w\in\mathcal{D}=\mathcal{D}(A^{1/2}),$$

where the λ_j and φ_j are the eigenvalues and eigenfunctions of A.

Replacing u in the weak formulation (12.2) by a function $U\in\mathcal{S}_k$ and integrating by parts in each J_n, we obtain for the first term on the left hand side of (12.2), with $w^n=w(t_n)$ (and $w^N=0$),

$$\begin{aligned}-\int_0^{t_N}(U,w')\,dt &= -\sum_{n=1}^{N}\Big((U,w)\big|_{t_{n-1}+0}^{t_n}-\int_{J_n}(U',w)\,dt\Big)\\ &= \int_0^{t_N}(U',w)\,dt+\sum_{n=1}^{N-1}([U]_n,w^n)+(U_+^0,w^0),\end{aligned}$$

where $[U]_n=U_+^n-U^n$ denotes the jump of U at t_n, and where U' is the piecewise polynomial of degree $q-2$, which agrees with dU/dt on each J_n. In particular, if $q=1$, we have $U'\equiv 0$ so that the integrand vanishes.

Recalling (12.2), we now define our discrete scheme by requiring that $U\in\mathcal{S}_k$ satisfies

$$\begin{aligned}\int_0^{t_N}\big((U',X)+A(U,X)\big)\,dt &+ \sum_{n=1}^{N-1}([U]_n,X_+^n)+(U_+^0,X_+^0)\\ &= (v,X_+^0)+\int_0^{t_N}(f,X)\,dt, \quad \forall X\in\mathcal{S}_k,\\ U^0 &= v.\end{aligned} \tag{12.3}$$

Since the function X in $\mathcal{S}_k$ is not required to be continuous at the t_n, we may choose its values on the different time intervals independently. By choosing X to vanish outside J_n we therefore see that the equation reduces to one equation for each J_n with $n\le N$ so that the discrete scheme requires us to determine $U\in\mathcal{S}_k$ such that

$$\begin{aligned}\int_{J_n}\big((U',X)+A(U,X)\big)\,dt &+ (U_+^{n-1},X_+^{n-1})\\ &= (U^{n-1},X_+^{n-1})+\int_{J_n}(f,X)\,dt, \quad \forall X\in\mathcal{S}_k^n,\ 1\le n\le N,\\ U^0 &= v.\end{aligned} \tag{12.4}$$

This also shows that the definition of the discrete solution is independent of the choice of the final nodal point t_N. We remark that the exact solution of (12.1) also satisfies this equation.

We now show that the local problem (12.4) has a unique solution in $\mathcal{S}_k$ on J_n for U^{n-1} and $f\big|_{J_n}$ given. We first note that to show uniqueness it suffices to see that the corresponding homogeneous equation,

$$\int_{J_n} ((U', X) + A(U, X))\, dt + (U_+^{n-1}, X_+^{n-1}) = 0, \quad \forall X \in \mathcal{S}_k^n,$$

only has the trivial solution $U \equiv 0$. For this purpose, assume U is a solution, and choose $X = U$ in J_n. Then, since $2(U', U) = \frac{d}{dt}\|U\|^2$, we find that

$$\|U^n\|^2 - \|U_+^{n-1}\|^2 + 2\int_{J_n} A(U, U)\, dt + 2\|U_+^{n-1}\|^2 = 0,$$

or

$$\|U^n\|^2 + \|U_+^{n-1}\|^2 + 2\int_{J_n} |U|_1^2\, dt = 0. \tag{12.5}$$

Here and below we again use the norm $|v|_s = \|A^{s/2}v\|$ in $\dot{H}^s$. In particular, $A(U, U) = \|A^{1/2}U\|^2 = |U|_1^2$. It follows from (12.5) that $A(U, U) = 0$ in J_n and hence $U(t) \equiv 0$ in J_n, which proves our claim. Note that we may also conclude directly from (12.5) that $U^n = U_+^{n-1} = 0$ which implies that $U(t) \equiv 0$ in J_n if $q = 1$ or 2, but not for higher values of q.

The existence of a solution to (12.4) follows from the uniqueness since, using the eigenspaces of A, (12.4) can be reduced to a set of finite dimensional problems, for each of which obviously uniqueness implies existence.

In the case $q = 1$, i.e., when the approximating functions are piecewise constant in time, then $U' \equiv 0$ and $U(t) = U^n = U_+^{n-1}$ in J_n, and the method reduces to the modified backward Euler method

$$(U^n, \psi) + k_n A(U^n, \psi) = (U^{n-1}, \psi) + \Big(\int_{J_n} f(t)\, dt, \psi\Big), \quad \forall \psi \in \mathcal{D},$$

or

$$(I + k_n A)U^n = U^{n-1} + \int_{J_n} f(t)\, dt. \tag{12.6}$$

Clearly then $U^n \in \mathcal{D}(A)$. Equation (12.6) may also be written

$$\bar{\partial}_n U^n + AU^n = \frac{1}{k_n}\int_{J_n} f(t)\, dt, \quad \text{where } \bar{\partial}_n U^n = \frac{U^n - U^{n-1}}{k_n}.$$

Note that the $f^n = f(t_n)$ occurring in the standard error estimate for the backward Euler method studied earlier has been replaced by an average of

f over J_n; the standard method may thus be interpreted as resulting from (12.6) after quadrature.

In the case $q = 2$, i.e., for piecewise linear functions of t, we may write $U(t) = \widetilde{U}_0^n + \widetilde{U}_1^n(t - t_{n-1})/k_n$ on J_n, and obtain for the determination of $\widetilde{U}_0^n = \widetilde{U}_0$ and $\widetilde{U}_1^n = \widetilde{U}_1$ the system

$$\begin{aligned}(\widetilde{U}_0, \psi) + k_n A(\widetilde{U}_0, \psi) + (\widetilde{U}_1, \psi) + \tfrac{1}{2}k_n A(\widetilde{U}_1, \psi) \\ = (U^{n-1}, \psi) + \Big(\int_{J_n} f(t)dt, \psi\Big), \\ \tfrac{1}{2}k_n A(\widetilde{U}_0, \eta) + \tfrac{1}{2}(\widetilde{U}_1, \eta) + \tfrac{1}{3}k_n A(\widetilde{U}_1, \eta) \\ = \Big(k_n^{-1}\int_{J_n}(t - t_n)f(t)\,dt, \eta\Big), \quad \text{for } \psi, \eta \in \mathcal{D},\ n \geq 1.\end{aligned}$$

Once $\widetilde{U}_0$ and $\widetilde{U}_1$ are determined, we have $U^n = \widetilde{U}_0 + \widetilde{U}_1$. This system may also be written

$$\begin{aligned}(I + k_n A)\widetilde{U}_0 + (I + \tfrac{1}{2}k_n A)\widetilde{U}_1 &= U^{n-1} + \int_{J_n} f(t)\,dt, \\ \tfrac{1}{2}k_n A\widetilde{U}_0 + (\tfrac{1}{2}I + \tfrac{1}{3}k_n A)\widetilde{U}_1 &= k_n^{-1}\int_{J_n}(t - t_n)f(t)\,dt.\end{aligned}$$

In the case of the homogeneous equation, i.e., when $f \equiv 0$, it is easy to show that, with the notation of Chapter 7, $U^n = r_{21}(k_n A)U^{n-1}$ where $r_{21}(\lambda)$ is the $(2,1)$-Padé approximation of $e^{-\lambda}$.

Before we turn to the analysis of the method introduced, we pause to discuss briefly some alternative approaches. It could perhaps appear more natural to seek the approximate solution as a piecewise polynomial in t of degree $q-1$, which is continuous at the nodes of the partition, thus avoiding the jump terms in (12.4). For suitable test functions X the defining equation would then be

$$\int_{J_n} \big((U', X) + A(U, X)\big)\,dt = \int_{J_n} (f, X)\,dt, \quad \text{for } n \geq 1, \tag{12.7}$$

again with $U^0 = v$. Given $U^{n-1} = U(t_{n-1})$, only $q-1$ conditions are now needed to determine U on J_n, and the local test space therefore should only be of dimension $q-1$ in time. We consider two such possibilities:

$$\mathcal{S}_{k,I}^n = \{X \in \Pi_{q-1} \otimes \mathcal{D}; X(t_{n-1}) = X^{n-1} = 0\}$$

and

$$\mathcal{S}_{k,II}^n = \{X \in \Pi_{q-2} \otimes \mathcal{D}\}.$$

Let us demonstrate that in both cases the solution of (12.7) is uniquely defined. As above, it suffices for this to show uniqueness, i.e., that if (12.7) holds with $U^{n-1} = 0$ and $f \equiv 0$ on J_n, then $U \equiv 0$ on J_n. In case I we may choose $X = U$, since $U \in \mathcal{S}^n_{k,I}$, and obtain

$$\tfrac{1}{2}\|U^n\|^2 + \int_{J_n} |U|_1^2\,dt = 0,$$

which implies $U \equiv 0$ on J_n. In case II we choose instead $X = U' \in \mathcal{S}^n_{k,II}$ to find

$$\int_{J_n} \|U'\|^2\,dt + \tfrac{1}{2}|U^n|_1^2 = 0,$$

from which we again conclude $U \equiv 0$ on J_n.

For $q = 2$, i.e., for U piecewise linear, the methods reduce to

$$\frac{U^n - U^{n-1}}{k_n} + A\Big(\frac{U^{n-1} + 2U^n}{3}\Big) = \frac{2}{k_n^2}\int_{J_n} (t - t_{n-1})\, f(t)\,dt,$$

and

$$\frac{U^n - U^{n-1}}{k_n} + A\Big(\frac{U^{n-1} + U^n}{2}\Big) = \frac{1}{k_n}\int_{J_n} f(t)\,dt,$$

respectively. The first of these is only first order accurate, and the second is a modified Crank-Nicolson method; the case II method is sometimes referred to as the continuous Galerkin method. Because it has less advantageous smoothing properties than the discontinuous Galerkin method (cf. Chapter 7), we shall refrain from a detailed analysis here.

The following theorem gives our first error estimate for the time stepping method (12.4). Here and below $u^{(l)} = (d/dt)^l u$.

Theorem 12.1 *We have, for the solutions of* (12.4) *(with $q \ge 1$) and* (12.1),

$$\|U^N - u(t_N)\| \le C\Big(\sum_{n=1}^{N} k_n^{2q}\int_{J_n} |u^{(q)}|_1^2\,dt\Big)^{1/2}, \quad \textit{for } t_N \ge 0. \tag{12.8}$$

Proof. We define an interpolant $\widetilde{u}(t) \in \mathcal{S}_k$ of the exact solution $u(t)$ of (12.1) by demanding, for each $n \ge 1$,

$$\begin{aligned} \widetilde{u}(t_n) &= u(t_n), \quad \text{for } n \ge 0,\\ \int_{J_n} (\widetilde{u}(t) - u(t))t^l\,dt &= 0, \quad \text{for } l \le q-2,\ n \ge 1, \end{aligned} \tag{12.9}$$

i.e., $\widetilde{u}$ interpolates at the nodal points, and the interpolation error is orthogonal to Π_{q-2} on J_n. (For $q = 1$ the latter condition is void.) In order to see that these equations define a unique $\widetilde{u} \in \Pi_{q-1}$ on J_n, it suffices, by expansion in $\mathcal{H}$ with respect to an orthonormal basis, to consider the scalar case, and, since the number of equations and the number of unknowns in (12.9) then both equal q, to show that $u(t) \equiv 0$ on J_n implies $\widetilde{u}(t) \equiv 0$ there. Transforming to the unit interval $(0,1)$, with t_n corresponding to 0, we thus need to see that if $\widetilde{u}(t) = t\sum_{j=0}^{q-2} a_j t^j$ is orthogonal to Π_{q-2} on $(0,1)$, then $\widetilde{u}(t) \equiv 0$ there. But this follows from

$$\int_0^1 \widetilde{u}(t) \sum_{j=0}^{q-2} a_j t^j \, dt = \int_0^1 t\,\Big(\sum_{j=0}^{q-2} a_j t^j\Big)^2 dt = 0.$$

This also shows that $\widetilde{u}$ agrees with u on J_n when $u \in \Pi_{q-1}$ so that the interpolation is accurate of order q. By standard arguments we then have (with $|\cdot|_0 = \|\cdot\|$)

$$|\widetilde{u}(t) - u(t)|_j^2 \le C k_n^{2q-1} \int_{J_n} |u^{(q)}|_j^2 \, dt, \quad \text{for } t \in J_n,\ j = 0, 1. \tag{12.10}$$

We now decompose the error as

$$U - u = (U - \widetilde{u}) + (\widetilde{u} - u) = \theta + \rho, \tag{12.11}$$

and note that $\rho^n = \rho(t_n) = 0$ for all $n \ge 0$. It therefore suffices to bound θ^N by the right hand side of (12.8). We have by (12.4) and (12.1)

$$\begin{aligned} &\int_{J_n} \big((\theta', X) + A(\theta, X)\big)\, dt + ([\theta]_{n-1}, X_+^{n-1}) \\ &\quad = -\int_{J_n} \big((\rho', X) + A(\rho, X)\big)\, dt - ([\rho]_{n-1}, X_+^{n-1}), \quad \forall X \in \mathcal{S}_k^n. \end{aligned} \tag{12.12}$$

Here, by the defining properties (12.9) of $\widetilde{u}$,

$$\begin{aligned} &\int_{J_n} (\rho', X)\, dt + ([\rho]_{n-1}, X_+^{n-1}) \\ &\quad = (\rho, X)\Big|_{t_{n-1}+0}^{t_n} - \int_{J_n} (\rho, X')\, dt + ([\rho]_{n-1}, X_+^{n-1}) \\ &\quad = -(\rho_+^{n-1}, X_+^{n-1}) + (\rho_+^{n-1}, X_+^{n-1}) = 0, \quad \forall X \in \mathcal{S}_k^n. \end{aligned} \tag{12.13}$$

Choosing $X = 2\theta$ in (12.12) and noting that

$$\begin{aligned} 2\int_{J_n} (\theta', \theta)\, dt &+ 2([\theta]_{n-1}, \theta_+^{n-1}) \\ &= \|\theta^n\|^2 - \|\theta_+^{n-1}\|^2 + 2\|\theta_+^{n-1}\|^2 - 2(\theta^{n-1}, \theta_+^{n-1}) \\ &\ge \|\theta^n\|^2 - \|\theta_+^{n-1}\|^2 + \|\theta_+^{n-1}\|^2 - \|\theta^{n-1}\|^2 \\ &= \|\theta^n\|^2 - \|\theta^{n-1}\|^2, \end{aligned}$$

we have

$$\|\theta^n\|^2 + 2\int_{J_n} |\theta|_1^2\,dt \le \|\theta^{n-1}\|^2 + 2\int_{J_n} |A(\rho,\theta)|\,dt$$
$$\le \|\theta^{n-1}\|^2 + \int_{J_n} |\theta|_1^2\,dt + \int_{J_n} |\rho|_1^2\,dt.$$

Hence

$$\|\theta^n\|^2 + \int_{J_n} |\theta|_1^2\,dt \le \|\theta^{n-1}\|^2 + \int_{J_n} |\rho|_1^2\,dt, \tag{12.14}$$

and, by summation, since $\theta^0 = U^0 - \widetilde{u}^0 = v - v = 0$,

$$\|\theta^N\|^2 + \int_0^{t_N} |\theta|_1^2\,dt \le \int_0^{t_N} |\rho|_1^2\,dt. \tag{12.15}$$

Using (12.10) we conclude from (12.15) that

$$\|\theta^N\|^2 \le \sum_{n=1}^N \int_{J_n} |\rho|_1^2\,dt \le C\sum_{n=1}^N k_n^{2q}\int_{J_n} |u^{(q)}|_1^2\,dt, \tag{12.16}$$

which completes the proof. □

In the case of constant time steps $k_n = k$, the error estimate of Theorem 12.1 reduces to

$$\|U^N - u(t_N)\| \le Ck^q\Big(\int_0^{t_N} |u^{(q)}|_1^2\,dt\Big)^{1/2}.$$

We note that in the case of the backward Euler method (12.6) the error bound contains only first derivatives with respect to time, in contrast to the standard error estimate for the backward Euler method for which u'' enters; it is natural that more regularity is required when the integral in (12.6) is evaluated by a point-value quadrature formula.

Although the above error estimate concerns only the nodal values, estimates of the same optimal order may be derived also in the interior of the intervals J_n, as follows from the next theorem. Here and below we use the notation

$$\|\varphi\|_{J_n} = \sup_{t\in J_n} \|\varphi(t)\|.$$

Theorem 12.2 *We have, for the solutions of* (12.4) *(with $q \ge 1$) and* (12.1), *for $1 \le n \le N$,*

$$\|U - u\|_{J_n} \le \|U^n - u(t_n)\| + C\|U^{n-1} - u(t_{n-1})\| + Ck_n^q\|u^{(q)}\|_{J_n}.$$

Proof. This time we write the error

$$e = U - u = (U - P_k u) + (P_k u - u) = \xi + \eta,$$

where P_k denotes the L_2-projection in time onto $\mathcal{S}_k^n$. Clearly then

$$\|\eta\|_{J_n} + k_n \|\eta'\|_{J_n} \le C k_n^q \|u^{(q)}\|_{J_n}.$$

Hence, in order to prove our result it remains to bound ξ. But

$$\|\xi(t)\| \le \|\xi^n\| + \int_t^{t_n} \|\xi'\|\, ds \le \|e^n\| + \|\eta^n\| + \int_{J_n} \|\xi'\|\, dt,$$

so that it now remains to bound the latter integral. We shall prove

$$\int_{J_n} \|\xi'\|\, dt \le C(\|e^{n-1}\| + \|\eta^n\|), \tag{12.17}$$

which will imply our claim.

We first note that, using for the second inequality a transformation to a unit size interval and the finite dimensionality of the polynomial spaces involved,

$$\Big(\int_{J_n} \|\xi'\|\, dt\Big)^2 \le k_n \int_{J_n} \|\xi'\|^2\, dt \le C \int_{J_n} (t - t_{n-1}) \|\xi'\|^2\, dt. \tag{12.18}$$

To estimate the latter integral we note that, using the orthogonality of η to $\mathcal{S}_k^n$,

$$\begin{aligned}
&\int_{J_n} \big((\xi', X) + A(\xi, X)\big)\, dt + (\xi_+^{n-1}, X_+^{n-1}) \\
&= (U^{n-1}, X_+^{n-1}) + \int_{J_n} (f, X)\, dt \\
&\qquad - \int_{J_n} \big(((P_k u)', X) + A(P_k u, X)\big)\, dt - ((P_k u)_+^{n-1}, X_+^{n-1}) \\
&= (U^{n-1} - (P_k u)_+^{n-1}, X_+^{n-1}) - \int_{J_n} (\eta', X)\, dt \\
&= (e^{n-1} - \eta_+^{n-1}, X_+^{n-1}) \\
&\qquad - \Big((\eta^n, X^n) - (\eta_+^{n-1}, X_+^{n-1}) - \int_{J_n} (\eta, X')\, dt\Big) \\
&= (e^{n-1}, X_+^{n-1}) - (\eta^n, X^n).
\end{aligned} \tag{12.19}$$

Choosing $X(t) = (t - t_{n-1})\xi'(t)$ in (12.19) we find, since $X_+^{n-1} = 0$,

$$\int_{J_n} (t - t_{n-1}) \|\xi'\|^2\, dt + \tfrac{1}{2} k_n |\xi^n|_1^2 \le \tfrac{1}{2} \int_{J_n} |\xi|_1^2\, dt + k_n \|\eta^n\|\, \|\xi'(t_n)\|.$$

In the same way as in (12.18), we have

$$k_n^2\|\xi'(t_n)\|^2 \le C\int_{J_n}(t-t_{n-1})\|\xi'\|^2\,dt,$$

so that we may now conclude

$$\int_{J_n}(t-t_{n-1})\|\xi'\|^2\,dt \le C\int_{J_n}|\xi|_1^2\,dt + C\|\eta^n\|^2. \tag{12.20}$$

To estimate the integral on the right we now choose $X=2\xi$ in (12.19) to obtain

$$\|\xi^n\|^2 + \|\xi_+^{n-1}\|^2 + 2\int_{J_n}|\xi|_1^2\,dt \le 2\|e^{n-1}\|\,\|\xi_+^{n-1}\| + 2\|\eta^n\|\,\|\xi^n\|,$$

which yields

$$\int_{J_n}|\xi|_1^2\,dt \le C(\|e^{n-1}\|^2 + \|\eta^n\|^2).$$

Together with (12.20) and (12.18) this completes the proof of (12.17) and thus of the theorem. □

We have thus shown a global error estimate of order $O(k^q)$, which is the optimal order using polynomials of degree $q-1$. We recall, however, that in the case $q=2$, the approximation of the homogeneous equation is associated with the subdiagonal (2,1)-Padé approximation of $e^{-\lambda}$. Since this is accurate of order $O(k^3)$, this raises the question of the optimality at the nodal points of the error bound derived, which is only second order for $q=2$. In our next result we shall see that, at the nodes, the error in the discontinuous Galerkin method is actually of order $O(k^{2q-1})$, which is of superconvergent order for $q\ge 2$.

Theorem 12.3 *We have, for the solutions of* (12.4), *with* $q\ge 2$, *and* (12.1),

$$\|U^N - u(t_N)\| \le Ck^{q-1}\Big(\sum_{n=1}^N k_n^{2q}\int_{J_n}|u^{(q)}|_{2q-1}^2\,dt\Big)^{1/2}, \quad \text{for } t_N\ge 0,$$

where $k=\max_n k_n$.

We note that this thus shows

$$\|U^N - u(t_N)\| \le Ck^{2q-1}\Big(\int_0^{t_N}|u^{(q)}|_{2q-1}^2\,dt\Big)^{1/2}, \quad \text{for } t_N\ge 0.$$

We remark that in application to partial differential operators A, severe boundary conditions need to be imposed on the solution of the continuous problem for $q\ge 2$ since $u^{(q)}(t)$ is required to be in $\dot H^{2q-1}$ for $t>0$.

The proof of this theorem will require some preparation. Because we are interested in bounding the error in the solution of (12.4) at $t = t_N$ we introduce the global bilinear form

$$(12.21)\qquad \begin{aligned} B_N(V,W) = &\int_0^{t_N} \big((V',W) + A(V,W)\big)\,dt \\ &+ \sum_{n=1}^{N-1}([V]_n, W_+^n) + (V_+^0, W_+^0). \end{aligned}$$

Here, for V discontinuous at the points of the partition, we understand by V' the piecewise smooth function obtained by differentiation on each J_n. With this definition the discrete equations (12.3) may be written

$$B_N(U,X) = (v, X_+^0) + \int_0^{t_N} (f,X)\,dt, \quad \forall\, X \in \mathcal{S}_k.$$

Since clearly the solution u of the continuous problem satisfies, for any appropriately regular W, in particular for $W = X \in \mathcal{S}_k$, the equation

$$(12.22)\qquad B_N(u,W) = (v, W_+^0) + \int_0^{t_N} (f,W)\,dt,$$

we have, for the error $e = U - u$,

$$(12.23)\qquad B_N(e,X) = 0, \quad \forall\, X \in \mathcal{S}_k.$$

In our analysis we shall also consider the *backward* homogeneous problem

$$(12.24)\qquad -z' + Az = 0, \quad \text{for } t < t_N, \quad \text{with } z(t_N) = \varphi.$$

We note that, if z is the solution of (12.24), then

$$(12.25)\qquad B_N(u,z) = (u(t_N), \varphi).$$

Replacing the variable t by $t_N - t$ we find that the natural analogue of the discrete problem (12.4) for (12.24) is to find $Z \in \mathcal{S}_k$ such that

$$(12.26)\qquad \begin{aligned} &\int_{J_n} \big(-(X,Z') + A(X,Z)\big)\,dt + (X^n, Z^n) = (X^n, Z_+^n), \\ &\qquad\qquad\qquad\qquad \forall X \in \mathcal{S}_k^n,\ n \le N, \\ &Z_+^N = \varphi. \end{aligned}$$

By integration by parts in (12.21) our bilinear form $B_N(V,W)$ may also be represented as

$$(12.27)\qquad \begin{aligned} B_N(V,W) = &\int_0^{t_N} \big(-(V,W') + A(V,W)\big)\,dt \\ &- \sum_{n=1}^{N-1}(V^n, [W]_n) + (V^N, W^N). \end{aligned}$$

As a result of this it is clear that the discrete analogue (12.26) of (12.24) may also be stated as to find $Z \in \mathcal{S}_k$ such that

$$B_N(X, Z) = (X^N, \varphi), \quad \forall\, X \in \mathcal{S}_k. \tag{12.28}$$

Thus, in particular, this problem has a unique solution and also other results obtained for the forward problem translate to this case. In particular,

$$B_N(X, Z - z) = 0, \quad \forall X \in \mathcal{S}_k. \tag{12.29}$$

Proof of Theorem 12.3. Let z and Z be the solutions of (12.24) and (12.28), with $\varphi \in \mathcal{H}$, and let $e^N = U^N - u(t_N)$. Then, by (12.25) and (12.28),

$$\begin{aligned}(e^N, \varphi) &= (U^N, \varphi) - (u(t_N), \varphi) = B_N(U, Z) - B_N(u, z)\\ &= B_N(U - u, z) + B_N(U, Z - z).\end{aligned}$$

Thus, using also (12.23) and (12.29), and setting $Z - z = \zeta$ and $\rho = \widetilde{u} - u$, where $\widetilde{u}$ is the interpolant defined in (12.9), we find

$$\begin{aligned}(e^N, \varphi) &= B_N(U - u, z - Z) = B_N(\widetilde{u} - u, z - Z)\\ &= -B_N(\rho, \zeta) = -\int_0^{t_N} \big(-(\rho, \zeta') + A(\rho, \zeta)\big)\, dt,\end{aligned}$$

where in the last step we have used (12.27) and the fact that $\rho^n = 0$ for $n \geq 0$. Since $|(v, w)| \leq |v|_s |w|_{-s}$ it then follows that

$$|(e^N, \varphi)| \leq \Big(\int_0^{t_N} |\rho|_{2q-1}^2\, dt\Big)^{1/2} \Big(\int_0^{t_N} \big(|\zeta'|_{-2q+1}^2 + |\zeta|_{-2q+3}^2\big)\, dt\Big)^{1/2}.$$

Here, cf. (12.10),

$$\int_0^{t_N} |\rho|_{2q-1}^2\, dt = \sum_{n=1}^N \int_{J_n} |\rho|_{2q-1}^2\, dt \leq C \sum_{n=1}^N k_n^{2q} \int_{J_n} |u^{(q)}|_{2q-1}^2\, dt.$$

We shall also show below that

$$\int_0^{t_N} \big(|\zeta'|_{-2q+1}^2 + |\zeta|_{-2q+3}^2\big)\, dt \leq C k^{2q-2} \|\varphi\|^2. \tag{12.30}$$

Assuming this for a moment, we find

$$(e^N, \varphi) \leq C k^{q-1} \Big(\sum_{n=1}^N k_n^{2q} \int_{J_n} |u^{(q)}|_{2q-1}^2\, dt\Big)^{1/2} \|\varphi\|,$$

which completes the proof of the theorem.

In order to show (12.30), we consider the corresponding forward problem (12.1) and (12.4), with $f \equiv 0$, and show, for $e = U - u$,

$$\int_0^{t_N} \left(|e'|_{-2q+1}^2 + |e|_{-2q+3}^2\right) dt \le Ck^{2q-2}\|v\|^2. \tag{12.31}$$

We now note that this will follow from

$$\int_0^{t_N} \left(|e'|_{-1}^2 + |e|_1^2\right) dt \le Ck^{2q-2}\|A^{q-1}v\|^2,$$

by replacing v by $A^{-q+1}v$ in the latter error estimate.

We write as in (12.11) $e = (U - \widetilde{u}) + (\widetilde{u} - u) = \theta + \rho$, and begin by bounding ρ and ρ'. In the same way as in (12.10) we have, for any j,

$$\int_{J_n} |\rho^{(s)}|_j^2\, dt \le Ck_n^{2(r-s)} \int_{J_n} |u^{(r)}u|_j^2\, dt, \quad \text{for } t \in J_n, \quad 0 \le s \le 1 \le r \le q.$$

This implies, as in (12.16),

$$\begin{aligned} &\int_0^{t_N} \left(|\rho'|_{-1}^2 + |\rho|_1^2\right) dt \\ &\le Ck^{2q-2} \int_0^{t_N} \left(|u^{(q)}|_{-1}^2 + |u^{(q-1)}|_1^2\right) dt \\ &= Ck^{2q-2} \int_0^{t_N} |A^{q-1}u|_1^2\, dt \le Ck^{2q-2}\|A^{q-1}v\|^2. \end{aligned} \tag{12.32}$$

Note that in order to be in parity with the optimal order estimate for ρ' we have used a suboptimal order estimate for ρ.

It remains to bound θ and θ'. We have by (12.15) and (12.32)

$$\int_0^{t_N} |\theta|_1^2\, dt \le \int_0^{t_N} |\rho|_1^2\, dt \le Ck^{2q-2}\|A^{q-1}v\|^2, \tag{12.33}$$

which is the desired estimate for θ. For θ', we note that by (12.12) and (12.13), with $X(t) = (t - t_{n-1})\theta'(t)$, we have

$$\int_{J_n} (t - t_{n-1})\left(\|\theta'\|^2 + A(\theta, \theta')\right) dt = -\int_{J_n} (t - t_{n-1})A(\rho, \theta')\, dt,$$

from which we conclude

$$\int_{J_n} (t - t_{n-1})\|\theta'\|^2\, dt \le C\int_{J_n} (t - t_{n-1})\left(\|A\theta\|^2 + \|A\rho\|^2\right) dt,$$

or, using a local inverse estimate on J_n,

$$\int_{J_n} \|\theta'\|^2\, dt \le C\int_{J_n} \left(\|A\theta\|^2 + \|A\rho\|^2\right) dt.$$

Application of this estimate with initial values $A^{-1/2}v$ rather than v, and summation, together with (12.32) and (12.33) shows

$$\int_0^{t_N} |\theta'|_{-1}^2\,dt \le C\int_0^{t_N} \big(|\theta|_1^2 + |\rho|_1^2\big)\,dt \le Ck^{2q-2}\|A^{q-1}v\|^2.$$

This completes the proof of (12.31) and thus of the theorem. □

We shall now turn to a different type of error estimates in which the L_2-type norm in time of the error bound is replaced by a maximum-norm. In this regard we state the following theorem; for $q = 1$, cf. Theorem 7.6.

Theorem 12.4 *Assume that $k_{n+1}/k_n \ge c > 0$ for $n \ge 0$. Then we have, for the solutions of* (12.4) *with $q \ge 1$ and of* (12.1),

$$\|U - u\|_{J_N} \le CL_N \max_{n\le N} \big(k_n^q\|u^{(q)}\|_{J_n}\big), \quad \textit{where } L_N = (\log\frac{t_N}{k_N})^{1/2} + 1.$$

This result suggests, e.g., that to keep the error uniformly small, we should choose the time steps inversely proportional to $\|u^{(q)}\|_{J_n}^{1/q}$. Note that L_N is of moderate size and does not effect the error bound in an essential way.

In the proof we shall use the following representation of the error, which contains the approximate solution of the backward problem (12.24).

Lemma 12.1 *Let U and u be the solution of* (12.4) *and* (12.1), *and Z that of* (12.28), *with $\varphi \in \mathcal{H}$. Then, with $u^N = u(t_N)$ we have for the error $e^N = U^N - u^N$*

$$(e^N, \varphi) = B_N(u - X, Z) + (X^N - u^N, \varphi), \quad \forall X \in \mathcal{S}_k.$$

Proof. We have by (12.28) and the error relation (12.23)

$$\begin{aligned}(e^N,\varphi) &= (U^N - X^N, \varphi) + (X^N - u^N, \varphi)\\ &= B_N(U - X, Z) + (X^N - u^N, \varphi)\\ &= B_N(u - X, Z) + (X^N - u^N, \varphi).\end{aligned} \tag{12.34}$$

which shows our claim. □

We will also need the following two stability results, the second of which is the main technical step in our analysis.

Lemma 12.2 *When $f = 0$ we have for the solution of* (12.4)

$$\|U^N\|^2 + 2\int_0^{t_N} |U|_1^2\,dt + \sum_{n=0}^{N-1} \|[U]_n\|^2 = \|v\|^2.$$

Proof. We choose $X = 2U$ in (12.4) (with $f \equiv 0$) to obtain

$$2\int_{J_n} \big((U', U) + A(U, U)\big)\,dt + 2([U]_{n-1}, U_+^{n-1}) = 0. \tag{12.35}$$

Here

$$2\int_{J_n}(U',U)\,dt = \int_{J_n}\frac{d}{dt}\|U\|^2\,dt = \|U^n\|^2 - \|U_+^{n-1}\|^2,$$

and

$$\begin{aligned}2([U]_{n-1},U_+^{n-1}) &= ([U]_{n-1},U_+^{n-1}+U^{n-1}+[U]_{n-1})\\ &= \|U_+^{n-1}\|^2 - \|U^{n-1}\|^2 + \|[U]_{n-1}\|^2,\end{aligned}$$

so that (12.35) yields

$$\|U^n\|^2 + 2\int_{J_n}|U|_1^2\,dt + \|[U]_{n-1}\|^2 = \|U^{n-1}\|^2. \tag{12.36}$$

Summation from $n=1$ to N now shows the lemma. □

Lemma 12.3 *Assume that* $k_{n+1}/k_n \geq c > 0$. *Then we have for the solution of* (12.28)

$$\int_0^{t_N}(\|Z'\| + \|AZ\|)\,dt + \sum_{n=1}^N \|[Z]_n\| \leq CL_N\|\varphi\|.$$

Proof. We shall show the corresponding estimate for the forward problem, i.e., assuming now that $k_{n-1}/k_n \geq c > 0$, and with $U^0 = v$, $L_N^* = (\log(t_N/k_1))^{1/2} + 1$ we show for the solution of (12.4) with $f \equiv 0$

$$\int_0^{t_N}(\|U'\| + \|AU\|)\,dt + \sum_{n=1}^N \|[U]_{n-1}\| \leq CL_N^*\|v\|. \tag{12.37}$$

For this purpose we shall establish

$$\sum_{n=1}^N\Big(t_n\int_{J_n}(\|U'\|^2 + \|AU\|^2)\,dt + t_nk_n^{-1}\|[U]_{n-1}\|^2\Big) \leq C\|v\|^2, \tag{12.38}$$

which easily shows (12.37). In fact, by Schwarz' inequality and (12.38)

$$\int_0^{t_N}\|U'\|\,dt = \sum_{n=1}^N\int_{J_n}\|U'\|\,dt \leq \sum_{n=1}^N k_n^{1/2}\Big(\int_{J_n}\|U'\|^2\,dt\Big)^{1/2},$$

and hence

$$\Big(\int_0^{t_N}\|U'\|\,dt\Big)^2 \leq \sum_{n=1}^N k_nt_n^{-1}\sum_{n=1}^N t_n\int_{J_n}\|U'\|^2\,dt \leq C(L_N^*)^2\|v\|^2,$$

where we have used

$$\sum_{n=1}^N k_nt_n^{-1} \leq 1 + \int_{k_1}^{t_N}\frac{dt}{t} = 1 + \log\frac{t_N}{k_1} \leq (L_N^*)^2.$$

The term in $\|AU\|$ is treated in a similar way, and finally

$$\Big(\sum_{n=1}^{N}\|[U]_{n-1}\|\Big)^2 \le \sum_{n=1}^{N} k_n t_n^{-1} \sum_{n=1}^{N} t_n k_n^{-1}\|[U]_{n-1}\|^2 \le C(L_N^*)^2\|v\|^2,$$

which completes the proof of (12.37).

We begin the proof of (12.38) with the estimate for AU, and choose $X = 2AU$ in (12.4) (with $f = 0$), to obtain, similarly to (12.36),

$$|U^n|_1^2 + 2\int_{J_n}\|AU\|^2\,dt + |[U]_{n-1}|_1^2 \le |U^{n-1}|_1^2, \quad \text{for } n \ge 2,$$

and, after multiplication by t_n, since $k_n \le Ck_{n-1}$,

$$\begin{aligned} t_n|U^n|_1^2 + 2t_n\int_{J_n}\|AU\|^2\,dt + t_n|[U]_{n-1}|_1^2 \\ \le t_{n-1}|U^{n-1}|_1^2 + k_n|U^{n-1}|_1^2 \le t_{n-1}|U^{n-1}|_1^2 + Ck_{n-1}|U^{n-1}|_1^2. \end{aligned}$$

Summation from $n = 2$ to N shows

$$2\sum_{n=2}^{N} t_n\int_{J_n}\|AU\|^2\,dt \le C\sum_{n=1}^{N-1} k_n|U^n|_1^2. \tag{12.39}$$

Here, using an inverse inequality on each J_n and Lemma 12.2,

$$\sum_{n=1}^{N-1} k_n|U^n|_1^2 \le C\int_0^{t_{N-1}}|U|_1^2\,dt \le C\|v\|^2. \tag{12.40}$$

To estimate $\|AU\|$ on J_1 we set again $X = 2AU$ in (12.4) (with $n = 1, f = 0$) to obtain

$$\begin{aligned} |U^1|_1^2 + |U_+^0|_1^2 + 2\int_{J_1}\|AU\|^2\,dt &= 2(v, AU_+^0) \\ &\le \varepsilon k_1\|AU_+^0\|^2 + (\varepsilon k_1)^{-1}\|v\|^2. \end{aligned}$$

Here, since $\|AX\|^2$ is a polynomial of degree $\le 2q$ on J_1,

$$k_1\|AX_+^0\|^2 \le C_q\int_{J_1}\|AX\|^2\,dt, \quad \forall X \in \mathcal{S}_k^1,$$

and hence, with $X = U$ and by choosing $\varepsilon \le C_q^{-1}$, we conclude

$$k_1\int_{J_1}\|AU\|^2\,dt \le C\|v\|^2.$$

Together with (12.39) and (12.40) this shows the desired bound for $\|AU\|$.

To estimate $\|U'\|$ we choose $X = (t - t_{n-1})U'$ in (12.4) (with $f = 0$) to obtain

$$\int_{J_n} (t - t_{n-1})\|U'\|^2\,dt = -\int_{J_n} (t - t_{n-1})(AU, U')\,dt$$
$$\leq \frac{1}{2}\int_{J_n} (t - t_{n-1})\|AU\|^2\,dt + \frac{1}{2}\int_{J_n} (t - t_{n-1})\|U'\|^2\,dt,$$

and hence

$$\int_{J_n} (t - t_{n-1})\|U'\|^2\,dt \leq \int_{J_n} (t - t_{n-1})\|AU\|^2\,dt \leq k_n \int_{J_n} \|AU\|^2\,dt.$$

Again a local inverse estimate gives

$$\int_{J_n} \|U'\|^2\,dt \leq Ck_n^{-1} \int_{J_n} (t - t_{n-1})\|U'\|^2\,dt \leq C \int_{J_n} \|AU\|^2\,dt,$$

and the desired inequality for $\|U'\|$ now follows from that for $\|AU\|$.

To estimate $[U]_{n-1}$, finally, we choose $X = [U]_{n-1}$ in (12.4) (with $f = 0$) to obtain, for $n \geq 1$,

$$\|[U]_{n-1}\|^2 = -\int_{J_n} \big((U', [U]_{n-1}) + (AU, [U]_{n-1})\big)\,dt$$
$$\leq \frac{1}{2}\|[U]_{n-1}\|^2 + \frac{1}{2}k_n \int_{J_n} \big(\|U'\|^2 + \|AU\|^2\big)\,dt,$$

or

$$k_n^{-1}\|[U]_{n-1}\|^2 \leq \int_{J_n} \big(\|U'\|^2 + \|AU\|^2\big)\,dt.$$

The desired result again follows by multiplication by t_n and summation using the results already obtained for $\|U'\|$ and $\|AU\|$. □

Proof of Theorem 12.4. We shall first bound $e^N = U^N - u^N$. We apply Lemma 12.1, choosing $X = \widetilde{u}$, where $\widetilde{u}$ is the interpolant defined in (12.9). With $\rho = \widetilde{u} - u$ we then have, using (12.27) and the properties in (12.9),

$$(12.41) \qquad \begin{aligned} (e^N, \varphi) &= -B_N(\rho, Z) + (\rho^N, \varphi) \\ &= \int_0^{t_N} \big((\rho, Z') - A(\rho, Z)\big)\,dt + \sum_{n=0}^{N-1} (\rho^n, [Z]_n) \\ &= -\int_0^{t_N} (\rho, AZ)\,dt, \end{aligned}$$

and hence, using (12.10) and Lemma 12.3,

$$|(e^N, \varphi)| \leq \max_{n \leq N} \|\rho\|_{J_n} \int_0^{t_N} \|AZ\|\,dt \leq CL_N\|\varphi\| \max_{n \leq N} \big(k_n^q \|u^{(q)}\|_{J_n}\big).$$

This implies

$$\|e^N\| \le CL_N \max_{n\le N} \big(k_n^q \|u^{(q)}\|_{J_n}\big).$$

It remains to show the estimate at the interior points of J_N. But with the nodal estimates now proven, this follows from Theorem 12.2. The proof is therefore now complete. □

For $q = 2$ we shall also show the following superconvergent third order error estimate at the nodal points, with a maximum-norm error bound.

Theorem 12.5 *Assume that* $k_{n+1}/k_n \ge c > 0$ *for all* n. *Then, for* $q = 2$, *we have for the solutions of* (12.4) *and* (12.1)

$$\|U^N - u(t_N)\| \le CL_N \max_{n\le N} \big(k_n^3 \|Au_{tt}\|_{J_n}\big).$$

Proof. Using (12.41) we have

$$(e^N, \varphi) = -\sum_{n=1}^{N} \int_{J_n} (A\rho, Z)\,dt = -\sum_{n=1}^{N} K_n.$$

Here, since ρ is orthogonal to constants for $q = 2$, we find

$$K_n = \int_{J_n} \Big(A\rho, Z_+^{n-1} + \int_{t_{n-1}}^{t} Z'(s)\,ds\Big)\,dt = \int_{J_n} \Big(A\rho, \int_{t_{n-1}}^{t} Z'(s)ds\Big)\,dt,$$

and hence

(12.42) $$|K_n| \le k_n \|A\rho\|_{J_n} \int_{J_n} \|Z'\|\,dt.$$

We conclude, using (12.10) with $j = 2$, that

$$\begin{aligned}|(e^N, \varphi)| &\le \max_{n\le N} \Big(k_n \|A\rho\|_{J_n}\Big) \int_0^{t_N} \|Z'\|\,dt \\ &\le CL_N \max_{n\le N} \Big(k_n^3 \|Au_{tt}\|_{J_n}\Big) \|\varphi\|,\end{aligned}$$

which completes the proof. □

We now turn to the application of the discontinuous Galerkin method to the solution of the partial differential equation problem

(12.43) $$u_t - \Delta u = f \quad \text{in } \Omega, \qquad \text{for } t > 0,$$
$$u = 0 \quad \text{on } \partial\Omega, \quad \text{for } t > 0, \quad u(\cdot, 0) = v \quad \text{in } \Omega.$$

For simplicity we now assume Ω to be a convex polygonal plane domain, and recall that the standard elliptic regularity estimate (1.7) holds in this case for $m = 0$. We restrict the discussion to the standard family of continuous, piecewise linear finite element spaces, and consider the semidiscrete problem to find $u_h(t) \in S_h$ for $t \geq 0$ such that

$$(12.44) \qquad \begin{aligned} (u_{h,t}, \chi) + (\nabla u_h, \nabla \chi) &= (f, \chi), \quad \forall \chi \in S_h, \ t > 0, \\ u_h(0) &= v_h, \end{aligned}$$

where v_h is an approximation of v.

It is to this problem that we now want to apply the discontinuous Galerkin time stepping method, so that the Hilbert space $\mathcal{H}$ will be S_h, equipped with the L_2 inner product, and the discrete Laplacian Δ_h defined in (1.33) will play the role of the operator A. In order to discretize (12.44) in time, we shall thus use the finite dimensional space

$$\mathcal{S}_{kh} = \{X : [0, \infty) \to S_h; \ X\big|_{J_n} = \sum_{j=0}^{q-1} X_j t^j, \ X_j \in S_h\},$$

and our fully discrete method is now to find $U_h \in \mathcal{S}_{kh}$ such that

$$(12.45) \qquad B_N(U_h, X) = (v_h, X_+^0) + \int_0^{t_N} (f, X)\, dt, \quad \forall\, X \in \mathcal{S}_{kh}, \ N \geq 0,$$

where this time

$$(12.46) \qquad \begin{aligned} &B_N(V, W) \\ &= \int_0^{t_N} \big((V_t, W) + (\nabla V, \nabla W)\big)\, dt + \sum_{n=1}^{N-1} ([V]_n, W_+^n) + (V_+^0, W_+^0) \\ &= \int_0^{t_N} \big(-(V, W_t) + (\nabla V, \nabla W)\big)\, dt - \sum_{n=1}^{N-1} (V^n, [W]_n) + (V^N, W^N). \end{aligned}$$

The equation satisfied by the error now takes the form

$$B_N(e, X) = (v_h - v, X_+^0), \quad \forall\, X \in \mathcal{S}_{kh}.$$

Note that the right hand side vanishes when $v_h = P_h v$.

We shall only show the nodal error estimates corresponding to Theorem 12.4 with $q = 1$ and to Theorem 12.5 where $q = 2$. We first have the following result where the approximating functions are piecewise constant in time. Here and below we use the notation

$$\|\varphi\|_{J_n} = \sup_{t \in J_n} \|\varphi(t)\|, \quad \text{and} \quad \|\varphi\|_{2,J_n} = \sup_{t \in J_n} \|\varphi(t)\|_2,$$

where now $\|\cdot\|$ is the norm in $L_2 = L_2(\Omega)$ and $\|\cdot\|_2$ that in $H^2 = H^2(\Omega)$.

Theorem 12.6 *Assume that* $k_{n+1}/k_n \ge c > 0$ *for* $n \ge 0$ *and let* $q = 1$. *Then we have for the solutions of* (12.45) *and* (12.43), *with* $v_h = P_h v$,

$$\|U_h^N - u(t_N)\| \le CL_N \max_{n\le N}(h^2\|u\|_{2,J_n} + k_n\|u_t\|_{J_n}).$$

Proof. Let $\widetilde{u}$ denote the piecewise constant function (with respect to t) defined by $\widetilde{u}(t) = u(t_n)$ for $t \in J_n$, and write

$$e = U_h - u = (U_h - R_h\widetilde{u}) + (R_h\widetilde{u} - u) = \theta + \rho, \tag{12.47}$$

where $R_h : H_0^1 \to S_h$ is the Ritz projection defined by (1.22). Since $\widetilde{u}(t_N) = u(t_N)$, we have $\|\rho^N\| = \|(R_h u - u)(t_N)\| \le Ch^2\|u(t_N)\|_2$. To bound θ^N, let $\varphi \in L_2$ and let Z_h be the fully discrete analogue of our previous Z, i.e., the solution of

$$B_N(X, Z_h) = (X^N, P_h\varphi) = (X^N, \varphi), \quad \forall X \in \mathcal{S}_{kh}.$$

Then, since $Z_{h,t}(t) \equiv 0$ on each J_n,

$$\begin{aligned}(\theta^N, \varphi) &= B_N(\theta, Z_h) = -B_N(\rho, Z_h)\\ &= -\sum_{n=1}^{N}\int_{J_n}(\nabla\rho, \nabla Z_h)\,dt + \sum_{n=1}^{N-1}(\rho^n, [Z_h]_n) - (\rho^N, P_h\varphi),\end{aligned} \tag{12.48}$$

and hence, since $(\nabla\rho, \nabla Z_h) = (R_h\rho, \Delta_h Z_h)$ we have

$$|(\theta^N, \varphi)| \le \max_{n\le N}(\|\rho\|_{J_n} + \|R_h\rho\|_{J_n})\Big(\int_0^{t_N}\|\Delta_h Z_h\|\,dt + \sum_{n=1}^{N-1}\|[Z_h]_n\| + \|\varphi\|\Big).$$

By the stability result of Lemma 12.3, applied in the discrete context, we thus have

$$\|\theta^N\| \le CL_N \max_{n\le N}\|\rho\|_{J_n}.$$

Now

$$\begin{aligned}\|\rho\|_{J_n} &= \|R_h\widetilde{u} - u\|_{J_n} \le \|(R_h - I)\widetilde{u}\|_{J_n} + \|\widetilde{u} - u\|_{J_n}\\ &\le Ch^2\|u\|_{2,J_n} + Ck_n\|u_t\|_{J_n},\end{aligned} \tag{12.49}$$

and since $R_h\rho = R_h\widetilde{u} - R_h u = \rho - (R_h u - u)$, this function admits the same bound. This completes the proof. □

Theorem 12.7 *Let* $q = 2$, *and assume that* $k_{n+1}/k_n \ge c > 0$, *for* $n \ge 0$. *We have, for the solutions of* (12.45), *with* $v_h = P_h v$, *and* (12.43),

$$\|U_h^N - u(t_N)\| \le CL_N \max_{n\le N}\big(h^2\|u\|_{2,J_n} + k_n^3\|u_{tt}\|_{2,J_n}\big).$$

Proof. We again split the error according to (12.47), where now $\widetilde{u}$ is the piecewise linear interpolant defined by the case $q = 2$ of (12.9). This time we find instead of (12.48)

$$(\theta^N, \varphi) = -\sum_{n=1}^{N} \int_{J_n} \big(-(\rho, Z_{h,t}) + (\nabla\rho, \nabla Z_h)\big)\, dt \\ + \sum_{n=1}^{N-1} (\rho^n, [Z_h]_n) - (\rho^N, P_h\varphi).$$

Here we have, using the definition of $\widetilde{u}$,

$$\int_{J_n} (\rho, Z_{h,t})\, dt = \int_{J_n} (R_h\widetilde{u} - u, Z_{h,t})\, dt = \int_{J_n} (R_h u - u, Z_{h,t})\, dt,$$

and, by Lemma 12.3,

$$\Big| \sum_{n=1}^{N} \int_{J_n} (R_h u - u, Z_{h,t})\, dt \Big| \le \max_{n\le N} \|R_h u - u\|_{J_n} \int_0^{t_N} \|Z_{h,t}\|\, dt \\ \le C L_N h^2 \max_{n\le N} \|u\|_{2,J_n} \|\varphi\|,$$

and similarly

$$\Big| \sum_{n=1}^{N-1} (\rho^n, [Z_h]_n) \Big| + |(\rho^N, P_h\varphi)| \\ \le \max_{n\le N} \|(R_h u - u)(t_n)\| \Big(\sum_{n=1}^{N-1} \|[Z_h]_n\| + \|P_h\varphi\| \Big) \\ \le C L_N h^2 \max_{n\le N} \|u\|_{2,J_n} \|\varphi\|.$$

Finally, by the definition of R_h,

$$\sum_{n=1}^{N} \int_{J_n} (\nabla\rho, \nabla Z_h)\, dt = \sum_{n=1}^{N} \int_{J_n} (\nabla(\widetilde{u} - u), \nabla Z_h)\, dt \\ = -\sum_{n=1}^{N} \int_{J_n} (\Delta(\widetilde{u} - u), Z_h)\, dt = \sum_{n=1}^{N} K_n,$$

and we conclude as before in (12.42) that

$$|K_n| \le k_n \|\widetilde{u} - u\|_{2,J_n} \int_{J_n} \|Z_{h,t}\|\, dt,$$

and hence that

$$\begin{aligned}\sum_{n=1}^{N}|K_n| &\le \max_{n\le N}\big(k_n\|\widetilde{u}-u\|_{2,J_n}\big)\sum_{n=1}^{N}\int_{J_n}\|Z_{h,t}\|\,dt\\ &\le CL_N\max_{n\le N}\big(k_n^3\|u_{tt}\|_{2,J_n}\big)\,\|\varphi\|.\end{aligned}$$

Together these estimates show

$$|(\theta^N,\varphi)|\le CL_N\max_{n\le N}\Big(k_n^3\|u_{tt}\|_{2,J_n}+h^2\|u\|_{2,J_n}\Big)\|\varphi\|,$$

which bounds $\|\theta^N\|$ as desired. The proof is now complete. □

Our earlier error bounds contain quantities which depend on the exact solution and are of the desired order of magnitude provided this exact solution has specified regularity properties. Such error bounds are referred to as *a priori* error bounds. However, since the exact solution is unknown, such estimates do not provide precise quantitative upper bounds for the error. We shall therefore now show an *a posteriori* bound, which gives an error estimate expressed in terms of only the data of the problem and of the computed solution. Such estimates may be used to design adaptive methods for solving our initial value problem, thus defining the successive time steps of the method so that the error is guaranteed to be below some fixed tolerance.

We shall restrict our discussion here to the discontinuous Galerkin method studied above in the case of piecewise constant approximating functions in time, i.e., with $q=1$. We shall again begin to do so in our Hilbert space framework, so that only the discretization in time is involved, and then apply this to the spatially discrete version of the heat equation, for simplicity here only with piecewise linear finite elements, i.e., with $r=2$.

We consider thus first the initial value problem (12.1) and an approximate solution in $\mathcal{S}_k=\{X:[0,\infty)\to\mathcal{H};\ X\big|_{J_n}=\psi\in\mathcal{H}\}$, defined by

$$B_N(U,X)=(v,X_+^0)+\int_0^{t_N}(f,X)\,dt,\quad \forall X\in\mathcal{S}_k,$$

where $B_N(U,X)$ is defined in (12.21). As noted in (12.6), this may be written as

$$U^n+k_nAU^n=U^{n-1}+\int_{J_n}f\,dt,\quad \text{for } n\ge 1,\quad U^0=v. \tag{12.50}$$

Our *a posteriori* error estimate is then the following. Recall that $L_N=(\log(t_N/k_N))^{1/2}+1$.

Theorem 12.8 *We have, for the solutions of* (12.50) *and* (12.1),

$$\|U^N-u(t_N)\|\le CL_N\max_{n\le N}\big(k_n\|f\|_{J_n}+k_n\|\bar\partial_nU^n\|\big).$$

We remark that, by Theorem 12.4, if $k_{n+1}/k_n \ge c > 0$ for $n \ge 0$,

$$\begin{aligned} k_n\|\bar\partial_n U^n\| &= \|U^n - U^{n-1}\| \\ &\le \|U^n - u(t_n)\| + \|U^{n-1} - u(t_{n-1})\| + \|u(t_n) - u(t_{n-1})\| \\ &\le CL_n \max_{j\le n}\big(k_j\|u_t\|_{J_j}\big), \quad \text{for } n \le N, \end{aligned}$$

so that, modulo the logarithmic factor L_N, the contribution of this term to the error bound is bounded by the earlier derived *a priori* error bound.

The proof requires some preparation. It will use the solution of the backward problem (12.24). Note that in the proof of the *a priori* error estimate of Theorem 12.4, it was the discrete analogue of the solution of this problem that entered.

We shall need the following representation of the error.

Lemma 12.4 *With U and u the solutions of* (12.50) *and* (12.1), *and z that of* (12.24), *we have for $e = U - u$*

$$\begin{aligned} (e^N, \varphi) &= \int_0^{t_N} A(U, z - X)\, dt \\ &\quad + \sum_{n=0}^{N-1} ([U]_n, z^n - X_+^n) - \int_0^{t_N} (f, z - X)\, dt, \quad \forall\, X \in \mathcal{S}_k. \end{aligned}$$

Proof. We recall that the error e satisfies (12.23). Using first (12.27) with $V = e, W = z$ and then (12.21) and (12.22) with $W = z - X$ we therefore have at once

$$\begin{aligned} (e^N, \varphi) &= B_N(e, z) = B_N(e, z - X) = B_N(U, z - X) - B_N(u, z - X) \\ &= \int_0^{t_N} A(U, z - X)\, dt + \sum_{n=1}^{N-1} ([U]_n, (z - X)_+^n) + (U_+^0, (z - X)_+^0) \\ &\quad - (v, (z - X)_+^0) - \int_0^{t_N} (f, z - X)\, dt, \end{aligned}$$

which shows our claim. □

We shall also need a stability estimate for the exact solution of (12.24).

Lemma 12.5 *We have for the solution of the backward problem (12.24)*

$$\int_0^{t_{N-1}} \|z_t\|\, dt + \|z\|_{J_N} \le CL_N\|\varphi\|.$$

Proof. This follows from the corresponding result for the forward problem (12.1) with $f = 0$, which reads

$$\int_{k_1}^{t_N} \|u_t\|\, dt + \|u\|_{J_1} \le CL_N^*\|v\|, \quad \text{with } L_N^* = \Big(\log \frac{t_N}{k_1}\Big)^{1/2} + 1.$$

To show the latter estimate we note that $\|u(t)\| \leq \|v\|$ for $t \geq 0$, so that, in particular, $\|u\|_{J_1} \leq \|v\|$, and, by a simple energy argument,

$$\int_0^\infty t\|u_t\|^2\,dt = \tfrac{1}{4}\|v\|^2,$$

from which we conclude

$$\Big(\int_{k_1}^{t_N} \|u_t\|\,dt\Big)^2 \leq \int_{k_1}^{t_N} \frac{dt}{t} \int_{k_1}^{t_N} t\|u_t\|^2 dt \leq \tfrac{1}{4}\log\frac{t_N}{k_1}\,\|v\|^2,$$

which completes the proof. □

We are now ready for the proof of Theorem 12.8.

Proof of Theorem 12.8. We denote the three terms in the representation in Lemma 12.4 by *I, II* and *III*. We choose

$$X(t) = \bar{z}^n = k_n^{-1}\int_{J_n} z(t)\,dt, \quad \text{for } t \in J_n,\ n \geq 1.$$

Then, since $U(t)$ is constant on J_n we have

$$\int_{J_n} A(U, z - \bar{z}^n)\,dt = 0, \quad \text{for } n \geq 1,$$

and hence $I = 0$. For *II* we have since $X_+^n = \bar{z}^{n+1}$

$$|II| \leq \max_{n\leq N-1} \|[U]_n\| \sum_{n=1}^N \|\bar{z}^n - z^{n-1}\|.$$

Here

$$\|\bar{z}^n - z^{n-1}\| = \|k_n^{-1}\int_{J_n}(z - z^{n-1})\,dt\| \leq \int_{J_n} \|z_t\|\,dt, \quad \text{for } n < N,$$

and $\|\bar{z}^N - z^{N-1}\| \leq 2\|z\|_{J_N}$, so that, by Lemma 12.5,

$$\sum_{n=1}^N \|\bar{z}^n - z^{n-1}\| \leq \int_0^{t_{N-1}} \|z_t\|\,dt + 2\|z\|_{J_N} \leq CL_N\|\varphi\|.$$

Since $[U]_n = k_n\bar{\partial}_n U^n$ this shows the desired estimate for II.

For *III* we have similarly

$$\begin{aligned}|III| &\leq \max_{n\leq N}\big(k_n\|f\|_{J_n}\big)\sum_{n=1}^N \Big(k_n^{-1}\int_{J_n}\|\bar{z}^n - z\|\,dt\Big)\\ &\leq \max_{n\leq N}\big(k_n\|f\|_{J_n}\big)\Big(\int_0^{t_{N-1}}\|z_t\|\,dt + 2\|z\|_{J_N}\Big)\\ &\leq CL_N\max_{n\leq N}\big(k_n\|f\|_{J_n}\big)\,\|\varphi\|.\end{aligned}$$

This completes the proof. □

We shall close with a discussion of an *a posteriori* error estimates for the discontinuous Galerkin method in the case of the heat equation in a bounded convex polygonal domain $\Omega \subset \mathbb{R}^2$, with Dirichlet boundary conditions. The continuous problem we want to solve is (12.43) and its fully discrete analogue is now

$$B_N(U_h, X) = (v, X_+^0) + \int_0^{t_N} (f, X)\, dt, \quad \forall\, X \in \mathcal{S}_{kh},\ N \geq 1, \tag{12.51}$$

with $B_N(\cdot,\cdot)$ again defined by (12.46), and with S_h the basic family of continuous, piecewise linear finite element functions, which, for simplicity, we now assume to be associated with a quasiuniform family of triangulations. Note that (12.51) implies that we assume that the discrete initial values are chosen as $v_h = P_h v$.

We emphasize that we thus restrict the considerations to the case when S_h is independent of time and only the time steps vary. For more refined estimates, allowing different approximating spaces S_{h_n} on different time intervals, thus resulting in more precise adaptive schemes, see the references below. The method is thus to find $U_h \in \mathcal{S}_{kh}$ such that

$$\int_{J_n} (\nabla U_h, \nabla X)\, dt + ([U_h]_{n-1}, X_+^{n-1}) = \int_{J_n} (f, X)\, dt, \quad \forall X \in \mathcal{S}_{kh}, \quad n \geq 1,$$

or

$$(U_h^n, \chi) + k_n(\nabla U_h^n, \nabla \chi) = (U_h^{n-1}, \chi) + \Big(\int_{J_n} f(t)\, dt, \chi \Big), \quad \forall \chi \in S_h, \quad n \geq 1,$$

and with $U^0 = P_h v$.

We recall that Theorem 12.6 shows, assuming k_{n+1}/k_n is bounded away from zero,

$$\|U_h - u\|_{J_N} \leq C L_N \max_{n \leq N} (k_n \|u_t\|_{J_n} + h^2 \|u\|_{2,J_n}).$$

For an *a posteriori* error estimate we thus have to replace the right hand side of this estimate by quantities which are known at the time of the computation. For this purpose it is natural to try to replace u_t on J_n by $\bar{\partial}_n U_h^n$. We also need to use an approximation for the second order spatial derivative norm. We therefore introduce the interior edges $\{\gamma\}$ of the triangulation $\mathcal{T}_h$, and denote, for $\chi \in S_h$, by $[\partial\chi/\partial n]_\gamma$ the jump in the normal derivative across γ and set

$$\|\chi\|_{2,h} = \Big(\sum_\gamma \big| \big[\frac{\partial \chi}{\partial n} \big]_\gamma \big|^2 \Big)^{1/2}.$$

Note that because $\nabla\chi$ is constant in each $\tau \in \mathcal{T}_h$ so is the normal derivative along γ within τ. We may therefore also think of the jump in $\partial\chi/\partial n$ as $\partial\chi/\partial n(P_1) - \partial\chi/\partial n(P_2)$ where P_1 and P_2 are the points of gravity of the two triangles involved, or as a multiple of order $O(h)$ times the difference quotient

$$\Big(\frac{\partial \chi}{\partial n}(P_1) - \frac{\partial \chi}{\partial n}(P_2)\Big)/|P_1 - P_2|,$$

which latter has the character of an approximation of a second order derivative.

The *a posteriori* error estimate we shall show may now be stated as follows:

Theorem 12.9 *We have for the solutions of* (12.51) *and* (12.43)

$$\|U_h^N - u(t_N)\| \le CL_N \max_{n\le N} \big((h^2 + k_n)\|f\|_{J_n} + h^2\|U_h^n\|_{2,h} + k_n\|\bar\partial_n U_h^n\|\big).$$

For the proof we shall need the following auxiliary estimates.

Lemma 12.6 *If* $W \in S_h, v \in H_0^1 \cap H^2$, *then*

$$|(\nabla W, \nabla(P_h - I)v)| \le Ch^2\|W\|_{2,h}\ \|v\|_2.$$

Proof. We first show that if $W \in S_h, v \in H_0^1$, then

$$|(\nabla W, \nabla v)| \le C\|W\|_{2,h}\ (\|v\| + h\|\nabla v\|). \tag{12.52}$$

In fact, for each triangle τ in $\mathcal{T}_h$ with edges $\gamma_{\tau,j}, j = 1, 2, 3$, we have by Green's formula

$$\int_\tau \nabla W \cdot \nabla v\, dx = \sum_{j=1}^3 \frac{\partial W}{\partial n}\Big|_{\gamma_{\tau,j}} \int_{\gamma_{\tau,j}} v\, ds.$$

Summing over the triangles τ we find that each edge γ occurs twice and thus the coefficient for $\int_\gamma v\, ds$ is $[\partial W/\partial n]_\gamma$. Using the Cauchy-Schwarz inequality we thus have

$$|(\nabla W, \nabla v)| \le C\|W\|_{2,h}\ \Big(\sum_\gamma \big(\int_\gamma v\, ds\big)^2\Big)^{1/2}. \tag{12.53}$$

Here, using the trace inequality (2.10) scaled down to each triangle, when γ is one of the sides of τ (cf. (2.11)),

$$\Big(\int_\gamma v\, ds\Big)^2 \le Ch\int_\gamma v^2\, ds \le Ch\Big(h\int_\tau |\nabla v|^2\, dx + h^{-1}\int_\tau v^2\, dx\Big),$$

and hence

$$\sum_\gamma \Big(\int_\gamma v\, ds\Big)^2 \le C(\|v\|^2 + h^2\|\nabla v\|^2).$$

Together with (12.53) this completes the proof of (12.52). The proof of the lemma is now concluded by noting that, when the triangulation is quasiuniform,

$$\|(P_h - I)v\| + h\|\nabla(P_h - I)v\| \le Ch^2\ \|v\|_2. \qquad \square$$

Proof of Theorem 12.9. The representation of the error $e = U_h - u$ of Lemma 12.4 remains valid and we thus have

$$
\begin{aligned}
(e^N, \varphi) = &- \int_0^{t_N} (\nabla U_h, \nabla(X - z))\, dt - \sum_{n=0}^{N-1} ([U_h]_n, X_+^n - z^n) \\
&+ \int_0^{t_N} (f, X - z)\, dt = I + II + III, \quad \text{for } X \in \mathcal{S}_{kh},
\end{aligned} \tag{12.54}
$$

where $U_h^0 = P_h v$. We now choose $X \in \mathcal{S}_{kh}$ as the orthogonal projection onto $L_2(\Omega \times J_n)$ of z, for $n \geq 1$, i.e., $X = P_h \bar{z}$, where P_h is the L_2-projection onto S_h and $\bar{z}|_{J_n} = k_n^{-1} \int_{J_n} z\, dt$. We then write $X - z = (P_h \bar{z} - P_h z) + (P_h z - z)$. Now

$$
\int_{J_n} (\nabla U_h, \nabla(P_h \bar{z} - P_h z))\, dt = - \int_{J_n} (\Delta_h U_h, \bar{z} - z)\, dt = 0,
$$

whereas by Lemma 12.6

$$
\begin{aligned}
\Big| \int_{J_n} (\nabla U_h, \nabla(P_h z - z))\, dt \Big| &= \Big| \Big(\nabla U_h^n, \nabla\big((P_h - I) \int_{J_n} z\, dt\big)\Big) \Big| \\
&\leq C h^2 \|U_h^n\|_{2,h} \Big\| \int_{J_n} z\, dt \Big\|_2 \leq C h^2 \|U_h^n\|_{2,h} \Big\| \Delta \int_{J_n} z\, dt \Big\|.
\end{aligned}
$$

Since $\int_{J_N} \Delta z\, dt = \int_{J_N} z_t\, dt = z(t_N) - z(t_{N-1})$, Lemma 12.5 shows

$$
\begin{aligned}
|I| &\leq C h^2 \max_{n \leq N} \|U_h^n\|_{2,h} \sum_{n=1}^{N} \Big\| \int_{J_n} \Delta z\, dt \Big\| \\
&\leq C h^2 \max_{n \leq N} \|U_h^n\|_{2,h} \Big(\int_0^{t_{N-1}} \|z_t\|\, dt + \|z\|_{J_N} \Big) \\
&\leq C L_N h^2 \max_{n \leq N} \|U_h^n\|_{2,h}\, \|\varphi\|.
\end{aligned}
$$

We have since $[U_h]_{n-1} = U_h^n - U_h^{n-1} = k_n \bar{\partial}_n U_h^n$

$$
\begin{aligned}
II &= - \sum_{n=1}^{N} ([U_h]_{n-1}, X^n - z^{n-1}) = \sum_{n=1}^{N} ([U_h]_{n-1}, X^n - P_h z^{n-1}) \\
&\leq \sum_{n=1}^{N} k_n \|\bar{\partial}_n U_h^n\|\, \|\bar{z}^n - z^{n-1}\|,
\end{aligned}
$$

so, again by Lemma 12.5,

$$
\begin{aligned}
|II| &\leq C \max_{n \leq N} \big(k_n \|\bar{\partial}_n U_h^n\|\big) \Big(\int_0^{t_{N-1}} \|z_t\|\, dt + \|z\|_{J_N} \Big) \\
&\leq C L_N \max_{n \leq N} \big(k_n \|\bar{\partial}_n U_h^n\|\big) \|\varphi\|.
\end{aligned}
$$

For *III* finally we have

$$\Big|\int_{J_n} (f, X - z)\, dt\Big| \le \|f\|_{J_n} \int_{J_n} \|P_h\bar{z} - z\|\, dt.$$

Here, by adding and subtracting $P_h z$, we have on J_n

$$\|P_h\bar{z} - z\| \le \|P_h z - z\| + \|\bar{z} - z\| \le Ch^2\|z\|_2 + \int_{J_n} \|z_t\|\, dt,$$

and hence,

$$\int_{J_n} \|z - P_h\bar{z}\|\, dt \le C(h^2 + k_n)\int_{J_n} \|z_t\|\, dt, \quad \text{for } n < N.$$

Further, since $\|z - P_h\bar{z}\| \le 2\|z\|_{J_N}$ on J_N, we find

$$\int_{J_N} \|z - P_h\bar{z}\|\, dt \le 2k_N\|z\|_{J_N},$$

so that altogether, using as before Lemma 12.5,

$$\begin{aligned} |III| &\le C \max_{n\le N} \big((h^2 + k_n)\|f\|_{J_n}\big)\Big(\int_0^{t_{N-1}} \|z_t\|\, dt + \|z\|_{J_N}\Big) \\ &\le CL_N \max_{n\le N} \big((h^2 + k_n)\|f\|_{J_n}\big)\|\varphi\|. \end{aligned} \qquad (12.55)$$

Our three estimates for *I, II* and *III*, together with (12.54) complete the proof. □

In order to see that the error bound in Theorem 12.9 is not excessively large we shall demonstrate in the next theorem that the quantities in the error bound which depend on the computed solution may, in fact, be bounded by the *a priori* error bound of Theorem 12.6.

Theorem 12.10 *Assume that $k_{n+1}/k_n \ge c > 0$ for $n \ge 0$. Then we have for the solutions of* (12.51) *and* (12.43)

$$h^2\|U_h^N\|_{2,h} + k_N\|\bar{\partial}_N U_h^N\| \le CL_N \max_{n\le N}(h^2\|u\|_{2,J_n} + k_n\|u_t\|_{J_n}). \qquad (12.56)$$

Proof. With $\widetilde{u}_h = I_h u$ the standard interpolant of u we have

$$h^2\|U_h^N\|_{2,h} \le h^2\|U_h^N - \widetilde{u}_h^N\|_{2,h} + h^2\|\widetilde{u}_h^N\|_{2,h}.$$

Using the Bramble-Hilbert lemma one easily shows $\|\widetilde{u}_h^N\|_{2,h} \le C\|u^N\|_2$. We now note that quasiuniformity implies the inverse estimate $\|\chi\|_{2,h} \le$

$Ch^{-2}\|\chi\|$ for $\chi \in S_h$. In fact, since $\text{area}(\tau) \geq ch^2$, with $c > 0$, we have

$$\|\chi\|_{2,h} = \Big(\sum_\gamma |[\frac{\partial \chi}{\partial n}]|_\gamma^2\Big)^{1/2} \leq C\Big(\sum_\tau \|\nabla\chi\|^2_{L_\infty(\tau)}\Big)^{1/2}$$
$$\leq C\Big(\sum_\tau (\text{area}(\tau))^{-1}\|\nabla\chi\|^2_{L_2(\tau)}\Big)^{1/2} = Ch^{-1}\|\nabla\chi\| \leq Ch^{-2}\|\chi\|.$$

Using Theorem 12.6 and the standard estimate for the interpolation error, we therefore have

$$h^2\|U_h^N - \widetilde{u}_h^N\|_{2,h} \leq C\|U_h^N - \widetilde{u}^N\| \leq \|U_h^N - u^N\| + \|u^N - \widetilde{u}^N\|$$
$$\leq CL_N \max_{n\leq N}\big(h^2\|u\|_{2,J_n} + k_n\|u_t\|_{J_n}\big),$$

so that we have shown the estimate claimed for the first term in (12.56). For the second term, we have

$$k_N\|\bar\partial_N U_h^N\| \leq \|U_h^N - u^N\| + \|U_h^{N-1} - u^{N-1}\| + k_N\|u_t\|_{J_N},$$

which is bounded as desired, by Theorem 12.6. □

The discontinuous Galerkin method was introduced and analyzed for ordinary differential equations in Delfour, Hager and Trochu [67], and applied to partial differential equations in, e.g., Lesaint and Raviart [155] and Jamet [128]. In the context of parabolic equations it was first studied in Eriksson, Johnson and Thomée [93]. A posteriori error analysis and adaptive time step control was initiated in Johnson, Nie and Thomée [131]. The approach taken here was essentially proposed by Lippold [157], and further developed in a sequence of papers by Eriksson and Johnson, in the linear case in [88], [89], [91] and Eriksson, Johnson and Larsson [92]. The variant (12.6) of the backward Euler method, which appears here as a special case, was analyzed in Luskin and Rannachaer [167]. The continuous Galerkin method was investigated by Aziz and Monk [8].

13. A Nonlinear Problem

In this chapter we shall consider the application of our previous methods of analysis to a nonlinear model problem. For simplicity and concreteness, we restrict our attention to the situation in the beginning of Chapter 1, with a convex plane domain and with piecewise linear approximating functions. We also consider the problem on a finite interval $J = (0, \bar{t}]$ in time; some of the constants in our estimates will depend on $\bar{t}$, without explicit mention.

Let thus Ω be a plane convex domain with smooth boundary and consider the parabolic problem

$$(13.1) \qquad u_t - \nabla \cdot (a(u)\nabla u) = f(u) \quad \text{in } \Omega,\ t \in J,$$
$$u = 0 \quad \text{on } \partial\Omega, \quad t \in J, \quad u(\cdot, 0) = v \text{ in } \Omega,$$

where a and f are smooth functions defined on $\mathbb{R}$ such that

$$(13.2) \qquad 0 < \mu \le a(u) \le M, \quad |a'(u)| + |f'(u)| \le B, \quad \text{for } u \in \mathbb{R}.$$

We assume that the above problem admits a unique solution which is sufficiently smooth for our purposes.

Let now, as in Chapter 1, $\mathcal{T}_h$ be a member of a family of quasiuniform triangulations of Ω with $\max_{\tau \in \mathcal{T}_h} \operatorname{diam} \tau \le h$ and let S_h be the corresponding finite dimensional space of continuous functions on Ω which reduce to linear functions in each of the triangles of $\mathcal{T}_h$, and which vanish on $\partial\Omega$. We may then pose the semidiscrete problem to find $u_h : \bar{J} \to S_h$ such that

$$(13.3) \qquad \begin{aligned} (u_{h,t}, \chi) + (a(u_h)\nabla u_h, \nabla\chi) &= (f(u_h), \chi), \quad \forall \chi \in S_h,\ t \in J, \\ u_h(0) &= v_h, \end{aligned}$$

where v_h is an approximation of v in S_h. Representing the solution as $u_h(x,t) = \sum_{j=1}^{N_h} \alpha_j(t)\Phi_j(x)$, where $\{\Phi_j\}_{j=1}^{N_h}$ is the standard basis of pyramid functions, this may be written

$$(13.4) \qquad \begin{aligned} \sum_{j=1}^{N_h} \alpha_j'(t)(\Phi_j, \Phi_k) + \sum_{j=1}^{N_h} \alpha_j(t)\Big(a\big(\sum_{l=1}^{N_h} \alpha_l(t)\Phi_l\big)\nabla\Phi_j, \nabla\Phi_k\Big) \\ = \Big(f\big(\sum_{l=1}^{N_h} \alpha_l(t)\Phi_l\big), \Phi_k\Big), \quad k = 1, \dots, N_h. \end{aligned}$$

Setting $\alpha = \alpha(t) = (\alpha_1(t), \dots, \alpha_{N_h}(t))^T$ and introducing the matrices $\mathcal{B} = (b_{jk})$ and $\mathcal{A}(\alpha) = (a_{jk}(\alpha))$ with elements

$$b_{jk} = (\Phi_j, \Phi_k) \quad \text{and } a_{jk}(\alpha) = \Big(a\big(\sum_{l=1}^{N_h} \alpha_l \Phi_l\big) \nabla \Phi_j, \nabla \Phi_k\Big),$$

respectively, and the vector $\widetilde{f}(\alpha) = (\widetilde{f}_1(\alpha), \dots, \widetilde{f}_{N_h}(\alpha))^T$, with $\widetilde{f}_j(\alpha) = (f(\sum_{l=1}^{N_h} \alpha_l \Phi_l), \Phi_j)$, the system (13.4) may also be written in matrix form as

$$(13.5) \qquad \mathcal{B}\alpha' + \mathcal{A}(\alpha)\alpha = \widetilde{f}(\alpha), \quad \text{for } t \in J, \quad \text{with } \alpha(0) = \gamma,$$

where γ is the vector of nodal values of v_h.

By our assumptions (13.2), the matrices $\mathcal{B}$ and $\mathcal{A}(\alpha)$ are positive definite, and $\mathcal{A}(\alpha)$ and $\widetilde{f}(\alpha)$ are globally Lipschitz continuous on $\mathbb{R}^{N_h}$. It follows easily that the system has a unique solution for $t \in J$, which is bounded there; it may be obtained, e.g., by determining the $\alpha_n = \alpha_n(t), n = 0, 1, \dots,$ from the iterative scheme

$$\mathcal{B}\alpha'_{n+1} + \mathcal{A}(\alpha_n)\alpha_{n+1} = \widetilde{f}(\alpha_n), \quad \text{for } t \in J, \quad \alpha_{n+1}(0) = \gamma, \quad \text{for } n \ge 0,$$
$$\alpha_0(t) \equiv \gamma \quad \text{on } \bar{J}.$$

Our first purpose is to estimate the error in the semidiscrete problem (13.3). As earlier we shall write the error as a sum of two terms,

$$(13.6) \qquad u_h - u = (u_h - w_h) + (w_h - u) = \theta + \rho,$$

where w_h is an elliptic projection in S_h of the exact solution u. This time we shall use the projection $w_h = w_h(t)$ defined by

$$(13.7) \qquad \big(a(u(t))\nabla(w_h(t) - u(t)), \nabla\chi\big) = 0, \quad \forall \chi \in S_h, \ t \ge 0,$$

and we shall therefore need some estimates for the error in this projection. Note that the inner product defining $\widetilde{u}_h$ depends on the exact solution u. We begin with the following auxiliary result.

Lemma 13.1 *Let $b = b(x)$ be a smooth function in Ω with $0 < \mu \le b(x) \le M$ for $x \in \Omega$. Assume that $u \in H^2 \cap H_0^1$ and let w_h be defined by*

$$(b\nabla(w_h - u), \nabla\chi) = 0, \quad \forall \chi \in S_h.$$

Then

$$(13.8) \qquad \|\nabla(w_h - u)\| \le C_1 h \|u\|_2$$

and

$$(13.9) \qquad \|w_h - u\| \le C_0 h^2 \|u\|_2.$$

Here C_1 depends on the family of triangulations $\mathcal{T}_h$, and on μ and M, and C_0 in addition on an upper bound for ∇b.

Proof. We have, for $\chi \in S_h$,

$$\mu\|\nabla(w_h - u)\|^2 \le (b\nabla(w_h - u), \nabla(w_h - u))$$
$$= (b\nabla(w_h - u), \nabla(\chi - u)) \le M\|\nabla(w_h - u)\|\,\|\nabla(\chi - u)\|,$$

and hence, with $I_h u$ the standard interpolant of w,

$$\|\nabla(w_h - u)\| \le (M/\mu)\|\nabla(I_h w - u)\| \le C_1 h\|u\|_2,$$

which is (13.8). To show (13.9) by duality, we solve the problem

$$(13.10) \qquad -\nabla\cdot(b\nabla\psi) \equiv -b\Delta\psi - \nabla b\cdot\nabla\psi = \varphi \quad \text{in } \Omega, \quad \psi = 0 \quad \text{on } \partial\Omega,$$

and note that, since $\|\psi\| \le C\|\nabla\psi\|$ for $\psi \in H_0^1$,

$$\mu\|\nabla\psi\|^2 \le (b\nabla\psi, \nabla\psi) = (\varphi, \psi) \le \|\varphi\|\,\|\psi\| \le C\|\varphi\|\,\|\nabla\psi\|,$$

so that $\|\nabla\psi\| \le C\|\varphi\|$. Hence, for ∇b bounded,

$$\|\psi\|_2 \le C\|\Delta\psi\| \le C\|b\Delta\psi\| = C\|\varphi + \nabla b\cdot\nabla\psi\| \le C\|\varphi\|.$$

Therefore, with $\chi = I_h\psi$,

$$(w_h - u, \varphi) = (b\nabla(w_h - u), \nabla\psi) = (b\nabla(w_h - u), \nabla(\psi - \chi))$$
$$\le M\|\nabla(w_h - u)\|\,\|\nabla(\psi - \chi)\| \le (Ch\,\|u\|_2)(Ch\,\|\psi\|_2) \le C_0 h^2\|u\|_2\,\|\varphi\|,$$

which completes the proof. □

We can now show the following result for the error in the elliptic projection $\widetilde{u}_h$, under the appropriate regularity assumptions for u. Here, and in the rest of this chapter, we refrain for brevity from specifying the dependence of the constants in the error estimates on the regularity of the exact solution.

Lemma 13.2 *With w_h defined by (13.7) and $\rho = w_h - u$ we have under the appropriate regularity assumptions on u, with $C(u)$ independent of $t \in J$,*

$$\|\rho(t)\| + h\|\nabla\rho(t)\| \le C(u)h^2, \quad \textit{for } t \in J,$$
$$\|\rho_t(t)\| + h\|\nabla\rho_t(t)\| \le C(u)h^2, \quad \textit{for } t \in J.$$

Proof. Since $\nabla a(u) = a'(u)\nabla u$ the first estimate follows at once by application of Lemma 13.1 with $b(x) = a(u(x,t))$.

By differentiation of (13.7) we have

$$(a(u)\nabla\rho_t, \nabla\chi) + (a(u)_t\nabla\rho, \nabla\chi) = 0, \quad \forall\chi \in S_h.$$

Hence, assuming $a(u)$ and $a(u)_t$ uniformly bounded,

$$\begin{aligned}\mu\|\nabla\rho_t\|^2 &\le (a(u)\nabla\rho_t, \nabla\rho_t)\\ &= (a(u)\nabla\rho_t, \nabla(\chi - u_t)) + (a(u)\nabla\rho_t, \nabla(w_{h,t} - \chi))\\ &= (a(u)\nabla\rho_t, \nabla(\chi - u_t)) + (a(u)_t\nabla\rho, \nabla(\chi - w_{h,t}))\\ &\le C\big(\|\nabla\rho_t\|\,\|\nabla(\chi - u_t)\| + \|\nabla\rho\|\,\|\nabla(\chi - w_{h,t})\|\big),\end{aligned}$$

and with $\chi = I_h u_t$,

$$\begin{aligned}\mu\|\nabla\rho_t\|^2 &\le Ch\|u_t\|_2\|\nabla\rho_t\| + \|\nabla\rho\|(Ch\|u_t\|_2 + \|\nabla\rho_t\|)\\ &\le \frac{\mu}{2}\|\nabla\rho_t\|^2 + C(\|\nabla\rho\|^2 + h^2\|u_t\|_2^2).\end{aligned}$$

In view of the estimate for $\nabla\rho$ already shown this yields $\|\nabla\rho_t\| \le C(u)h$.

For the L_2 estimate we use again the duality argument of the proof of Lemma 13.1. We have with ψ as in (13.10) (with $b = a(u)$),

$$\begin{aligned}(\rho_t, \varphi) =&(a(u)\nabla\rho_t, \nabla\psi) = (a(u)\nabla\rho_t, \nabla(\psi - \chi))\\ &+ (a(u)_t\nabla\rho, \nabla(\psi - \chi)) - (\nabla\rho, a(u)_t\nabla\psi),\end{aligned}$$

and hence, choosing $\chi = I_h\psi$ and using integration by parts in the last term,

$$|(\rho_t, \varphi)| \le C\big(\|\nabla\rho_t\|\,h\,\|\psi\|_2 + \|\nabla\rho\|\,h\,\|\psi\|_2 + \|\rho\|\,\|\psi\|_2\big),$$

whence, by the estimates already shown for $\rho, \nabla\rho$ and $\nabla\rho_t$,

$$|(\rho_t, \varphi)| \le C(u)h^2\|\psi\|_2 \le C(u)h^2\|\varphi\|,$$

so that $\|\rho_t\| \le C(u)h^2$. This completes the proof of the lemma. □

We shall also need the boundedness of $\nabla\widetilde{u}_h$:

Lemma 13.3 *We have, with w_h defined in* (13.7),

$$\|\nabla w_h(t)\|_{L_\infty} \le C(u), \quad \textit{for } t \in J.$$

Proof. Using the inverse estimate (which is trivial in this case since $\nabla\chi$ is constant on each triangle)

$$\|\nabla\chi\|_{L_\infty} \le Ch^{-1}\|\nabla\chi\|, \quad \text{for } \chi \in S_h, \tag{13.11}$$

together with Lemma 13.2 and the known error estimate for $I_h u$, we have

$$\begin{aligned}\|\nabla(w_h - I_h u)\|_{L_\infty} &\le Ch^{-1}\|\nabla(w_h - I_h u)\|\\ &\le Ch^{-1}(\|\nabla\rho\| + \|\nabla(I_h u - u)\|) \le C(u).\end{aligned}$$

Since it is easy to see that $\|\nabla I_h u\|_{L_\infty} \le C\|\nabla u\|_{L_\infty}$, the result follows. □

We are now ready for the L_2 error estimate for the semidiscrete problem.

Theorem 13.1 *Let u_h and u be the solutions of* (13.3) *and* (13.1), *respectively. Then, under the appropriate regularity assumptions for u, we have*

$$\|u_h(t) - u(t)\| \le C\|v_h - v\| + C(u)h^2, \quad \textit{for } t \in \bar{J}.$$

Proof. With the error written as in (13.6) it suffices, in view of Lemma 13.2, to bound $\theta = u_h - w_h$. We have, using (13.7), for $\chi \in S_h$,

$$\begin{aligned}
(\theta_t, \chi) &+ (a(u_h)\nabla\theta, \nabla\chi) \\
&= (u_{h,t}, \chi) + (a(u_h)\nabla u_h, \nabla\chi) - (w_{h,t}, \chi) - (a(u_h)\nabla w_h, \nabla\chi) \\
&= (f(u_h), \chi) - (\rho_t, \chi) - (u_t, \chi) - (a(u)\nabla w_h, \nabla\chi) \\
&\qquad + ((a(u) - a(u_h))\nabla w_h, \nabla\chi) \\
&= (f(u_h), \chi) - (\rho_t, \chi) - (u_t, \chi) - (a(u)\nabla u, \nabla\chi) \\
&\qquad + ((a(u) - a(u_h))\nabla w_h, \nabla\chi) \\
&= (f(u_h) - f(u), \chi) + ((a(u) - a(u_h))\nabla w_h, \nabla\chi) - (\rho_t, \chi).
\end{aligned}$$

Hence with $\chi = \theta$, using (13.2), Lemma 13.3 and (1.4)

$$\begin{aligned}
\tfrac{1}{2}\frac{d}{dt}\|\theta\|^2 + \mu\|\nabla\theta\|^2 &\le C\big(\|u_h - u\|(\|\theta\| + \|\nabla\theta\|) + \|\rho_t\|\,\|\theta\|\big) \\
&\le \mu\|\nabla\theta\|^2 + C(\|\theta\|^2 + \|\rho\|^2 + \|\rho_t\|^2).
\end{aligned}$$

After integration, this shows

$$\|\theta(t)\|^2 \le \|\theta(0)\|^2 + C\int_0^t (\|\theta\|^2 + \|\rho\|^2 + \|\rho_t\|^2)\,ds,$$

and, using Gronwall's lemma (with C now depending on $\bar{t}$),

$$\|\theta(t)\|^2 \le C\|\theta(0)\|^2 + C\int_0^t (\|\rho\|^2 + \|\rho_t\|^2)\,ds.$$

Using Lemma 13.2 together with

$$\|\theta(0)\| \le \|v_h - v\| + \|w_h(0) - v\| \le \|v_h - v\| + Ch^2\|v\|_2, \tag{13.12}$$

this shows $\|\theta(t)\| \le C\|v_h - v\| + C(u)h^2$, and thus completes the proof. □

The corresponding estimate for the gradient follows easily by an inverse estimate as in Theorem 2.4. We refrain from giving the details.

We shall now pause to make a comment concerning the global nature of our assumption (13.2) for the functions a and f. It should be clear from our analysis that as long as u_h is close to u, the assumptions referred to only come into play in a neighborhood of the range of u. It is therefore natural to make the less stringent assumption that a and f are defined in such a neighborhood and satisfy (13.2) there. It has to be kept in mind, however, that closeness now has to be interpreted as being valid at each point, or that u_h be close to u in the uniform norm.

Let thus I_0 be the range of $u, I_0 = [m_0, m_1] = \{u(x,t); x \in \bar{\Omega}, t \in \bar{J}\}$, and consider for a fixed $\delta > 0$ the interval $I_\delta = [m_0 - \delta, m_1 + \delta]$. Assume now that f and a belong to $\mathcal{C}^1(I_\delta)$, so that a' and f' are bounded on I_δ, and that a is positive and bounded away from 0 and ∞ on I_δ. Then, if v_h is sufficiently close to v, or $v_h(x) \in I_{\delta/2}$, say, for $x \in \Omega$, we have that the problem (13.3), or (13.5), is well defined and has a solution in I_δ, at least for t in an interval $[0, t_h]$ with $0 < t_h \le \bar{t}$. Assume, for instance, that v_h is chosen so that $\|v_h - v\| \le C_0(v)h^2$. Then, using the easily proven inverse estimate

$$\|\chi\|_{L_\infty} \le Ch^{-1}\|\chi\|, \quad \text{for } \chi \in S_h,$$

taking η to be the interpolant of v, say,

$$\begin{aligned} \|v_h - v\|_{L_\infty} &\le Ch^{-1}\|v_h - \eta\| + \|\eta - v\|_{L_\infty} \\ &\le Ch^{-1}\|v_h - v\| + Ch^{-1}\|\eta - v\| + \|\eta - v\|_{L_\infty} \le C_1(v)h, \end{aligned} \tag{13.13}$$

so that $v_h \in I_{\delta/2}$ for small h. As long as $u_h(t) \in I_\delta$, however, the above error analysis remains valid, and we conclude from the proof of Theorem 13.1 that

$$\|u_h(t) - u(t)\| \le C\|v_h - v\| + C(u)h^2 \le C_2(u)h^2, \quad \text{for } t \le t_h,$$

and thus, again for $t \le t_h$, as in (13.13),

$$\|u_h(t) - u(t)\|_{L_\infty} \le C(u)h < \delta/2, \quad \text{if } h \le h_0,$$

where the latter inequality defines h_0 independently of t_h. Thus $u_h(t_h) \in I_{\delta/2}$, and hence the solution continues to exist beyond t_h if $t_h < \bar{t}$. We may therefore conclude that t_h may be chosen as $\bar{t}$.

Thus the local assumptions for a and f suffice in the proof of Theorem 13.1, for h small. These hold, in particular, if a and f are in $\mathcal{C}^1(\mathbb{R})$, without any requirement of boundedness of a' and f'. On the other hand, since $a(u)$ and $f(u)$ only enter in the semidiscrete problem for values of the argument in the interval I_δ, these functions may be modified outside I_δ, so that the more stringent condition (13.2) may be assumed without restriction of generality.

We shall now turn to fully discrete schemes. As usual, let k be the time step, $t_n = nk$, and let now U^n be the approximation of $u(t_n)$ in S_h; as in Chapter 1 we shall omit the subscript h in the notation for the fully discrete solution. We begin with the backward Euler Galerkin scheme which in this case reads

(13.14) $$(\bar\partial U^n,\chi)+(a(U^n)\nabla U^n,\nabla\chi)=(f(U^n),\chi),\quad \forall\chi\in S_h,\ t_n\in J,$$

with $U^0=v_h$, where as earlier $\bar\partial U^n=(U^n-U^{n-1})/k$.

Introducing the vector α^n by $U^n=\sum_{j=1}^{N_h}\alpha_j^n\Phi_j$, the equation (13.14) may be written in matrix form as

$$\mathcal{B}\frac{\alpha^n-\alpha^{n-1}}{k}+\mathcal{A}(\alpha^n)\alpha^n=\widetilde{f}(\alpha^n),$$

or

$$(\mathcal{B}+k\mathcal{A}(\alpha^n))\alpha^n=\mathcal{B}\alpha^{n-1}+k\widetilde{f}(\alpha^n),\quad \text{for } t_n\in J,$$

with $\alpha^0=\gamma$ given by v_h, where $\mathcal{B},\mathcal{A}(\alpha)$, and $\widetilde{f}(\alpha)$ are as above.

In order to show that there exists a solution of this equation we multiply (13.14) by $2k$ and write it as $(G_h(U^n),\chi)=0$, where $G_h:S_h\to S_h$ is continuous. It is a well-known simple consequence of Brouwer's fixed point theorem that the equation $G_h(X)=0$ has a solution $X\in B_q=\{\chi\in S_h;\ \|\chi\|\le q\}$ if $(G_h(\chi),\chi)>0$ for $\|\chi\|=q$. In fact, if we assume that $G_h(\chi)\neq 0$ in B_q, then the mapping $\Phi_h(\chi)=-qG_h(\chi)/\|G_h(\chi)\|:B_q\to B_q$ is continuous, and therefore has a fixed point $\bar\chi\in B_q$, with $q^2=\|\bar\chi\|^2=-q(G_h(\bar\chi),\bar\chi)/\|G_h(\bar\chi)\|$, which contradicts $(G_h(\bar\chi),\bar\chi)>0$.

To show the condition needed for $(G_h(\chi),\chi)$, we use (13.2) to obtain

$$\begin{aligned}(G_h(\chi),\chi)&=2(\chi-U^{n-1},\chi)+2k(a(\chi)\nabla\chi,\nabla\chi)-2k(f(\chi),\chi)\\&\ge\|\chi\|^2-\|U^{n-1}\|^2-Ck(1+\|\chi\|)\|\chi\|,\end{aligned}$$

which is positive if $\|\chi\|$ is large enough, provided $k\le k_0<1/C$.

Uniqueness is less obvious, but in the following theorem we show an error estimate which is valid for any solution of (13.14). After this theorem we shall comment again on uniqueness.

Theorem 13.2 *Let U^n and u be solutions of* (13.14) *and* (13.1), *respectively. Then, under the appropriate regularity assumptions for u, we have*

$$\|U^n-u(t_n)\|\le C\|v_h-v\|+C(u)(h^2+k),\quad \textit{for } t_n\in\bar J.$$

Proof. We write as before, with $u^n=u(t_n)$,

(13.15) $$U^n-u^n=(U^n-w_h^n)+(w_h^n-u^n)=\theta^n+\rho^n,$$

where w_h^n is the elliptic projection of u^n, defined in (13.7). In view of Lemma 13.2, it remains to bound θ^n. We have, for $\chi\in S_h$,

$$\begin{aligned}&(\bar\partial\theta^n,\chi)+(a(U^n)\nabla\theta^n,\nabla\chi)\\&\quad=(\bar\partial U^n,\chi)+(a(U^n)\nabla U^n,\nabla\chi)-(\bar\partial w_h^n,\chi)-(a(U^n)\nabla w_h^n,\nabla\chi)\\&\quad=(f(U^n),\chi)-(u_t^n,\chi)-(\bar\partial w_h^n-u_t^n,\chi)\\&\qquad-(a(u^n)\nabla w_h^n,\nabla\chi)-((a(U^n)-a(u^n))\nabla w_h^n,\nabla\chi).\end{aligned}$$

Using (13.7) in the second to last term and the weak form of the continuous problem, we find

$$(\bar\partial\theta^n,\chi)+(a(U^n)\nabla\theta^n,\nabla\chi)=(f(U^n)-f(u^n),\chi)-(\bar\partial\rho^n,\chi)$$
$$-(\bar\partial u^n-u_t^n,\chi)-((a(U^n)-a(u^n))\nabla w_h^n,\nabla\chi).$$

Taking $\chi=\theta^n$ this yields, by (13.2) and the boundedness of ∇w_h^n shown in Lemma 13.3,

$$\tfrac12\bar\partial\|\theta^n\|^2+\mu\|\nabla\theta^n\|^2$$
$$\le C\|U^n-u^n\|(\|\theta^n\|+\|\nabla\theta^n\|)+(\|\bar\partial\rho^n\|+\|\bar\partial u^n-u_t^n\|)\|\theta^n\|,$$

and hence, after kicking back $\|\nabla\theta^n\|$,

$$\bar\partial\|\theta^n\|^2+\mu\|\nabla\theta^n\|\le C(\|\theta^n\|^2+R_n), \tag{13.16}$$

where

$$R_n=\|\rho^n\|^2+\|\bar\partial\rho^n\|^2+\|\bar\partial u^n-u_t^n\|^2).$$

This shows

$$(1-Ck)\|\theta^n\|^2\le\|\theta^{n-1}\|^2+CkR_n,$$

or, for small k,

$$\|\theta^n\|^2\le(1+Ck)\|\theta^{n-1}\|^2+CkR_n,$$

whence, by repeated application,

$$\begin{aligned}\|\theta^n\|^2&\le(1+Ck)^n\|\theta^0\|^2+Ck\sum_{j=1}^n(1+Ck)^{n-j}R_j\\&\le C\|\theta^0\|^2+Ck\sum_{j=1}^nR_j,\quad\text{for } t_n\in J.\end{aligned}\tag{13.17}$$

By Lemma 13.2 we have $\|\rho^j\|\le C(u)h^2$,

$$\|\bar\partial\rho^j\|=\|k^{-1}\int_{t_{j-1}}^{t_j}\rho_t\,ds\|\le C(u)h^2,$$

and, cf. the estimate of ω_1^j in the proof of Theorem 1.5,

$$\|\bar\partial u^j-u_t^j\|=\|k^{-1}\int_{t_{j-1}}^{t_j}(s-t_{j-1})u_{tt}(s)\,ds\|\le C(u)k.$$

This shows $R_j\le C(u)(h^2+k)^2$, and using also (13.12), (13.17) yields

$$\|\theta^n\|\le C\|v_h-v\|+C(u)(h^2+k), \tag{13.18}$$

which completes the proof. □

We return briefly to the question of uniqueness of the solution of (13.14), and show that this holds when the solution of the continuous problem is smooth and when $\|v_h - v\| \le Ch^2$, provided that h and the mesh-ratio k/h, and thus also k, are sufficiently small. In fact, let X and Y be two solutions of (13.14) with U^{n-1} given. Then by subtraction

$$(X - Y, \chi) + k(a(X)\nabla X - a(Y)\nabla Y, \nabla\chi) = k(f(X) - f(Y), \chi), \quad \forall \chi \in S_h.$$

Choosing $\chi = X - Y$ we find

$$\begin{aligned}&\|X - Y\|^2 + k(a(Y)\nabla(X - Y), \nabla(X - Y))\\ &\quad = k(f(X) - f(Y), X - Y) - k((a(X) - a(Y))\nabla X, \nabla(X - Y)),\end{aligned}$$

and hence, after obvious estimates and a kickback of $\|\nabla(X - Y)\|$,

$$\|X - Y\|^2 + \tfrac{1}{2}k\mu\|\nabla(X - Y)\|^2 \le C\|X - Y\|^2(k + k\|\nabla X\|_{L_\infty}^2).$$

Thus, if $k\|\nabla X\|_{L_\infty}^2$ may be bounded by an arbitrarily small constant and k is small, we conclude that $\|X - Y\| = 0$, which shows uniqueness. But, by Lemma 13.3 and (13.11),

$$\|\nabla X\|_{L_\infty} \le \|\nabla w_h^n\|_{L_\infty} + \|\nabla\theta^n\|_{L_\infty} \le C + Ch^{-1}\|\nabla\theta^n\|.$$

Here by (13.16), (13.18) and the estimate for R_n before (13.18),

$$k\|\nabla\theta^n\|^2 \le C(\|\theta^{n-1}\|^2 + k\|\theta^n\|^2 + kR_n) \le C(h^2 + k)^2,$$

and hence $k\|\nabla X\|_{L_\infty}^2 \le C(k + h^2 + (k/h)^2)$, which shows our claim.

For the solution of (13.14) we could employ the iterative scheme

$$(X^{j+1} - U^{n-1}, \chi) + k(a(X^j)\nabla X^{j+1}, \nabla\chi) = (f(X^j), \chi), \quad \forall \chi \in S_h, \ j \ge 0,$$

with $X^0 = U^{n-1}$, say. Multiplying (13.14) by k and subtracting we obtain by similar calculations as the above

$$\|X^{j+1} - U^n\|^2 \le C(k + k\|U^n\|_{L_\infty}^2)\|X^j - U^n\|^2 \le \gamma\|X^j - U^n\|^2,$$

with $\gamma < 1$ if h and k/h are small. Thus the iterative scheme converges to the solution of (13.14).

We remark finally that when a is independent of u, so that the only source of nonlinearity is $f(u)$, no mesh-ratio condition is required for uniqueness or convergence of the iterative scheme.

The above method thus has the disadvantage that a nonlinear system of algebraic equations has to be solved at each time step, as a result of the presence of $a(U^n)$ and $f(U^n)$ in (13.14). We shall therefore now consider a linearized modification of the method in which this difficulty is avoided by replacing U^n by U^{n-1} in these two places, so that we now have, for $t_n \in J$,

(13.19) $$(\bar\partial U^n,\chi)+(a(U^{n-1})\nabla U^n,\nabla\chi)=(f(U^{n-1}),\chi),\quad \forall\chi\in S_h,$$

where $U^0=v_h$. With $\mathcal{B}$ and $\mathcal{A}(\alpha)$ as before, this equation may be written

$$(\mathcal{B}+k\mathcal{A}(\alpha^{n-1}))\alpha^n=\mathcal{B}\alpha^{n-1}+k\widetilde{f}(\alpha^{n-1}),\quad \text{for } t_n\in J.$$

Note that these linear systems may always be solved for α^n.

We shall show that the result of Theorem 13.2 remains valid for this linearized form of the backward Euler Galerkin method.

Theorem 13.3 *Let U^n and u be the solutions of* (13.19) *and* (13.1), *respectively. Then, under the appropriate regularity assumptions for u, we have*

$$\|U^n-u(t_n)\|\le C\|v_h-v\|+C(u)(h^2+k),\quad \textit{for } t_n\in\bar J.$$

Proof. Using again the splitting (13.15) we only have to consider the modification in the estimation of θ^n. Similarly to above we have now

$$\begin{aligned}(\bar\partial\theta^n,\chi)+(a(U^{n-1})\nabla\theta,\nabla\chi)&=(f(U^{n-1})-f(u^n),\chi)\\ &-((a(U^{n-1})-a(u^n))\nabla w^n,\nabla\chi)-(\bar\partial\rho^n,\chi)-(\bar\partial u^n-u_t^n,\chi).\end{aligned}$$

Here

$$\|f(U^{n-1})-f(u^n)\|\le C\|U^{n-1}-u^n\|\le C(\|\theta^{n-1}\|+\|\rho^{n-1}\|+k\|\bar\partial u^n\|),$$

and, bounding the term in $a(\cdot)$ similarly, we obtain now, with $\chi=\theta^n$,

$$\begin{aligned}&\tfrac12\bar\partial\|\theta^n\|^2+\mu\|\nabla\theta^n\|^2\\ &\quad\le C\big(\|\theta^{n-1}\|+\|\rho^{n-1}\|+k\|\bar\partial u^n\|+\|\bar\partial\rho^n\|+\|\bar\partial u^n-u_t^n\|\big)\|\nabla\theta^n\|,\end{aligned}$$

where we have used Friedrichs' inequality $\|\theta^n\|\le C\|\nabla\theta^n\|$. Hence, arguing as before,

$$\bar\partial\|\theta^n\|^2\le C\|\theta^{n-1}\|^2+C(u)(h^2+k)^2,$$

or

$$\|\theta^n\|^2\le(1+Ck)\|\theta^{n-1}\|^2+C(u)k(h^2+k)^2.$$

Hence, by repeated application,

$$\|\theta^n\|^2\le C\|\theta^0\|^2+C(u)(h^2+k)^2,$$

which shows (13.18) and thus completes the proof. □

For the purpose of obtaining higher accuracy in time we shall now consider the Crank-Nicolson Galerkin scheme, or, with $\widehat U^n=(U^n+U^{n-1})/2$,

(13.20) $$(\bar\partial U^n,\chi)+(a(\widehat U^n)\nabla\widehat U^n,\nabla\chi)=(f(\widehat U^n),\chi),\ \forall\chi\in S_h,\ t_n\in J,$$

with $U^0 = v_h$. This equation is symmetric around the point $t = t_{n-1/2}$, and one should therefore expect second order accuracy in time. It shares, however, with the first backward Euler method discussed above, the disadvantage of producing a nonlinear system of equations at each time level. For this reason we shall consider also below a linearized modification, in which the argument of a and f is obtained by extrapolation from U^{n-1} and U^{n-2}, or, more precisely, with $\bar{U}^n = \frac{3}{2}U^{n-1} - \frac{1}{2}U^{n-2}$, for $n \geq 2, t_n \in J$,

$$(\bar{\partial}U^n, \chi) + (a(\bar{U}^n)\nabla\widehat{U}^n, \nabla\chi) = (f(\bar{U}^n), \chi), \quad \forall \chi \in S_h. \tag{13.21}$$

As was the case for the backward Euler scheme, the nonlinear equation (13.20) will be solvable for U^n in terms of U^{n-1} for k small, whereas the linearized equation (13.21) is always solvable for U^n when U^{n-1} and U^{n-2} are given.

Note that taking a and f at U^{n-1}, as we did for the backward Euler scheme, will not be satisfactory here since this choice would be only first order accurate, whereas since

$$\bar{u}^n = \tfrac{3}{2}u^{n-1} - \tfrac{1}{2}u^{n-2} = u^{n-1/2} + O(k^2), \quad \text{as } k \to 0, \tag{13.22}$$

the extrapolation just proposed will give second order accuracy.

We observe that since the equation now contains U^{n-2} it may only be used for $n \geq 2$, and we have to supplement it with some other method for determining U^1. We shall discuss such a choice later.

We shall now present the error analysis for the basic Crank-Nicolson-Galerkin method. We shall then need another auxiliary estimate:

Lemma 13.4 *Assuming the appropriate regularity for u we have, for the elliptic projection defined by* (13.7),

$$\|\nabla w_{h,tt}(t)\| \leq C(u), \quad \textit{for } t \in \bar{J}.$$

Proof. Differentiation of (13.7) with respect to t twice gives

$$(a(u)\nabla w_{h,tt}, \nabla\chi) = (a(u)\nabla u_{tt}, \nabla\chi) - 2(a(u)_t\nabla\rho_t, \nabla\chi) - (a(u)_{tt}\nabla\rho, \nabla\chi),$$

and hence, with $\chi = w_{h,tt}$,

$$\mu\|\nabla w_{h,tt}\|^2 \leq C(u)\big(\|\nabla u_{tt}\| + \|\nabla\rho_t\| + \|\nabla\rho\|\big)\|\nabla w_{h,tt}\|,$$

from which the result follows, in view of Lemma 13.2. □

Theorem 13.4 *Let U^n and u be solutions of* (13.19) *and* (13.1), *respectively. Then, under the appropriate regularity assumptions for u, we have, for small k,*

$$\|U^n - u(t_n)\| \leq C\|v_h - v\| + C(u)(h^2 + k^2), \quad \textit{for } t_n \in \bar{J}.$$

Proof. Partitioning the error as usual according to (13.15), ρ^n is bounded as desired, and it remains to consider θ^n. We have this time

$$
\begin{aligned}
&(\bar\partial\theta^n,\chi) + (a(\widehat{U}^n)\nabla\widehat{\theta}^n,\nabla\chi)\\
=&(\bar\partial U^n,\chi) + (a(\widehat{U}^n)\nabla\widehat{U}^n,\nabla\chi) - (\bar\partial w_h^n,\chi) - (a(\widehat{U}^n)\nabla\widehat{w}_h^n,\nabla\chi)\\
=&(f(\widehat{U}^n),\chi) - (u_t^{n-\frac12},\chi) - (\bar\partial w_h^n - u_t^{n-\frac12},\chi)\\
&- \big(a(u^{n-\frac12})\nabla w_h^{n-\frac12},\nabla\chi\big) - \big(a(\widehat{U}^n)\nabla\widehat{w}_h^n - a(u^{n-\frac12})\nabla w_h^{n-\frac12},\nabla\chi\big)\\
=&\big(f(\widehat{U}^n) - f(u^{n-\frac12}),\chi\big) - (\bar\partial w_h^n - u_t^{n-\frac12},\chi)\\
&- \big((a(\widehat{U}^n) - a(u^{n-\frac12}))\nabla\widehat{w}_h^n + a(u^{n-\frac12})\nabla(\widehat{w}_h^n - w_h^{n-\frac12}),\nabla\chi\big).
\end{aligned}
\tag{13.23}
$$

Setting $\chi = \bar\theta^n$ and using $(\bar\partial\theta^n,\bar\theta^n) = \frac12\bar\partial\|\theta^n\|^2$ and (13.2), we find

$$
\begin{aligned}
&\tfrac12\bar\partial\|\theta^n\|^2 + \mu\|\nabla\bar\theta^n\|^2\\
&\le C\big(\|\bar U^n - u^{n-\frac12}\| + \|\bar\partial w_h^n - u_t^{n-\frac12}\| + \|\nabla(\widehat{w}_h^n - w_h^{n-\frac12})\|\big)\|\nabla\bar\theta^n\|,
\end{aligned}
$$

and hence

$$
\bar\partial\|\theta^n\|^2 \le C(\|\widehat{U}^n - u^{n-\frac12}\|^2 + \|\bar\partial\widetilde{U}^n - u_t^{n-\frac12}\|^2 + \|\nabla(\widehat{w_h}^n - w_h^{n-\frac12})\|^2).
\tag{13.24}
$$

Here applying Lemma 13.2

$$
\|\bar U^n - u^{n-\frac12}\| \le \|\bar\theta^n\| + \|\bar\rho^n\| + \|\bar u^n - u^{n-\frac12}\| \le \|\bar\theta^n\| + C(u)(h^2+k^2).
$$

Similarly

$$
\|\bar\partial w_h^n - u_t^{n-\frac12}\| \le \|\bar\partial\rho^n\| + \|\bar\partial u^n - u_t^{n-\frac12}\| \le C(u)(h^2+k^2),
$$

and, by Lemma 13.4,

$$
\|\nabla(\widehat{w}_h^n - w_h^{n-\frac12})\| \le Ck\int_{t_{n-1}}^{t_n}\|\nabla w_{h,tt}\|\,ds \le C(u)k^2.
$$

Altogether, this shows $\bar\partial\|\theta^n\|^2 \le C\|\bar\theta^n\|^2 + C(h^2+k^2)^2$, or

$$
(1-Ck)\|\theta^n\|^2 \le (1+Ck)\|\theta^{n-1}\|^2 + C(u)k(h^2+k^2)^2,
$$

whence, for small k, by repeated application,

$$
\|\theta^n\| \le C\|v_h - v\| + C(u)(h^2+k^2), \quad \text{for } t_n \in J,
$$

which completes the proof. □

We now turn our attention to the linearized Crank-Nicolson Galerkin method. As we mentioned earlier, this method will require a separate prescription for calculating U^1. We shall analyze here a predictor corrector method for this purpose, using as a first approximation the value $U^{1,0}$ determined by the case $n = 1$ of equation (13.21) with $\bar{U}^1$ replaced by U^0 and then as the final approximation the result of the same equation with $\bar{U}^1$ replaced by $\frac{1}{2}(U^{1,0} + U^0)$, so that thus our starting procedure is defined by

$$U^0 = v_h, \tag{13.25}$$

followed by

$$\Big(\frac{U^{1,0} - U^0}{k}, \chi\Big) + \big(a(U^0)\nabla\big(\frac{U^{1,0} + U^0}{2}\big), \nabla\chi\big) = (f(U^0), \chi), \tag{13.26}$$

and then

$$(\bar{\partial} U^1, \chi) + \big(a\big(\frac{U^{1,0} + U^0}{2}\big)\nabla \bar{U}^1, \nabla\chi\big) = \big(f\big(\frac{U^{1,0} + U^0}{2}\big), \chi\big), \tag{13.27}$$

with $\chi \in S_h$. For this method we shall show the following:

Theorem 13.5 *Let U^n be the solution of* (13.21), *with U^0 and U^1 defined by* (13.25) *and* (13.26), (13.27), *and let u be the solution of* (13.1). *Then, under the appropriate regularity assumptions for u, we have*

$$\|U^n - u(t_n)\| \le C\|v_h - v\| + C(u)(h^2 + k^2), \quad \textit{for } t_n \in \bar{J}.$$

Proof. This time we have instead of (13.23), for $n \ge 2$,

$$\begin{aligned}(\bar{\partial}\theta^n, \chi) + (a(\bar{U}^n)\nabla\bar{\theta}^n, \nabla\chi) &= \big(f(\bar{U}^n) - f(u^{n-\frac{1}{2}}), \chi\big) - (\bar{\partial} w_h^n - u_t^{n-\frac{1}{2}}, \chi) \\ &- \big((a(\bar{U}^n) - a(u^{n-\frac{1}{2}}))\nabla \widehat{w}_h^n + a(u^{n-\frac{1}{2}})\nabla(\widehat{w}_h^n - w_h^{n-\frac{1}{2}}), \nabla\chi\big),\end{aligned}$$

and therefore this time

$$\bar{\partial}\|\theta^n\|^2 \le C(\|\bar{U}^n - u^{n-\frac{1}{2}}\|^2 + \|\bar{\partial} w_h^n - u_t^{n-\frac{1}{2}}\|^2 + \|\nabla(\widehat{w}_h^n - w_h^{n-\frac{1}{2}})\|^2).$$

Here, using our definitions and (13.22),

$$\begin{aligned}\|\bar{U}^n - u^{n-\frac{1}{2}}\| &\le \|\bar{\theta}^n\| + \|\bar{\rho}^n\| + \|\bar{u}^n - u^{n-\frac{1}{2}}\| \\ &\le C(\|\theta^{n-1}\| + \|\theta^{n-2}\|) + C(u)(h^2 + k^2),\end{aligned}$$

and we obtain

$$\|\theta^n\|^2 \le (1 + Ck)\|\theta^{n-1}\|^2 + Ck\|\theta^{n-2}\|^2 + C(u)k(h^2 + k^2)^2,$$

or

$$\begin{aligned}&\|\theta^n\|^2 + Ck\|\theta^{n-1}\|^2 \\ &\qquad \le (1 + 2Ck)(\|\theta^{n-1}\|^2 + Ck\|\theta^{n-2}\|^2) + C(u)k(h^2 + k^2)^2.\end{aligned}$$

This shows

$$\|\theta^n\|^2 \le C(\|\theta^1\|^2 + k\|\theta^0\|^2) + C(u)(h^2+k^2)^2, \quad \text{for } n \ge 2. \tag{13.28}$$

We now estimate $\|\theta^1\|$ from the equations (13.26) and (13.27). In the same way as above we obtain from (13.26), with $\theta^{1,0} = U^{1,0} - w_h^1$, $\theta^{0,0} = \theta^0$,

$$\bar\partial\|\theta^{1,0}\|^2 \le C\|U^0 - u^{\frac12}\|^2 + C(u)(h^2+k^2)^2.$$

Since

$$\|U^0 - u^{\frac12}\| \le \|\theta^0\| + \|\rho^0\| + \|u^0 - u^{1/2}\| \le \|\theta^0\| + C(u)(h^2+k),$$

this shows $\bar\partial\|\theta^{1,0}\|^2 \le C\|\theta^0\|^2 + C(u)(h^4+k^2)$, and hence

$$\|\theta^{1,0}\|^2 \le (1+Ck)\|\theta^0\|^2 + C(u)k(h^4+k^2) \le C\|\theta^0\|^2 + C(u)(h^4+k^3).$$

We now apply equation (13.27) to obtain this time, instead of (13.24),

$$\bar\partial\|\theta^1\|^2 \le C(\|\tfrac12(U^{1,0}+U^0) - u^{\frac12}\|^2 + C(u)(h^2+k^2)^2). \tag{13.29}$$

Here, by above,

$$\begin{aligned}\|\tfrac12(U^{1,0}+U^0) - u^{\frac12}\| &\le \|\tfrac12(\theta^{1,0}+\theta^0)\| + \|\bar{\bar U}^1 - u^{\frac12}\| \\ &\le \tfrac12(\|\theta^{1,0}\| + \|\theta^0\|) + C(u)(k^2+h^2) \le C\|\theta^0\| + C(u)(h^2+k^{3/2})\end{aligned}$$

and hence from (13.29),

$$\|\theta^1\|^2 \le (1+Ck)\|\theta^0\|^2 + C(u)k(h^4+k^3) \le C\|\theta^0\|^2 + C(u)(h^2+k^2)^2.$$

Together with our previous estimate (13.28), this yields

$$\|\theta^n\| \le C\|\theta^0\| + C(u)(h^2+k^2) \le C\|v_h - v\| + C(u)(h^2+k^2).$$

The proof is now complete. □

The material of this chapter is already essentially covered in the work of Douglas and Dupont [74] and Wheeler [246] cited in Chapter 1.

Among a large number of later related works we quote Douglas and Dupont [75], [76], Rachford [198], Dendy [69], Douglas [73], Luskin [165], Lubich and Ostermann [163], [164], Zlàmal [252], [253], Cermak and Zlàmal [46], Chen, Larsson and Zhang [48], and Larsson, Thomée and Zhang [148].

The discontinuous Galerkin method was studied for nonlinear equations in Eriksson and Johnson [90]. For maximum-norm analyses, see Dobrowolski [71], [72]. Analysis of finite element methods for the Navier-Stokes equations has been pursued by Heywood and Rannacher [120], [121], [122], [123].

14. Semilinear Parabolic Equations

In the last chapter we considered discretization in both space and time of a model nonlinear parabolic equation. The discretization with respect to space was done by piecewise linear finite elements and in time we applied the backward Euler and Crank-Nicolson methods. In this chapter we shall restrict the consideration to the case when only the forcing term is nonlinear, but discuss more general approximations in the spatial variable. We shall begin with the spatially semidiscrete problem and first briefly study global conditions on the forcing term and the finite element spaces under which optimal order error estimates can be derived for smooth data, uniformly down to $t = 0$, and then turn our attention to the analysis for nonsmooth initial data. We then discuss discretization in time by the backward Euler method, in particular with reference to nonsmooth initial data.

We shall thus be concerned with spatially and fully discrete approximate solutions of the semilinear initial-boundary value problem

$$\begin{aligned} u_t - \Delta u &= f(u) \quad \text{in } \Omega, \qquad \text{for } t \in J = (0, \bar{t}\,], \\ u &= 0 \qquad\quad \text{on } \partial\Omega, \quad \text{for } t \in J, \qquad \text{with } u(0) = v \quad \text{in } \Omega. \end{aligned} \tag{14.1}$$

Here Ω is a bounded domain in $\mathbb{R}^d$ with a sufficiently smooth boundary $\partial\Omega$, and f is a smooth function on $\mathbb{R}$, for which we assume provisionally that

$$|f'(u)| \le B, \quad \text{for } u \in \mathbb{R}. \tag{14.2}$$

We shall now permit finite element spaces also of higher order than linear, and let thus $S_h \subset H_0^1 = H_0^1(\Omega)$ be a family of finite dimensional spaces satisfying our standard $O(h^r)$ approximation assumption (1.10) for some integer $r \ge 2$ and for $v \in H^r \cap H_0^1$.

We first study the semidiscrete solution $u_h : \bar{J} \to S_h$ defined by

$$(u_{h,t}, \chi) + (\nabla u_h, \nabla \chi) = (f(u_h), \chi), \quad \forall \chi \in S_h, \ t \in J, \tag{14.3}$$

with $u_h(0) = v_h$, where $v_h \in S_h$ is an approximation of v. It is easy to see that under our present assumptions this semilinear system of ordinary differential equations has a unique solution.

We first note that the argument of last chapter immediately shows the following result. Here and below we omit the dependence of constants on B.

Theorem 14.1 *Assume that* (14.2) *and* (1.10) *hold, and let* u_h *and* u *be solutions of* (14.3) *and* (14.1), *respectively. Then, if* v_h *is appropriately chosen and* u *sufficiently smooth, we have, with* $C = C(u, \bar{t})$,

$$\|u_h(t) - u(t)\| + h\|u_h(t) - u(t)\|_1 \le Ch^r, \quad \text{for } t \in \bar{J}. \tag{14.4}$$

In Chapter 13 we noted that in applications, f might not satisfy the global condition (14.2), but for the problem studied there it was sufficient to assume such a condition in a neighborhood of the range of the solution u considered. The analysis then required us to show closeness of u_h to u in maximum-norm, and this was accomplished by using the inverse property (1.12), satisfied when S_h consists of piecewise linear functions on quasiuniform triangulations. For the more general elements satisfying (1.10) it suffices to assume the inverse property

$$\|\chi\|_{L_\infty} \le Ch^{-\nu}\|\chi\|, \quad \forall \chi \in S_h, \quad \text{for some } \nu < r;$$

in the d–dimensional case with quasiuniform partitions this holds with $\nu = d/2$ and hence is always satisfied for $r > d/2$.

In the one-dimensional case the desired closeness may be shown, without requiring any inverse properties, from the fact that $\|v\|_{L_\infty} \le C\|v\|_1$. In fact, for as long as $u_h(t)$ belongs to a neighborhood $I_\delta = [m_0 - \delta, m_1 + \delta]$ with $\delta > 0$ of the range $I_0 = [m_0, m_1]$ of the solution u, in which f' is bounded, we have

$$\|u_h(t) - u(t)\|_{L_\infty} \le C\|u_h(t) - u(t)\|_1 \le C(u, \bar{t})h^{r-1} < \delta/2,$$

for small h, and $u_h(t)$ therefore remains in I_δ.

We shall now show that also when $d \ge 2$ the error estimate of Theorem 14.1 remains valid without inverse assumptions, provided that $f'(u)$ only grows mildly with u. We shall thus assume that there is a positive number p, with $p \le 2/(d-2)$ when $d \ge 3$ and with p arbitrary when $d = 2$, such that

$$|f'(u)| \le C(1 + |u|^p), \quad \text{for } u \in \mathbb{R}. \tag{14.5}$$

Theorem 14.2 *Let* $d \ge 2$ *and assume that* f *satisifies* (14.5) *with* p *appropriate, and that* (1.10) *holds. Let* u_h *and* u *be solutions of* (14.3) *and* (14.1), *respectively. Then the error estimates of Theorem 14.1 hold if* u *is sufficiently smooth and* v_h *is suitably chosen.*

Proof. In the standard way we write $u_h - u = (u_h - R_h u) + (R_h u - u) = \theta + \rho$, with R_h the elliptic projection onto S_h defined by (1.22), and recall that $\|\rho\| = O(h^r)$. For θ we have this time

$$\begin{aligned}(\theta_t, \chi) + (\nabla\theta, \nabla\chi) &= (f(u_h), \chi) - (R_h u_t, \chi) - (\nabla R_h u, \nabla\chi) \\ &= (f(u_h) - f(u), \chi) - (\rho_t, \chi).\end{aligned} \tag{14.6}$$

We shall use (14.5) to show that

$$(14.7) \qquad |(f(u_h) - f(u), \theta)| \le C\|u_h - u\| \, \|\nabla\theta\|.$$

Applying (14.6) with $\chi = \theta$, this implies

$$\tfrac{1}{2}\frac{d}{dt}\|\theta\|^2 + \|\nabla\theta\|^2 \le C(\|\theta\|^2 + \|\rho\|^2 + \|\rho_t\|^2) + \|\nabla\theta\|^2$$

and hence, using the standard estimates for ρ and ρ_t,

$$\frac{d}{dt}\|\theta\|^2 \le C\|\theta\|^2 + Ch^{2r}.$$

Choosing, e.g., $v_h = R_h v$, this implies $\|\theta(t)\| \le Ch^r$ on J, and thus completes the proof of the L_2-estimate for $u_h - u$.

To show (14.7) we consider first $d = 2$. Choosing q with $2 < q < \infty$, we have $\|\theta\|_{L_q} \le C\|\nabla\theta\|$, and Hölder's inequality shows, with $q^{-1} + (q')^{-1} = 1$,

$$(14.8) \qquad |(f(u_h) - f(u), \theta)| \le C\|f(u_h) - f(u)\|_{L_{q'}} \|\nabla\theta\|.$$

Here, using (14.5) and Hölder's inequality once more, now with exponents $2/q'$ and $2/(2-q')$,

$$(14.9) \qquad \begin{aligned} \|f(u_h) - f(u)\|_{L_{q'}}^{q'} &\le C\int_\Omega |u_h - u|^{q'}(1 + |u_h| + |u|)^{pq'}\,dx \\ &\le C\|u_h - u\|_{L_2}^{q'}(1 + \|u_h\|_{L_s} + \|u\|_{L_s})^{pq'}, \quad \text{with } s = 2pq/(q-2). \end{aligned}$$

Since $s < \infty$, we have $\|u_h\|_{L_s} \le C\|\nabla u_h\|$ and since u is smooth, we find

$$\|f(u_h) - f(u)\|_{L_{q'}} \le C\|u_h - u\|(1 + \|\nabla u_h\|)^p.$$

In view of (14.8), the proof of (14.7) may now be completed by showing that $\|\nabla\theta\|$ and hence also $\|\nabla u_h\|$ is bounded for small h. For this purpose we use (14.6) with $\chi = 2\theta_t$ to obtain, after kickback of $2\|\theta_t\|^2$,

$$(14.10) \qquad \begin{aligned} \frac{d}{dt}\|\nabla\theta\|^2 &\le \|f(u_h) - f(u)\|^2 + \|\rho_t\|^2 \\ &\le 2\|f(u_h) - f(R_h u)\|^2 + 2\|f(R_h u) - f(u)\|^2 + Ch^{2r}. \end{aligned}$$

Here, similarly to the above estimation of $f(u_h) - f(u)$,

$$\begin{aligned} \|f(R_h u) - f(u)\|^2 &\le C\int_\Omega \rho^2(1 + |R_h u|)^{2p}\,dx \\ &\le C\Big(\int_\Omega \rho^q\,dx\Big)^{2/q}\Big(\int_\Omega (1 + |R_h u|)^s\,dx\Big)^{(q-2)/q} \\ &\le C\|\rho\|_{L_q}^2(1 + \|R_h u\|_{L_s})^{2p} \le C\|\nabla\rho\|^2 \le Ch^{2r-2}, \end{aligned}$$

since $\|R_h u\|_{L_s} \le C\|\nabla R_h u\| \le C\|\nabla u\| \le C$. In the same way we have

$$\|f(u_h)-f(R_hu)\|^2 \le C\|\nabla\theta\|^2\big(1+\|\nabla u_h\|\big)^{2p} \le C\|\nabla\theta\|^2\big(1+\|\nabla\theta\|\big)^{2p}.$$

Let $\bar{t}_h \in \bar{J}$ be as large as possible with $\|\nabla\theta\| \le 1$ on $[0,\bar{t}_h]$. Then, for $t \le \bar{t}_h$, we have, by (14.10),

$$\frac{d}{dt}\|\nabla\theta\|^2 \le C\|\nabla\theta\|^2 + Ch^{2r-2}.$$

Thus, with C independent of $\bar{t}_h$,

$$\|\nabla\theta\| \le Ce^{C\bar{t}}h^{r-1} \le 1/2, \quad \text{for } h \le h_0.$$

It follows that $\bar{t}_h = \bar{t}$ for $h \le h_0$, so that $\|\nabla\theta\| \le 1$ on $\bar{J}$ for these h, and thus $\|\nabla u_h\| \le \|\nabla u\| + 1$ on $\bar{J}$. This completes the proof of (14.7) for $d = 2$.

For $d \ge 3$ we choose $q = 2d/(d-2)$. Then $\|v\|_{L_s} \le C\|\nabla v\|$ for $s \le q$, so that (14.8) and (14.9) remain valid. Since $p \le 2/(d-2)$ we have $s = 2pq/(q-2) \le q$, and the proof proceeds as for $d = 2$. □

To guarantee that u is smooth enough for Theorem 14.1 to apply, both smoothness of v and compatibility conditions between v and the differential equation at $\partial\Omega$ for $t = 0$ are needed. For instance, in the linear homogeneous case ($f = 0$ in (14.1)) we know from Chapter 7 that, with $|v|_r = \|(-\Delta)^{r/2}v\|$,

$$\|u_h(t) - u(t)\| \le Ch^r|v|_r, \quad \text{for } v \in \dot{H}^r = \dot{H}^r(\Omega), \quad t \ge 0,$$

and we recall that this requires $\Delta^j v = 0$ on $\partial\Omega$ for $j < r/2$.

We note that the solution of (14.1) will always be smooth for positive time; in the case of the linear homogeneous equation this was expressed in Lemma 3.2 as the fact that the solution operator $E(t)$ of the initial value problem is an analytic semigroup and that

$$|E(t)v|_\beta \le Ct^{-(\beta-\alpha)/2}|v|_\alpha, \quad \text{for } t > 0, \quad \text{if } 0 \le \alpha \le \beta. \tag{14.11}$$

Using this, and a similar property of the solution operator $E_h(t)$ of the corresponding semidiscrete problem, we showed that if the discrete initial data v_h are chosen as the L_2-projection P_hv of v, then

$$\|u_h(t) - u(t)\| = \|E_h(t)P_hv - E(t)v\| \le Ch^rt^{-r/2}\|v\|, \quad \text{for } t > 0. \tag{14.12}$$

With $F_h(t) = E_h(t)P_h - E(t)$ we have by Theorem 3.5 the whole scale of estimates

$$\|F_h(t)v\| \le Ch^\mu t^{-(\mu-\alpha)/2}|v|_\alpha, \quad \text{for } 0 \le \alpha \le \mu \le r. \tag{14.13}$$

We now show a somewhat weaker result of this type for the semilinear equation (14.1); in the case of piecewise linear finite elements, i.e., when $r = 2$, this implies that (14.12) essentially remains valid. Note that because of the nonlinear character of the problem, the norm $\|v\|$ of the initial data does not enter as a factor on the right, but instead the constant depends on a bound for $\|v\|$.

Theorem 14.3 *Assume that* (14.2) *and* (1.10) *hold. Then there is a constant* $C = C(\kappa, \bar{t})$ *such that, for all solutions* u_h *and* u *of* (14.3) *and* (14.1) *with* $v \in L_2$ *and* $v_h = P_h v$, *we have*

$$\|u_h(t) - u(t)\| \le Ch^2(t^{-1} + \max(0, \log(t/h^2))), \quad \text{for } \|v\| \le \kappa, \ t \in J. \tag{14.14}$$

Proof. Simple energy arguments together with Gronwall's lemma show that $u(t)$ and $u_h(t)$ are bounded in L_2 for $t \in \bar{J}$ so that (14.14) trivially holds for $t \le h^2$. With our above notation we have by Duhamel's principle, for the solutions of (14.1) and (14.3), that

$$u(t) = E(t)v + \int_0^t E(t-s) f(u(s))\, ds$$

and

$$u_h(t) = E_h(t)v_h + \int_0^t E_h(t-s) P_h f(u_h(s))\, ds,$$

respectively. Hence, with $F_h(t)$ as above, $e = u_h - u$ satisfies

$$\begin{aligned} e(t) = F_h(t)v &+ \int_0^t E_h(t-s) P_h\big(f(u_h(s)) - f(u(s))\big)\, ds \\ &+ \int_0^t F_h(t-s) f(u(s))\, ds. \end{aligned} \tag{14.15}$$

Using the cases $\mu = 2$ and 0, $\alpha = 0$ of (14.13) and (14.2) and the boundedness of $\|f(u(s))\|$ we thus find

$$\begin{aligned} \|e(t)\| &\le C\kappa h^2 t^{-1} + CB\Big(\int_0^{h^2} + \int_{h^2}^t\Big) \|e(s)\|\, ds \\ &\quad + \Big(\int_0^{t-h^2} + \int_{t-h^2}^t\Big) \|F_h(t-s) f(u(s))\|\, ds \\ &\le Ch^2 t^{-1} + Ch^2 + C\int_{h^2}^t \|e(s)\|\, ds + Ch^2 \int_0^{t-h^2} \frac{ds}{t-s} + Ch^2 \\ &\le Ch^2 t^{-1} + Ch^2 \log(t/h^2) + C\int_{h^2}^t \|e(s)\|\, ds, \quad \text{for } t \ge h^2. \end{aligned} \tag{14.16}$$

Letting $\varphi(t) = \int_{h^2}^t \|e\|\, ds$, we conclude that

$$\varphi'(t) - C\varphi(t) \le Ch^2 t^{-1} + Ch^2 \log(t/h^2), \quad \text{for } h^2 \le t \le \bar{t},$$

with $\varphi(h^2) = 0$, whence

$$\varphi(t) \le C \int_{h^2}^t e^{C(t-s)} (h^2 s^{-1} + h^2 \log(s/h^2))\, ds \le Ch^2 \log(t/h^2).$$

Inserted into (14.16), this completes the proof. □

It turns out that the nonsmooth data error estimate (14.12) for the linear problem, with optimal order convergence for positive time, without regularity restrictions on initial data, does not quite generalize to semilinear equations when $r > 2$. However, we shall demonstrate a reduced smoothness convergence result which will show an $O(h^r)$ error for $r > 2$ under the assumption of initial regularity and compatibility of essentially order $r-2$. We note that the argument of the proof of Theorem 14.3, using a superposition of the estimate (14.13) for the linear homogeneous problem, does not carry over to the present case. In fact, in order to apply (14.13) with $r = \mu > 5/2$ to the expression $F_h(t-s)f(u(s))$ in (14.15), this would require $f(u(s))$ to be in some $\dot{H}^\alpha$ space with $\alpha > 1/2$. In particular, this would demand $f(0) = 0$, which we do not want to assume. We shall therefore give a direct proof which does not depend on (14.13). We shall now assume that $\Omega \subset \mathbb{R}^d$ with $d \le 3$.

In order to express our assumptions on the initial data, we define the set $\mathcal{F}_\alpha$ of compatible data of order α, for simplicity only with $\alpha \le 4$, by $\mathcal{F}_\alpha = L_\infty \cap \dot{H}^\alpha$ if $0 \le \alpha \le 2$, and $\mathcal{F}_\alpha = \{v \in \mathcal{F}_2; f(v) + \Delta v \in \dot{H}^{\alpha-2}\}$ if $2 < \alpha \le 4$. (Note that for a smooth solution $u_t(0) = f(v) + \Delta v$ has to vanish on $\partial\Omega$.) To measure the regularity we also introduce the functional

$$F_\alpha(v) = \begin{cases} \|v\|_{L_\infty} + |v|_\alpha, & \text{for } 0 \le \alpha \le 2, \\ \|v\|_{L_\infty} + |v|_2 + |f(v) + \Delta v|_{\alpha-2}, & \text{for } 2 < \alpha \le 4. \end{cases}$$

We first state the following special case of a regularity result from Johnson, Larsson, Thomée and Wahlbin [130], which generalizes (14.11) for the values of α and β considered. The restriction in β derives from technical difficulties associated with the nonlinearity of the equation.

Theorem 14.4 *Let $d \le 3$ and assume that* (14.2) *holds, and let $0 \le \alpha \le 4$ and $\beta \le \alpha + 5$. Then there is a constant $C = C(\kappa, \bar{t})$ such that for all solutions u of* (14.1) *with $v \in \mathcal{F}_\alpha$ we have*

$$|u(t)|_\beta + |u_t(t)|_{\beta-2} \le Ct^{-(\beta-\alpha)/2}, \quad \textit{for } t \in J, \quad \textit{if } F_\alpha(v) \le \kappa.$$

In the elementary but somewhat lengthy proof one first estimates successive time derivatives of u in spaces $\dot{H}^\beta$ with $\beta \le 2$, and then uses elliptic regularity to translate regularity with respect to time into regularity in space. We refer to [130] for details.

We are now ready for the convergence result indicated above, which shows $O(h^\mu)$ convergence with $\mu \le r$ if the initial data are in $\mathcal{F}_\alpha$ with $\alpha > \mu - 2$. As in (14.13), a negative power of t is required in the error bound if $\alpha < \mu$. It follows that if α is given with $0 \le \alpha \le 4$ then for positive t the convergence rate is essentially $O(h^{2+\alpha})$.

We shall also demonstrate below that for $\alpha = 0$ this result is best possible, so that a convergence rate of higher order than $O(h^2)$ is not possible without regularity restrictions on initial data. In [62] this maximal order convergence for t positive was improved for $0 \le \alpha < 2$ from $O(h^{2+\alpha})$ to $O(h^{2+2\alpha})$, and this was also shown to be best possible for these α. For $r = 4$, e.g., this essentially brings down the regularity requirements for $O(h^4)$ convergence from $v \in \mathcal{F}_2$ to $v \in \mathcal{F}_1$.

Theorem 14.5 *Let $d \le 3$ and assume that* (14.2) *and* (1.10) *hold, and let $0 \le \alpha \le 4, 1 \le \mu \le r$, and $\alpha \le \mu < \alpha + 2$. Then there exists a constant $C = C(\kappa, \bar{t})$ such that, if u_h and u are solutions of* (14.3) *and* (14.1) *with initial values $v \in \mathcal{F}_\alpha$ and $v_h = P_h v$, respectively, we have*

$$\|u_h(t) - u(t)\| \le C h^\mu t^{-(\mu-\alpha)/2}, \quad \text{for } t \in J, \ F_\alpha(v) \le \kappa.$$

Proof. Let $T = (-\Delta)^{-1} : L_2 \to \dot{H}^2$, and let $T_h : L_2 \to S_h$ be the approximation defined by (3.10). We recall that the operator T_h is bounded, symmetric, positive semidefinite on L_2 and positive definite on S_h, and that the elliptic projection satisfies $R_h v = T_h(-\Delta)v$. Application of T to (14.1) yields

$$T u_t + u = T f(u), \quad \text{for } t \in J, \quad \text{with } u(0) = v,$$

and the semidiscrete problem (14.3) may similarly be written

$$T_h u_{h,t} + u_h = T_h f(u_h), \quad \text{for } t \in J, \quad \text{with } u_h(0) = v_h. \tag{14.17}$$

Let $e = u_h - u$ be the error. We have

$$\begin{aligned} T_h e_t + e &= T_h u_{h,t} + u_h - T_h u_t - u = T_h f(u_h) - T f(u) + (T - T_h) u_t \\ &= T_h(f(u_h) - f(u)) + (T_h - T)(f(u) - u_t) \end{aligned}$$

or

$$T_h e_t + e = T_h(\omega e) + \rho, \tag{14.18}$$

where $\omega e = f(u_h) - f(u)$ so that

$$\omega = \int_0^1 f'(\eta u_h + (1-\eta)u)\, d\eta \quad \text{and } \rho = (T_h - T)(-\Delta)u = (R_h - I)u.$$

Multiplication of (14.18) by e_t yields

$$(T_h e_t, e_t) + \tfrac{1}{2}\frac{d}{dt}\|e\|^2 = (T_h(\omega e), e_t) + \frac{d}{dt}(\rho, e) - (\rho_t, e).$$

Since T_h is positive semidefinite, we have

$$|(T_h v, w)| \le (T_h v, v)^{1/2} (T_h w, w)^{1/2}, \tag{14.19}$$

and hence, using also the geometric-arithmetic mean inequality,

$$(T_h e_t, e_t) + \tfrac{1}{2}\frac{d}{dt}\|e\|^2 \le \tfrac{1}{2}(T_h e_t, e_t) + \tfrac{1}{2}(T_h(\omega e), \omega e) + \frac{d}{dt}(\rho, e) - (\rho_t, e).$$

Employing the boundedness of T_h and of ω, this shows

$$\frac{d}{dt}\|e\|^2 \le C\|e\|^2 + 2\frac{d}{dt}(\rho, e) - 2(\rho_t, e).$$

Multiplication by t^2 now gives, recalling that $t \le \bar{t}$,

$$\begin{aligned}\frac{d}{dt}(t^2\|e\|^2) &\le 2t\|e\|^2 + Ct^2\|e\|^2 + 2\frac{d}{dt}(t^2(\rho, e)) - 4t(\rho, e) - 2t^2(\rho_t, e)\\ &\le 2\frac{d}{dt}(t^2(\rho, e)) + C(t\|\rho\|^2 + t^3\|\rho_t\|^2 + t\|e\|^2),\end{aligned}$$

whence, by integration and a trivial kickback argument,

$$t^2\|e\|^2 \le Ct^2\|\rho\|^2 + C\int_0^t (s\|\rho\|^2 + s^3\|\rho_t\|^2)\,ds + C\int_0^t s\|e\|^2 ds. \tag{14.20}$$

In order to bound the last integral, we return to the error equation (14.18), which we now multiply by $2te$ to obtain

$$\frac{d}{dt}(t(T_h e, e)) + 2t\|e\|^2 \le 2t(T_h(\omega e), e) + 2t(\rho, e) + (T_h e, e).$$

Here, by (14.19), for ε suitable, since T_h and ω are bounded,

$$(T_h(\omega e), e) \le \varepsilon(T_h(\omega e), \omega e) + \frac{1}{4\varepsilon}(T_h e, e) \le \frac{1}{4}\|e\|^2 + C(T_h e, e),$$

so that

$$\frac{d}{dt}(t(T_h e, e)) + t\|e\|^2 \le C(t\|\rho\|^2 + (T_h e, e)), \quad \text{for } t \le \bar{t},$$

and hence by integration

$$\int_0^t s\|e\|^2\,ds \le C\int_0^t s\|\rho\|^2\,ds + C\int_0^t (T_h e, e)\,ds. \tag{14.21}$$

For the last integral we set $\tilde{e}(t) = \int_0^t e(s)\,ds$ and integrate (14.18) to obtain

$$T_h(e(t) - e(0)) + \tilde{e}(t) = T_h\int_0^t \omega e\,ds + \int_0^t \rho\,ds.$$

Recalling from the proof of Theorem 2.5 that $T_h e(0) = 0$ we obtain after multiplication by $2\tilde{e}' = 2e$,

$$\begin{aligned}2(T_h e, e) + \frac{d}{dt}\|\tilde{e}\|^2 &= 2(T_h\int_0^t \omega e\,ds, e) + 2(\int_0^t \rho\,ds, e)\\ &\le (T_h e, e) + (T_h\int_0^t \omega e\,ds, \int_0^t \omega e\,ds) + 2\int_0^t \|\rho\|\,ds\,\|e\|,\end{aligned}$$

or, by integration, since $\widetilde{e}(0) = 0$,

$$\int_0^t (T_h e, e)\, ds \le C \int_0^t \Big(\int_0^s \|e(\tau)\|\, d\tau\Big)^2 ds + 2\int_0^t \|e(s)\| \int_0^s \|\rho(\tau)\|\, d\tau\, ds.$$

Together with (14.20) and (14.21), this yields

$$\begin{aligned} t^2\|e\|^2 \le & C\Big(t^2\|\rho\|^2 + \int_0^t (s\|\rho\|^2 + s^3\|\rho_t\|^2)\, ds \\ & + \int_0^t \Big(\int_0^s \|e(\tau)\| d\tau\Big)^2 ds + \int_0^t \|e(s)\| \int_0^s \|\rho(\tau)\|\, d\tau\, ds\Big). \end{aligned}$$

Now by our assumptions on u we have using Theorem 14.4

$$\|\rho(t)\| = \|(R_h - I)u(t)\| \le Ch^\mu |u(t)|_\mu \le Ch^\mu t^{-\sigma/2}, \quad \sigma = \mu - \alpha,$$

and similarly $\|\rho_t(t)\| \le Ch^\mu |u_t(t)|_\mu \le Ch^\mu t^{-1-\sigma/2}$. Hence, since $\sigma < 2$,

$$t^2\|e\|^2 \le C\Big(h^{2\mu} t^{2-\sigma} + \int_0^t \Big(\int_0^s \|e\| d\tau\Big)^2 ds + h^\mu \int_0^t s^{1-\sigma/2}\|e\|\, ds\Big).$$

For $\varphi(t) = t^{\sigma/2}\|e(t)\|$ this shows

$$\begin{aligned} \varphi(t)^2 \le C\Big(h^{2\mu} + t^{-(2-\sigma)} & \int_0^t \Big(\int_0^s \tau^{-\sigma/2}\varphi(\tau)\, d\tau\Big)^2 ds \\ & + h^\mu t^{-(2-\sigma)} \int_0^t s^{1-\sigma}\varphi(s)\, ds\Big). \end{aligned}$$

With $\psi(t) = \max_{0\le s\le t}\varphi(s)$, and choosing $t_0 = t_0(t)$ such that $\varphi(t_0) = \psi(t)$, we have

$$\varphi(t)^2 \le \psi(t)^2 \le C\Big(h^{2\mu} + t_0^{-(2-\sigma)} \int_0^{t_0} s^{2-\sigma}\psi(s)^2\, ds + h^\mu \psi(t)\Big)$$

whence, for small h,

$$\psi(t)^2 \le C\Big(h^{2\mu} + \int_0^t \psi(s)^2\, ds\Big).$$

Gronwall's lemma shows $\psi(t) \le Ch^\mu$, and since $t^{\sigma/2}\|e(t)\| = \varphi(t) \le \psi(t)$, this completes the proof. □

We remark that the proof of Theorem 14.5 immediately extends to the case that the semidiscrete problem is defined by (14.17) where T_h satisfies conditions (i) and (ii) of Chapter 2, with the only change that μ now has to satisfy $2 \le \mu \le r$.

For the special case $\alpha = 0$, Theorem 14.5 shows that for any $\sigma < 2$ there is a $C = C(\kappa, t_0, \bar{t})$ such that, for the solutions of (14.3) and (14.1) with $v_h = P_h v$, we have

$$\|u_h(t) - u(t)\| \leq Ch^\sigma, \quad \text{for } 0 < t_0 \leq t \leq \bar{t}, \quad \text{if } \|v\|_{L_\infty} \leq \kappa; \tag{14.22}$$

we recall from Theorem 14.3 that in this case $O(h^\sigma)$ may be replaced by $O(h^2 \log(1/h))$ and that only boundedness of v in L_2 is required. We shall now give an example which shows that, in contrast to the linear case, this result is essentially sharp, in the sense that (14.22) cannot hold for any $\sigma > 2$, even if $r > 2$.

Consider thus the spatially one-dimensional problem

$$\begin{aligned} u_t &= u_{xx} + u^2 \quad \text{in } [0, \pi], \quad \text{for } t > 0, \\ u(0,t) &= u(\pi, t) = 0 \quad \text{for } t > 0, \quad \text{with } u(\cdot, 0) = v, \end{aligned} \tag{14.23}$$

and let S_h consist of continuous piecewise polynomials of degree $< r$ on a uniform partition, i.e., $S_h = \{\chi \in C[0,\pi], \chi|_{I_j} \in \Pi_{r-1}, j = 1, \dots, n\}$, where $I_j = (x_{j-1}, x_j)$, with $x_j = jh, h = \pi/n, n$ integer. We assume that $r > 2$.

We shall construct a sequence of solutions $u = u_n$ of (14.23), with initial data $v = v_n$ depending on n, such that the corresponding semidiscrete solutions $u_h = u_{n,h} \in S_h$, with $h = \pi/n$, violate (14.22) when $\sigma > 2$. The construction will start by choosing $v_n = u_n(\cdot, 0)$ orthogonal to S_h. Since then for the discrete initial data $v_{n,h} = P_h v_n = 0$, the semidiscrete solution $u_{n,h}(t)$ vanishes for $t \geq 0$, and thus the error equals $-u_n$. The desired contradiction is therefore achieved by choosing v_n bounded in L_∞, uniformly in n, and such that, for some $n_0, t_0 > 0$,

$$\|u_n(t_0)\| \geq cn^{-2}, \quad \text{with } c > 0, \ n \geq n_0. \tag{14.24}$$

To accomplish this, let $\psi(y) = \sum_{j=1}^{r+1} \psi_j \sin(jy) \not\equiv 0$ be orthogonal to Π_{r-1} on $[0, \pi]$ (which is possible since the number of ψ_j is greater than r). The function $v_n(x) = \psi(nx)$ is then orthogonal to Π_{r-1} on each I_j, and hence orthogonal to S_h. Further, independently of n, $\|v_n\|_{L_\infty} \leq \sum_{j=1}^{r+1} |\psi_j| \equiv \kappa$.

By comparison with the initial value problems

$$z_t = z^2, \quad \text{for } t > 0, \quad \text{with } z(0) = \pm\kappa, \tag{14.25}$$

it follows that there exist $\bar{t} > 0$ and M such that $\|u_n(t)\|_{L_\infty} \leq M$ for $t \in J = (0, \bar{t}]$, uniformly in n. In fact, the solutions of (14.25) may be thought of as solutions of the differential equation in (14.23) which are independent of x and with boundary values dominating those in (14.23), so that the maximum principle may be used to achieve the comparison. Since thus u_n is bounded, we may regard u_n as the solution of an equation in which the forcing term u^2 in (14.23) has been replaced by a function $f(u)$ with $f(u) = u^2$ for $|u| \leq M$, and with f' bounded on $\mathbb{R}$, thus satisfying the assumptions of Theorem 14.4.

Letting $c_n = c_n(t) = \int_0^\pi u_n(x,t)\sin x\,dx$ denote the first Fourier sine coefficient of u_n, it sufficies for (14.24) to demonstrate that

$$c_n(t_0) \geq c_0 n^{-2}, \quad \text{for } n \geq n_0, \quad \text{for some } t_0 > 0, \tag{14.26}$$

for by Parseval's relation $\|u_n(t_0)\| \geq (2/\pi)^{1/2}|c_n(t_0)|$. Here $c_n(0) = 0$ since $v_n(x) = \psi(nx)$ is orthogonal to $\sin x$ for $n > 1$, and thus from (14.23)

$$c_n' + c_n = g_n(t) := \int_0^\pi u_n^2(x,t)\sin x\,dx, \quad \text{for } t \geq 0, \quad \text{with } c_n(0) = 0.$$

We shall show that with positive constants k_0, μ, and ω,

$$g_n(t) \geq \mu e^{-\omega n^2 t}, \quad \text{for } n^2 t \geq k_0 > 0. \tag{14.27}$$

Choosing t_0 such that $2k_0 n^{-2} \leq t_0 \leq \bar{t}$, this implies

$$c_n(t_0) \geq \mu \int_{k_0 n^{-2}}^{t_0} e^{-(t_0 - s)} e^{-\omega n^2 s}\,ds \geq \mu e^{-t_0} \int_{k_0 n^{-2}}^{t_0} e^{-\omega n^2 s}\,ds \geq c_0 n^{-2},$$

with $c_0 > 0$, and thus proves (14.26).

For (14.27) we first note that if w_n is the solution of (14.23) with the forcing term u^2 replaced by 0, then $u_n \geq w_n$, and hence $u_n^2 \geq w_{n,+}^2$, where $w_{n,+} = \max(w_n, 0)$. With ψ_m the first non-vanishing coefficient in $\psi(y)$, which we normalize so that $\psi_m = 1$, we have

$$w_n(x,t) = e^{-m^2 n^2 t}\sin(mnx) + \sum_{j=m+1}^{r+1} \psi_j e^{-j^2 n^2 t}\sin(jnx).$$

Denoting the first term on the right by $\widetilde{w}$, it is clear that

$$\int_0^\pi \widetilde{w}_+(x,t)^2 \sin x\,dx \geq c e^{-2m^2 n^2 t}, \quad \text{with } c > 0.$$

Since $\widetilde{w}_+ \leq w_+ + |w - \widetilde{w}|$ we have $\widetilde{w}_+^2 \leq 2w_+^2 + 2|w - \widetilde{w}|^2$, and hence

$$w_+^2 \geq \tfrac{1}{2}\widetilde{w}_+^2 - |w - \widetilde{w}|^2 \geq \tfrac{1}{2}\widetilde{w}_+^2 - Ce^{-2(m+1)^2 n^2 t}.$$

Thus, for $n^2 t \geq k_0 > 0$, with k_0 large enough,

$$g_n(t) \geq \int_0^\pi w_+(x,t)^2 \sin x\,dx \geq ce^{-2m^2 n^2 t} - Ce^{-2(m+1)^2 n^2 t} \geq ce^{-Cn^2 t},$$

which shows (14.27), and thus establishes our counter-example.

We now turn to a discussion of the fully discrete backward Euler Galerkin method for (14.1), to find $U_h^n \in S_h$ for $n \geq 0$ such that

$$(\bar{\partial}U_h^n,\chi)+(\nabla U_h^n,\nabla\chi)=(f(U_h^n),\chi),\quad \forall\chi\in S_h,\ n\geq 1. \tag{14.28}$$

We also consider the linearized version of (14.28) defined by

$$(\bar{\partial}U_h^n,\chi)+(\nabla U_h^n,\nabla\chi)=(f(U_h^{n-1}),\chi),\quad \forall\chi\in S_h,\ n\geq 1. \tag{14.29}$$

Since we are also going to discuss an abstract problem in a Hilbert space below, we now use the subscript h in the notation for the fully discrete solution.

As in Chapter 13 one shows at once the following smooth data results for these two methods.

Theorem 14.6 *Assume that* (14.2) *and* (1.10) *hold and let* U_h^n *and* u *be the solutions of* (14.28) *or* (14.29), *and* (14.1). *Assume that* $U_h^0=v_h$ *is appropriately chosen and that* u *is sufficiently smooth. Then there is a* $C=C(u,\bar{t})$ *such that (in case of* (14.28) *for* k *small)*

$$\|U_h^n-u(t_n)\|\leq C(h^r+k),\quad \textit{for } t_n\in J.$$

We now turn to the case of nonsmooth initial data and concentrate on the linearized method (14.29). We begin by considering the problem in the Hilbert space framework and consider thus the semilinear problem

$$u'+Au=f(u),\quad \text{for } t\in J,\quad \text{with } u(0)=v, \tag{14.30}$$

where A is a positive definite selfadjoint operator with a compact inverse in the Hilbert space $\mathcal{H}$, and $f:\mathcal{H}\to\mathcal{H}$ is continuous and such that

$$\|f'(u)\|\leq B,\quad \text{for } u\in\mathcal{H}, \tag{14.31}$$

where f' denotes the Fréchet derivative of f. The analogue of (14.29) is the linearized backward Euler scheme

$$U^n=E_kU^{n-1}+kE_kf(U^{n-1}),\quad \text{for } t_n\in J,\quad \text{with } U^0=v, \tag{14.32}$$

where $E_k=(I+kA)^{-1}$. We shall show the following:

Theorem 14.7 *Assume that* (14.31) *holds, and let* U^n *and* u *be the solutions of* (14.32) *and* (14.30), *respectively. Then there is a constant* $C=C(\kappa,\bar{t})$ *such that*

$$\|U^n-u(t_n)\|\leq Ck\Big(\frac{1}{t_n}+\log\frac{t_n}{k}\Big),\quad \textit{for } t_n\in J,\quad \textit{if } \|v\|\leq\kappa.$$

Proof. We find at once

$$U^n=E_k^nv+k\sum_{j=0}^{n-1}E_k^{n-j}f(U^j).$$

Similarly, with $J_j=(t_j,t_{j+1})$ and $u^n=u(t_n)$,

$$u^n = E(t_n)v + \sum_{j=0}^{n-1} \int_{J_j} E(t_n - s)f(u(s))\,ds.$$

Hence, for the error $e^n = U^n - u^n$,

$$e^n = (E_k^n - E(t_n))v + \sum_{j=0}^{n-1} \int_{J_j} d_j^n(s)\,ds \tag{14.33}$$

where

$$d_j^n(s) = E_k^{n-j} f(U^j) - E(t_n - s)f(u(s)).$$

We shall estimate the terms in (14.33). We first have, using the known nonsmooth data error estimate in the case of a linear homogeneous equation,

$$\|(E_k^n - E(t_n))v\| \le C\frac{k}{t_n}\|v\| \le C\kappa\frac{k}{t_n}.$$

We proceed with the terms in the sum in (14.33). We write

$$\begin{aligned} d_j^n(s) &= E_k^{n-j}(f(U^j) - f(u^j)) + (E_k^{n-j} - E(t_{n-j}))f(u^j) \\ &\quad + E(t_{n-j})(f(u^j) - f(u(s))) + (E(t_{n-j}) - E(t_n - s))f(u(s)) \\ &= \sum_{l=1}^{4} d_{jl}^n(s). \end{aligned}$$

For the first term we have by the stability of E_k and by (14.2)

$$\|d_{j1}^n\| \le C\|U^j - u^j\| = C\|e^j\|, \quad \text{for } j \le n-1$$

and, for the second term, again by the standard linear nonsmooth data estimate, since $\|f(u(s))\|$ is bounded,

$$\|d_{j2}^n\| \le C\frac{k}{t_{n-j}}, \quad \text{for } j \le n-1.$$

For the third term, we find, using the analogue of Theorem 14.4 with $\alpha = 0,\ \beta = 2$,

$$\begin{aligned} \|d_{j3}^n(s)\| &\le \|f(u^j) - f(u(s))\| \le C\|u^j - u(s)\| \\ &\le Ck \sup_{s\in J_j} \|u'(s)\| \le C\frac{k}{t_j}, \quad \text{for } s \in J_j,\ 1 \le j \le n-1, \end{aligned}$$

and since d_{03}^n is bounded we conclude

$$\|d_{j3}^n(s)\| \le C\frac{k}{t_{j+1}}, \quad \text{for } s \in J_j,\ 0 \le j \le n-1.$$

Finally, applying now also a spectral argument, since $E(t) = e^{-At}$,

$$\|d_{j4}^n(s)\| = \|AE(t_n - s)A^{-1}(E(s-t_j) - I)f(u(s))\|$$
$$\le C\frac{1}{t_n - s}\sup_{\lambda>0}\Big|\frac{e^{-\lambda(s-t_j)}-1}{\lambda}\Big| \le C\frac{k}{t_{n-j-1}}, \quad \text{for } s \in J_j,\ 0 \le j < n-1.$$

Since $\|d_{n-1,4}^n\|$ is bounded, we have

$$\|d_{j4}^n(s)\| \le C\frac{k}{t_{n-j}}, \quad \text{for } s \in J_j,\ 0 \le j \le n-1.$$

Altogether we obtain

$$\int_{J_j} \|d_j^n(s)\|\,ds \le Ck\|e^j\| + Ck\big(\frac{k}{t_{j+1}} + \frac{k}{t_{n-j}}\big), \quad \text{for } 0 \le j \le n-1,$$

and hence after summation, from (14.33),

$$\|e^n\| \le Ck\big(\frac{\kappa}{t_n} + \log n\big) + Ck\sum_{j=0}^{n-1}\|e^j\|. \tag{14.34}$$

Setting $\sigma^n = k\sum_{j=0}^n \|e^j\|$, we thus have

$$(\sigma^n - \sigma^{n-1})/k \le Ck\big(\frac{\kappa}{t_n} + \log n\big) + C\sigma^{n-1}, \quad \text{for } n \ge 1,$$

and hence

$$\sigma^n \le (1 + Ck)\sigma^{n-1} + Ck^2\big(\frac{\kappa}{t_n} + \log n\big),$$

and, since the time interval is bounded,

$$\sigma^n \le Ck\sum_{j=1}^n (1+Ck)^{n-1-j}\big(\frac{\kappa}{j} + k\log j\big) \le C(\kappa+1)k\log n.$$

By (14.34) this completes the proof. □

The above result may be applied to derive a nonsmooth data error estimate for (14.29). As an illustration we consider the solution $u_h : \bar{J} \to S_h$ of (14.3) in the case that S_h is a standard piecewise linear finite element space, so that Theorem 14.3 holds. For this problem we have the following.

Theorem 14.8 *Assume that* (14.2) *and* (1.10) *(with* $r = 2$*) hold, and let* U_h^n *and* u *be the solutions of* (14.29) *and* (14.1), *with* $U_h^0 = P_h v$ *Then there is a constant* $C = C(\kappa, \bar{t})$ *such that, for* $t_n \in J$,

$$\|U_h^n - u(t_n)\| \le Ck\big(\frac{1}{t_n} + \log\frac{t_n}{k}\big) + Ch^2\big(\frac{1}{t_n} + \max(0, \log\frac{t_n}{h^2})\big), \quad \textit{if } \|v\| \le \kappa.$$

Proof. We write $U_h^n - u(t_n) = (U^n - u_h(t_n)) + (u_h(t_n) - u(t_n))$. The second term is bounded by Theorem 14.3, and the first by Theorem 14.7, now applied with $\mathcal{H} = S_h$, and $A = -\Delta_h$, where Δ_h is the discrete Laplacian: With $E_{kh} = (I - k\Delta_h)^{-1}$ and P_h the L_2-projection onto S_h, (14.29) may be written

$$U_h^n = E_{kh}U^{n-1} + kE_{kh}P_h f(U_h^{n-1}), \quad \text{for } t_n \in J, \quad \text{with } U^0 = P_h v.$$

Since the assumptions of Theorem 14.7 are satisfied for the data $P_h f(u)$ and $P_h v$, uniformly in h, this theorem applies, and shows

$$\|U_h^n - u_h(t_n)\| \le Ck\Big(\frac{1}{t_n} + \log\frac{t_n}{k}\Big), \quad \text{for } \|P_h v\| \le \|v\| \le \kappa.$$

This completes the proof. □

In Chapter 13 we also considered the Crank-Nicolson method, and it is clear that the results shown there are valid in the particular case of a semilinear equation, and also generalize to the more general finite element spaces considered here to give $O(h^r + k^2)$ error bounds for smooth solutions.

Another class of methods for the semilinear equation are the Runge-Kutta methods which were introduced in Chapter 9 in the case of a linear inhomogeneous equation. Applied to the semilinear equation (14.30) such a method takes the form

$$\begin{aligned} U^{n+1} &= U^n + k\sum_{j=1}^m b_j(-AU_{nj} + f(U_{nj})), \\ U_{ni} &= U^n + k\sum_{j=1}^m a_{ij}(-AU_{nj} + f(U_{nj})), \quad i = 1, \dots, m, \end{aligned}$$

where the coefficients are associated with the quadrature formulas in (8.9). It may also be written

$$\begin{aligned} U^{n+1} &= r(kA)U^n + k\sum_{j=1}^m p_j(kA)f(U_{nj}), \\ U_{ni} &= s_i(kA)U^n + k\sum_{j=1}^m s_{ij}(kA)f(U_{nj}), \quad i = 1, \dots, m, \end{aligned}$$

where the $r(\lambda)$ and $p_j(\lambda)$ are as in (8.10) and, with $\bar{e} = (1, \dots, 1)$,

$$(s_1(\lambda), \dots, s_m(\lambda))^T = \sigma(\lambda)\bar{e} \quad \text{and } (s_{ij}(\lambda)) = \sigma(\lambda)\mathcal{A}, \quad \sigma(\lambda) = (I + \lambda\mathcal{A})^{-1}.$$

It is not difficult to show that if the method is stable, so that (7.10) holds, and such that the quadrature formulas in (8.9) are exact of orders $q-1$ and $q-2$, respectively, then the error is of order $O(k^q)$ if the exact solution is sufficiently smooth.

In the same way as for the spatial discretization discussed in connection with Theorem 14.5, it may be shown that it is not possible to generalize the nonsmooth data error estimate

$$\|U^n - u(t_n)\| \leq Ck^p t_n^{-p} \|v\|,$$

for the abstract linear homogeneous equation to the semilinear case when $p > 1$. In fact, in [58] using the ideas in our above counter-example in the spatially semidiscrete case, a simple semilinear system of the form (14.30), and with uniformly bounded initial values, was exhibited such that for any Runge-Kutta method corresponding to a rational function of type III, i.e., such that $|r(\lambda)| < 1$ for $\lambda > 0$, and $|r(\infty)| < 1$, one has for any $t \in J$

$$\limsup_{n=t/k\to\infty} \|U^n - u(t_n)\| \geq ck, \quad \text{with } c = c(t) > 0.$$

However, similarly to the situation in Theorem 14.5 one may show $O(k^p)$ convergence for $t_n > 0$ for such methods under regularity assumptions which are reduced compared to the smooth data case by essentially two orders. For instance, full second order convergence is achieved for positive time for a second order method for $v \in \dot{H}^2$, see [58].

The discussion in the beginning of the chapter concerning nonlinear forcing terms which are not globally Lipschitz is from Thomée and Wahlbin [231]. The results for nonsmooth data were derived in Helfrich [118], Johnson, Larsson, Thomée and Wahlbin [130] and Crouzeix, Thomée and Wahlbin [62] in the spatially semidiscrete case and in Crouzeix and Thomée [58] for fully discrete methods.

For similar analyses on other types of semilinear problems we mention, e.g., Elliott and Larsson [85], [86] and Akrivis, Crouzeix and Makridakis [3]. The long-time behavior of finite element solutions was studied in, e.g., Khalsa [139], Larsson [143], [144], Larsson and Sanz-Serna [145], [146], Elliott and Stuart [87]. Application of the discontinuous Galerkin method to semilinear equations was studied in Eriksson and Johnson [90], [91] and Estep and Larsson [95]. For work related to blow-up of solutions, see Nakagawa and Ushijima [174].

For a recent development concerning so called nonlinear Galerkin methods we refer to Marion and Temam [168], Temam [223] and Marion and Xu [169].

The continuous semilinear problem has been discussed in, e.g., Amann [4] and Henry [119].

15. The Method of Lumped Masses

In this chapter we shall consider a modification of the standard Galerkin method using piecewise linear trial functions, the so-called method of lumped masses. In this method the mass matrix is replaced by a diagonal matrix with the row sums of the original mass matrix as its diagonal elements. This can also be interpreted as using a quadrature rule for the corresponding L_2 inner product.

We consider the simple initial-boundary value problem

$$\begin{aligned} u_t - \Delta u &= f \quad \text{in } \Omega, \quad t > 0, \\ u &= 0 \quad \text{on } \partial\Omega,\ t > 0, \quad \text{with } u(\cdot, 0) = v \quad \text{in } \Omega, \end{aligned}$$

where again for simplicity Ω is a smooth convex domain in the plane.

Let $S_h \subset H_0^1 = H_0^1(\Omega)$ consist of continuous, piecewise linear functions on a quasiuniform family of triangulations $\mathcal{T}_h = \{\tau\}$ of Ω with its boundary vertices on $\partial\Omega$ and which vanish outside the polygonal domain Ω_h determined by $\mathcal{T}_h$. Let $\{P_j\}_{j=1}^{N_h}$ denote the interior vertices of $\mathcal{T}_h$ and let $\{\Phi_j\}_{j=1}^{N_h}$ be the standard basis for S_h consisting of the pyramid functions defined by $\Phi_j(P_k) = \delta_{jk}$.

Recall that the basic semidiscrete Galerkin method is to find $u_h : [0, \infty) \to S_h$ such that

$$(u_{h,t}, \chi) + (\nabla u_h, \nabla \chi) = (f, \chi), \quad \forall \chi \in S_h, \ t > 0, \quad u_h(0) = v_h, \tag{15.1}$$

where v_h is some approximation of v in S_h. Recall also that this method may be written in matrix form as

$$\mathcal{B}\alpha'(t) + \mathcal{A}\alpha(t) = \widetilde{f}(t), \quad \text{for } t > 0, \quad \text{with } \alpha(0) = \gamma, \tag{15.2}$$

where $\mathcal{B} = (b_{jk})$ and $\mathcal{A} = (a_{jk})$ are the mass and stiffness matrices with elements $b_{jk} = (\Phi_j, \Phi_k)$ and $a_{jk} = (\nabla\Phi_j, \nabla\Phi_k)$, respectively, where $\alpha_j(t)$ and γ_j are the coefficients of $u_h(t)$ and v_h with respect to $\{\Phi_j\}_{j=1}^{N_h}$ and where $\widetilde{f}$ is the vector with components (f, Φ_k).

The lumped mass method consists in replacing the mass matrix $\mathcal{B}$ in (15.2) by the diagonal matrix $\bar{\mathcal{B}}$ obtained by taking for its diagonal elements the numbers $\bar{b}_{jj} = \sum_{k=1}^{N_h} b_{jk}$, i.e., by lumping all masses in one row into the diagonal entry. This makes the inversion of the matrix coefficient of $\alpha'(t)$ a triviality.

We shall thus study the matrix problem

$$\tilde{\mathcal{B}}\alpha'(t) + \mathcal{A}\alpha(t) = \widetilde{f}(t), \quad \text{for } t > 0, \quad \text{with } \alpha(0) = \gamma. \tag{15.3}$$

We shall now describe two alternative interpretations of this procedure, and then use the first of these to show some error estimates for it.

Our first interpretation will be to think of (15.3) as being obtained by evaluating the first term in (15.1) by numerical quadrature. Let τ be a triangle of the triangulation $\mathcal{T}_h$, let $P_{\tau,j}$, $j = 1, 2, 3$, be its vertices, and consider the quadrature formula

$$Q_{\tau,h}(f) = \tfrac{1}{3} \text{ area } (\tau) \sum_{j=1}^{3} f(P_{\tau,j}) \approx \int_\tau f\,dx. \tag{15.4}$$

We may then define an approximation of the inner product in S_h by

$$(\psi, \chi)_h = \sum_{\tau \in \mathcal{T}_h} Q_{\tau,h}(\psi\chi). \tag{15.5}$$

We claim now that the lumped mass method defined by (15.3) above is equivalent to

$$(u_{h,t}, \chi)_h + (\nabla u_h, \nabla\chi) = (f, \chi), \quad \forall \chi \in S_h, \ t > 0, \quad u_h(0) = v_h. \tag{15.6}$$

In fact, setting $u_h(t) = \sum_{j=1}^{N_h} \alpha_j(t)\Phi_j$ this system may be written

$$\sum_{j=1}^{N_h} \alpha_j'(t)(\Phi_j, \Phi_k)_h + \sum_{j=1}^{N_h} \alpha_j(t)(\nabla\Phi_j, \nabla\Phi_k) = (f, \Phi_k), \quad k = 1, \dots, N_h,$$

and to show the equivalence it remains only to observe that $(\Phi_j, \Phi_k)_h = 0$ for $j \neq k$, as $\Phi_j(x)\Phi_k(x)$ vanishes at all vertices of $\mathcal{T}_h$, and to show that

$$\|\Phi_j\|_h^2 = (\Phi_j, \Phi_j)_h = \sum_{k=1}^{N_h} (\Phi_j, \Phi_k). \tag{15.7}$$

To prove this latter fact, note that (Φ_j, Φ_k) is only non-zero for $j \neq k$ if P_j and P_k are neighbors, and observe that in such a case, if τ is a triangle with P_j and P_k as vertices, simple calculations, for instance after transformation to a reference triangle, show that

$$\int_\tau \Phi_j\Phi_k\,dx = \tfrac{1}{12} \text{ area } (\tau) \quad \text{and} \quad \int_\tau \Phi_j^2\,dx = \tfrac{1}{6} \text{ area } (\tau).$$

It follows, since for each pair of neighbors P_j, P_k there are two such triangles τ, that with D_j the union of the triangles which have P_j as a vertex,

$$\sum_{k=1}^{N_h}(\Phi_j, \Phi_k) = \tfrac{1}{3} \text{ area } (D_j).$$

Since clearly

$$\|\Phi_j\|_h^2 = \sum_\tau Q_{\tau,h}(\Phi_j^2) = \tfrac{1}{3} \text{ area } (D_j),$$

this completes the proof of (15.7).

We now turn to the other formulation of the method under consideration. Let again τ be a triangle of the triangulation and P_j one of its vertices. Now draw the straight lines connecting each vertex of τ to the midpoint of the opposite side of τ. These straight lines intersect at the barycenter of τ and divide τ into six triangles of equal area. Let $B_{j,\tau}$ be the union of the two of these that have P_j as a vertex. Clearly, then, the area of $B_{j,\tau}$ is a third of that of τ. For each interior vertex P_j, let B_j be the union of the $B_{j,\tau}$ for which τ has P_j as a vertex.

Now let $\bar{S}_h$ denote the functions which are constant on each B_j and vanish outside the union of the B_j. We note that the elements $\bar{\chi}$ of $\bar{S}_h$ are uniquely defined by the values at the vertices P_j and we may write

$$\bar{\chi}(x) = \sum_{j=1}^{N_h} \bar{\chi}(P_j)\bar{\Phi}_j(x),$$

where $\bar{\Phi}_j = 1$ on B_j and vanishes elsewhere. Since the functions of S_h are also uniquely determined by their values at the P_j there is a one-to-one correspondence between the functions of S_h and those of $\bar{S}_h$, and for χ in S_h we denote by $\bar{\chi}$ the associated function in $\bar{S}_h$ which agrees with χ at the P_j.

With this notation the semidiscrete equation (15.3) or (15.6) may also be formulated as

$$(\bar{u}_{h,t}, \bar{\chi}) + (\nabla u_h, \nabla \chi) = (f, \chi), \ \forall \chi \in S_h, \ t > 0, \quad u_h(0) = v_h.$$

In fact, this follows similarly to above if we observe that trivially $(\bar{\Phi}_j, \bar{\Phi}_k) = 0$ for $j \neq k$ and that $\|\bar{\Phi}_j\|^2 = \text{ area } (B_j) = \text{ area } (D_j)/3 = \|\Phi_j\|_h^2$.

One may think of this latter formulation as being obtained by reducing the H^1 regularity requirements for the functions in S_h in the first term of (15.1), where they are not needed for the products to make sense. This latter approach was taken in [203] and [49].

We now turn to the error analysis and return to the formulation (15.6). We begin with the following lemma.

Lemma 15.1 *Let $\varepsilon_h(v, w) = (v, w)_h - (v, w)$ denote the quadrature error in (15.5). We then have*

$$|\varepsilon_h(\psi, \chi)| \leq Ch^2 \|\nabla\psi\| \, \|\nabla\chi\|, \quad \textit{for } \psi, \chi \in S_h.$$

Proof. Since the quadrature formula (15.4) is exact for f linear we have, by transformation to a fixed reference triangle τ_0 and using the Bramble-Hilbert lemma and the Sobolev inequality $\|f\|_{L_\infty(\tau_0)} \le C\|f\|_{W_1^2(\tau_0)}$, that

$$|Q_{\tau,h}(f) - \int_\tau f\,dx| \le Ch^2 \sum_{|\alpha|=2} \|D^\alpha f\|_{L_1(\tau)}.$$

After application to $f = \psi\chi$ this implies, since both ψ and χ are linear in τ, that

$$|Q_{\tau,h}(\psi\chi) - \int_\tau \psi\chi\,dx| \le Ch^2 \|\nabla\psi\|_{L_2(\tau)} \|\nabla\chi\|_{L_2(\tau)}.$$

Using the Cauchy-Schwarz inequality we conclude that

$$|\varepsilon_h(\psi,\chi)| \le Ch^2 \sum_{\tau\in\mathcal{T}_h} \|\nabla\psi\|_{L_2(\tau)} \|\nabla\chi\|_{L_2(\tau)} \le Ch^2 \|\nabla\psi\|\,\|\nabla\chi\|,$$

which is the desired estimate. □

We shall now show the following error estimate:

Theorem 15.1 *We have for the error in the semidiscrete lumped mass method* (15.6), *for* $t \ge 0$,

$$\|u_h(t) - u(t)\| \le C\|v_h - v\| + Ch^2\Big(\|v\|_2 + \|u(t)\|_2 + \big(\int_0^t \|u_t\|_2^2\,ds\big)^{1/2}\Big).$$

Proof. We write, with R_h the standard Ritz projection, $u_h - u = (u_h - R_h u) + (R_h u - u) = \theta + \rho$, and $\rho(t)$ is bounded in the desired way. In order to estimate θ, we write

$$\begin{aligned}(15.8)\qquad &(\theta_t,\chi)_h + (\nabla\theta,\nabla\chi)\\ &= (u_{h,t},\chi)_h + (\nabla u_h,\nabla\chi) - (R_h u_t,\chi)_h - (\nabla R_h u,\nabla\chi)\\ &= (f,\chi) - (R_h u_t,\chi)_h - (\nabla u,\nabla\chi) = (u_t,\chi) - (R_h u_t,\chi)_h\\ &= -(\rho_t,\chi) - \varepsilon_h(R_h u_t,\chi).\end{aligned}$$

Setting $\chi = \theta$ we obtain

$$(15.9)\qquad \tfrac12 \frac{d}{dt}\|\theta\|_h^2 + \|\nabla\theta\|^2 = -(\rho_t,\theta) - \varepsilon_h(R_h u_t,\theta).$$

Here we have at once

$$|(\rho_t,\theta)| \le \|u_t - R_h u_t\|\,\|\theta\| \le Ch^2\|u_t\|_2\|\theta\| \le Ch^2\|u_t\|_2\|\nabla\theta\|,$$

and, using Lemma 15.1,

$$|\varepsilon_h(R_h u_t,\theta)| \le Ch^2\|\nabla R_h u_t\|\,\|\nabla\theta\| \le Ch^2\|u_t\|_2\|\nabla\theta\|.$$

It follows thus that

$$\tfrac{1}{2}\frac{d}{dt}\|\theta\|_h^2 + \|\nabla\theta\|^2 \le Ch^2\|u_t\|_2\|\nabla\theta\| \le \|\nabla\theta\|^2 + Ch^4\|u_t\|_2^2,$$

from which we infer

$$\|\theta(t)\|_h^2 \le \|\theta(0)\|_h^2 + Ch^4\int_0^t \|u_t\|_2^2\,ds.$$

We now note that $\|\cdot\|_h$ and $\|\cdot\|$ are equivalent norms on S_h, uniformly in h (this follows easily by considering each triangle separately), and that hence

$$\|\theta(t)\| \le C\|\theta(0)\| + Ch^2\Big(\int_0^t \|u_t\|_2^2\,ds\Big)^{1/2}.$$

Here $\|\theta(0)\| = \|v_h - R_h v\| \le \|v_h - v\| + Ch^2\|v\|_2$, whence $\theta(t)$ is bounded as desired. The proof is complete. □

We now turn to an estimate for the gradient.

Theorem 15.2 *We have for the error in the semidiscrete method* (15.6), *for* $t \ge 0$,

$$\|\nabla(u_h - u)(t)\| \le \|\nabla(v_h - v)\| + Ch\Big(\|v\|_2 + \|u(t)\|_2 + \Big(\int_0^t \|\nabla u_t\|^2\,ds\Big)^{1/2}\Big).$$

Proof. We now set $\chi = \theta_t$ in the equation (15.8) for θ to obtain

$$\|\theta_t\|_h^2 + \tfrac{1}{2}\frac{d}{dt}\|\nabla\theta\|^2 = -(\rho_t, \theta_t) - \varepsilon_h(R_h u_t, \theta_t). \tag{15.10}$$

Here, as in the proof of Theorem 15.1,

$$|(\rho_t, \theta_t)| \le \|u_t - R_h u_t\|\,\|\theta_t\| \le Ch\|\nabla u_t\|\,\|\theta_t\|.$$

Further, by Lemma 15.1,

$$|\varepsilon_h(R_h u_t, \theta_t)| \le Ch^2\|\nabla R_h u_t\|\,\|\nabla\theta_t\| \le Ch\|\nabla u_t\|\,\|\theta_t\|,$$

where in the last step we have applied the inverse estimate (1.12). (The use of the inverse estimate may be avoided by a slight modification of Lemma 15.1.) Using again the equivalence between the norms $\|\cdot\|_h$ and $\|\cdot\|$ on S_h we conclude

$$\|\theta_t\|_h^2 + \tfrac{1}{2}\frac{d}{dt}\|\nabla\theta\|^2 \le Ch\|\nabla u_t\|\,\|\theta_t\|_h \le \|\theta_t\|_h^2 + Ch^2\|\nabla u_t\|^2,$$

so that, after integration,

$$\begin{aligned}\|\nabla\theta(t)\| &\le \|\nabla\theta(0)\| + Ch\Big(\int_0^t \|\nabla u_t\|^2\,ds\Big)^{1/2}\\ &\le \|\nabla(v_h - v)\| + Ch\Big(\|v\|_2 + \Big(\int_0^t \|\nabla u_t\|^2\,ds\Big)^{1/2}\Big).\end{aligned}$$

Together with the standard estimate for $\nabla\rho(t)$ this completes the proof. □

This demonstration does not immediately yield the superconvergent order $O(h^2)$ estimate for $\nabla\theta$ which holds for the standard Galerkin method. However, as is shown in the following lemma, a slight modification of the proof shows such a result.

Lemma 15.2 *For each $\bar{t} > 0$ there is a constant $C = C(\bar{t})$ such that for $\theta = u_h - R_h u$ and $0 \le t \le \bar{t}$,*

$$\|\nabla\theta(t)\| \le \|\nabla\theta(0)\| + Ch^2\Big(\|u_t(t)\|_2 + \big(\int_0^t (\|u_t\|_2^2 + \|u_{tt}\|_1^2)\,ds\big)^{1/2}\Big).$$

Proof. It suffices to consider the case $v_h = R_h v$, or $\theta(0) = 0$. For the solution $\widetilde{u}_h$ of the homogeneous equation with initial data $\widetilde{u}_h(0) = v_h - R_h v = \theta(0)$ satisfies

$$\|\widetilde{u}_{h,t}\|_h^2 + \tfrac{1}{2}\frac{d}{dt}\|\nabla\widetilde{u}_h\|^2 = 0,$$

and hence

$$\|\nabla\widetilde{u}_h(t)\|^2 \le \|\nabla\widetilde{u}_h(0)\|^2 = \|\nabla\theta(0)\|^2.$$

We have as before (15.10), which we now write in the form

$$\|\theta_t\|_h^2 + \tfrac{1}{2}\frac{d}{dt}\|\nabla\theta\|^2 = -(\rho_t,\theta_t) - \frac{d}{dt}\varepsilon_h(R_h u_t,\theta) + \varepsilon_h(R_h u_{tt},\theta). \tag{15.11}$$

Here

$$|(\rho_t,\theta_t)| \le \|\rho_t\|\,\|\theta_t\| \le Ch^2\|u_t\|_2\|\theta_t\|_h \le Ch^4\|u_t\|_2^2 + \|\theta_t\|_h^2.$$

Further, by Lemma 15.1,

$$|\varepsilon_h(R_h u_t,\theta)| \le Ch^2\|\nabla R_h u_t\|\,\|\nabla\theta\| \le Ch^4\|u_t\|_1^2 + \tfrac{1}{4}\|\nabla\theta\|^2,$$

and similarly with u_t replaced by u_{tt}. By integration of (15.11) we therefore obtain, since $\theta(0) = 0$,

$$\|\nabla\theta(t)\|^2 \le Ch^4\Big(\|u_t(t)\|_1^2 + \int_0^t (\|u_t\|_2^2 + \|u_{tt}\|_1^2)\,ds\Big) + \int_0^t \|\nabla\theta\|^2\,ds.$$

The result now follows by Gronwall's lemma. □

As one application of the lemma we shall prove the following maximum-norm error estimate:

Theorem 15.3 *Let v_h be chosen so that $\|\nabla(v_h - R_h v)\| \le Ch^2$. Then under the appropriate regularity assumptions we have for the error in* (15.6)

$$\|u_h(t) - u(t)\|_{L_\infty} \le C(\bar{t};u)h^2\ell_h, \quad \text{where } \ell_h = \max(1,\log(1/h)), \quad \text{for } t \le \bar{t}.$$

Proof. We recall that since the triangulation is quasiuniform we may apply the "almost" Sobolev inequality of Lemma 6.4 together with Lemma 15.2 to obtain

$$\|\theta(t)\|_{L_\infty} \le C(\bar{t};u)h^2\ell_h^{1/2}.$$

In view of the maximum-norm error estimate of Theorem 1.4 for the elliptic problem, this shows the result. □

We observe that because of the use of quadrature, our above error analyses of Theorem 15.1 and Lemma 15.2 require more regularity of the solution than was the case for the standard Galerkin method. For the homogeneous equation, for instance, Theorem 15.1 shows by standard calculations using the definition of the norm in $\dot{H}^s = \dot{H}^s(\Omega)$ (cf. Chapter 3) that, for $v_h = R_h v$, say,

$$\|u_h(t) - u(t)\| \le Ch^2|v|_3, \quad \text{for } v \in \dot{H}^3,$$

and Lemma 15.2 shows similarly

$$\|\nabla\theta(t)\| \le Ch^2|v|_4, \quad \text{for } v \in \dot{H}^4.$$

In addition to smoothness these estimates require $v = \Delta v = 0$ on $\partial\Omega$. We shall demonstrate now how at least the latter boundary condition may be removed for t positive, by using our previous techniques for nonsmooth data error estimates.

Lemma 15.3 *Consider the homogeneous equation* ($f = 0$) *and let* $\theta = u_h - R_h u$. *Then for each* $\bar{t} > 0$ *there is a constant* C *such that if* $\theta(0) = 0$ *then for* $0 < t \le \bar{t}$ *and* $v \in \dot{H}^2$,

$$\|\theta(t)\| \le Ch^2 t^{-1/2}|v|_2 \quad \textit{and} \quad \|\nabla\theta(t)\| \le Ch^2 t^{-1}|v|_2.$$

Proof. Multiplying (15.9) by t we have

$$\tfrac{1}{2}\frac{d}{dt}(t\|\theta\|_h^2) + t\|\nabla\theta\|^2 = -t(\rho_t,\theta) - t\varepsilon_h(R_h u_t,\theta) + \tfrac{1}{2}\|\theta\|_h^2.$$

Hence by integration and routine estimates

$$\begin{aligned} t\|\theta\|_h^2 + \int_0^t s\|\nabla\theta\|^2\,ds &\le Ch^4\int_0^t (s^2\|u_t\|_2^2 + s\|u_t\|_1^2)\,ds \\ &\quad + C\int_0^t \|\theta\|_h^2\,ds \le Ch^4\|v\|_1^2 + C\int_0^t \|\theta\|_h^2\,ds. \end{aligned} \tag{15.12}$$

In order to estimate the latter integral we set $\widetilde{\theta}(t) = \int_0^t \theta(s)\,ds$ and integrate the error equation (15.8) from 0 to t to obtain

$$(\theta,\chi)_h + (\nabla\widetilde{\theta},\nabla\chi) = (\rho(0) - \rho(t),\chi) - \varepsilon_h(R_h(u(t) - v),\chi), \quad \forall\chi \in S_h.$$

Setting $\chi = \theta = \widetilde{\theta}_t$ this yields

$$\|\theta\|_h^2 + \tfrac{1}{2}\frac{d}{dt}\|\nabla\widetilde{\theta}\|^2$$
$$= (\rho(0) - \rho(t), \theta) - \frac{d}{dt}\varepsilon_h(R_h(u(t)-v), \widetilde{\theta}) + \varepsilon_h(R_h u_t, \widetilde{\theta}),$$

and hence, by obvious estimates,

$$\int_0^t \|\theta\|_h^2\,ds + \|\nabla\widetilde{\theta}\|^2 \le Ch^4 \int_0^t (\|u(s)\|_2 + \|v\|_2)^2\,ds$$
$$+ Ch^4\|\nabla R_h(u(t)-v)\|^2 + Ch^4 \int_0^t \|\nabla R_h u_t\|^2\,ds + \int_0^t \|\nabla\widetilde{\theta}\|^2\,ds,$$

so that, using also Gronwall's lemma, for $t \le \bar{t}$,

$$\int_0^t \|\theta\|_h^2\,ds \le Ch^4(\|v\|_2^2 + \int_0^t \|u\|_3^2\,ds) \le Ch^4|v|_2^2. \tag{15.13}$$

Together with (15.12) this proves the first estimate of the lemma.

In order to bound $\nabla\theta$ we multiply (15.11) by t^2 and obtain easily

$$\tfrac{1}{2}\frac{d}{dt}(t^2\|\nabla\theta\|^2) \le -\frac{d}{dt}(t^2\varepsilon_h(R_h u_t, \theta)) + Ct^2\|\rho_t\|^2$$
$$+ t^2\varepsilon_h(R_h u_{tt}, \theta) + 2t\varepsilon_h(R_h u_t, \theta) + t\|\nabla\theta\|^2,$$

or

$$t^2\|\nabla\theta(t)\|^2 \le Ch^4(t^2\|\nabla u_t(t)\|^2 + \int_0^t (s^3\|u_{tt}\|_1^2 + s^2\|u_t\|_2^2 + s\|u_t\|_1^2)\,ds)$$
$$+ C\int_0^t s\|\nabla\theta\|^2\,ds \le Ch^4 t\|v\|_2^2 + C\int_0^t s\|\nabla\theta\|^2\,ds.$$

Hence we have using (15.12) and (15.13), since $t \le \bar{t}$,

$$t^2\|\nabla\theta(t)\|^2 \le Ch^4|v|_2^2,$$

which completes the proof. □

It is obvious how Lemma 15.3 may be combined with our different estimates for the error ρ in the elliptic projection to yield L_2 and L_∞ norm bounds for the error in the homogeneous semidiscrete equation with initial data $v \in \dot{H}^2$. We shall not insist on the details.

The method of lumped masses may, of course, be used in fully discrete methods. With $\bar{\partial}$ as usual denoting the backward difference quotient and $0 \le \kappa \le 1$ one could, for instance, consider the method defining $U^n = U_h^n \in S_h$ by,

$$\begin{aligned}(\bar{\partial}U^n, \chi)_h + \kappa(\nabla U^n, \nabla\chi) + (1-\kappa)(\nabla U^{n-1}, \nabla\chi)\\ = (f(t_{n-1} + \kappa k), \chi), \quad \forall \chi \in S_h,\ n \ge 1, \quad \text{with } U^0 = v_h,\end{aligned} \tag{15.14}$$

or in matrix form, with α^n the vector of the components of U^n with respect to the basis $\{\Phi_j\}_{j=1}^{N_h}$, and $F^{n-1+\kappa}$ that with components $(f(t_{n-1}+k\kappa),\Phi_j)$,

$$\bar{\mathcal{B}}\bar{\partial}\alpha^n+\kappa\mathcal{A}\alpha^n+(1-\kappa)\mathcal{A}\alpha^{n-1}=F^{n-1+\kappa},$$

or since $\bar{\mathcal{B}}+\kappa k\mathcal{A}$ is obviously positive definite,

$$\alpha^n=(\bar{\mathcal{B}}+\kappa k\mathcal{A})^{-1}(\bar{\mathcal{B}}-(1-\kappa)k\mathcal{A})\alpha^{n-1}+(\bar{\mathcal{B}}+\kappa k\mathcal{A})^{-1}kF^{n-1+\kappa}.$$

The backward Euler method corresponds to $\kappa=1$, the Crank-Nicolson method to $\kappa=\frac{1}{2}$, and for $\kappa=0$ we now have a method which is purely explicit since $\bar{\mathcal{B}}$ is diagonal.

As an example, let us briefly analyze the backward Euler method and show the following.

Theorem 15.4 *We have for the backward Euler Galerkin method* (15.14) *with $\kappa=1$, for $t_n\ge 0$,*

$$\begin{aligned}&\|U^n-u(t_n)\|\le C\|v_h-v\|\\&+Ch^2\Big(\|v\|_2+\|u(t_n)\|_2+\big(\int_0^{t_n}\|u_t\|_2^2\,ds\big)^{1/2}\Big)+Ck\big(\int_0^{t_n}\|u_{tt}\|^2\,ds\big)^{1/2}.\end{aligned}$$

Proof. Writing as usual $U^n-u^n=\theta^n+\rho^n$, we need only bound θ^n. We have

$$\begin{aligned}&(\bar{\partial}\theta^n,\chi)_h+(\nabla\theta^n,\nabla\chi)\\&=(\bar{\partial}U^n,\chi)_h+(\nabla U^n,\nabla\chi)-(\bar{\partial}R_hu^n,\chi)_h-(\nabla R_nu^n,\nabla\chi)\\&=(f^n,\chi)-(\bar{\partial}R_hu^n,\chi)_h-(\nabla u^n,\nabla\chi)=(u_t^n,\chi)-(\bar{\partial}R_hu^n,\chi)_h.\end{aligned}$$

Choosing $\chi=\theta^n$ we find after some manipulation

$$\begin{aligned}&\frac{1}{2k}(\|\theta^n\|_h^2-\|\theta^{n-1}\|_h^2)+\frac{1}{2k}\|\theta^n-\theta^{n-1}\|_h^2+\|\nabla\theta^n\|^2\\&=(u_t^n-\bar{\partial}u^n,\theta^n)+(\bar{\partial}u^n-\bar{\partial}R_hu^n,\theta^n)-\varepsilon_h(\bar{\partial}R_hu^n,\theta^n)=R_1+R_2+R_3.\end{aligned}$$

We have for the contribution of the discretization in time

$$\begin{aligned}|R_1|&\le\|u_t^n-\bar{\partial}u^n\|\,\|\theta^n\|\le C\int_{t_{n-1}}^{t_n}\|u_{tt}\|\,ds\,\|\nabla\theta^n\|\\&\le Ck^{1/2}\Big(\int_{t_{n-1}}^{t_n}\|u_{tt}\|^2\,ds\Big)^{1/2}\|\nabla\theta^n\|.\end{aligned}$$

Further

$$\begin{aligned}&|R_2|\le\|(I-R_h)\bar{\partial}u^n\|\,\|\theta^n\|\le Ch^2\|\bar{\partial}u^n\|_2\|\theta^n\|\\&\le Ch^2k^{-1}\int_{t_{n-1}}^{t_n}\|u_t\|_2\,ds\|\nabla\theta^n\|\le Ch^2k^{-1/2}\Big(\int_{t_{n-1}}^{t_n}\|u_t\|_2^2\,ds\Big)^{1/2}\|\nabla\theta^n\|,\end{aligned}$$

and finally, using again Lemma 15.1,

$$|R_3| \le Ch^2 \|\nabla R_h \bar\partial u^n\| \, \|\nabla \theta^n\| \le Ch^2 \|\bar\partial \nabla u^n\| \, \|\nabla \theta^n\|$$
$$\le Ch^2 k^{-1} \int_{t_{n-1}}^{t_n} \|\nabla u_t\| \, ds \, \|\nabla \theta^n\| \le Ch^2 k^{-1/2} \Big(\int_{t_{n-1}}^{t_n} \|u_t\|_2^2 \, ds \Big)^{1/2} \|\nabla \theta^n\|.$$

Altogether we conclude, after a kickback of $k\|\nabla\theta^n\|^2$,

$$\|\theta^n\|_h^2 \le \|\theta^{n-1}\|_h^2 + Ch^4 \int_{t_{n-1}}^{t_n} \|u_t\|_2^2 \, ds + Ck^2 \int_{t_{n-1}}^{t_n} \|u_{tt}\|^2 \, ds,$$

and hence

$$\|\theta^n\|_h^2 \le \|\theta^0\|_h^2 + Ch^4 \int_0^{t_n} \|u_t\|_2^2 \, ds + Ck^2 \int_0^{t_n} \|u_{tt}\|^2 \, ds.$$

By the equivalence of $\|\cdot\|_h$ and $\|\cdot\|$ on S_h and the standard estimate for $\|\theta^0\|$ this concludes the proof of the theorem. □

We shall now show that if the triangulations used are of Delaunay type, then there is a maximum-principle associated with the lumped mass method. Recall from the beginning of Chapter 6 that this is not the case for the standard Galerkin method. A triangulation is of Delaunay type if for all edges e of $\mathcal{T}_h$, with α_1 and α_2 the two angles opposite to e in the two triangles τ_1 and τ_2 determined by e, respectively, we have $\alpha_1 + \alpha_2 \le \pi$. This condition is satisfied, in particular, if all angles α of $\mathcal{T}_h$ are acute. Note that this does not require $\mathcal{T}_h$ to be quasiuniform.

We consider the homogeneous equation

$$(u_{h,t}, \chi)_h + (\nabla u_h, \nabla \chi) = 0, \quad \forall \chi \in S_h, \ t > 0, \quad \text{with } u_h(0) = v_h,$$

and we denote by $\bar E_h(t) : S_h \to S_h$ the solution operator of this problem. This problem may also be written

$$u_{h,t} - \Delta_h u_h = 0, \quad \text{for } t \ge 0, \quad \text{with } u_h(0) = v_h,$$

where $\Delta_h : S_h \to S_h$ is now defined by

$$-(\Delta_h \psi, \chi)_h = (\nabla \psi, \nabla \chi), \quad \psi, \chi \in S_h. \tag{15.15}$$

With this notation $\bar E_h(t)$ is the semigroup on S_h generated by Δ_h.

Note that we may write the complex form of the discrete inner product $(\cdot,\cdot)_h$ defined in (15.5) as

$$(\psi, \chi)_h = \sum_{j=1}^{N_h} \omega_j \psi_j \bar\chi_j, \quad \text{where } \omega_j = \frac{1}{3} \sum_{P_j \in \bar\tau} \text{area}(\tau), \quad \psi_j = \psi(P_j). \tag{15.16}$$

From this we may easily see that

$$-(\Delta_h\psi)_j = \omega_j^{-1}\sum_{i=1}^{N_h}\alpha_{ij}\psi_i, \quad \text{where } \alpha_{ij} = (\nabla\Phi_i, \nabla\Phi_j), \quad \text{for } j = 1, \dots, N_h.$$

We shall begin with a characterization of Delaunay triangulations.

Lemma 15.4 *The triangulation $\mathcal{T}_h$ is of Delaunay type if and only if*

$$(\nabla\Phi_i, \nabla\Phi_j) \le 0, \quad \textit{for all } P_i \ne P_j.$$

Proof. Let $e = P_iP_j$ be an edge of $\mathcal{T}_h$, let τ be one of the two triangles determined by e, and let α be the angle in τ opposite e. Then $\nabla\Phi_i\big|_\tau$ is in the direction of the normal to the side of τ opposite P_i and $|\nabla\Phi_i|\big|_\tau = 1/\delta_{i,\tau}$, where $\delta_{i,\tau}$ is the distance from P_i to the opposite side of τ. One sees at once that the angle between the two normals is $\pi - \alpha$, and hence

$$(\nabla\Phi_i, \nabla\Phi_j)_\tau = -\cos\alpha\,|\nabla\Phi_i|\big|_\tau\,|\nabla\Phi_j|\big|_\tau\,\text{area}(\tau) = -\cos\alpha\,\delta_{i,\tau}^{-1}\,\delta_{j,\tau}^{-1}\,\text{area}(\tau).$$

But we also have

$$\text{area}(\tau) = \ell_{i,\tau}\ell_{j,\tau}\,\sin\alpha = \ell_{i,\tau}\delta_{i,\tau}/2,$$

where $\ell_{i,\tau}$ is the length of the side opposite to P_i. Hence altogether

$$(\nabla\Phi_i, \nabla\Phi_j)_\tau = -\tfrac{1}{4}\cot\alpha.$$

We finally have

$$(\nabla\Phi_i, \nabla\Phi_j) = \sum_{j=1}^{2}(\nabla\Phi_i, \nabla\Phi_j)_{\tau_j} = -\tfrac{1}{4}(\cot\alpha_1 + \cot\alpha_2) = -\frac{\sin(\alpha_1+\alpha_2)}{4\sin\alpha_1\sin\alpha_2},$$

from which the conclusion of the lemma immediately follows. □

We shall show the following discrete maximum-principle.

Theorem 15.5 *Assume that the triangulations $\mathcal{T}_h$ are of Delaunay type. Then*

$$\min(0, \min_{x\in\Omega} v_h(x)) \le (\bar{E}_h(t)v_h)(x) \le \max(0, \max_{x\in\Omega} v_h(x)), \quad \forall v_h \in S_h.$$

In particular, $\bar{E}_h(t)$ is stable with respect to the maximum-norm, and

$$\|\bar{E}_h(t)v_h\|_{L_\infty} \le \|v_h\|_{L_\infty}, \quad \textit{for } t \ge 0.$$

Proof. We write as above the system in matrix form

$$\bar{\mathcal{B}}\alpha'(t) + \mathcal{A}\alpha(t) = 0, \quad \text{for } t > 0, \quad \text{with } \alpha(0) = \gamma,$$

where $\alpha(t)$ and γ are the vectors whose components are the coefficients of $u_h(t) = \bar{E}_h(t)v_h$ and v_h with respect to the basis $\{\Phi_j\}_{j=1}^{N_h}$ of S_h, and where $\bar{\mathcal{B}} = ((\Phi_j, \Phi_l)_h)$ is diagonal and $\mathcal{A} = ((\nabla\Phi_j, \nabla\Phi_l))$ is the stiffness matrix. Clearly, the maxima and minima of $u_h(t)$ and v_h coincide with those of the components of $\alpha(t)$ and γ, respectively. Since, with $\widetilde{\mathcal{A}}$ and $\mathcal{G}(t)$ defined by the latter two equalities,

$$\alpha(t) = e^{-\bar{\mathcal{B}}^{-1}\mathcal{A}t}\gamma = e^{-\widetilde{\mathcal{A}}t}\gamma = \mathcal{G}(t)\gamma, \quad \text{for } t \geq 0,$$

it suffices for the first statement of the theorem to show that the matrix $\mathcal{G}(t) \geq 0$ (in the sense that its elements $g_{jl}(t)$ are nonnegative) and that for each j,

$$\sum_{l=1}^{N_h} g_{jl}(t) \leq 1. \tag{15.17}$$

For the purpose of showing $\mathcal{G}(t) \geq 0$ we observe that the off-diagonal elements of the stiffness matrix $\mathcal{A}$ are nonpositive by Lemma 15.4. Therefore we have use for the following simple matrix lemma.

Lemma 15.5 *Let $\mathcal{M} = (m_{jl})$ be a positive definite symmetric matrix with $m_{jl} \leq 0$ for $j \neq l$. Then $\mathcal{M}^{-1} \geq 0$.*

Proof. Let $\mu \geq \max_j m_{jj}$ be such that all eigenvalues of $\mathcal{K} = \mu\mathcal{I} - \mathcal{M}$ are positive. Then the largest eigenvalue of $\mathcal{K}$ and thus also its norm are smaller than μ. Hence

$$\mathcal{M}^{-1} = (\mu\mathcal{I} - \mathcal{K})^{-1} = \mu^{-1}(\mathcal{I} - \mu^{-1}\mathcal{K})^{-1} = \sum_{j=0}^{\infty} \mu^{-j-1}\mathcal{K}^j \geq 0,$$

since $\mathcal{K}$ has nonnegative elements. □

It follows from the lemma that $(\mathcal{I} + k\widetilde{\mathcal{A}})^{-1} \geq 0$ for $k > 0$. In fact, $\bar{\mathcal{B}} + k\mathcal{A}$ satisfies the assumptions of the lemma so that $(\bar{\mathcal{B}} + k\mathcal{A})^{-1} \geq 0$, and hence

$$(\mathcal{I} + k\widetilde{\mathcal{A}})^{-1} = (\bar{\mathcal{B}}^{-1}(\bar{\mathcal{B}} + k\mathcal{A}))^{-1} = (\bar{\mathcal{B}} + k\mathcal{A})^{-1}\bar{\mathcal{B}} \geq 0.$$

Since the powers of nonnegative matrices are nonnegative, we conclude

$$\mathcal{G}(t) = e^{-t\widetilde{\mathcal{A}}} = \lim_{n\to\infty} (\mathcal{I} + \frac{t}{n}\widetilde{\mathcal{A}})^{-n} \geq 0.$$

We now complete the proof by showing (15.17), that is, with $\underline{1}$ the N_h-vector with components 1, that (element-wise) $\mathcal{G}(t)\underline{1} \leq \underline{1}$. We shall show below that $\mathcal{A}\underline{1} \geq 0$. Assuming this for a moment we have $(\bar{\mathcal{B}} + k\mathcal{A})\underline{1} \geq \bar{\mathcal{B}}\underline{1}$. It follows that $(\bar{\mathcal{B}} + k\mathcal{A})^{-1}\bar{\mathcal{B}}\underline{1} = (I + k\widetilde{\mathcal{A}})^{-1}\underline{1} \leq \underline{1}$, and hence as above

$$\mathcal{G}(t)\underline{1} = e^{-t\widetilde{\mathcal{A}}}\underline{1} = \lim_{n\to\infty}(I + \frac{t}{n}\widetilde{\mathcal{A}})^{-n}\underline{1} \le \underline{1}.$$

For the purpose of showing that $\mathcal{A}\underline{1} \ge 0$, we extend the basis $\{\Phi_j\}_{j=1}^{N_h}$ with additional pyramid functions $\{\Phi_{N_h+l}\}_{l=1}^{M_h}$ corresponding to the boundary vertices. In fact, we only need to consider these defined on the polygonal domain Ω_h defined by $\mathcal{T}_h$, so no extension of $\mathcal{T}_h$ is needed. In the same way as before, we have for P_j an interior vertex and P_{N_h+l} a boundary vertex that $(\nabla\Phi_j, \nabla\Phi_{N_h+l}) \le 0$. Hence, since $\sum_{l=1}^{N_h+M_h} \Phi_l \equiv 1$ in Ω_h,

$$\sum_{l=1}^{N_h} a_{jl} = (\nabla\Phi_j, \nabla \sum_{l=1}^{N_h+M_h} \Phi_l) - \sum_{l=1}^{M_h}(\nabla\Phi_j, \nabla\Phi_{N_h+l}) \ge 0.$$

This shows $\mathcal{A}\underline{1} \ge 0$ and thus completes the proof of the maximum-principle. The second part of the theorem is an obvious consequence of the first. □

Maximum-principles are also valid under certain conditions for the homogeneous case ($f = 0$) of the fully discrete schemes (15.14) with $\kappa \in [0,1]$. We show the following:

Theorem 15.6 *Assume that $\mathcal{T}_h$ is of Delaunay type, and that $(1-\kappa)k \le \delta_{\min}^2/3$, where $\delta_{\min} = \min_{j,\tau}\delta_{j,\tau}$. Then the solution of (15.14) with $f = 0$ satisfies, for $x \in \Omega$,*

$$\min(0, \min_{x\in\Omega} v_h(x)) \le U^n(x) \le \max(0, \max_{x\in\Omega} v_h(x)), \quad \textit{for } n \ge 0.$$

In particular,

$$\|U^n\|_{L_\infty} \le \|v_h\|_{L_\infty}.$$

Proof. We write the scheme (15.14) with $f = 0$ as above in matrix form,

$$\alpha^n = (\bar{\mathcal{B}} + \kappa k\mathcal{A})^{-1}(\bar{\mathcal{B}} - (1-\kappa)k\mathcal{A})\alpha^{n-1} = \bar{\mathcal{G}}_{k,\kappa}\alpha^{n-1}.$$

We need to show as before that $\bar{\mathcal{G}}_{k,\kappa} \ge 0$ and $\bar{\mathcal{G}}_{k,\kappa}\underline{1} \le \underline{1}$. For the backward Euler scheme, corresponding to $\kappa = 1$, our above proof of Theorem 15.5 shows the result. For more general $\kappa \in [0,1]$ we still have $(\bar{\mathcal{B}}+\kappa k\mathcal{A})^{-1} \ge 0$ by Lemma 15.5. In order to guarantee $\bar{\mathcal{G}}_{k,\kappa} \ge 0$ we now demand $\bar{\mathcal{B}} - (1-\kappa)k\mathcal{A} \ge 0$. Since $a_{jl} \le 0$ for $j \ne l$ it suffices for this to require $\bar{b}_{jj} - (1-\kappa)ka_{jj} \ge 0$ or $(1-\kappa)k\|\nabla\Phi_j\|^2 \le \|\Phi_j\|_h^2$ for $j = 1, \dots, N_h$. But

$$\|\nabla\Phi_j\|^2 = \sum_{\tau\subset \operatorname{supp}\Phi_j} \delta_{j,\tau}^{-2} \text{ area } (\tau),$$

and, recalling that $D_j = \operatorname{supp}\Phi_j$,

$$\|\Phi_j\|_h^2 = \tfrac{1}{3} \sum_{\tau\in \operatorname{supp}\Phi_j} \text{area } (\tau) = \tfrac{1}{3} \text{ area } (D_j),$$

so that the condition is valid if $(1-\kappa)k \le \delta_{j,\tau}^2/3$, for all j, τ, which is satisfied under the assumptions of the theorem. Since $\mathcal{A}\underline{1} \ge 0$ as before we have $(\bar{\mathcal{B}} + k\kappa\mathcal{A})\underline{1} \ge (\bar{\mathcal{B}} - k(1-\kappa)\mathcal{A})\underline{1}$, and thus

$$\bar{\mathcal{G}}_{k,\kappa}\underline{1} = (\bar{\mathcal{B}} + k\kappa\mathcal{A})^{-1}(\bar{\mathcal{B}} - k(1-\kappa)\mathcal{A})\underline{1} \le \underline{1}.$$

This completes the proof of the theorem. □

Note that, except when $\kappa = 1$, a mesh-ratio condition of type $k \le Ch^2$ is required in this result.

We shall end this chapter by showing that the semigroup $\bar{E}_h(t)$ discussed above is, in fact, an analytic semigroup with respect to the maximum-norm. We shall then use this fact to conclude that it has a smoothing property and also to demonstrate a stability estimate for the fully discrete method (15.14). The analyticity of $\bar{E}_h(t)$ is a consequence of the following resolvent estimate, where we again assume that the family $\{\mathcal{T}_h\}$ is quasiuniform.

Theorem 15.7 *With Δ_h defined by* (15.15) *we have*

$$(15.18) \qquad \|R(z; -\Delta_h)\|_{L_\infty} \le C\ell_h^{1/2}|z|^{-1}, \quad \textit{for } z \in \Sigma_{\delta_h}, \ \delta_h = \tfrac{1}{2}\pi - c\ell_h^{-1/2}.$$

We begin by stating a resolvent estimate in the discrete L_p-norm which we define in analogy with (15.16) as

$$\|\chi\|_{L_p,h} = \Big(\sum_j \omega_j |\chi_j|^p\Big)^{1/p}, \quad \text{for } \chi \in S_h.$$

Theorem 15.8 *With Δ_h defined by* (15.15) *we have*

$$(15.19) \qquad \|R(z; -\Delta_h)\|_{L_p,h} \le \sqrt{p}\,|z|^{-1}, \quad \textit{for } z \in \Sigma_{\delta_p}, \ \delta_p = \tfrac{1}{2}\pi - p^{-1/2}.$$

We now use this result to give the

Proof of Theorem 15.7. Setting $U = R(z; -\Delta_h)F$ we have, with j appropriate and $p < \infty$, since $\omega_j \ge ch^2$,

$$\|U\|_{L_\infty} = |U_j| \le \omega_j^{-1/p}\|U\|_{L_p,h} \le Ch^{-2/p}\sqrt{p}\,|z|^{-1}\|F\|_{L_p,h},$$

for $z \in \Sigma_{\delta_p}$, $\delta_p = \frac{1}{2}\pi - p^{-1/2}$. Choosing $p = \ell_h = \log(1/h)$ for small h now completes the proof. □

The basis of our L_p analysis is the following lemma where for an edge e of $\mathcal{T}_h$ defined by two neighbors P_i and P_j, $\partial_j U = U_{j_1} - U_{j_2}$.

Lemma 15.6 *For every edge e of $\mathcal{T}_h$ there is a real-valued constant γ_e such that*

$$(\nabla\psi, \nabla\chi) = \sum_j \gamma_e \partial_e\psi \cdot \overline{\partial_e\chi}, \quad \forall \psi, \chi \in S_h.$$

Proof. We note that $\sum_{i=1}^{N_h+M_h} \alpha_{ji} = 0$ since $\sum_{i=1}^{N_h+M_h} \Phi_i = 1$. It therefore suffices to remark that, noting that $\psi_j = \chi_j = 0$ for $N_h + 1 \le j \le N_h + M_h$,

$$(\nabla\psi, \nabla\chi) = \sum_{i,j=1}^{N_h+M_h} \alpha_{ij}\psi_i\bar{\chi}_j = \sum_{i\neq j} \alpha_{ij}(\psi_i - \psi_j)(\bar{\chi}_j - \bar{\chi}_i).$$

This shows the lemma with $\gamma_e = -a_{ij}$ for $e = P_iP_j$. □

We also need the following lemma:

Lemma 15.7 *Let z and w be two complex numbers and set*

$$H_p = (w - z)(\bar{w}|w|^{p-2} - \bar{z}|z|^{p-2}), \quad \textit{where } p > 2.$$

Then

$$|\arg H_p| \le \arcsin(1 - 2/p).$$

Proof. Setting $d = w - z$ and $\varphi(t) = d\overline{(z + td)}\,|z + td|^{p-2}$ we may write

$$H_p = d\,\overline{(z + d)}\,|z + d|^{p-2} - d\,\bar{z}|z|^{p-2} = \varphi(1) - \varphi(0) = \int_0^1 \varphi'(t)\,dt,$$

and it hence suffices to show $|\arg\varphi'(t)| \le \arcsin|1 - 2/p|$. For this we write $d^2\,\overline{(z + td)}^2 = re^{i\omega}$ we have

$$\begin{aligned}\varphi'(t) &= \frac{p}{2}|d|^2|z + td|^{p-2} + \frac{p-2}{2}d^2\,\overline{(z + td)}^2\,|z + td|^{p-4}\\ &= \tfrac{1}{2}|z + td|^{p-4}r^2(p + (p-2)e^{2i\omega}).\end{aligned}$$

We now easily find

$$|\arg\varphi'(t)| = |\arg(p + (p-2)e^{2i\omega})| \le \arcsin(1 - 2/p),$$

which completes the proof. □

Proof of Theorem 15.8. We first show, with $\theta_p = \frac{1}{2}\pi + \arcsin(1 - 2/p)$,

$$\|R(z; -\Delta_h)\|_{L_p,h} \le |z|^{-1}, \quad \text{for } z \in \Sigma_{\theta_p}. \tag{15.20}$$

Letting $U \in S_h$ be the solution of the discrete elliptic problem

$$zU + \Delta_h U = F, \tag{15.21}$$

we have $U = R(z; -\Delta_h)F$ so that the statement (15.20) will follow from

$$\|U\|_{L_p,h} \le |z|^{-1}\|F\|_{L_p,h}, \quad \text{for } z \in \Sigma_{\theta_p}. \tag{15.22}$$

We obtain from (15.21)

$$(15.23) \qquad -z(U,\chi)_h + (\nabla U, \nabla \chi) = -(F,\chi)_h, \quad \forall \chi \in S_h.$$

We choose $\chi = \widetilde{\chi} = I_h(U|U|^{p-2})$ and note that, by Lemma 15.6,

$$(\nabla U, \nabla \widetilde{\chi}) = \sum_e \gamma_e \partial_e U \partial_e(\bar{U}|U|^{p-2}) = \sum_e \gamma_e H_{p,e},$$

where, for $e = P_i P_j$,

$$H_{p,e} = (U_j - U_i)(\bar{U}_j|U_j|^{p-2} - \bar{U}_i|U_i|^{p-2}).$$

Note that each $H_{p,e}$ is of the form of H_p in Lemma 15.7, and this lemma therefore shows $|\arg(\nabla U, \nabla \widetilde{\chi})| \le \arcsin(1-2/p)$.

We may then write (15.23), with $\chi = \widetilde{\chi}$, as

$$(15.24) \qquad -z\|U\|_{L_p,h}^p + (\nabla U, \nabla \widetilde{\chi}) = -(F,\chi)_h.$$

and think of this as a relation of the form

$$(15.25) \qquad ae^{i\varphi} + be^{i\psi} = c, \quad \text{with } a, b > 0, \ \varphi, \ \psi \in \mathbb{R},$$

where $\varphi = \arg(-z)$ and where $|\arg \psi| \le \arcsin(1-2/p)$. By multiplication by $e^{-i\varphi}$ and taking real parts, this implies

$$(15.26) \qquad a \le |c|, \quad \text{if} \quad |\arg(-z)| \le \tfrac{1}{2}\pi - \arcsin(1-2/p),$$

since then $\cos(\psi - \varphi) \ge 0$. Hence

$$(15.27) \qquad |z| \, \|U\|_{L_p,h}^p \le \|F\|_{L_p,h} \|U\|_{L_p,h}^{p-1}, \quad \text{for } z \in \Sigma_{\theta_p},$$

from which (15.20) follows.

Noting that $\theta_p > \frac{1}{2}\pi$, we now want to derive a bound for the resolvent in a wider sector which extends to the right half-plane. For this we use (15.20) (with λ replaced by ζ) to obtain

$$\begin{aligned} \|R(z; -\Delta_h)\|_{L_p,h} &\le \|R(\zeta; -\Delta_h)\|_{L_p,h} / (1 - |z-\zeta| \|R(\zeta; -\Delta_h)\|_{L_p,h}) \\ &\le \frac{1}{|\zeta| - |z - \zeta|}, \quad \text{if } |\arg \zeta| = \theta_p, \quad |z-\zeta|/|\zeta| < 1. \end{aligned}$$

Letting $|\zeta| \to \infty$ we find $|\zeta| - |z-\zeta| \to |z| \cos(\theta_p - |\arg z|)$ and hence, with $M_p(\varphi) = 1/\cos(\theta_p - |\varphi|)$,

$$\|R(z; -\Delta_h)\|_{L_p,h} \le \frac{M_p(\arg z)}{|z|}, \quad \text{for } \theta_p - \tfrac{1}{2}\pi < |\arg z| \le \theta_p.$$

In particular, if we assume that $z \in \Sigma_{\pi/2 - \arcsin(1/\sqrt{p})}$, then

$$\cos(\theta_p - |\arg z|) \ge \cos(\arcsin(1/\sqrt{p}) + \arcsin(1-2/p)) = 1/\sqrt{p},$$

and hence

$$\|R(z; -\Delta_h)\|_{L_p,h} \le \sqrt{p}\,|z|^{-1}, \qquad \text{for } |\arg z| \ge \tfrac{1}{2}\pi - \arcsin(1/\sqrt{p}).$$

Since $\frac{1}{2}\pi - \arcsin(1/\sqrt{p}) \ge \frac{1}{2}\pi - 1/\sqrt{p}$, this shows (15.19) and thus completes the proof. □

In the same way as in Chapter 6, Theorem 15.7 can be translated into properties for the semigroup $\bar{E}_h(t) = e^{t\Delta_h}$. In particular, we have the following smoothing property in maximum-norm.

Theorem 15.9 *Assume that the $\mathcal{T}_h$ are of Delaunay type. Then we have*

$$\|\bar{E}_h'(t)\|_{L_\infty} \le C\ell_h t^{-1}, \quad \textit{for } t > 0.$$

Proof. This follow at once from Theorem 15.7 and Lemma 6.6, with $M = M_h = C\ell_h^{1/2}$, $\delta = \delta_h = \frac{1}{2}\pi - c\ell_h^{-1/2}$, since

$$(15.28) \qquad \cos\delta_h = \cos(\tfrac{1}{2}\pi - c\ell_h^{-1/2}) = \sin(c\ell_h^{-1/2}) \ge c\ell_h^{-1/2}, \quad c > 0. \qquad \square$$

Using the techniques of Chapter 9 one may also use the resolvent estimate of Theorem 15.7 to show stability of fully discrete methods. We illustrate with the homogeneous case of (15.14), which we now write

$$(15.29) \quad \bar{\partial}U^n - \kappa\Delta_h U^n + (1-\kappa)\Delta_h U^{n-1} = 0, \quad \text{for } n \ge 1, \quad \text{with } U^0 = v_h.$$

The solution of this problem is then

$$(15.30) \qquad U^n = E_{kh}^n v_h = r_\kappa(-\Delta_h)^n v_h, \quad \text{where } r_\kappa(z) = \frac{1-(1-\kappa)z}{1+\kappa z}.$$

We first need a somewhat more precise result than that of Theorem 9.1 in the special case of the rational function in (15.30). Recalling from (6.48) that $\ell(t) = \max(1, \log(1/t))$ we have the following.

Lemma 15.8 *Let A be an operator in the Banach space $\mathcal{B}$ satisfying* (9.2) *and* (9.3), *and let $r_\kappa(z)$ be the rational function in* (15.30) *with $\frac{1}{2} \le \kappa \le 1$. Then*

$$\|E_k^n\| \le CM\ell(\cos\delta), \quad \textit{for } n \ge 0, \quad \textit{with } E_k = r_\kappa(kA).$$

Proof. We follow the proof of Theorem 9.1. Consider first the case $\kappa > \frac{1}{2}$, so that $|r_\kappa(\infty)| < 1$. Choosing $\psi = \delta$ the estimates for the integrals over γ^R and $\gamma^{\varepsilon/n}$ are unchanged. For the integral over $\Gamma_{\varepsilon/n}^R$ we note that, as is readily proved,

$$|r_\kappa(z)| \le e^{-c\,\mathrm{Re}\,z} \le e^{-c\cos\delta\,|z|}, \quad \text{for } z \in \Sigma_\delta, \ |z| \le R,$$

and hence the bound in (9.11) is replaced by

$$\frac{M}{\pi}\int_{\varepsilon/n}^\infty e^{-cn\cos\delta\,\rho}\frac{d\rho}{\rho} \le CM\ell(\cos\delta).$$

For $\kappa = \frac{1}{2}$ we have $|r_\kappa(\infty)| = 1$, which case is handled correspondingly as in the proof of Theorem 9.1. $\square$

The following is now our stability result for (15.30).

Theorem 15.10 *Assume that the $\mathcal{T}_h$ are of Delaunay type. Then we have for the solution of the fully discrete scheme* (15.30), *for* $\frac{1}{2} \le \kappa \le 1$,

$$\|U^n\|_{L_\infty} \le C\ell_h^{1/2}\ell(\ell_h)\|v_h\|_{L_\infty}, \quad \text{for } n \ge 0.$$

Proof. Since $r_\kappa(z)$ is A-stable (cf. (9.5)) for $\frac{1}{2} \le \kappa \le 1$, this is an immediate consequence of Theorem 15.7 and Lemma 15.8, together with (15.28). □

The lumped mass method described here is a special case of a family of methods involving quadrature analyzed in Raviart [203]. The superconvergence result of Lemma 15.2 and the corresponding maximum-norm error estimate as well as the reduced smoothness estimates are from Chen and Thomée [49]. The maximum-principles of Theorems 15.5 and 15.6 are contained in Fujii [102], and applied in Ushijima [237], [238] to derive uniform convergence, which, except for the case of uniform triangulations, was only shown to be of first order in h. The resolvent estimate of Theorem 15.7 is from Crouzeix and Thomée [60], where also nonquasiuniform families of triangulations are considered. In Nie and Thomée [177] a lumped mass method with quadrature also in the other terms in the variational formulation was discussed for a nonlinear parabolic problem.

16. The H^1 and H^{-1} Methods

In this chapter we briefly discuss some alternatives to the Galerkin methods considered above which use other inner products than that in $L_2(\Omega)$ to formulate the discrete problem. For simplicity we shall content ourselves with describing the situation in the case of a simple selfadjoint parabolic equation in one space dimension, and only study spatially semidiscrete methods.

We begin with the H^1 method in which Galerkin's method is applied with respect to an inner product in H^1. We consider the initial-boundary value problem

$$(16.1)\qquad \begin{aligned} &u_t + Au = f \quad \text{in } I, &&\text{for } t > 0, \quad \text{where } I = (0,1),\\ &u(0,t) = u(1,t) = 0, &&\text{for } t > 0, \qquad \text{with } u(\cdot,0) = v \quad \text{in } I, \end{aligned}$$

where $Au := -(au')' + bu$, with a and b smooth on $\bar{I}$, $a > 0, b \geq 0$.

Let r and k be integers with $r \geq 4$ and $1 \leq k \leq r-2$, and consider a family of partitions $0 = x_0 < x_1 < \cdots < x_M = 1$ of I into subintervals $I_j = (x_{j-1}, x_j)$. Set $h = \max(x_j - x_{j-1})$ and

$$S_h = \{\chi \in \mathcal{C}^k(\bar{I}); \chi|_{I_j} \in \Pi_{r-1} \quad \text{for } 1 \leq j \leq M;\ \chi(0) = \chi(1) = 0\}.$$

Since $k \geq 1$ we have $S_h \subset H^2 \cap H_0^1$ (in this chapter all spaces are with respect to I), and we have with our standard notation

$$(16.2)\qquad \inf_{\chi \in S_h} \sum_{j=0}^{2} h^j \|v - \chi\|_j \leq Ch^s \|v\|_s, \quad \text{for } 2 \leq s \leq r,\ v(0) = v(1) = 0.$$

Introducing the bilinear form corresponding to A,

$$A(v,w) = \int_0^1 (av'w' + bvw)\, dx,$$

the semidiscrete H^1 method for our parabolic problem is then to find $u_h : [0,\infty) \to S_h$ such that

$$(16.3)\qquad A(u_{h,t}, \chi) + (Au_h, A\chi) = (f, A\chi),\ \forall \chi \in S_h,\ t > 0,\ u_h(0) = v_h,$$

with $v_h \in S_h$ a suitable approximation of v. This is based on the corresponding weak formulation of (16.1) obtained by multiplying the parabolic equation by $A\varphi$, integrating over I, and integrating by parts in the first term. It may also be thought of as resulting from a weak formulation with respect to the inner product $A(\cdot,\cdot)$, or

$$A(u_t,\varphi) + A(Au,\varphi) = A(f,\varphi);$$

since $f - Au = u_t$ vanishes at $x = 0$ and 1, an integration by parts brings it to a form analogous to (16.3).

With $\{\Phi_j\}_{j=1}^{N_h}$ a basis for S_h, the semidiscrete problem (16.3) may be written in matrix form as

$$\mathcal{B}\alpha'(t) + \mathcal{A}\alpha(t) = \widetilde{f}(t) \quad \text{for } t > 0, \quad \text{with } \alpha(0) = \gamma, \tag{16.4}$$

where the elements of $\mathcal{B}$ and $\mathcal{A}$ are $A(\Phi_j,\Phi_l)$ and $(A\Phi_j, A\Phi_l)$, respectively. Both $\mathcal{B}$ and $\mathcal{A}$ are thus symmetric and positive definite, and it is therefore clear that a unique solution of (16.4) exists for $t \geq 0$.

As usual in the analysis of a parabolic problem we shall need to study separately the corresponding stationary problem, in this case the two-point boundary value problem

$$Au = f \quad \text{in } I, \quad \text{with } u(0) = u(1) = 0. \tag{16.5}$$

The corresponding discrete problem is then to find $u_h \in S_h$ such that

$$(Au_h, A\chi) = (f, A\chi), \quad \forall \chi \in S_h. \tag{16.6}$$

As is easily checked, this Galerkin formulation is, in fact, equivalent with the least squares problem to find $u_h \in S_h$ such that it minimizes $\|Au_h - f\|$.

We shall begin by demonstrating the following result in which we note that an error estimate in H^2 is also included. We use the negative norm

$$\|v\|_{-q} = \sup\{(v,\varphi)/\|\varphi\|_q;\ \varphi \in H^q\}, \quad \text{for } q \geq 0.$$

Lemma 16.1 *If u_h and u are the solution of* (16.6) *and* (16.5), *then*

$$\sum_{j=0}^{2} h^j \|u_h - u\|_j \leq Ch^s \|u\|_s, \quad \text{for } 2 \leq s \leq r,$$

and

$$\|u_h - u\|_{-q} \leq Ch^{s+q}\|u\|_s, \quad \text{for } 2 \leq s \leq r,\ 0 \leq q \leq r-4.$$

Proof. We have for the error, $e = u_h - u$,

$$(Ae, A\chi) = 0, \quad \forall \chi \in S_h, \tag{16.7}$$

and hence

$$\|Ae\|^2 = (Ae, A(\chi - u)) \le \|Ae\| \, \|A(\chi - u)\|, \quad \text{for } \chi \in S_h,$$

so that by (16.2)

$$\|Ae\| \le \inf_{\chi \in S_h} \|A(u - \chi)\| \le Ch^{s-2}\|u\|_s, \quad \text{for } 2 \le s \le r. \tag{16.8}$$

Since $\|e''\| \le C\|Ae\|$, the desired estimate in H^2 follows.

We now turn to the negative norm estimate; this includes the L_2-norm error bound as a special case. We shall show that

$$|(e, \varphi)| \le Ch^{s+q}\|u\|_s\|\varphi\|_q, \quad \text{for } \varphi \in H^q.$$

For this purpose we associate with φ the solution ψ of the two-point boundary value problem

$$A^2\psi = \varphi \quad \text{in } I, \quad \text{with } \psi(0) = A\psi(0) = \psi(1) = A\psi(1) = 0,$$

and observe that for any $q \ge 0$, $\|\psi\|_{q+4} \le C\|\varphi\|_q$. By integration by parts we have, in view of the boundary conditions, $(e, \varphi) = (e, A^2\psi) = (Ae, A\psi)$, and hence, by (16.7), (16.8) and (16.2),

$$\begin{aligned}(e, \varphi) &= (Ae, A(\psi - \chi)) \le \|Ae\| \inf_{\chi \in S_h} \|A(\psi - \chi)\| \\ &\le (Ch^{s-2}\|u\|_s)(Ch^{q+2}\|\psi\|_{q+4}) \le Ch^{s+q}\|u\|_s\|\varphi\|_q,\end{aligned}$$

which is the desired estimate.

Finally, for the first derivatives, we have by the results already obtained

$$\|e'\|^2 \le CA(e, e) \le C\|Ae\| \, \|e\| \le Ch^{2s-2}\|u\|_s^2,$$

which completes the proof. □

For the case that the approximating functions are at most twice differentiable at the nodal points, i.e., for $k = 1$ or 2, we shall also show some superconvergence results for the error at the points of the partition; for $k = 1$ we have such a result also for the error in the derivative at these points.

Lemma 16.2 *Let $\bar{x}$ be a point of the partition and let u_h and u be the solutions of* (16.6) *and* (16.5)*, respectively. Then*

$$|u_h(\bar{x}) - u(\bar{x})| \le Ch^{2r-4}\|u\|_r, \quad \textit{if } k = 1 \textit{ or } 2,$$

and

$$|u_h'(\bar{x}) - u'(\bar{x})| \le Ch^{2r-4}\|u\|_r, \quad \textit{if } k = 1. \tag{16.9}$$

Proof. Letting $G^x = G^x(y) = G(x, y)$ be the Green's function for the two-point boundary value problem (16.5), we have for any $v \in H^2 \cap H_0^1$,

$$v(x) = (Av, G^x). \tag{16.10}$$

In particular, with T the exact solution operator of (16.5),

$$v(\bar{x}) = (Av, ATG^{\bar{x}}) = (Av, Ag_0), \quad \text{with } g_0 = TG^{\bar{x}}. \tag{16.11}$$

We note that since $G^{\bar{x}}$ is smooth except at $\bar{x}$, but continuous there, g_0 is in $\mathcal{C}^2$ and smooth outside $\bar{x}$, so that for $k = 1, 2$

$$\inf_{\chi \in S_h} \|g_0 - \chi\|_2 \le Ch^{r-2}. \tag{16.12}$$

We now apply (16.11) to $e = u_h - u$ and find in view of (16.7)

$$e(\bar{x}) = (Ae, Ag_0) = (Ae, A(g_0 - \chi)), \quad \forall \chi \in S_h,$$

and hence, using (16.12) and Lemma 16.1,

$$|e(\bar{x})| \le C\|e\|_2 \inf_{\chi \in S_h} \|g_0 - \chi\|_2 \le Ch^{2r-4}\|u\|_r,$$

which is the first estimate of the lemma.

Let now $k = 1$. By differentiation of (16.10) we obtain $u'(x) = (Au, G^x_x)$, and setting $g_1 = TG^x_x|_{x=\bar{x}}$ we have similarly to the above

$$e'(\bar{x}) = (Ae, Ag_1) = (Ae, A(g_1 - \chi)), \quad \forall \chi \in S_h.$$

Since $G^x_x(y)$ has a simple discontinuity at $y = x$ we have $g_1 \in \mathcal{C}^1$ and thus $\inf_{\chi \in S_h} \|g_1 - \chi\|_2 \le Ch^{r-2}$. This implies (16.9), and completes the proof. □

We are now ready to analyze the error in the semidiscrete parabolic problem (16.3). We shall then use the elliptic projection $\widetilde{R}_h : H^2 \cap H^1_0 \to S_h$ corresponding to the method (16.6) for the stationary problem, i.e., $(A(\widetilde{R}_h u - u), A\chi) = 0$ for $\chi \in S_h$, for which thus, by Lemma 16.1,

$$\|\widetilde{R}_h u - u\|_q \le Ch^{s-q}\|u\|_s, \quad \text{for } -(r-4) \le q \le 2 \le s \le r. \tag{16.13}$$

We begin with the following:

Theorem 16.1 *Let u_h and u be the solutions of* (16.3) *and* (16.1), *respectively. Then, if $v_h = \widetilde{R}_h v$, we have, for $t \ge 0$,*

$$\|u_h(t) - u(t)\| \le Ch^r\Big(\|u(t)\|_r + \big(\int_0^t \|u_t\|_r^2\,ds\big)^{1/2}\Big)$$

and

$$\|u_h(t) - u(t)\|_j \le Ch^{r-j}\Big(\|u(t)\|_r + \big(\int_0^t \|u_t\|_{r-1}^2\,ds\big)^{1/2}\Big), \quad \text{for } j = 1, 2.$$

Proof. We write $u_h - u = (u_h - \widetilde{R}_h u) + (\widetilde{R}_h u - u) = \theta + \rho$, and find at once from (16.13),

$$\|\rho(t)\|_j \le Ch^{r-j}\|u(t)\|_r, \quad \text{for } j = 0, 1, 2.$$

In order to estimate θ, we note that

$$A(\theta_t, \chi) + (A\theta, A\chi) = -A(\rho_t, \chi) = -(\rho_t, A\chi), \quad \forall \chi \in S_h, \ t > 0.$$

Setting $\chi = \theta_t$ yields

$$A(\theta_t, \theta_t) + \tfrac{1}{2}\frac{d}{dt}\|A\theta\|^2 = -A(\rho_t, \theta_t) \le A(\rho_t, \rho_t)^{1/2} A(\theta_t, \theta_t)^{1/2},$$

whence, since $\theta(0) = 0$,

$$\|A\theta\|^2 \le \int_0^t A(\rho_t, \rho_t)\, ds \le C \int_0^t \|\rho_t\|_1^2\, ds,$$

and thus, by Lemma 16.1,

$$\|\theta(t)\|_2 \le Ch^{r-2}\Big(\int_0^t \|u_t\|_{r-1}^2\, ds\Big)^{1/2}.$$

Similarly, using $\chi = \theta$, we find

$$\tfrac{1}{2}\frac{d}{dt}A(\theta, \theta) + \|A\theta\|^2 = -(\rho_t, A\theta),$$

and hence

$$\|\theta\|_1 \le C\Big(\int_0^t \|\rho_t\|^2\, ds\Big)^{1/2} \le Ch^{r-j}\Big(\int_0^t \|u_t\|_{r-j}^2\, ds\Big)^{1/2}, \quad \text{for } j = 0, 1.$$

Together these estimates show the theorem. □

In order to indicate how negative norm estimates and superconvergence results at nodes may be derived, we shall briefly sketch the adaptation of the methods employed in Chapter 6 to the present context. Let thus $T_h : L_2 \to S_h$ be the solution operator of the discrete problem (16.6) so that

$$(AT_h f, A\chi) = (f, A\chi), \quad \forall \chi \in S_h. \tag{16.14}$$

With T as above the solution operator of the continuous problem, the estimates of Lemma 16.1 may then be stated as

$$\|T_h f - Tf\|_q \le Ch^{s-q}\|f\|_{s-2}, \quad \text{for } -(r-4) \le q \le 2 \le s \le r.$$

For $f \in H_0^1$ our definition (16.14) may also be written

$$(AT_h f, A\chi) = A(f, \chi), \quad \forall \chi \in S_h.$$

In particular, $(AT_h f, AT_h g) = A(f, T_h g)$ for $f, g \in H_0^1$, from which one easily shows that the restriction of T_h to H_0^1 is selfadjoint and positive semidefinite with respect to the inner product $A(\cdot,\cdot)$, and positive definite when further restricted to S_h.

With this notation our parabolic problem (16.3) may be stated as

$$T_h u_{h,t} + u_h = T_h f, \quad \text{for } t > 0, \quad \text{with } u_h(0) = v_h,$$

and the machinery developed in Chapters 2, 3 and 6 may be applied. The equation for the error $e = u_h - u$ takes the form

$$T_h e_t + e = \rho := (T_h - T)Au = (\widetilde{R}_h - I)u,$$

and recalling that the basic inner product for the analysis is now $A(\cdot,\cdot)$, we have by Lemma 2.4 that

$$\|u_h(t) - u(t)\|_1 \le C\|v_h - v\|_1 + Ch^{r-1}\Big(\|v\|_r + \int_0^t \|u_t\|_r\, ds\Big),$$

and for the homogeneous equation, the technique of Theorem 3.4 shows, now with $v_h = R_h v$ defined by $A(\cdot,\cdot)$, that

$$\|u_h(t) - u(t)\|_1 \le Ch^{r-1} t^{-(r-1)/2} \|v\|_1.$$

We may also define discrete negative norms and corresponding inner products as in Chapter 6, this time by

$$\|v\|_{-s,h} = (v, v)_{-s,h}^{1/2}, \quad \text{with } (v, w)_{-s,h} = A(T_h^{s+1} v, w),$$

and we find easily as in Lemma 5.3, for $0 \le s \le r-2, v \in H_0^1$, and with $\|v\|_{-s} = (T^s v, v)^{1/2}$,

$$\|v\|_{-s,h} \le C(\|v\|_{-s} + h^s\|v\|) \quad \text{and} \quad \|v\|_{-s} \le C(\|v\|_{-s,h} + h^s\|v\|).$$

For example, we have for $s = 0$, with $v \in H_0^1$,

$$\begin{aligned} \|v\|_{0,h}^2 &= A(T_h v, v) = (AT_h v, v) \le C\|T_h v\|_2 \, \|v\| \\ &\le \big(C\|T_h v - Tv\|_2 + \|v\|\big)\|v\| \le C\|v\|^2. \end{aligned}$$

In the same way as in the proof of Theorem 5.2 it is possible to use these discrete negative norms to show the following negative norm estimates for the parabolic problem.

Theorem 16.2 *Let u_h and u be the solutions of* (16.3) *and* (16.1), *and let $0 \le s \le r-4$. Assume that v_h is chosen so that $\|v_h - v\|_{-s} + h^s\|v_h - v\| \le Ch^{r+s}\|v\|_r$. Then we have*

$$\|u_h(t) - u(t)\|_{-s} \le Ch^{r+s}\Big(\|v\|_r + \int_0^t \|u_t\|_r\, ds\Big), \quad \textit{for } t \ge 0.$$

We shall not give the details of the proof.

Similarly to the situation in Chapter 6 one may also demonstrate estimates of the type

$$(16.15)\qquad \|D_t^j(u_h(t)-u(t))\|_{-s}\le C(t,u)h^{r+s},\quad \text{for } -2\le s\le r-4.$$

Such estimates are again useful for deriving superconvergent order error estimates at the nodes for low order k of continuity of S_h. This time we have:

Theorem 16.3 *Let $\bar{x}$ be one of the points of the partition, and let u_h and u be solutions of* (16.3) *and* (16.1). *Then if $k=1$ or 2 we have for $e=u_h-u$, and any $n\ge 0$,*

$$|e(\bar{x},t)|\le C\Big(h^{r-1}\|D_t^{n+1}e\|_1+h^{r-2}\sum_{j=0}^{n}\|D_t^j e\|_2+\|D_t^{n+1}e\|_{-2n}\Big),\quad t\ge 0.$$

If $k=1$ the quantity $\big|\frac{\partial e}{\partial x}(\bar{x},t)\big|$ is bounded by the same expression.

Proof. We have as in the proof of Lemma 16.2 that $e(\bar{x},t)=(Ae,Ag_0)$. Setting

$$L(v,w)=A(v_t,w)+(Av,Aw),$$

we have, as in the proof of Theorem 5.6,

$$e(\bar{x},t)=(Ae,Ag_0)=\sum_{j=0}^{n}(-1)^jL(D_t^je,T_{g_0}^j)+(-1)^{n+1}A(D_t^{n+1}e,T^ng_0).$$

Noting that $L(D_t^je,\chi)=0$ for $\chi\in S_h$, we obtain

$$|L(D_t^je,T^jg_0)|=|L(D_t^je,T^jg_0-\chi)|\le C\Big(h^{r-1}\|D_t^{j+1}e\|_1+h^{r-2}\|D_t^je\|_2\Big),$$

where we have used the fact that T^jg_0 is twice continuously differentiable at $\bar{x}$. Further

$$|A(D_t^{n-1}e,T^ng_0)|=|(T^nD_t^{n-1}e,Ag_0)|\le C\|D_t^{n-1}e\|_{-2n},$$

which completes the proof for $e(\bar{x},t)$. The proof for $(\partial/\partial x)e(\bar{x},t)$ is similar. □

Combined with the appropriate estimates of the form (16.15) we may conclude that

$$|e(\bar{x},t)|\le C(t,u)h^{2r-4},\quad \text{if } k=1 \text{ or } 2,$$

and

$$\Big|\frac{\partial e}{\partial x}(\bar{x},t)\Big|\le C(t,u)h^{2r-4},\quad \text{if } k=1.$$

For piecewise cubic elements ($r=4$) both these estimates are $O(h^4)$.

We shall next turn to the H^{-1} method for (16.1). This method may be described as a Petrov-Galerkin method, which term is used to indicate that the trial and test functions are selected from two different spaces. With the above partition of I, and r and k integers with $r \geq 1$ and $-1 \leq k \leq r-2$, we shall then use as the trial space

$$S_h = \{\chi \in \mathcal{C}^k(I); \chi|_{I_j} \in \Pi_{r-1}, \quad \text{for } 1 \leq j \leq M\}$$

(where $\mathcal{C}^{-1}(I)$ is interpreted as not requiring any continuity at the nodal points) and as the test space

$$V_h = \{\omega \in \mathcal{C}^{k+2}(I); \omega|_{I_j} \in \Pi_{r+1}, \quad \text{for } 1 \leq j \leq M;\ \omega(0) = \omega(1) = 0\}.$$

Note that no boundary conditions are prescribed for S_h and that the order of continuity and the degree of the polynomials are two orders higher for V_h than for S_h. An interesting choice is $k = -1$ for which the functions of S_h may have discontinuities at the nodes of the partition and the functions in V_h are continuously differentiable.

The semidiscrete method we shall study is then to find $u_h : [0, \infty) \to S_h$ such that

$$(u_{h,t}, \omega) + (u_h, A\omega) = (f, \omega), \quad \forall \omega \in V_h,\ t > 0,\ u_h(0) = v_h, \tag{16.16}$$

where v_h is a given approximation of v in S_h and where, as usual, $(\cdot, \cdot)$ is the inner product in L_2. As we shall see below, the present method may also be interpreted as an ordinary Galerkin method, now with respect to an inner product in the dual space to H_0^1, which is the reason the method is referred to as the H^{-1} method.

For the purpose stated, we introduce the solution operator T_0 of the two-point boundary value problem

$$-u'' = f \quad \text{in } I, \quad \text{with } u(0) = u(1) = 0,$$

and observe that $V_h = T_0 S_h$. The operator T_0 is positive definite on L_2, and we may therefore define the inner product $\langle v, w \rangle = (v, T_0 w)$ and the corresponding norm $|v| = \langle v, v \rangle^{1/2}$. In fact,

$$|v|^2 = (v, T_0 v) = -((T_0 v)'', T_0 v) = \|(T_0 v)'\|^2, \tag{16.17}$$

and it follows easily (cf. the discussion following (5.14)) that

$$c|v| \leq \sup_{w \in H_0^1(I)} \frac{(v, w)}{\|w\|_1} = \sup_{w \in H_0^1(I)} \frac{((T_0 v)', w')}{\|w\|_1} \leq C|v| \quad \text{with } c > 0,$$

so that $|\cdot|$ is a norm on the dual space to H_0^1. Note that $|v| \leq C\|v\|$. Setting also

$$B(v, w) = (v, AT_0 w) = (v, aw) + (v, A_1 T_0 w),$$

where $A_1 v = -a'v' + bv$, we may now write (16.16), with $\omega = T_0\chi \in V_h$, in the form of the ordinary Galerkin method

$$\langle u_{h,t}, \chi\rangle + B(u_h, \chi) = \langle f, \chi\rangle, \quad \forall \chi \in S_h,\ t > 0. \tag{16.18}$$

Since in view of (16.17)

$$|(v, A_1 T_0 w)| \le C\|v\|\,\|T_0 w\|_1 \le C\|v\|\,\|(T_0 w)'\| \le C\|v\|\,|w|, \tag{16.19}$$

we have

$$B(u,u) = \|a^{1/2}u\|^2 + (u, A_1 T_0 u) \ge c_0\|u\|^2 - \kappa|u|^2.$$

It is also clear from (16.19) that

$$|B(u,v)| \le C\|u\|\,\|v\|. \tag{16.20}$$

After a transformation of variables $\widetilde{u} = e^{-\kappa t}u$, the equation (16.18) takes the form

$$\langle \widetilde{u}_{h,t}, \chi\rangle + B_\kappa(\widetilde{u}_h, \chi) = \langle \widetilde{f}, \chi\rangle, \quad \forall \chi \in S_h,$$

where $B_\kappa(v,w) = B(v,w) + \kappa\langle v, w\rangle$ is positive definite. We shall assume that this transformation has been performed from the outset so that we may keep the equation in the original form (16.18), where now

$$B(u,u) \ge c_0\|u\|^2. \tag{16.21}$$

For the analysis we now introduce the elliptic projection $Q_h : L_2 \to S_h$ defined by

$$B(Q_h u - u, \chi) = 0, \quad \forall \chi \in S_h; \tag{16.22}$$

the existence and uniqueness of $Q_h u$ is guaranteed by the positivity of $B(\cdot,\cdot)$. We shall have use for the following lemma, where

$$\|u\|_{-q} = \sup_{\varphi \in H^q} \frac{(u,\varphi)}{\|\varphi\|_q}.$$

Lemma 16.3 *With $Q_h : L_2 \to S_h$ defined by (16.22) we have*

$$\|Q_h u - u\|_{-q} \le Ch^{s+q}\|u\|_s, \quad \textit{for } 0 \le q,\ s \le r. \tag{16.23}$$

Proof. Recall that for the standard L_2-projection P_h onto S_h we have

$$\|P_h u - u\| = \inf_{\chi \in S_h} \|u - \chi\| \le Ch^s\|u\|_s, \quad \text{for } 0 \le s \le r. \tag{16.24}$$

From (16.20), (16.21) and (16.22) we infer

$$\begin{aligned} c_0\|Q_h u - u\|^2 &\le B(Q_h u - u, Q_h u - u) \\ &= B(Q_h u - u, P_h u - u) \le C\|Q_h u - u\|\,\|P_h u - u\|, \end{aligned}$$

whence, by (16.24), $\|Q_h u - u\| \le C\|P_h u - u\| \le Ch^s\|u\|_s$.

In order to show (16.23) for $q > 0$, we define for $\varphi \in L_2$ the function $\psi \in L_2$ as the unique solution of the equation $AT_0\psi = \varphi$; it may be found by determining $\omega = T_0\psi$ from $A\omega = \varphi$ in I, with $\omega(0) = \omega(1) = 0$ and then setting $\psi = -\omega''$. We note that

$$\|\psi\|_q = \|(T_0\psi)''\|_q \le C\|T_0\psi\|_{q+2} \le C\|\varphi\|_q.$$

We have now

$$|(Q_hu - u, \varphi)| = |(Q_hu - u, AT_0\psi)| = |B(Q_hu - u, \psi)|$$
$$= |B(Q_hu - u, \psi - P_h\psi)| \le C\|Q_hu - u\|\,\|\psi - P_h\psi\| \le Ch^{s+q}\|u\|_s\|\psi\|_q,$$

which completes the proof. □

We shall now begin our error analysis for the parabolic problem and start by an error estimate for the case of a smooth solution.

Theorem 16.4 *Let u_h and u be the solutions of* (16.16) *and* (16.1), *respectively, with $\|v_h - v\| \le Ch^r\|v\|_r$. Then for each $\bar{t} > 0$ there is a constant $C = C_{\bar{t}}$ such that, for $t \in [0, \bar{t}]$,*

$$\|u_h(t) - u(t)\| \le Ch^r\Big(\|v\|_r + \|u(t)\|_r + \big(\int_0^t \|u_t(s)\|_{r-1}^2\,ds\big)^{1/2}\Big).$$

Proof. We write $u_h - u = (u_h - Q_hu) + (Q_hu - u) = \theta + \rho$, and find at once by Lemma 16.3, $\|\rho(t)\| \le Ch^r\|u(t)\|_r$. From our definitions we have in the standard fashion

$$\langle\theta_t, \chi\rangle + B(\theta, \chi) = -\langle\rho_t, \chi\rangle, \quad \forall\chi \in S_h,\ t > 0. \tag{16.25}$$

We set $\chi = \theta_t$ and note that, using (16.19),

$$B(\theta, \theta_t) = (\theta, a\theta_t) + (\theta, A_1T_0\theta_t) = \tfrac{1}{2}\frac{d}{dt}\|a^{1/2}\theta\|^2 + (\theta, A_1T_0\theta_t)$$
$$\ge \tfrac{1}{2}\frac{d}{dt}\|a^{1/2}\theta\|^2 - C\|a^{1/2}\theta\|\,|\theta_t|.$$

This yields

$$|\theta_t|^2 + \tfrac{1}{2}\frac{d}{dt}\|a^{1/2}\theta\|^2 = -\langle\rho_t, \theta_t\rangle - (\theta, A_1T_0\theta_t) \le C(|\rho_t|^2 + \|a^{1/2}\theta\|^2) + |\theta_t|^2,$$

or

$$\frac{d}{dt}\|a^{1/2}\theta\|^2 \le C(|\rho_t|^2 + \|a^{1/2}\theta\|^2). \tag{16.26}$$

Gronwall's lemma now shows

$$\|a^{1/2}\theta(t)\|^2 \le e^{Ct}\|a^{1/2}\theta(0)\|^2 + C\int_0^t e^{C(t-s)}|\rho_t(s)|^2\,ds,$$

or, for t bounded,

$$\|\theta(t)\| \le C\Big(\|\theta(0)\| + \big(\int_0^t |\rho_t|^2\,ds\big)^{1/2}\Big).$$

Here, using Lemma 16.3,

$$\|\theta(0)\| = \|v_h - Q_h v\| \le Ch^r\|v\|_r,$$

and $|\rho_t| \le C\|\rho_t\|_{-1} \le Ch^r\|u_t\|_{r-1}$, so that

$$\Big(\int_0^t |\rho_t|^2\,ds\Big)^{1/2} \le Ch^r\Big(\int_0^t \|u_t\|_{r-1}^2\,ds\Big)^{1/2}.$$

Together these estimates show the theorem. □

For the special case of the homogeneous equation we have the following result, where in the same way as in Chapter 3, $\dot{H}^r = \dot{H}^r(I)$ denotes the space defined by the norm

$$|v|_r = \|v\|_{\dot{H}^r} = \big(\sum_{j=1}^{\infty}\lambda_j^r(v,\varphi_j)^2\big)^{1/2},$$

where $\{\lambda_j\}_{j=1}^{\infty}$ and $\{\varphi_j\}_{j=1}^{\infty}$ are the eigenvalues and eigenfunctions of A, with boundary conditions $\varphi_j(0) = \varphi_j(1) = 0$.

Theorem 16.5 *Let u_h and u be the solutions of* (16.16) *and* (16.1), *respectively. Assume that $v \in \dot{H}^r$ and $f = 0$. We then have, with $C = C_{\bar{t}}$,*

$$\|u_h(t) - u(t)\| \le C\|v_h - v\| + Ch^r|v|_r, \quad \text{for } 0 \le t \le \bar{t}.$$

Proof. This follows at once from Theorem 16.4 upon noting that as in Chapter 3, $\|u(t)\|_r \le C|u(t)|_r \le C|v|_r$, and similarly

$$\int_0^t \|u_t\|_{r-1}^2\,ds \le C\int_0^t \|u\|_{r+1}^2\,ds \le C|v|_r^2. \qquad \square$$

We shall end this chapter by showing the following nonsmooth data error estimate.

Theorem 16.6 *Let u_h and u be the solutions of* (16.16) *and* (16.1), *respectively, with $v_h = P_h v$ and $f = 0$. Then, with $C = C_{\bar{t}}$,*

$$\|u_h(t) - u(t)\| \le Ch^r t^{-r/2}\|v\|, \quad \text{for } 0 < t \le \bar{t}.$$

Proof. We shall show that

$$\|u_h(t) - u(t)\| \le Cht^{-1/2}\|v\|, \quad \text{for } 0 < t \le \bar{t}.$$

The result claimed then follows by an integration argument, exactly as in Chapter 3.

As in Theorem 16.4 we write the error $e = u_h - u = \theta + \rho$, and note first that by Lemma 16.3 and a standard smoothing estimate,

$$\|\rho(t)\| = \|Q_h u(t) - u(t)\| \le Ch\|u(t)\|_1 \le Cht^{-1/2}\|v\|, \quad \text{for } 0 < t \le \bar{t}.$$

In order to derive the estimate needed for $\theta = u_h - Q_h u$, we shall first use (16.25) to show

$$t^2\|\theta(t)\|^2 \le C\big(t|e(0)|^2 + \int_0^t (s^2|\rho_t|^2 + |\rho|^2)\,ds\big), \tag{16.27}$$

and then observe that this implies

$$t^{1/2}\|\theta(t)\| \le Ch\|v\|, \quad \text{for } 0 < t \le \bar{t}, \tag{16.28}$$

thus completing the proof.

To prove (16.27) we first multiply (16.26) by t^2 to obtain

$$\frac{d}{dt}(t^2\|a^{1/2}\theta\|^2) \le Ct^2|\rho_t|^2 + Ct\|a^{1/2}\theta\|^2,$$

or, after integration,

$$t^2\|\theta(t)\|^2 \le C\int_0^t s^2|\rho_t|^2\,ds + C\int_0^t s\|\theta\|^2\,ds. \tag{16.29}$$

To estimate the latter integral, we choose $\chi = \theta$ in (16.25) and multiply by t, which gives

$$\tfrac{1}{2}\frac{d}{dt}(t|\theta|^2) + tB(\theta,\theta) = -t\langle\rho_t,\theta\rangle + \tfrac{1}{2}|\theta|^2 \le Ct^2|\rho_t|^2 + C|\theta|^2$$

and hence, using (16.21),

$$\int_0^t s\|\theta\|^2\,ds \le C\int_0^t s^2|\rho_t|^2\,ds + C\int_0^t |\theta|^2\,ds. \tag{16.30}$$

For the latter term we now integrate equation (16.25) to get, with $R(t) = \int_0^t \theta(s)\,ds$ and $\psi(t) = e(0) - \rho(t)$,

$$\langle\theta(t),\chi\rangle + B(R,\chi) = \langle\theta(0),\chi\rangle + \langle\rho(0) - \rho(t),\chi\rangle = \langle\psi(t),\chi\rangle.$$

Choosing $\chi = \theta(t) = R_t(t)$ we obtain using (16.19)

$$|\theta|^2 + \tfrac{1}{2}\frac{d}{dt}\|a^{1/2}R\|^2 = \langle \psi, \theta\rangle - (R, A_1 T_0 \theta) \le \tfrac{1}{2}|\theta|^2 + C(|\psi|^2 + \|a^{1/2}R\|^2).$$

Since $R(0) = 0$, Gronwall's lemma now shows, for $0 \le t \le \bar{t}$,

$$\int_0^t |\theta|^2\, ds \le C \int_0^t |\psi|^2\, ds \le Ct|e(0)|^2 + C\int_0^t |\rho|^2\, ds,$$

Together with (16.29) and (16.30) this shows (16.27).

Now $|e(0)| = |v_h - v| \le C\|P_h v - v\|_{-1}$, and since P_h is selfadjoint,

$$\|P_h v - v\|_{-1} = \sup_{\varphi \in H^1} \frac{(v, (P_h - I)\varphi)}{\|\varphi\|_1} \le Ch\|v\|,$$

so that $|e(0)| \le Ch\|v\|$. Further, applying Lemma 16.3 once more and also the stability and smoothing property of the solution operator of (16.1), we have

$$|\rho(s)| + s|\rho_t(s)| \le Ch(\|u(s)\| + s\|u_t(s)\|) \le Ch\|v\|.$$

Hence

$$\int_0^t (s^2|\rho_t(s)|^2 + |\rho(s)|^2)\, ds \le Ch^2 t\|v\|^2,$$

so that (16.28) follows from (16.27). The proof is now complete. □

We conclude by remarking that the L_2-norm error estimates of our last three theorems are different in character from our earlier L_2 estimates for the standard Galerkin method and would correspond to H^1 estimates for those methods.

The H^1 method was first proposed in Thomée and Wahlbin [231] for a semilinear problem in two and three space dimensions where the fact that the H^2-norm majorizes the maximum-norm was used to show optimal order error estimates without inverse assumptions, under local regularity assumptions on the nonlinear forcing term. It was further studied in Douglas, Dupont and Wheeler [79] and [80], in the latter reference also in several space dimensions. The method may also be designed to employ approximating subspaces whose elements do not necessarily satisfy the homogeneous Dirichlet boundary conditions of the continuous problem, which is an advantage when the boundary is curved. In Bramble and Thomée [38] a somewhat similar fully discrete method is studied in which the parabolic equation is first discretized in time by the backward Euler method, after which the resulting elliptic equations at the time levels are solved in the approximating finite dimensional space by a least squares method. Again the approximating functions in the spatial variables are not required to satisfy the homogeneous boundary conditions exactly. The approach in [38] is developed further in Bramble and Thomée [39] to include higher order time discretization methods.

The H^{-1} method was introduced for two-point boundary value problems by Rachford and Wheeler [199], and for corresponding parabolic problems by Wheeler [247] and Kendall and Wheeler [138]. The use of approximating subspaces of discontinuous functions combined with a judicious choice of discrete initial data, in a manner discussed in Chapter 6 and referred to as quasi-projections, was shown to lead to superconvergent $O(h^{r+1})$ error estimates at certain Gaussian points and, after a posteriori local quadratures, to $O(h^{2r})$ nodal estimates for u and u_x. Douglas and Dupont [77] contains certain generalizations to more than one space dimension. The above presentation is based on Huang and Thomée [126].

17. A Mixed Method

In this chapter we shall consider a finite element method for our model parabolic equation which is based on a mixed formulation of the problem. In this formulation the gradient of the solution is introduced as a separate dependent variable, the approximation of which is sought in a different finite element space than the solution itself. One advantage of this procedure is that the gradient of the solution may be approximated to the same order of accuracy as the solution itself.

Letting thus Ω be a convex plane domain with smooth boundary, we shall consider first the stationary problem

$$-\Delta u = f \quad \text{in } \Omega, \quad \text{with } u = 0 \quad \text{on } \partial\Omega. \tag{17.1}$$

Introducing the gradient of u as a new variable this may also be formulated

$$-\operatorname{div} \sigma = f \quad \text{in } \Omega, \quad \sigma = \nabla u \quad \text{in } \Omega, \quad \text{with } u = 0 \quad \text{on } \partial\Omega. \tag{17.2}$$

With $L_2 = L_2(\Omega)$ and $H = \{\omega = (\omega_1, \omega_2) \in L_2 \times L_2;\ \operatorname{div} \omega \in L_2\}$ we note that the solution $(u, \sigma) \in L_2 \times H$ also solves the variational problem

$$\begin{aligned} (\operatorname{div} \sigma, \varphi) + (f, \varphi) &= 0, \quad \forall \varphi \in L_2, \\ (\sigma, \omega) + (u, \operatorname{div} \omega) &= 0, \quad \forall \omega \in H, \end{aligned} \tag{17.3}$$

where the $(\cdot,\cdot)$ denote the appropriate L_2 inner products. Note that the boundary condition $u = 0$ is implicitly contained in (17.3); using Green's formula in the second equation we have, with n the exterior normal to $\partial\Omega$,

$$(\sigma, \omega) = -(u, \operatorname{div} \omega) = -\int_{\partial\Omega} u\, \omega \cdot n\, ds + (\nabla u, \omega), \quad \forall\, \omega \in H,$$

and hence, formally, $\sigma = \nabla u$ in Ω and $u = 0$ on $\partial\Omega$.

With S_h and H_h finite-dimensional subspaces of L_2 and H to be specified below, we shall consider the following discrete analogue of (17.2) (or (17.3)), namely to find $(u_h, \sigma_h) \in S_h \times H_h$ such that

$$\begin{aligned} (\operatorname{div} \sigma_h, \chi) + (f, \chi) &= 0, \quad \forall \chi \in S_h, \\ (\sigma_h, \psi) + (u_h, \operatorname{div} \psi) &= 0, \quad \forall \psi \in H_h. \end{aligned} \tag{17.4}$$

We shall now describe our choice of subspaces S_h and H_h; they belong to a family of such pairs introduced by Raviart and Thomas [204].

Let $\mathcal{T}_h$ be a quasiuniform triangulation of Ω of the type we have used repeatedly before, e.g., in Chapter 1, and set, with Ω_h the polygonal domain determined by the union of the triangles of $\mathcal{T}_h$,

$$S_h = \{\chi \in L_2; \chi|_\tau \text{ linear } \forall \tau \in \mathcal{T}_h,\ \chi = 0 \text{ in } \Omega \setminus \Omega_h\},$$

where no continuity is required across inter-element boundaries. In order to define H_h, let $\hat{\tau}$ be the standard reference triangle in the ξ-plane, with vertices $P_0 = (0,0)$, $P_1 = (1,0)$, and $P_2 = (0,1)$, and let $\hat{H}$ denote the space of $\hat{\psi} = (\hat{\psi}_1, \hat{\psi}_2) \in \Pi_2^2$ on $\hat{\tau}$ of the form

$$\begin{aligned} \hat{\psi}_1(\xi) &= \alpha_0 + \alpha_1\xi_1 + \alpha_2\xi_2 + \alpha_3(\xi_1^2 + \xi_1\xi_2), \\ \hat{\psi}_2(\xi) &= \beta_0 + \beta_1\xi_1 + \beta_2\xi_2 + \beta_3(\xi_1\xi_2 + \xi_2^2), \end{aligned} \tag{17.5}$$

where α_j, β_j, $j = 0,1,2,3$, are real numbers, and $\xi = (\xi_1, \xi_2)$. For $\tau \in \mathcal{T}_h$ let F_τ be an affine mapping of $\hat{\tau}$ onto τ, so that $x = F_\tau(\xi) = B_\tau\xi + b_\tau$, where B_τ is a 2×2 matrix and $b_\tau \in \mathbb{R}^2$, and set

$$H(\tau) = \{\psi = B_\tau\hat{\psi} \circ F_\tau^{-1} :\ \hat{\psi} \in \hat{H}\}.$$

For a triangle $\tau \in \mathcal{T}_h$ with two vertices on $\partial\Omega$ we define $\widetilde{\tau}$ to be the obvious extension of τ to a triangle with one curved edge, and set for convenience $\widetilde{\tau} = \tau$ for other triangles τ in $\mathcal{T}_h$. We then define

$$H_h = \{\psi = (\psi_1, \psi_2) \in H;\ \psi|_{\widetilde{\tau}} \in \Pi_2^2,\ \psi|_\tau \in H(\tau), \quad \forall \tau \in \mathcal{T}_h\}.$$

This space thus consists of piecewise quadratics on the triangulation $\mathcal{T}_h$ which are of the specific form implied by the definition of $H(\tau)$, and for boundary triangles these polynomials are extended to the curved boundary.

Let us note that if $\varphi = (\varphi_1, \varphi_2)$ and $\hat{\varphi} = (\hat{\varphi}_1, \hat{\varphi}_2)$ are defined on τ and $\hat{\tau}$, respectively, and related as in the definition of $H(\tau)$, so that $\varphi(F_\tau(\xi)) = B_\tau\hat{\varphi}(\xi)$, then their normal components at their corresponding segments of the boundary are proportional. In fact, if $\hat{n}$ is the normal to a side $\hat{\delta}$ of $\hat{\tau}$, then $\hat{\varphi} \cdot \hat{n} = B_\tau^{-1}\varphi \cdot \hat{n} = \varphi \cdot \widetilde{n}$ with $\widetilde{n} = (B_\tau^{-1})^T\hat{n}$. Further, if $\hat{v}$ is a vector along $\hat{\delta}$, then its image along the corresponding side δ of τ is $B_\tau\hat{v}$, and $(B_\tau\hat{v}) \cdot \widetilde{n} = \hat{v} \cdot (B_\tau^T\widetilde{n}) = \hat{v} \cdot \hat{n} = 0$, so that $\widetilde{n}$ is a normal to δ.

We see from (17.5) that the dimension of $H(\tau)$ is 8. As degrees of freedom for this space we may use the values of $\psi \cdot n$ at two points on each side of τ (6 conditions) and in addition the mean-values of ψ_1 and ψ_2 over τ (2 conditions). In the usual way, in order to show that these values determine a unique element of $H(\tau)$ it suffices to show uniqueness. For this purpose, we first note that the normal component of ψ is linear on each side of τ. For, by the above, it suffices to see that this is the case of the reference triangle, and there we have

$$\hat{\psi} \cdot \hat{n} = \begin{cases} -\hat{\psi}_2 = -\beta_0 - \beta_1 \xi_1, & \text{on } P_0 P_1 \\ -\hat{\psi}_1 = -\alpha_0 - \alpha_2 \xi_2, & \text{on } P_0 P_2, \\ \frac{1}{\sqrt{2}}(\hat{\psi}_1 + \hat{\psi}_2) = \frac{1}{\sqrt{2}}\big((\alpha_0 + \beta_0 + \alpha_2 + \beta_2 + \beta_3) \\ \qquad + (\alpha_1 + \beta_1 - \alpha_2 - \beta_2 + \alpha_3 - \beta_3)\xi_1\big), & \text{on } P_1 P_2. \end{cases}$$

In particular, if $\psi \cdot n$ vanishes at two points on each of the three sides of τ, then the same holds for $\hat{\psi} \cdot \hat{n}$ on $\hat{\tau}$, and we have $\alpha_0 = \beta_0 = \alpha_2 = \beta_1 = \alpha_1 + \alpha_3 = \beta_2 + \beta_3 = 0$, so that $\hat{\psi}$ reduces to

$$\hat{\psi}_1(\xi) = \alpha_1 \xi_1 (1 - \xi_1 - \xi_2), \quad \hat{\psi}_2(\xi) = \beta_2 \xi_2 (1 - \xi_1 - \xi_2).$$

Since ξ_1, ξ_2 and $1 - \xi_1 - \xi_2$ are positive in $\hat{\tau}$ it follows that if the averages of $\hat{\psi}_1$ and $\hat{\psi}_2$ over $\hat{\tau}$ vanish, then we also have $\alpha_1 = \beta_2 = 0$ and hence $\hat{\psi} \equiv 0$ in $\hat{\tau}$ and $\psi \equiv 0$ in τ.

In order to further elucidate the definition of H_h we recall that the condition $\chi \in H$ in the definition of H_h requires that $\operatorname{div} \chi \in L_2$, and observe that this in turn is equivalent to requiring $\chi \cdot n$ to be continuous across inter-element boundaries. In fact, if $\operatorname{div} \chi \in L_2$ we have

$$(\operatorname{div} \chi, \varphi) = -(\chi, \nabla \varphi), \quad \forall \varphi \in C_0^\infty(\Omega).$$

On the other hand, considering two neighboring triangles τ_1 and τ_2 with a common side, and φ with its support in the interior of their union, using Green's formula on each of the triangles separately yields

$$(\operatorname{div} \chi, \varphi) = \int_{\partial \tau_1} (\chi \cdot n) \varphi \, ds + \int_{\partial \tau_2} (\chi \cdot n) \varphi \, ds - (\chi, \nabla \varphi),$$

which shows that, modulo its sign, $\chi \cdot n$ is the same on both sides of the common side of τ_1 and τ_2.

In conclusion we may thus state that the values of $\psi \cdot n$ at two points on each side and the averages over all triangles of the triangulation $\mathcal{T}_h$ uniquely determine an element ψ of H_h.

Our first goal is now to prove the following error estimates for our mixed method for the stationary problem.

Theorem 17.1 *The discrete problem* (17.4) *has a unique solution* $(u_h, \sigma_h) \in S_h \times H_h$. *With* $(u, \sigma) = (u, \nabla u)$ *the solution of* (17.2) *we have*

$$\|u_h - u\| \le C h^2 \|u\|_2 \quad \textit{and} \ \|\sigma_h - \sigma\| \le C h^s \|u\|_{s+1}, \quad s = 1, 2.$$

The proof will require some preparation. In our first lemma we construct an interpolation operator which will be useful in the analysis. Here $H^1 = H^1(\Omega)$.

Lemma 17.1 *There exists a linear operator* $Q_h : H^1 \times H^1 \to H_h$ *such that*

$$(\operatorname{div} Q_h\omega, \chi) = (\operatorname{div} \omega, \chi), \quad \forall \chi \in S_h,\ \omega \in H, \tag{17.6}$$

$$\|Q_h\omega - \omega\| \le Ch^s\|\omega\|_s, \quad \textit{for } s = 1, 2, \tag{17.7}$$

and

$$\|Q_h\omega\| \le C\|\omega\|_1. \tag{17.8}$$

Proof. We define Q_h by requiring that, with $\partial\mathcal{T}_h$ denoting the set of sides of the triangles $\tau \in \mathcal{T}_h$ (note that ω is defined on $\partial\mathcal{T}_h$ when $\omega \in H^1 \times H^1$),

$$\int_\delta (Q_h\omega - \omega) \cdot n\, ds = \int_\delta s(Q_h\omega - \omega) \cdot n\, ds = 0, \quad \forall \delta \in \partial\mathcal{T}_h, \tag{17.9}$$

and

$$\int_\tau (Q_h\omega - \omega)\, dx = 0, \qquad \text{for each } \tau \in \mathcal{T}_h. \tag{17.10}$$

It follows easily from our above discussion that this defines $Q_h\omega$ on each $\tau \in \mathcal{T}_h$, and hence by extension on each $\widetilde{\tau}$, and that the resulting $Q_h\omega$ belongs to H_h.

The first property (17.6) follows by Green's formula applied to each τ: Since χ is linear and $\nabla\chi$ constant, conditions (17.9) and (17.10) yield

$$\int_\tau \operatorname{div}(Q_h\omega - \omega)\chi\, dx = \int_{\partial\tau} \chi(Q_h\omega - \omega) \cdot n ds - \int_\tau \nabla\chi \cdot (Q_h\omega - \omega)\, dx = 0,$$

and (17.6) follows since $\chi = 0$ outside Ω_h. The second statement (17.7) follows by the Bramble-Hilbert lemma, since clearly Q_h reproduces linear functions on each τ, and since the appropriate boundedness condition needed is valid on the reference triangle, namely

$$\|\omega\|_{L_1(\partial\hat{\tau})} + \|\omega\|_{L_1(\hat{\tau})} \le C\|\omega\|_{H^s(\hat{\tau})}, \quad \text{for } s = 1, 2.$$

The inequality (17.8) follows at once from (17.7). □

In our second lemma we show a stability result which will be needed in the existence and uniqueness proof below.

Lemma 17.2 *There is a constant* C *such that if* $v_h \in S_h$ *and* $\omega = (\omega_1, \omega_2) \in L_2^2$ *satisfy*

$$(\omega, \psi) + (v_h, \operatorname{div} \psi) = 0, \quad \forall \psi \in H_h, \tag{17.11}$$

then

$$\|v_h\| \le C\|\omega\|. \tag{17.12}$$

Proof. Let $\varphi \in L_2$ and let g be the solution of

$$(17.13) \qquad -\Delta g = \varphi \quad \text{in } \Omega, \quad \text{with } g = 0 \quad \text{on } \partial\Omega.$$

Then, using (17.6) and (17.11), we have

$$(17.14) \qquad (v_h, \varphi) = -(v_h, \operatorname{div} \nabla g) = -(v_h, \operatorname{div} Q_h \nabla g) = (\omega, Q_h \nabla g),$$

and hence, by (17.8) and the standard elliptic regularity estimate,

$$|(v_h, \varphi)| \le \|\omega\| \, \|Q_h \nabla g\| \le C\|\omega\| \, \|\nabla g\|_1 \le C\|\omega\| \, \|g\|_2 \le C\|\omega\| \, \|\varphi\|,$$

which shows (17.12) and thus proves the lemma. □

We note that locally, on each $\tau \in \mathcal{T}_h$, we have $\operatorname{div} \psi \in \Pi_1$ for $\psi \in H_h$ and thus the restriction of $\operatorname{div} \psi$ to Ω_h agrees there with an element of S_h. However, since $\Omega \neq \Omega_h$, $\operatorname{div} \psi$ does not in general belong to S_h for $\psi \in H_h$, but rather to the space

$$\widetilde{S}_h = \{\widetilde{\chi} \in L_2 : \widetilde{\chi}|_{\widetilde{\tau}} \in \Pi_1, \quad \forall \tau \in \mathcal{T}_h\}.$$

In the following lemma we shall consider a modification $\widetilde{P}_h$ of the L_2-projection P_h onto S_h which uses $\widetilde{S}_h$ as the test space.

Lemma 17.3 *Let $\widetilde{P}_h : L_2 \to S_h$ be defined by*

$$(17.15) \qquad (\widetilde{P}_h v, \widetilde{\chi}) = (v, \widetilde{\chi}), \quad \forall \widetilde{\chi} \in \widetilde{S}_h.$$

Then

$$\|\widetilde{P}_h v - v\| \le Ch^2 \|v\|_2, \quad \text{if } v = 0 \quad \text{on } \partial\Omega.$$

Proof. We first note that (17.15) defines $\widetilde{P}_h v$ uniquely in S_h. For if $v = 0$ we have $(\widetilde{P}_h v, \chi) = 0$ for $\chi \in S_h$, since $\widetilde{P}_h v = 0$ in $\Omega \setminus \Omega_h$.

As is well-known (note that P_h is locally defined on the triangles of $\mathcal{T}_h$),

$$(17.16) \qquad \|P_h v - v\| \le Ch^2 \|v\|_2,$$

and in order to prove the desired result, we shall compare $\widetilde{P}_h$ with P_h. For this purpose, let for $\chi \in S_h$ $\widetilde{\chi}$ denote the associated element in $\widetilde{S}_h$ which agrees with χ on Ω_h. We have by (17.15)

$$\begin{aligned}(\widetilde{P}_h v - P_h v, \chi) &= (\widetilde{P}_h v, \widetilde{\chi}) - (P_h v, \chi) = (v, \widetilde{\chi} - \chi) \\ &= \int_{\Omega \setminus \Omega_h} v \widetilde{\chi} \, dx \le \|v\|_{L_2(\Omega \setminus \Omega_h)} \|\widetilde{\chi}\|_{L_2(\Omega \setminus \Omega_h)}.\end{aligned}$$

Note that, uniformly in τ and h, $\|\widetilde{\chi}\|_{L_2(\widetilde{\tau} \setminus \tau)} \le C\|\widetilde{\chi}\|_{L_2(\tau)}$ for $\widetilde{\chi} \in \widetilde{S}_h$, which shows that $\|\widetilde{\chi}\|_{L_2(\Omega \setminus \Omega_h)} \le C\|\widetilde{\chi}\|_{L_2(\Omega)} = C\|\chi\|$ and hence

$$\|\widetilde{P}_h v - P_h v\| \le C\|v\|_{L_2(\Omega\setminus\Omega_h)}.$$

Since $\operatorname{dist}(x, \partial\Omega) \le Ch^2$ for each point of $\Omega \setminus \Omega_h$ we have, for v vanishing on $\partial\Omega$,

$$\|v\|_{L_2(\Omega\setminus\Omega_h)} \le Ch^2\|\nabla v\|_{L_2(\Omega\setminus\Omega_h)} \le Ch^2\|v\|_1,$$

and we thus conclude that for such v,

$$\|\widetilde{P}_h v - P_h v\| \le Ch^2\|v\|_1.$$

Together with (17.16) this completes the proof. □

The following final lemma is the main ingredient in the proof of the error estimate for u_h.

Lemma 17.4 *There is a constant C such that for $v_h \in S_h$ and $\omega \in H$ satisfying*

$$(17.17) \qquad \begin{aligned} (\omega, \psi) + (v_h, \operatorname{div}\psi) &= 0, \quad \forall \psi \in H_h, \\ (\operatorname{div}\omega, \chi) &= 0, \quad \forall \chi \in S_h, \end{aligned}$$

we have

$$(17.18) \qquad \|v_h\| \le C(h\|\omega\| + h^2\|\operatorname{div}\omega\|).$$

Proof. As in the proof of Lemma 17.2, let $\varphi \in L_2$ and let g be the solution of (17.13). Since (17.11) holds we then have (17.14) so that

$$(v_h, \varphi) = (\omega, Q_h\nabla g - \nabla g) + (\omega, \nabla g) = I_1 + I_2.$$

Here, by Lemma 17.1 and the elliptic regularity estimate,

$$|I_1| \le \|\omega\|\,\|Q_h\nabla g - \nabla g\| \le Ch\|\omega\|\,\|\nabla g\|_1 \le Ch\|\omega\|\,\|\varphi\|,$$

and using Green's formula and the second equation in (17.17),

$$I_2 = -(\operatorname{div}\omega, g) = (\operatorname{div}\omega, \widetilde{P}_h g - g),$$

so that by Lemma 17.3

$$|I_2| \le \|\operatorname{div}\omega\|\,\|\widetilde{P}_h g - g\| \le Ch^2\|\operatorname{div}\omega\|\,\|g\|_2 \le Ch^2\|\operatorname{div}\omega\|\,\|\varphi\|.$$

Altogether,

$$|(v_h, \varphi)| \le C(h\|\omega\| + h^2\|\operatorname{div}\omega\|)\|\varphi\|,$$

which shows (17.18) and completes the proof. □

Proof of Theorem 17.1. As usual in a linear finite dimensional problem with the same number of equations as unknowns, in order to show the existence, it suffices to prove uniqueness. Thus let $f = 0$. By setting $\chi = u_h, \psi = \sigma_h$ in (17.4) we obtain

$$\|\sigma_h\|^2 = -(u_h, \operatorname{div} \sigma_h) = 0,$$

so that $\sigma_h = 0$. By Lemma 17.2 we now conclude that $u_h = 0$ which shows the uniqueness.

In the error analysis we shall begin with the estimate for $\sigma_h - \sigma$. In view of (17.7) it suffices to show

$$\|\sigma_h - \sigma\| \le \|Q_h\sigma - \sigma\|. \tag{17.19}$$

For this purpose we note that by (17.6), (17.3) and (17.4) we have

$$\begin{aligned}(\operatorname{div}(Q_h\sigma - \sigma_h), \chi) &= (\operatorname{div} \sigma, \chi) - (\operatorname{div} \sigma_h, \chi)\\ &= -(f, \chi) + (f, \chi) = 0, \quad \text{for } \chi \in S_h.\end{aligned}$$

Thus $\operatorname{div}(Q_h\sigma - \sigma_h)$ vanishes on Ω_h, and hence, since it is linear in each $\widetilde{\tau}$, also in Ω. But, by (17.3) and (17.4),

$$(\sigma_h - \sigma, \psi) + (u_h - u, \operatorname{div} \psi) = 0, \quad \forall \psi \in H_h, \tag{17.20}$$

so that, in particular, with $\psi = \sigma_h - Q_h\sigma$, we have $(\sigma_h - \sigma, \sigma_h - Q_h\sigma) = 0$. Hence

$$\|\sigma_h - \sigma\|^2 = (\sigma_h - \sigma, Q_h\sigma - \sigma) \le \|\sigma_h - \sigma\| \, \|Q_h\sigma - \sigma\|,$$

which proves (17.19).

For the estimate for $u_h - u$ we note that since $\operatorname{div} \psi \in \widetilde{S}_h$ for $\psi \in H_h$, we have, by our definition (17.15),

$$(u, \operatorname{div} \psi) = (\widetilde{P}_h u, \operatorname{div} \psi), \quad \forall \psi \in H_h,$$

and hence, by (17.20),

$$(\sigma_h - \sigma, \psi) + (u_h - \widetilde{P}_h u, \operatorname{div} \psi) = 0, \quad \forall \psi \in H_h.$$

Since further, by (17.3) and (17.4)

$$(\operatorname{div}(\sigma_h - \sigma), \chi) = 0, \quad \forall \chi \in S_h, \tag{17.21}$$

we conclude from Lemma 17.4 that

$$\|u_h - \widetilde{P}_h u\| \le C\big(h\|\sigma_h - \sigma\| + h^2\|\operatorname{div}(\sigma_h - \sigma)\|\big). \tag{17.22}$$

Here, by above,

$$\|\sigma_h - \sigma\| \le Ch\|u\|_2.$$

We note now by considering each boundary triangle separately that

$$\|\operatorname{div}\sigma_h\| \le C\|\operatorname{div}\sigma_h\|_{L_2(\Omega_h)}.$$

Hence choosing $\chi = \operatorname{div}\sigma_h|_{\Omega_h} \in S_h$ in (17.21) we have

$$\|\operatorname{div}\sigma_h\| \le C\|\operatorname{div}\sigma\| \le C\|u\|_2,$$

so that altogether (17.22) yields

$$\|u_h - \widetilde{P}_h u\| \le Ch^2\|u\|_2.$$

In view of Lemma 17.3 this completes the proof. □

By a refinement of the above arguments it is also possible to show an almost optimal order maximum-norm error estimate for the first component of the solution, namely

$$\|u_h - u\|_{L_\infty} \le Ch^2\ell_h\|u\|_3, \quad \text{where } \ell_h = \max(1, \log(1/h)).$$

We shall not give the details here.

We may think of the solution $(u_h, \sigma_h) \in S_h \times H_h$ of (17.4) with $f \in L_2$ as the result of a pair of operators $(T_h, M_h) : L_2 \to S_h \times H_h$ defined by $T_h f = u_h, M_h f = \sigma_h$. With $T : L_2 \to H^2 \cap H_0^1$ the solution operator of the continuous problem (17.1) we may now state that the conditions (i) and (ii) of Chapter 2 are satisfied with $r = 2$:

Lemma 17.5 *Let u_h be the first component of the solution of* (17.4). *Then the operator $T_h : L_2 \to S_h$, defined by $T_h f = u_h$, is selfadjoint, positive semidefinite on L_2 and positive definite on S_h. Further,*

$$\|T_h f - Tf\| \le Ch^2\|f\|.$$

Proof. The discrete problem (17.4) may be written

$$\begin{aligned} (\operatorname{div} M_h f, \chi) + (f, \chi) &= 0, \quad \forall \chi \in S_h, \\ (M_h f, \psi) + (T_h f, \operatorname{div}\psi) &= 0, \quad \forall \psi \in H_h. \end{aligned} \tag{17.23}$$

By these relations we have

$$(f, T_h g) = -(\operatorname{div} M_h f, T_h g) = (M_h f, M_h g), \quad \forall f, g \in L_2,$$

which shows that T_h is selfadjoint and positive semidefinite on L_2. Let now $f_h \in S_h$ be such that $T_h f_h = 0$. Then $M_h f_h = 0$ by (17.23) and hence $\|f_h\|^2 = -(f_h, \operatorname{div} M_h f_h) = 0$, so that $f_h = 0$, which shows that T_h is positive definite on S_h. The error estimate follows at once by Theorem 17.1. □

We now turn to the parabolic problem

$$\begin{aligned} u_t - \Delta u &= f \quad \text{in } \Omega, \qquad \text{for } t > 0, \\ u &= 0 \quad \text{on } \partial\Omega, \quad \text{for } t > 0, \quad \text{with } u(\cdot, 0) = v \quad \text{in } \Omega. \end{aligned}$$

Introducing again $\sigma = \nabla u$, the pair $(u, \sigma) \in L_2 \times H$ satisfies

$$(17.24) \qquad \begin{aligned} &(u_t, \varphi) - (\operatorname{div} \sigma, \varphi) = (f, \varphi), \quad \forall \varphi \in L_2,\ t > 0, \\ &(\sigma, \omega) + (u, \operatorname{div} \omega) = 0, \qquad \forall \omega \in H,\ t > 0, \quad u(0) = v, \end{aligned}$$

and we are led to consider the semidiscrete problem to find $(u_h, \sigma_h) \in S_h \times H_h$ such that

$$(17.25) \qquad \begin{aligned} &(u_{h,t}, \chi) - (\operatorname{div} \sigma_h, \chi) = (f, \chi), \quad \forall \chi \in S_h,\ t > 0, \\ &(\sigma_h, \psi) + (u_h, \operatorname{div} \psi) = 0, \quad \forall \psi \in H_h,\ t > 0, \quad u_h(0) = v_h, \end{aligned}$$

where v_h is some approximation of v in S_h. Note that $u_h(0)$ determines $\sigma_h(0)$ by the second equation in (17.25).

Introducing bases in S_h and H_h this problem may be written in matrix form as

$$\mathcal{B}U_t - \mathcal{K}\Sigma = F, \quad \mathcal{K}^T U + \mathcal{L}\Sigma = 0, \quad \text{for } t > 0, \quad \text{with } U(0) \text{ given},$$

where U and Σ are the vectors corresponding to u_h and σ_h, respectively, and where A and D are positive definite. After elimination of Σ we get the linear system of ordinary differential equations

$$\mathcal{B}U_t + \mathcal{A}U = F, \quad \text{with } \mathcal{A} = \mathcal{K}\mathcal{L}^{-1}\mathcal{K}^T, \quad \text{for } t > 0, \quad \text{with } U(0) \text{ given},$$

which clearly has a unique solution.

Recalling the definition of the operator T_h above, our problem may also be written

$$(17.26) \qquad T_h u_{h,t} + u_h = T_h f, \quad \text{for } t > 0, \quad \text{with } u_h(0) = v_h,$$

and since T_h is positive definite on S_h, this again shows that (17.25) has a unique solution $u_h : [0, \infty) \to S_h$. Once u_h has been determined, σ_h may be found from the second equation of (17.25).

The representation (17.26) of the semidiscrete problem together with Lemma 17.5 puts the present problem into the framework introduced in Chapter 2, and the appropriate error estimates of Chapters 2, 3 and 6 may therefore apply. It may also be used to formulate fully discrete schemes and show error estimates corresponding to those in Chapters 7, 8 and 10.

In our first result below we shall derive error estimates for the inhomogeneous equation by means of the energy method. This has the advantage that we analyze simultaneously the errors in u_h and σ_h. In doing so we shall use an analogue in the present context of the elliptic projection of the exact solution which we define here to be the pair

$$(17.27) \qquad (\widetilde{u}_h, \widetilde{\sigma}_h) = (-T_h \Delta u, -M_h \Delta u) \in S_h \times H_h,$$

that is, the solution of the discrete elliptic problem with exact solution $(u, \nabla u)$. We shall use for our discrete initial data $\widetilde{u}_h(0)$, which we may think of as the ordinary elliptic projection $R_h v = -T_h \Delta v$ onto S_h.

Theorem 17.2 *Let* (u_h, σ_h) *and* $(u, \sigma) = (u, \nabla u)$ *the solutions of* (17.25) *and* (17.24), *with* $v_h = R_h v = -T_h \Delta v$. *Then, for* $t \geq 0$,

$$\|u_h(t) - u(t)\| \leq Ch^2 \Big(\|u(t)\|_2 + \int_0^t \|u_t\|_2 \, ds \Big) \tag{17.28}$$

and

$$\|\sigma_h(t) - \sigma(t)\| \leq Ch^2 \Big(\|u(t)\|_3 + \big(\int_0^t \|u_t\|_2^2 \, ds \big)^{1/2} \Big). \tag{17.29}$$

Proof. With $(\widetilde{u}_h, \widetilde{\sigma}_h)$ defined by (17.27), we set $\theta = u_h - \widetilde{u}_h, \rho = \widetilde{u}_h - u$, and $\varepsilon = \sigma_h - \widetilde{\sigma}_h$. Recall from Theorem 17.1 that

$$\begin{aligned} \|\rho(t)\| &= \|\widetilde{u}_h(t) - u(t)\| \leq Ch^2 \|u(t)\|_2, \\ \|\widetilde{\sigma}_h(t) - \sigma(t)\| &\leq Ch^2 \|u(t)\|_3, \end{aligned} \tag{17.30}$$

so that it remains to estimate θ and ε.

Using the variational formulation we have

$$\begin{aligned} (\theta_t, \chi) - (\operatorname{div} \varepsilon, \chi) &= -(\rho_t, \chi), \quad \forall \chi \in S_h, \ t > 0, \\ (\varepsilon, \psi) + (\theta, \operatorname{div} \psi) &= 0, \quad \forall \psi \in H_h, \ t > 0. \end{aligned} \tag{17.31}$$

Setting $\chi = \theta, \psi = \varepsilon$ and adding we obtain

$$\tfrac{1}{2} \frac{d}{dt} \|\theta\|^2 + \|\varepsilon\|^2 = -(\rho_t, \theta), \quad \text{for } t > 0,$$

and hence, since $\theta(0) = 0$, in the standard fashion,

$$\|\theta(t)\| \leq \int_0^t \|\rho_t\| \, ds \leq Ch^2 \int_0^t \|u_t\|_2 \, ds,$$

which completes the proof of (17.28).

In order to show (17.29) we first differentiate the second equation in (17.31) with respect to t, then set $\chi = 2\theta_t, \psi = 2\varepsilon$, and add to obtain

$$\frac{d}{dt} \|\varepsilon\|^2 + 2\|\theta_t\|^2 = -2(\rho_t, \theta_t) \leq \|\rho_t\|^2 + \|\theta_t\|^2. \tag{17.32}$$

We now note that since $\theta(0) = 0$ we have $\varepsilon(0) = 0$. Integration of (17.32) together with the standard estimate for ρ_t therefore shows that

$$\|\varepsilon(t)\|^2 \leq \int_0^t \|\rho_t\|^2 \, ds \leq Ch^4 \int_0^t \|u_t\|_2^2 \, ds,$$

which completes the proof of (17.29) and hence of the theorem. □

We shall now discuss some error estimates for the homogeneous equation and begin with a smooth data estimate. We shall use the spaces $\dot{H}^s = \dot{H}^s(\Omega)$ as in Chapter 3.

Theorem 17.3 *Let (u_h, σ_h) and (u, σ) be the solutions of the homogeneous cases ($f = 0$) of (17.25) and (17.24), with $v_h = R_h v$. Then we have, for $t \geq 0$,*

$$\|u_h(t) - u(t)\| \leq Ch^2 |v|_2, \quad \text{if } v \in \dot{H}^2$$

and

$$\|\sigma_h(t) - \sigma(t)\| \leq Ch^2 |v|_3, \quad \text{if } v \in \dot{H}^3.$$

Proof. In view of Lemma 17.5 and the representation (17.26), the first estimate follows at once from Theorem 3.1 and the second from Theorem 17.2 upon noticing that $\|u(t)\|_3 \leq C\|v\|_3$, and, with the notation of Chapter 3,

$$\int_0^t \|u_t\|_2^2\, ds \leq C \int_0^t \|u\|_4^2\, ds \leq C \int_0^t \sum_{j=1}^{\infty} \lambda_j^4 e^{-2\lambda_j s} (v, \varphi_j)^2\, ds$$
$$\leq C \sum_{j=1}^{\infty} \lambda_j^3 (v, \varphi_j)^2 \leq C|v|_3^2. \qquad \square$$

We shall end by showing a nonsmooth data estimate for the homogeneous equations.

Theorem 17.4 *Lt (u_h, σ_h) and (u, σ) be the solutions of the homogeneous cases ($f = 0$) of equations (17.25) and (17.24), now with $v_h = P_h v$. Then we have, for $t > 0$,*

$$\|u_h(t) - u(t)\| \leq Ch^2 t^{-1} \|v\| \tag{17.33}$$

and

$$\|\sigma_h(t) - \sigma(t)\| \leq Ch^2 t^{-3/2} \|v\|. \tag{17.34}$$

Proof. It follows from Theorems 3.3 and 3.4 that for $j \geq 0$,

$$\|D_t^j (u_h(t) - u(t))\| \leq Ch^2 t^{-1-j} \|v\|, \quad \text{for } t > 0,$$

which includes (17.33) as the special case $j = 0$. For the purpose of showing (17.34) we use again the elliptic projection $(\widetilde{u}_h, \widetilde{\sigma}_h)$ defined by (17.27) and find, as in the proof of Theorem 17.2,

$$(\theta_t, \theta) + \|\varepsilon\|^2 = -(\rho_t, \theta),$$

so that

$$\|\varepsilon\|^2 \leq (\|\rho_t\| + \|\theta_t\|)\|\theta\|. \tag{17.35}$$

Here

$$\|\theta(t)\| \le \|u_h(t) - u(t)\| + \|\rho(t)\| \le Ch^2 t^{-1}\|v\|,$$

$$\|\rho_t(t)\| \le Ch^2\|u_t(t)\|_2 \le Ch^2 t^{-2}\|v\|$$

and

$$\|\theta_t(t)\| \le \|u_{h,t}(t) - u_t(t)\| + \|\rho_t(t)\| \le \; Ch^2 t^{-2}\|v\|,$$

so that (17.35) shows

$$\|\varepsilon(t)\| \le Ch^2 t^{-3/2}\|v\|.$$

Since by (17.30),

$$\|\widetilde{\sigma}_h(t) - \sigma(t)\| \le Ch^2\|u(t)\|_3 \le Ch^2 t^{-3/2}\|v\|,$$

this competes the proof of (17.34) and thus of the theorem. □

As was the case for the stationary problem, our above error analysis may be refined to yield almost optimal order maximum-norm error estimates for $u_h(t)$. These error bounds for the error in the uniform norm corresponding to Theorems 17.2, 17.3 and 17.4 are all obtained by multiplication of the error bound given for $\sigma_h(t)$ by ℓ_h, as for instance in the case of Theorem 17.4,

$$\|u_h(t) - u(t)\|_{L_\infty} \le Ch^2 \ell_h t^{-3/2}\|v\|.$$

We shall not carry out the details.

The mixed method discussed above is a special case of a family of such methods introduced for the stationary problem in polygonal domains by Raviart and Thomas in [204] and further studied in, e.g., Falk and Osborn [97]. The present analysis, with the application to the parabolic problem, is from Johnson and Thomée [132], where the method was also adapted to the stationary and evolutionary Stokes equations.

For some more recent work on mixed methods for parabolic equations, see Scholz [211], where optimal order maximum-norm error estimates are shown, and Squeff [218] where asymptotic expansions are used to derive superconvergence results.

18. A Singular Problem

In this chapter we shall study the numerical solution of a singular parabolic equation in one space dimension which arises after reduction by polar coordinates of a radially symmetric parabolic equation in three space dimensions. We shall analyze and compare finite element discretizations based on two different variational formulations.

We consider thus the initial-boundary value problem

$$
\begin{aligned}
(18.1)\qquad & u_t - u_{xx} - 2x^{-1}u_x + q(x)u = f(x) \quad \text{for } x \in I = (0,1),\ t > 0,\\
& u_x(0,t) = u(1,t) = 0, \quad \text{for } t > 0, \quad \text{with } u(x,0) = v(x) \text{ for } x \in I,
\end{aligned}
$$

and, as a preparation, also its stationary analogue

$$
(18.2)\qquad -u'' - 2x^{-1}u' + qu = f \quad \text{in } I, \quad u'(0) = u(1) = 0,
$$

where q is a smooth bounded nonnegative function on I. If u is a solution of

$$
\begin{aligned}
u_t - \Delta u + qu = f \quad & \text{in } B, \qquad \text{for } t > 0,\\
u = 0 \quad & \text{on } \partial B, \quad \text{for } t > 0, \quad \text{with } u(\cdot,0) = v \quad \text{in } B,
\end{aligned}
$$

where B is the unit ball $B = B_1(0) \subset \mathbb{R}^3$, and where q, f and v depend only on $|x|$, then transformation by polar coordinates brings it into the form (18.1), with x denoting the radial coordinate. Note that if $u \in \mathcal{C}^2(\bar{I})$ and u satisfies the differential equation in (18.2), and if f is bounded at $x = 0$, then the boundary condition at $x = 0$ is automatically satisfied. In fact, it is easy to see that this conclusion holds if $u \in \mathcal{C}^2(I)$ and u and f are bounded near zero. Similar statements hold for (18.1).

We shall discuss finite element methods for solving these problems, using approximating functions of x from the space S_h of continuous functions on I, which vanish at $x = 1$ and reduce to polynomials of degree at most $r - 1$ on each interval $I_j = (x_{j-1}, x_j)$, $j = 1, \ldots, M$, with $x_j = jh, h = 1/M$, and where $r \geq 2$.

We begin with the stationary problem (18.2). A natural variational formulation of this problem arises from noting that the equation may be written

$$
-(x^2u')' + x^2q(x)u = x^2 f, \quad \text{for } x \in I,
$$

and thus a solution of (18.2) also solves

$$A(u, \varphi) := \int_0^1 (x^2 u' \varphi' + x^2 q \, u \, \varphi) \, dx = (x^2 f, \varphi), \quad \forall \varphi \in \dot{H}^1,$$

where $\dot{H}^1$ now denotes the functions in $H^1(I)$ which vanish at $x = 1$, and $(\cdot, \cdot)$ the inner product in $L_2 = L_2(I)$. We may therefore pose the discrete stationary problem to find $u_h \in S_h$ such that

$$A(u_h, \chi) = (x^2 f, \chi), \quad \forall \chi \in S_h. \tag{18.3}$$

We note at once that $A(\cdot, \cdot)$ is a positive definite symmetric bilinear form on $\dot{H}^1$, and that $S_h \subset \dot{H}^1$. In particular, our discrete problem (18.3) admits a unique solution in S_h for f given.

Before we proceed, we shall establish a simple Poincaré type inequality.

Lemma 18.1 *If $\alpha \geq 0$ and $d > 0$ we have*

$$\|x^\alpha v\|_{L_2(0,d)} \leq d \, \|x^\alpha v'\|_{L_2(0,d)}, \quad \textit{if } v(d) = 0.$$

Proof. For $x \in [0, d]$ we have

$$|x^\alpha v(x)| = |x^\alpha \int_x^d s^{-\alpha} s^\alpha v'(s) \, ds| \leq \|x^\alpha v'\|_{L_1(0,d)} \leq d^{1/2} \|x^\alpha v'\|_{L_2(0,d)},$$

from which the result at once follows by integration. □

Using the special case $\alpha = d = 1$, our lemma implies, in particular, that our bilinear form $A(\cdot, \cdot)$ is continuous with respect to the norm $\|xu'\|$ on $\dot{H}^1$, where $\|\cdot\| = \|\cdot\|_{L_2}$. For

$$|A(u, v)| \leq \|xu'\| \, \|xv'\| + \|q\|_{L_\infty} \|xu\| \, \|xv\| \leq (1 + \|q\|_{L_\infty}) \|xu'\| \, \|xv'\|.$$

We may now show the following error estimate for (18.3).

Theorem 18.1 *Under the above assumptions we have for the solutions u_h and u of* (18.3) *and* (18.2), *respectively, that*

$$\|x(u_h - u)\| \leq C h^r \|x u^{(r)}\|.$$

Proof. Setting $e = u_h - u$ we shall first prove directly from the variational formulation that

$$\|xe'\| \leq C h^{r-1} \|x u^{(r)}\|, \tag{18.4}$$

and then, by a duality argument, that

$$\|xe\| \leq Ch \|xe'\|. \tag{18.5}$$

Together these inequalities prove the theorem.

In order to show (18.4) we note that by our definitions

$$A(u_h, \chi) = (x^2 f, \chi) = A(u, \chi), \quad \forall \chi \in S_h,$$

so that

$$\text{(18.6)} \qquad A(e, \chi) = 0, \quad \forall \chi \in S_h.$$

Since q is nonnegative, we hence have

$$\|xe'\|^2 \leq A(e,e) = A(e, \chi - u) \leq C\|xe'\| \, \|x(\chi - u)'\|$$

so that

$$\|xe'\| \leq C \inf_{\chi \in S_h} \|x(\chi - u)'\|.$$

We now choose for χ the interpolant $\widetilde{u}_h$ of u in S_h defined locally on each interval $I_j, j = 2, \dots, M$, by

$$\begin{aligned} \widetilde{u}_h(x_j + kh/(r-1)) &= u(x_j + kh/(r-1)), \\ &\qquad \text{for } k = 0, \dots, r-2, \; j = 1, \dots, M-1, \\ \widetilde{u}_h(1) &= u(1) = 0, \end{aligned}$$

and such that, for the first interval I_1, $\widetilde{u}_h^{(k)}(x_1 - 0) = u^{(k)}(x_1)$ for $k = 0, \dots, r-1$. These conditions clearly determine $\widetilde{u}_h$ uniquely and

$$\|(\widetilde{u}_h - u)'\|_{L_2(I_j)} \leq Ch^{r-1}\|u^{(r)}\|_{L_2(I_j)}, \quad \text{for } j = 1, \dots, M.$$

Hence, excepting the first interval,

$$\begin{aligned} \|x(\widetilde{u}_h - u)'\|_{L_2(I_j)} &\leq x_j \|(\widetilde{u}_h - u)'\|_{L_2(I_j)} \leq Ch^{r-1} x_j \|u^{(r)}\|_{L_2(I_j)} \\ &\leq Ch^{r-1} x_j x_{j-1}^{-1} \|xu^{(r)}\|_{L_2(I_j)} \leq Ch^{r-1}\|xu^{(r)}\|_{L_2(I_j)}, \quad \text{for } j = 2, \dots, M. \end{aligned}$$

For the first interval we have, by repeated use of Lemma 18.1,

$$\|x(\widetilde{u}_h - u)'\|_{L_2(I_1)} \leq h\|x(\widetilde{u}_h - u)''\|_{L_2(I_1)} \leq \dots \leq h^{r-1}\|xu^{(r)}\|_{L_2(I_1)},$$

and we conclude

$$\inf_{\chi \in S_h} \|x(\chi - u)'\| \leq \|x(\widetilde{u}_h - u)'\| \leq Ch^{r-1}\|xu^{(r)}\|,$$

which completes the proof of (18.4).

We now turn to the proof of (18.5), and let ψ denote the solution of

$$\text{(18.7)} \qquad -\psi'' - 2x^{-1}\psi' + q\psi = \varphi \quad \text{in } I, \quad \text{with } \psi'(0) = \psi(1) = 0,$$

where φ is a given smooth function vanishing in a neighborhood of 0, say. Since (18.7) can be interpreted as a three-dimensional spherically symmetric elliptic problem, we may assume that ψ is smooth on $\bar{I}$. We have then, using the orthogonality relation (18.6),

$$|(x^2 e, \varphi)| = |A(e, \psi)| = |A(e, \psi - \chi)| \le C\|xe'\|\,\|x(\psi - \chi)'\|, \quad \forall \chi \in S_h.$$

With $\widetilde{\psi}_h$ a piecewise linear interpolant of ψ we have, as above,

$$\|x(\widetilde{\psi}_h - \psi)'\| \le Ch\|x\psi''\|.$$

We shall show presently that

$$\|x\psi''\| \le C\|x\varphi\|. \tag{18.8}$$

Assuming this for a moment, we conclude $(x^2 e, \varphi)| \le Ch\|xe'\|\,\|x\varphi\|$, from which (18.5) follows at once.

It remains only to show (18.8). By (18.7),

$$\|x\psi''\| \le C(\|\psi'\| + \|x\psi\| + \|x\varphi\|).$$

We have, by Lemma 18.1 and (18.7),

$$\|x\psi\|^2 \le A(\psi, \psi) = (x^2\varphi, \psi) \le \|x\varphi\|\,\|x\psi\|,$$

so that $\|x\psi\| \le \|x\varphi\|$. The proof is thus complete if we show

$$\|\psi'\| \le C\|x\varphi\|. \tag{18.9}$$

But multiplying (18.7) by $-x\psi'$ and integrating we have

$$\begin{aligned}(x\psi'', \psi') + 2\|\psi'\|^2 &= -(x(\varphi - q\psi), \psi')\\ &\le (\|x\varphi\| + \|q\|_{L_\infty}\|x\psi\|)\|\psi'\| \le C\|x\varphi\|\,\|\psi'\|.\end{aligned}$$

Here

$$(x\psi'', \psi') = \left[\tfrac{1}{2}x\psi'^2\right]_0^1 - \tfrac{1}{2}\|\psi'\|^2 \ge -\tfrac{1}{2}\|\psi'\|^2,$$

so that altogether $\frac{3}{2}\|\psi'\|^2 \le C\|x\varphi\|\,\|\psi'\|$. This completes the proof of (18.9) and thus of our theorem. □

We now address the time dependent problem (18.1) and define a spatially semidiscrete analogue by

$$(x^2 u_{h,t}, \chi) + A(u_h, \chi) = (x^2 f, \chi), \quad \forall \chi \in S_h, \quad \text{for } t > 0, \tag{18.10}$$

with $u_h(0) = v_h$. This problem clearly admits a unique solution and we have:

Theorem 18.2 *Let u be the solution of* (18.1) *and u_h that of* (18.10). *Then, with $v_h = u_h(0)$ appropriately chosen, we have*

$$\|x(u_h(t) - u(t))\| \le Ch^r\Big(\|xv^{(r)}\| + \int_0^t \|su_t^{(r)}(s)\|\,ds\Big), \quad \text{for } t \ge 0.$$

Proof. The proof will proceed along well established lines. We define an elliptic projection R_h^A onto S_h by

$$A(R_h^A u - u, \chi) = 0, \quad \forall \chi \in S_h,$$

and write $u_h - u = (u_h - R_h^A u) + (R_h^A u - u) = \theta + \rho$. From Theorem 18.1 we conclude at once that

$$\|x\rho(t)\| \le Ch^r \|xu^{(r)}(t)\| \le Ch^r \Big(\|xv^{(r)}\| + \int_0^t \|xu_t^{(r)}\| \, ds \Big),$$

and it remains to bound θ. We have

$$(x^2\theta_t, \chi) + A(\theta, \chi) = -(x^2\rho_t, \chi), \quad \forall \chi \in S_h, \ t > 0,$$

and hence, setting $\chi = 2\theta$, using the positivity of $A(\theta, \theta)$, and integration, we have, with $u_h(0) = R_h^A v$,

$$\|x\theta(t)\| \le \|x\theta(0)\| + \int_0^t \|s\rho_t\| \, ds \le Ch^r \Big(\|xv^{(r)}\| + \int_0^t \|xu_t^{(r)}\| \, ds \Big).$$

Together our estimates complete the proof. □

Numerical experiments show that the above methods for solving our singular problems produce approximate solutions for which the error is relatively large near $x = 0$. This is not surprising since our variational formulation contains the weight factor x^2 and thus the values of our functions have less influence when x is smaller. In order to modify the method so as to get a more even distribution of the error, we shall now consider an alternative weak formulation of our problem which gives more weight to these function values.

We begin with the stationary problem which we first write as

$$-xu'' - 2u' + xq(x)u = xf(x) \quad \text{for } x \in I, \quad \text{with } u'(0) = u(1) = 0.$$

Multiplication by φ, integration over I, and integration by parts in the first term shows that the solution of (18.1) satisfies

$$B(u, \varphi) = (xu', \varphi') - (u', \varphi) + (xqu, \varphi) = (xf, \varphi), \quad \forall \varphi \in \dot{H}^1. \tag{18.11}$$

This variational formulation thus uses a bilinear form $B(\cdot, \cdot)$ which is nonsymmetric, but it is still positive, as

$$B(v, v) = \|x^{1/2}v'\|^2 + \tfrac{1}{2}v(0)^2 + \|(xq)^{1/2}v\|^2, \quad \text{if } v(1) = 0.$$

We may now pose the discrete problem to find $u_h \in S_h$ such that

$$B(u_h, \chi) = (xf, \chi), \quad \forall \chi \in S_h. \tag{18.12}$$

By the positivity of $B(\cdot, \cdot)$ this problem admits a unique solution u_h and

$$B(u_h - u, \chi) = 0, \quad \forall \chi \in S_h. \tag{18.13}$$

The most natural norm for the analysis appears now to be $\|x^{1/2}v\| = (xv, v)^{1/2}$, and we should then expect a less marked increase of the error near the origin. Instead of pursuing the error analysis in this weighted norm, we shall directly derive a uniform error bound. For simplicity of presentation, we shall restrict our considerations to the case $r = 2$, that is, we shall consider piecewise linear approximations only. We set $\|\cdot\|_{L_\infty} = \|\cdot\|_{L_\infty(I)}$.

Theorem 18.3 *Let $r = 2$ and let u_h and u be the solutions of* (18.12) *and* (18.2), *respectively. Then we have*

$$\|u_h - u\|_{L_\infty} \leq Ch^2 \|u''\|_{L_\infty}.$$

Proof. Setting again $e = u_h - u$ we shall first show that

$$\|e\|_{L_\infty} \leq Ch\|e'\|_{L_\infty}, \tag{18.14}$$

and then that

$$\|e'\|_{L_\infty} \leq Ch\|u''\|_{L_\infty}. \tag{18.15}$$

Together these estimates prove the desired result.

We begin by showing (18.14). For φ given, let ψ be the solution of

$$-\psi'' + q\psi = \varphi \quad \text{in } I, \quad \text{with } \psi(0) = \psi(1) = 0. \tag{18.16}$$

We then have, for any $\chi \in S_h$,

$$(xe, \varphi) = (xe, -\psi'' + q\psi) = ((xe)', \psi') + (xqe, \psi) = B(e, \psi) = B(e, \psi - \chi),$$

where in the last step we have used (18.13), and hence

$$|(xe, \varphi)| \leq C\|e'\|_{L_\infty} \big(\|x(\psi - \chi)'\|_{L_1} + \|\psi - \chi\|_{L_1}\big).$$

We next show

$$\inf_{\chi \in S_h} (\|x(\psi - \chi)'\|_{L_1} + \|\psi - \chi\|_{L_1}) \leq Ch\|x\psi''\|_{L_1}, \tag{18.17}$$

and

$$\|x\psi''\|_{L_1} \leq C\|x\varphi\|_{L_1}. \tag{18.18}$$

Together, (18.17) and (18.18) yield $|(e, x\varphi)| \leq Ch\|e'\|_{L_\infty}\|x\varphi\|_{L_1}$, and hence (18.14).

For (18.17) we note that the piecewise linear interpolant $\widetilde{\psi}_h$ of ψ satisfies

$$h\|(\widetilde{\psi}_h - \psi)'\|_{L_1(I_i)} + \|\widetilde{\psi}_h - \psi\|_{L_1(I_i)} \leq Ch^2\|\psi''\|_{L_1(I_i)}, \quad \text{for } i = 1, \dots, M.$$

It follows easily for all intervals I_i except the first, that

$$\|x(\widetilde{\psi}_h - \psi)'\|_{L_1(I_i)} + \|\widetilde{\psi}_h - \psi\|_{L_1(I_i)} \le Ch\|x\psi''\|_{L_1(I_i)}. \tag{18.19}$$

Defining $\widetilde{\psi}_h$ on I_1 by $\widetilde{\psi}_h^{(l)}(x_1 - 0) = \psi^{(l)}(x_1)$, for $l = 0, 1$, we have

$$|(\widetilde{\psi}_h - \psi)(x)| = \left| \int_x^h s\psi''(s)\, ds \right| \le \|x\psi''\|_{L_1(I_1)}, \quad \text{for } x \in I_1,$$

and hence, after integration, $\|\widetilde{\psi}_h - \psi\|_{L_1(I_1)} \le h\|x\psi''\|_{L_1(I_1)}$. Similarly

$$|x(\widetilde{\psi}_h' - \psi')(x)| = \left| x \int_x^{x_1} \psi''(s)\, ds \right| \le \|x\psi''\|_{L_1(I_1)},$$

and hence $\|x(\widetilde{\psi}_h - \psi)'\|_{L_1(I_1)} \le h\|x\psi''\|_{L_1(I_1)}$. This shows (18.19) for $i = 1$ and thus completes the proof of (18.17).

Turning to (18.18), we note that with G^x the Green's function for (18.16), we may write $\psi(x) = \int_0^1 G^x(y)\varphi(y)\, dy$. It is easily seen that $|G^x(y)| \le C(q)y$, for $0 \le y \le 1$, and hence $\|\psi\|_{L_\infty} \le C\|x\varphi\|_{L_1}$, whence, using also the differential equation,

$$\|x\psi''\|_{L_1} \le \|x\varphi\|_{L_1} + C\|x\psi\|_{L_1} \le C\|x\varphi\|_{L_1}.$$

This proves (18.18) and thus completes the proof of (18.14).

In order to demonstrate (18.15), we introduce this time the elliptic projection R_h^B onto S_h defined by

$$B(R_h^B v - v, \chi) = 0, \quad \forall \chi \in S_h, \tag{18.20}$$

so that $u_h = R_h^B u$. We shall show

$$\|(R_h^B v)'\|_{L_\infty} \le C\|v'\|_{L_\infty}. \tag{18.21}$$

With this already proved we have with $\widetilde{u}_h$ a suitable interpolant of u,

$$\begin{aligned} \|e'\|_{L_\infty} &= \|(R_h^B u - u)'\|_{L_\infty} = \|((R_h^B - I)(u - \widetilde{u}_h))'\|_{L_\infty} \\ &\le C\|(\widetilde{u}_h - u)'\|_{L_\infty} \le Ch\|u''\|_{L_\infty}, \end{aligned}$$

which is (18.15).

To prove (18.21), we set $v_h = R_h^B v$ and write (18.20) in the form

$$B_0(v_h, \chi) = B_0(v, \chi) - (xq(v_h - v), \chi), \quad \forall \chi \in S_h, \tag{18.22}$$

where $B_0(v, w) = (v', xw' - w)$. We now introduce a basis $\{\varphi_i\}_{i=1}^M$ for the trial functions by

$$\varphi_i(x) = \begin{cases} -h, & \text{for } x \le x_{i-1}, \\ x - x_i, & \text{for } x_{i-1} < x < x_i, \\ 0, & \text{for } x \ge x_i, \end{cases}$$

and set $v_h = \sum_{i=1}^M w_i \varphi_i$. We have at once

$$\|v_h'\|_{L_\infty} = \max_i |w_i|, \tag{18.23}$$

and by (18.22) the w_i are determined by

$$\sum_{i=1}^M w_i B_0(\varphi_i, \chi) = B_0(v, \chi) - (xq(v_h - v), \chi), \quad \forall \chi \in S_h. \tag{18.24}$$

To bound them we shall choose for the χ the elements of the basis $\{\psi_j\}_{j=1}^M$ for the test functions defined by

$$\psi_j(x) = \begin{cases} \varphi_j(x), & \text{for } x \ge x_{j-1}, \\ -hx/x_{j-1}, & \text{for } 0 \le x \le x_{j-1}. \end{cases}$$

As a simple calculation shows, these functions are such that $B_0(\varphi_i, \psi_j) = \delta_{ij} h x_i$. Thus, setting $\chi = \psi_j$ in (18.24) we have, for $j = 1, \dots, M$,

$$\begin{aligned} |w_j h x_j| &= |B_0(v, \psi_j) - (xq(v_h - v), \psi_j)| \\ &\le \|v'\|_{L_\infty} \|x\psi_j' - \psi_j\|_{L_1} + C\|v_h - v\|_{L_\infty} \|\psi_j\|_{L_1} \\ &= \|v'\|_{L_\infty} h x_j + C\|v_h - v\|_{L_\infty} h x_j/2, \end{aligned}$$

and hence $|w_j| \le \|v'\|_{L_\infty} + C\|v_h - v\|_{L_\infty}$. By (18.23) and (18.14) this yields

$$\|v_h'\|_{L_\infty} \le \|v'\|_{L_\infty} + Ch\|v_h' - v'\|_{L_\infty} \le C\|v'\|_{L_\infty} + Ch\|v_h'\|_{L_\infty},$$

which implies (18.21) for h small. The proof of (18.15) and thus of Theorem 18.3 is now complete. □

We finally consider the time-dependent analogue of the nonsymmetric method (18.12), i.e.,

$$(xu_{h,t}, \chi) + B(u_h, \chi) = (xf, \chi), \quad \forall \chi \in S_h, \ t > 0, \tag{18.25}$$

with $u_h(0) = v_h$ and $B(\cdot,\cdot)$ defined in (18.11). Recall that R_h^B is defined by (18.20), and that $\ell_h = \max(1, \log(1/h))$. We show the following.

Theorem 18.4 *Assume $r = 2$, and let u be the solution of (18.1) and u_h that of (18.25), with $v_h = R_h^B v$. Then, for $t \ge 0$,*

$$\|u_h(t) - u(t)\|_{L_\infty} \le Ch^2 \ell_h^{1/2} \Big(\|u''(t)\|_{L_\infty} + \|u_t''(0)\|_{L_\infty} + \int_0^t \|u_{tt}''\|_{L_\infty} \, ds \Big).$$

Proof. We write $u_h - u = (u_h - R_h^B u) + (R_h^B u - u) = \theta + \rho$, and recall from Theorem 18.3 that $\|\rho(t)\|_{L_\infty} \le Ch^2\|u''(t)\|_{L_\infty}$, so that it only remains to estimate $\theta(t)$. Since $\theta \in S_h$ we have

$$\|\theta\|_{L_\infty} \le \|\theta'\|_{L_1} \le C\ell_h^{1/2}\|x^{1/2}\theta'\|. \tag{18.26}$$

In fact, using the finite dimensionality of S_h on I_1, we have $\|\theta'\|_{L_1(0,h)} \le C\|\theta'\|_{L_1(h/2,h)}$, and hence

$$\|\theta'\|_{L_1(0,1)} \le C\|\theta'\|_{L_1(h/2,1)} \le C\Big(\int_{h/2}^1 \frac{ds}{s}\Big)^{1/2}\|x^{1/2}\theta'\| \le C\ell_h^{1/2}\|x^{1/2}\theta'\|.$$

In view of (18.26) it remains to show

$$\|x^{1/2}\theta'(t)\| \le Ch^2\Big(\|u_t''(0)\|_{L_\infty} + \int_0^t \|u_{tt}''\|_{L_\infty}\,ds\Big). \tag{18.27}$$

By (18.25), its analogue for the solution of (18.1), and (18.20), we have

$$(x\theta_t, \chi) + B(\theta, \chi) = -(x\rho_t, \chi), \quad \forall \chi \in S_h,\ t > 0, \tag{18.28}$$

and setting $\chi = \theta$ we find

$$\|x^{1/2}\theta'\|^2 \le B(\theta,\theta) \le \big(\|x^{1/2}\theta_t\| + \|x^{1/2}\rho_t\|\big)\|x^{1/2}\theta\|,$$

and hence, after application of Lemma 18.1 to the last factor,

$$\|x^{1/2}\theta'\| \le \|x^{1/2}\theta_t\| + \|x^{1/2}\rho_t\|. \tag{18.29}$$

Here, using Theorem 18.3 once more,

$$\begin{aligned}\|x^{1/2}\rho_t(t)\| \le \|\rho_t(t)\|_{L_\infty} &\le Ch^2\|u_t''(t)\|_{L_\infty}\\ &\le Ch^2\Big(\|u_t''(0)\|_{L_\infty} + \int_0^t \|u_{tt}''\|_{L_\infty}\,ds\Big).\end{aligned}$$

In order to estimate the term in θ_t in (18.29), we differentiate (18.28) and set $\chi = \theta_t$ to obtain

$$(x\theta_{tt}, \theta_t) + B(\theta_t, \theta_t) = -(x\rho_{tt}, \theta_t), \quad \text{for } t > 0,$$

whence in the standard way

$$\|x^{1/2}\theta_t(t)\| \le \|x^{1/2}\theta_t(0)\| + \int_0^t \|x^{1/2}\rho_{tt}\|\,ds.$$

Since $\theta(0) = 0$ we obtain from (18.28), with $t = 0, \chi = \theta_t(t)$,

$$\|x^{1/2}\theta_t(0)\|^2 = -(x\rho_t(0), \theta_t(0)) \le \|x^{1/2}\rho_t(0)\|\,\|x^{1/2}\theta_t(0)\|,$$

so that

$$\|x^{1/2}\theta_t(0)\| \le \|\rho_t(0)\|_{L_\infty} \le Ch^2\|u_t''(0)\|_{L_\infty}.$$

Finally, by Theorem 18.3, $\|x^{1/2}\rho_{tt}\| \le \|\rho_{tt}\|_{L_\infty} \le Ch^2\|u_{tt}''\|_{L_\infty}$. Together our estimates show (18.27), and thus complete the proof of Theorem 18.4. □

In the above result the initial data v_h were chosen as the elliptic projection of v. We shall now show that any optimal order initial approximation will produce a discrete solution which is essentially optimal order in the uniform norm for t positive. In fact, with $\widetilde{u}_h$ the solution of Theorem 18.4 and u_h that of (18.25) with v_h arbitrary, this statement follows from the appropriate estimate for $\eta = u_h - \widetilde{u}_h$. Since η is in S_h and satisfies

$$(x\eta_t, \chi) + B(\eta, \chi) = 0, \quad \forall \chi \in S_h,\ t \geq 0,\ \eta(0) = v_h - v, \tag{18.30}$$

an estimate of the desired type is a consequence of the following:

Lemma 18.2 *Let* $\eta \in S_h$ *be a solution of* (18.30). *Then*

$$\|\eta(t)\|_{L_\infty} \leq Ct^{-1/2}\ell_h \|x^{1/2}\eta(0)\|, \quad \textit{for } t > 0.$$

Proof. Using the analogue of (18.26), this result follows from

$$\|x^{1/2}\eta'(t)\| \leq Ct^{-1/2}\ell_h^{1/2}\|x^{1/2}\eta(0)\|, \quad \text{for } t > 0,$$

which we now prove. By (18.30) we have

$$\|x^{1/2}\eta'\|^2 \leq B(\eta, \eta) = -(x\eta_t, \eta) \leq \|x^{1/2}\eta_t\|\,\|x^{1/2}\eta\|.$$

To bound the last factor we use (18.30) to obtain, in the obvious way,

$$\|x^{1/2}\eta(t)\|^2 + 2\int_0^t B(\eta, \eta)\,ds = \|x^{1/2}\eta(0)\|^2. \tag{18.31}$$

The proof will be completed by showing $\|x^{1/2}\eta_t(t)\| \leq Ct^{-1}\ell_h\|x^{1/2}\eta(0)\|$. But from (18.30) we find

$$\frac{d}{dt}(t^2\|x^{1/2}\eta_t\|^2) + 2t^2 B(\eta_t, \eta_t) = 2t\|x^{1/2}\eta_t\|^2. \tag{18.32}$$

To bound the right hand side in (18.32), we first note that for $\chi, \zeta \in S_h$

$$|B(\chi, \zeta)| = |(x\chi', \zeta') - (\chi', \zeta) + (xq\,\chi, \zeta)| \leq C\ell_h B(\chi, \chi)^{1/2} B(\zeta, \zeta)^{1/2},$$

which follows by observing that, by (18.26),

$$|(\chi', \zeta)| \leq \|\chi'\|_{L_1}\|\zeta\|_{L_\infty} \leq C\ell_h \|x^{1/2}\chi'\|\,\|x^{1/2}\zeta'\|.$$

Therefore, from (18.30),

$$2t\|x^{1/2}\eta_t\|^2 = -2tB(\eta, \eta_t) \leq 2t^2 B(\eta_t, \eta_t) + C\ell_h^2 B(\eta, \eta),$$

so that, by integration of (18.32) and using (18.31),

$$t^2\|x^{1/2}\eta_t\|^2 \leq C\ell_h^2 \int_0^t B(\eta, \eta)\,ds \leq C\ell_h^2\|x^{1/2}\eta(0)\|^2.$$

This shows the desired estimate for $\|x^{1/2}\eta_t\|$. □

The weighted norm estimate of Theorem 18.1 is from Schreiber and Eisenstat [212]. A maximum-norm estimate for the same problem may be found in Jespersen [129]. For the rest of the analysis, see Eriksson and Thomée [94], where the results are given in greater generality than here.

19. Problems in Polygonal Domains

In earlier parts of this book we have generally assumed the spatial domain Ω to have a smooth boundary $\partial\Omega$, which has made it possible to guarantee that the solution of the initial-boundary value problem is sufficiently regular for the purpose at hand, provided the data of the problem are sufficiently smooth and satisfy certain compatibility conditions at $t = 0$. In this chapter we shall consider the case when Ω is a plane polygonal domain. In this case singularities will in general appear in the solution even for smooth compatible data, and this will affect the convergence properties of the approximating finite element solution. We shall analyze in some detail the case of piecewise linear finite elements. In this case, no special difficulties arise when Ω is convex, but when Ω is nonconvex the singularities will normally reduce the rate of convergence both for elliptic and for parabolic problems.

We shall consider the model initial-boundary value problem for the heat equation,

$$
\begin{aligned}
(19.1)\qquad u_t - \Delta u &= f \quad \text{in } \Omega, \qquad \text{for } t > 0,\\
u &= 0 \quad \text{on } \partial\Omega, \quad \text{for } t > 0, \quad \text{with } u(\cdot,0) = v \quad \text{in } \Omega,
\end{aligned}
$$

where $\Omega \subset \mathbb{R}^2$ is a polygonal domain.

We begin by studying the corresponding stationary elliptic problem

$$
(19.2)\qquad -\Delta u = f \quad \text{in } \Omega, \quad \text{with } u = 0 \quad \text{on } \partial\Omega.
$$

It is known that when Ω is convex and $f \in L_2$, then the solution u belongs to $H^2 \cap H_0^1$ and

$$
(19.3)\qquad \|u\|_2 \le C\|f\| = C\|\Delta u\|.
$$

However, higher order regularity estimates do not hold, and, as we shall see, the situation is more complicated for Ω nonconvex. Regularity results for the solution of (19.2) will therefore be important for our discussion below. We refer to the standard references Grisvard [110], [111] for more details than given here.

Since we are going to be concerned with solutions of low regularity we will also consider weaker variational solutions $u \in H_0^1$ satisfying

$$
(19.4)\qquad (\nabla u, \nabla \varphi) = \langle f, \varphi \rangle, \quad \forall \varphi \in H_0^1,
$$

where the linear functional f belongs to $H^{-1} = H^{-1}(\Omega) = (H_0^1(\Omega))^*$. It is well-known that this problem has a unique solution, and that

$$\|u\|_{H_0^1} = \|\nabla u\| \le \|f\|_{-1} = \|f\|_{H^{-1}}.$$

For $u \in H_0^1$ we may think of (19.4) as defining $f \in H^{-1}$, and then the operator $\Delta : H_0^1 \to H^{-1}$ is defined by $\Delta u = -f$.

We denote by S_h the piecewise linear finite element spaces used in Chapter 1, and consider the discrete problem to find $u_h \in S_h$ such that

$$(19.5) \qquad (\nabla u_h, \nabla \chi) = (f, \chi), \quad \text{for } \chi \in S_h,$$

which, as earlier, has a unique solution for $f \in L_2$.

We begin by considering the case when Ω is convex. In the same way as in Theorem 1.1 we may use the regularity estimate (19.3) to show the following.

Theorem 19.1 *Let u_h and u be the solutions of* (19.5) *and* (19.4), *respectively. Then*

$$\|u_h - u\| \le Ch^2\|u\|_2 \quad \text{and} \quad \|\nabla u_h - \nabla u\| \le Ch\|u\|_2.$$

In view of the regularity estimate (19.3), the norms on the right hand sides are finite for $f \in L_2$.

Our first goal is now to show that these second order error estimates carry over to the semidiscrete parabolic problem, to find $u_h(t) \in S_h$ for $t \ge 0$ such that

$$(19.6) \qquad \begin{aligned} (u_{h,t}, \chi) + (\nabla u_h, \nabla \chi) &= (f, \chi), \quad \text{for } \chi \in S_h, \ t > 0, \\ u_h(0) &= v_h \approx v. \end{aligned}$$

We begin by noting that the error bound of Theorem 1.2 remains valid in the case of a convex polygonal domain.

Theorem 19.2 *Let u_h and u be the solutions of* (19.6), *with $v_h = R_h v$ and* (19.1), *respectively. Then we have*

$$\|u_h(t) - u(t)\| \le Ch^2\Big(\|v\|_2 + \int_0^t \|u_t\|_2\, d\tau\Big), \quad \text{for } t \ge 0.$$

In order to see that this indeed constitutes an $O(h^2)$ error bound for the semidiscrete solution, we need to know that under the appropriate smoothness and compatibility assumptions on data the expression within parentheses on the right hand side is finite. For this we note that, as a result of the elliptic regularity inequality (19.3), the case $m = 1$ of the parabolic regularity result (1.20) holds, not only for a domain with smooth boundary, but also for a convex polygonal domain, so that, in particular,

$$(19.7) \qquad \int_0^t \|u_t\|_2^2\, d\tau \le C_T\Big(\|v\|_3^2 + \int_0^T (\|f\|_2^2 + \|f_t\|^2) d\tau\Big), \quad \text{for } t \le T.$$

This estimate requires, however, also the compatibility conditions $v = g = 0$ on $\partial\Omega$, where we use the notation

$$g = u_t(0) = f(0) + \Delta v. \tag{19.8}$$

Using the Cauchy-Schwarz inequality in time, (19.7) yields a bound for $\int_0^t \|u_t\|_2 \, d\tau$ in terms of data. As we shall see below in Lemma 19.1, the regularity assumptions on data needed for this may be somewhat reduced.

Proof of Theorem 19.2. Using the Ritz projection $R_h : H_0^1 \to S_h$, defined as usual by (1.22), we split the error $u_h - u$ in the standard way

$$u_h - u = (u_h - R_h u) + (R_h u - u) = \theta + \rho. \tag{19.9}$$

In the same way as in the proof of Theorem 1.2 we have

$$\|\rho(t)\| \le Ch^2 \|u(t)\|_2 \le Ch^2 \Big(\|v\|_2 + \int_0^t \|u_t\|_2 \, d\tau \Big), \quad t \ge 0. \tag{19.10}$$

To bound θ, we recall from (1.27) that

$$(\theta_t, \chi) + (\nabla\theta, \nabla\chi) = -(\rho_t, \chi), \quad \forall \chi \in S_h. \tag{19.11}$$

and choosing $\chi = \theta$ and, using the fact that $\theta(0) = 0$, together with Theorem 19.1,

$$\|\theta(t)\| \le \int_0^t \|\rho_t\| \, d\tau \le Ch^2 \int_0^t \|u_t\|_2 d\tau, \quad \text{for } t \ge 0. \tag{19.12}$$

Thus together with (19.10) this shows the result stated. □

Also the error estimate for the gradient of the solution of the semidiscrete problem of Theorem 1.2 and its proof carry over to convex polygonal domains, and again the regularity inequality (19.7) shows that this yields an $O(h)$ error bound, under the appropriate assumptions on data:

Theorem 19.3 *Under the assumptions of Theorem 19.2 we have*

$$\|\nabla u_h(t) - \nabla u(t)\| \le Ch\Big(\|v\|_2 + \|u(t)\|_2 + \Big(\int_0^t \|u_t\|_1^2 \, ds \Big)^{1/2} \Big), \quad \textit{for } t \ge 0.$$

In order to discuss regularity results here and below, particularly for nonconvex domains, we shall need to use function spaces with a fractional number of derivatives. We therefore now briefly review some facts about such spaces, without proofs. For more details, we refer to the references at the end of this chapter.

For Ω with a piecewise smooth boundary let $H^m = H^m(\Omega)$ with norm $\|\cdot\|_m$ denote the standard Sobolev spaces of integer order $m \ge 0$. For $s = m + \sigma$, with $0 < \sigma < 1$, we then define $H^s = H^s(\Omega)$ by the norm

$$\|u\|_s = \Big(\|u\|_m^2 + \sum_{|\alpha|=m} \iint_{\Omega\times\Omega} \frac{|D^\alpha u(x) - D^\alpha u(y)|^2}{|x-y|^{2+2\sigma}}\,dx\,dy\Big)^{1/2}.$$

The space H^s may thought of as an intermediate space between H^m and H^{m+1} in the sense of interpolation of Banach spaces, as follows.

For two Banach spaces $\mathcal{B}_0$ and $\mathcal{B}_1$ with $\mathcal{B}_1 \subset \mathcal{B}_0$, the associated K-functional is defined by

$$K(t,u) = K(\mathcal{B}_0, \mathcal{B}_1; t, u) = \inf_{v\in\mathcal{B}_1} \big(\|u-v\|_{\mathcal{B}_0} + t\|v\|_{\mathcal{B}_1}\big).$$

We may then define the intermediate space $\mathcal{B} = [\mathcal{B}_0, \mathcal{B}_1]_{\sigma,q}$ for $0<\sigma<1$, $1 \le q \le \infty$, as the set of $u \in \mathcal{B}_0$ for which the norm defined by

$$\|u\|_{[\mathcal{B}_0,\mathcal{B}_1]_{\sigma,q}} = \begin{cases} \Big(\int_0^\infty t^{-\sigma q-1} K(t,u)^q\,dt\Big)^{1/q}, & \text{if } 1 \le q < \infty, \\ \sup_{t>0} t^{-\sigma} K(t,u), & \text{if } q = \infty, \end{cases}$$

is finite. Obviously $\mathcal{B}_1 \subset \mathcal{B} = [\mathcal{B}_0, \mathcal{B}_1]_{\sigma,q} \subset \mathcal{B}_0$.

With this notation one may show that the space H^s introduced above may also be defined as $H^s = [H^m, H^{m+1}]_{\sigma,2}$, with $\sigma = s-m$.

We shall also have reason to use fractional order spaces of functions satisfying homogeneous boundary conditions, and define

$$(19.13)\quad H_0^\sigma = [L_2, H_0^1]_{\sigma,2} \quad\text{and}\quad H_0^{1+\sigma} = [H_0^1, H^2\cap H_0^1]_{\sigma,2}, \quad \text{for } 0<\sigma<1,$$

as well as the negative order spaces

$$H^{-\sigma} = [H^{-1}, L_2]_{1-\sigma,2}, \quad \text{for } 0<\sigma<1.$$

We note that, by duality and (19.13), $H^{-\sigma} = (H_0^\sigma)^*$, for $0 \le \sigma \le 1$. We remark that $H_0^{1+\sigma} = H^{1+\sigma} \cap H_0^1$, for $0<\sigma<1$. In the statements of several of our error bounds below we shall have reason to know that H_0^σ does not require any boundary condition for small σ, or, more precisely,

$$(19.14)\qquad H_0^\sigma = H^\sigma, \quad \text{for } 0<\sigma<\tfrac12.$$

It will also be convenient to use the Hilbert spaces $\dot H^s = \dot H^s(\Omega)$ introduced in Chapter 3, defined by $|v|_s < \infty$ where

$$|v|_s = \Big(\sum_{j=1}^\infty \lambda_j^s \langle v, \varphi_j\rangle^2\Big)^{1/2}, \quad \text{for } s \ge -1, \quad v \in H^{-1},$$

where $\{\lambda_j\}_{j=1}^\infty$ are the eigenvalues and $\{\varphi_j\}_{j=1}^\infty$ the corresponding orthonormal eigenfunctions of $-\Delta$. As for the spaces H^s above, the spaces $\dot H^s$ have the interpolation property $\dot H^s = [\dot H^m, \dot H^{m+1}]_{\sigma,2}$, where $\sigma = s-m$. Then, for $0 < s < 1$, since both $\dot{H}^{-s}$ and H^{-s} is the uniquely defined interpolation space between L_2 and H^{-1}, we have $\dot{H}^{-s} = H^{-s}$. Also, $\dot{H}^s = H_0^s$ for $0 \le s \le 1$, and for $1 \le s \le 2$, $\dot{H}^s$ consists of the functions $u \in H_0^1$ such that Δu is in the negative order space H^{s-2}. In particular, with Δ considered as an operator in L_2, we have for its domain $D(A) = D(A; L_2) = \dot{H}^2$, and the range of Δ is L_2. Thus, if $f \in L_2$ the solution u of (19.4) belongs to $\dot{H}^2$.

The solution operator of the homogeneous case ($f = 0$) of (19.1) may be defined as

$$E(t)v = \sum_{j=1}^{\infty} e^{-\lambda_j t} \langle v, \varphi_j \rangle \varphi_j, \quad \text{for } v \in H^{-1}, \quad t > 0,$$

and it follows at once as in Chapter 3 by Parseval's relation that $E(t)$ is a contraction in L_2 and has the smoothing property

$$|E(t)v|_{s_2} \le Ct^{-(s_2-s_1)/2} |v|_{s_1}, \quad \text{for } -1 \le s_1 \le s_2. \tag{19.15}$$

In particular, $E(t)$ is an analytic semigroup in L_2, with Δ as its generator.

Recall that, by Duhamel's principle, we have, for the solution of the inhomogeneous equation (19.1), under the appropriate assumptions,

$$u(t) = E(t)v + \int_0^t E(t-s)f(s)\,ds, \tag{19.16}$$

We are now ready to show the following regularity result for Ω convex.

Lemma 19.1 *Let $\Omega \subset \mathbb{R}^2$ be convex and let u be the solution of* (19.1), *with g defined by* (19.8). *Then, for any $\varepsilon \in (0, \frac{1}{2})$, we have, with $C = C_{\varepsilon,T}$,*

$$\int_0^t (\|u_t\|_2 + \|u_{tt}\|)\,d\tau \le C\Big(\|g\|_\varepsilon + \int_0^t \|f_t\|_\varepsilon\,d\tau\Big), \quad \textit{for } t \le T.$$

Proof. We use the elliptic regularity estimate (19.3) and differentiate (19.1) to obtain

$$\|u_t\|_2 \le C\|\Delta u_t\| \le C(\|u_{tt}\| + \|f_t\|).$$

It therefore suffices to bound the integral of $\|u_{tt}\|$. For this purpose, we use the equation for u_t, and find by (19.16) that

$$u_t(t) = E(t)g + \int_0^t E(t-s)f_t(s)\,ds,$$

and, after differentiation,

$$u_{tt}(t) = E'(t)g + f_t(t) + \int_0^t E'(t-s)f_t(s)\,ds. \tag{19.17}$$

Using (19.15), we have, for $\varepsilon \in (0, \frac{1}{2})$,

$$\|E'(t)v\| = \|\Delta E(t)v\| = |E(t)v|_2 \le Ct^{-1+\varepsilon/2}|v|_\varepsilon, \quad \text{for } t > 0. \tag{19.18}$$

Applying this in (19.17) we find

$$\begin{aligned}\int_0^t \|u_{tt}\|\,d\tau &\le C\Big(\int_0^t \tau^{-1+\varepsilon/2}|g|_\varepsilon\,d\tau + \int_0^t \|f_t(\tau)\|\,d\tau \\ &\qquad + \int_0^t\int_0^\tau (\tau-s)^{-1+\varepsilon/2}|f_t(s)|_\varepsilon\,ds\,d\tau\Big) \\ &\le Ct^{\varepsilon/2}\varepsilon^{-1}\Big(|g|_\varepsilon + \int_0^t |f_t|_\varepsilon\,d\tau\Big), \quad \text{for } t \ge 0.\end{aligned}$$

Since the norms in $\dot{H}^\varepsilon$ and H^ε are equivalent by (19.14), this completes the proof. □

Applied in Theorem 19.2 this lemma shows the error estimate

$$\|u_h(t) - u(t)\| \le Ch^2\Big(\|v\|_2 + \|g\|_\varepsilon + \int_0^t \|f_t\|_\varepsilon\,d\tau\Big), \quad \text{for } t \le T,$$

where we assume $v = 0$ on $\partial\Omega$. Note that in view of (19.14) no boundary conditions are required for functions in H^ε with $\varepsilon \in (0, \frac{1}{2})$. Thus, in addition to allowing milder regularity requirements on g and f_t compared with the regularity on data assumed by (19.7), the smoothing property (19.18) makes it possible to avoid imposing unnatural boundary conditions for these functions. In the rest of this chapter we shall present our error estimates in this form, thus indicating the regularity requirements on data, rather than on the solution itself, which are needed for the order of convergence stated.

We now turn to the case that Ω is a nonconvex polygonal domain. For simplicity we assume that there is only one reentrant corner $O = (0,0)$, with interior angle ω, $\pi < \omega < 2\pi$, and set $\beta = \pi/\omega \in (\frac{1}{2}, 1)$. It is known that, in polar coordinates near this corner, the solution of (19.4) normally behaves like a multiple of the function $r^\beta \sin(\beta\theta)$, for $0 \le \theta \le \omega$, $r \le r_0$. We note that this is a harmonic function and that it vanishes on the edges of the sector, corresponding to $\theta = 0, \omega$. Letting $\eta = \eta(r)$ be a smooth cutoff function such that $\eta(r) \equiv 1$ near the nonconvex corner O and such that the support of η only meets the two edges emerging from O, we introduce the singular function

$$S(r,\theta) = \eta(r)r^\beta \sin(\beta\theta). \tag{19.19}$$

It is easy to see that $S \in \mathcal{C}^\beta$, but $S \notin \mathcal{C}^s$ for $s > \beta$. Also, $S \in H^{1+s}$ for $0 \le s < \beta$, but $S \notin H^{1+s}$ for $s \ge \beta$, in particular $S \notin H^2$. On the other hand, we find by a simple calculation that $\Delta S \in \mathcal{C}^\infty(\overline{\Omega})$, and vanishes near O. Hence, with $f = -\Delta S$, the problem (19.4) has smooth data, but, even though its variational solution is in $\dot{H}^2$, it does not belong to H^2.

However, it can be shown that for $f \in L_2$, there is a number $\kappa(f)$ such that the solution of (19.4) satisfies

$$u - \kappa(f)S \in V^2 = H^2 \cap H_0^1, \tag{19.20}$$

and hence may be written as

$$u = u_S + u_R, \quad \text{with } u_S = \kappa(f)S \quad \text{and } u_R \in V^2, \quad \|u_R\|_2 \le C\|f\|. \tag{19.21}$$

Here $\kappa(f)$ may be represented as

$$\kappa(f) = (f, q), \quad \text{where } q \in H_0^\sigma, \quad \text{for } 0 < \sigma < 1 - \beta < \tfrac{1}{2}.$$

In fact, $q \approx cS_*$, where $S_*(r,\theta) = r^{-\beta}\sin(\beta\theta)$ is the so-called dual singular function, and $q \notin H^{1-\beta}$. This means that $\kappa(f)$, and hence u_S, is well defined for $f \in H^{-1+s}$, with $\beta < s \le 1$, which is consistent with the following regularity results for the solution of (19.4).

We begin with the following shift theorem by Kellogg [137].

Lemma 19.2 *The solution u of (19.4) satisfies, with $C = C_s$,*

$$\|u\|_{1+s} \le C\|f\|_{-1+s} = C\|\Delta u\|_{-1+s}, \quad \textit{for } 0 \le s < \beta.$$

Since $H^{-1+s} = \dot{H}^{-1+s}$ for $0 \le s \le 1$, with equivalent norms, and since obviously $-\Delta$ is an isomorphism between $\dot{H}^{1+s}$ and $\dot{H}^{-1+s}$, this result implies

$$\|u\|_{1+s} \le C|\Delta u|_{-1+s} = C|u|_{1+s}, \quad \text{for } 0 \le s < \beta.$$

For the critical value $s = \beta$ we have the following regularity result by Bacuta, Bramble and Xu [14], which we shall depend on below. It is expressed in terms of the norm in the Besov space defined as the interpolation space

$$B_2^{1+\beta,\infty} = [H^1, H^2]_{\beta,\infty}.$$

Lemma 19.3 *For the solution of (19.4) we have, with $C = C_s$,*

$$\|u\|_{B_2^{1+\beta,\infty}} \le C\|f\|_{-1+s} = C\|\Delta u\|_{-1+s} \le C|u|_{1+s}, \quad \textit{for } \beta < s \le 1.$$

We consider now the finite element approximation u_h of u defined by (19.5). Using the regularity result of Lemma 19.3 one is then able to show the following error estimates, see [14].

Lemma 19.4 *We have, for the solutions of (19.5) and (19.4), with $C = C_s$,*

$$\|u_h - u\|_1 \le Ch^\beta |u|_{1+s}, \quad \textit{for } \beta < s \le 1. \tag{19.22}$$

Further,

$$\|u_h - u\| \le Ch^{2\beta} |u|_{1+s}, \quad \textit{for } \beta < s \le 1, \tag{19.23}$$

and

$$\|u_h - u\| \le Ch^\beta |u|_1. \tag{19.24}$$

Note that the L_2-estimate (19.23), which is obtained by the standard duality argument, is of double the order of the H^1-estimate (19.22).

We emphasize that the error bounds in (19.22) and (19.23) are expressed in terms of $\dot{H}^{1+s}$-norms, and that this requires less regularity than with the corresponding H^{1+s}-norms, which are not normally finite.

With the aid of these error estimates for the elliptic problem we are now ready to show an error estimate for the solution of the semidiscrete parabolic problem (19.6).

Theorem 19.4 *Let u_h and u be the solutions of* (19.6) *and* (19.1), *with $v_h = R_h v$. Then we have, with $C = C_T$,*

$$\|u_h(t) - u(t)\| \le Ch^{2\beta}\Big(\|\Delta v\| + \|g\| + \int_0^t \|f_t\|\,d\tau\Big), \quad \textit{for } t \le T.$$

Proof. Writing again the error as in (19.9), we have, by (19.12),

$$\|u_h(t) - u(t)\| \le \|\rho(t)\| + \|\theta(t)\| \le \|\rho(0)\| + 2\int_0^t \|\rho_t\|\,d\tau. \tag{19.25}$$

and hence, using the elliptic finite element estimate (19.23) to bound ρ,

$$\|u_h(t) - u(t)\| \le Ch^{2\beta}\Big(|v|_{1+s} + \int_0^t |u_t|_{1+s}\,d\tau\Big), \quad \text{for } t \ge 0, \quad \text{with } \beta < s < 1.$$

To complete the proof we need the first regularity estimate for the solution of (19.1) of the following lemma. □

Lemma 19.5 *Let $u(t)$ be the solution of* (19.1), *and let g be defined by* (19.8). *Then we have, for $0 \le s < 1$, with $C = C_{s,T}$,*

$$\int_0^t \big(|u_t|_{1+s} + |u_{tt}|_{-1+s}\big)\,d\tau \le C\Big(\|g\| + \int_0^t \|f_t\|\,d\tau\Big), \quad \textit{for } t \le T. \tag{19.26}$$

Further, for $\varepsilon \in (0, \frac{1}{2})$, with $C = C_{\varepsilon,T}$,

$$\int_0^t \big(|u_t|_2 + \|u_{tt}\|\big)\,d\tau \le C\Big(\|g\|_\varepsilon + \int_0^t \|f_t\|_\varepsilon\,d\tau\Big), \quad \textit{for } t \le T. \tag{19.27}$$

Proof. We first note that, for $0 < s \le 1$,

$$\begin{aligned} |u_t(t)|_{1+s} &= |\Delta u_t(t)|_{-1+s} \\ &\le |u_{tt}(t)|_{-1+s} + |f_t(t)|_{-1+s} \le |u_{tt}(t)|_{-1+s} + \|f_t(t)\|, \end{aligned} \tag{19.28}$$

so that it suffices to consider the second integrand on the left in (19.26). Let $\varepsilon = 0$ if $s < 1$ and $\varepsilon \in (0, \frac{1}{2})$ if $s = 1$. We use again (19.17), together with (19.15), to obtain this time

$$|E'(t)g|_{-1+s} = |E(t)g|_{1+s} \le Ct^{-\sigma}\|g\|, \quad \text{with } \sigma = (1+s-\varepsilon)/2,$$

and similarly for the integrand in (19.17). We conclude

$$|u_{tt}(t)|_{-1+s} \le C\Big(t^{-\sigma}\|g\| + \|f_t(t)\| + \int_0^t (t-\tau)^{-\sigma}\|f_t(\tau)\|\,d\tau\Big),$$

and hence after integration, since $\sigma = (s+1-\varepsilon)/2 < 1$,

$$\int_0^t |u_{tt}|_{-1+s}\,d\tau \le C(1+T^{1-\sigma})\Big(|g|_\varepsilon + \int_0^t |f_t|_\varepsilon\,d\tau\Big), \text{ for } t \le T.$$

Since the norms in $\dot{H}^\varepsilon$ and H^ε are equivalent, this completes the proof. □

Note that $|u_t|_{1+s}$ could not be replaced by $\|u_t\|_{1+s}$ in (19.26) (or (19.27)) because in the first inequality in (19.28), this would require an elliptic regularity result which does not hold for $\beta < s \le 1$. We also note that in (19.28) the differential equation is used to transfer the regularity requirement on the solution from the spatial to the time variable, which is easier to handle.

We next show an $O(h^\beta)$ estimate for the gradient of the error.

Theorem 19.5 *With u_h and u as in Theorem 19.2, and g as in* (19.8), *we have for $t \le T$, with $C = C_T$,*

$$\|\nabla(u_h(t) - u(t))\| \le Ch^\beta\Big(\|\Delta v\| + \|g\| + \int_0^t \|f_t\|\,d\tau + \big(\int_0^t \|f_t\|_{-1}^2 d\tau\big)^{1/2}\Big).$$

Proof. Using Lemma 19.4 we find, with $\beta < s < 1$,

$$\|\nabla\rho(t)\| \le Ch^\beta|u(t)|_{1+s} \le Ch^\beta\Big(|v|_{1+s} + \int_0^t |u_t|_{1+s}\,d\tau\Big).$$

The right hand side is bounded as desired by Lemma 19.5.

To bound $\nabla\theta(t)$ we proceed as in the proof of Theorem 1.3, and obtain, using (19.24),

$$\|\nabla\theta(t)\|^2 \le \int_0^t \|\rho_t\|^2 d\tau \le Ch^{2\beta}\int_0^t \|u_t\|_1^2 d\tau, \quad \text{for } t \ge 0. \tag{19.29}$$

The bound stated therefore follows from the first estimate of the following lemma. □

Lemma 19.6 *We have for the solution of* (19.1)

$$\int_0^t \|u_t\|_1^2 d\tau \le C\Big(\|g\|^2 + \int_0^t \|f_t\|_{-1}^2\,d\tau\Big)$$

and

$$\int_0^t (\|\Delta u_t\|^2 + \|u_{tt}\|^2)\,d\tau \le C\Big(|g|_1^2 + \int_0^t \|f_t\|^2 d\tau\Big), \quad \textit{for } t \ge 0.$$

Proof. By differentiation of (19.1), multiplication by u_t and integration, we obtain

$$\tfrac{1}{2}\frac{d}{dt}\|u_t\|^2 + \|\nabla u_t\|^2 = (f_t, u_t) \le C\|f_t\|_{-1}^2 + \tfrac{1}{2}\|\nabla u_t\|^2,$$

from which the first result follows by integration. Multiplication instead by u_{tt} shows

$$\|u_{tt}\|^2 + \tfrac{1}{2}\frac{d}{dt}\|\nabla u_t\|^2 = (f_t, u_{tt}) \le \tfrac{1}{2}\|u_{tt}\|^2 + \tfrac{1}{2}\|f_t\|^2,$$

which yields

$$\int_0^t \|u_{tt}(\tau)\|^2 d\tau \le \|\nabla g\|^2 + \int_0^t \|f_t(\tau)\|^2 d\tau.$$

Since $\|\Delta u_t\| \le \|u_{tt}\| + \|f_t\|$ the result stated follows. □

The above analysis of $\nabla\theta$ also yields the following superconvergence type result.

Lemma 19.7 *With the above notation we have*

$$\|\nabla\theta(t)\| \le Ch^{2\beta}\Big(|g|_1^2 + \int_0^t \|f_t(\tau)\|^2 d\tau\Big)^{1/2}, \quad \textit{for } t \ge 0.$$

Proof. This follows at once from (19.29) by using (19.23) with $s = 1$ instead of (19.24), together with Lemma 19.6. □

We shall now turn to some error estimates in maximum-norm, and begin with the elliptic problem. Our analysis will be based on the discrete Sobolev type inequality of Lemma 6.4, together with estimates for the gradient of the error. We begin with an essentially $O(h^\beta)$ error bound.

Lemma 19.8 *Assume the triangulations are such that $h_{\min} \ge Ch^\gamma$ for some $\gamma > 0$, and let u_h and u be the solutions of* (19.5) *and* (19.4). *Then, for any s, s_1 with $0 \le s < s_1 < \beta$, we have, with $C = C_{s,s_1}$,*

$$\|u_h - u\|_{L_\infty} \le Ch^s\|\Delta u\|_{-1+s_1}.$$

Proof. We have

$$\|u_h - u\|_{L_\infty} \le \|u_h - I_h u\|_{L_\infty} + \|I_h u - u\|_{L_\infty}.$$

Here, by Lemmas 6.4 and 19.2, we have, since u_h is the best approximation of u in $|\cdot|_1$, with $s < s_1 < \beta$,

$$\begin{aligned}\|u_h - I_h u\|_{L_\infty} &\le C\ell_h^{1/2}\|\nabla(u_h - u)\| + C\ell_h^{1/2}\|\nabla(u - I_h u)\| \\ &\le C\ell_h^{1/2}\|\nabla(I_h u - u)\| \le C\ell_h^{1/2}h^{s_1}\|u\|_{1+s_1} \le Ch^s\|\Delta u\|_{-1+s_1}.\end{aligned}$$

By Sobolev's inequality and Lemma 19.2, we find

$$\|I_h u - u\|_{L_\infty} \le Ch^s\|u\|_{W_\infty^s} \le Ch^s\|u\|_{1+s_1} \le Ch^s\|\Delta u\|_{-1+s_1}.$$

This shows the result stated. □

The global error bound derived is thus of lower order in maximum-norm than in L_2. Away from the corners of the domain, however, the convergence in maximum-norm is of the same order $O(h^{2\beta})$ as in the global L_2-error estimate. This follows from the following lemma.

Lemma 19.9 *Let $\Omega_0 \subset \Omega_1 \subset \Omega$ be such that Ω_1 does not contain any corner of Ω and the distance between $\partial\Omega_1 \cap \Omega$ and $\partial\Omega_0 \cap \Omega$ is positive, and let u_h and u be the solutions of* (19.5) *and* (19.2). *Assume that the triangulations associated with S_h are quasiuniform in Ω_1. Then, with $C = C_s$,*

$$\|u_h - u\|_{L_\infty(\Omega_0)} \le Ch^{2\beta}\big(\|u\|_{W^{2s}_\infty(\Omega_1)} + \|\Delta u\|_{-1+s}\big), \quad \text{for } \beta < s \le 1.$$

Proof. This is a consequence of the following interior estimate, cf. [244], valid up to the interiors of the sides of Ω, namely

$$\|u_h - u\|_{L_\infty(\Omega_0)} \le C\ell_h\|I_h u - u\|_{L_\infty(\Omega_1)} + C\|u_h - u\|,$$

together with Lemma 19.4 and the fact that $h^{2s}\ell_h \le Ch^{2\beta}$. □

We now turn to maximum-norm error estimates for the semidiscrete parabolic problem, and begin with an essentially $O(h^\beta)$ global estimate.

Theorem 19.6 *Assume that the family of triangulations underlying S_h is such that $h_{\min} \ge Ch^\gamma$ for some $\gamma > 0$, and let u_h and u be the solutions of* (19.6) *and* (19.1), *with $v_h = R_h v$. Then, for any s with $0 \le s < \beta$, we have for $t \le T$, with $C = C_{s,T}$,*

$$\|u_h(t) - u(t)\|_{L_\infty} \le Ch^s\Big(\|\Delta v\| + \|f(0)\| + \int_0^t \|f_t\|\,d\tau + \big(\int_0^t \|f_t\|_{-1}^2\,d\tau\big)^{1/2}\Big).$$

Proof. We have by Lemma 19.8, with $s_1 \in (s, \beta)$,

$$\|\rho(t)\|_{L_\infty} \le Ch^s\|\Delta u(t)\|_{-1+s_1} \le Ch^s|u(t)|_{1+s_1}.$$

Here

$$|u(t)|_{1+s_1} \le C\Big(|v|_{1+s_1} + \int_0^t |u_t(\tau)|_{1+s_1}\,d\tau\Big), \quad \text{for } t \ge 0.$$

which is bounded as desired by Lemma 19.5.

Using Lemma 6.4 together with (19.29) and Lemma 19.6 we have

$$\|\theta(t)\|_{L_\infty} \le C\ell_h^{1/2}\|\nabla\theta(t)\| \le C\ell_h^{1/2}h^\beta\Big(\|g\| + \Big(\int_0^t \|f_t\|_{-1}^2 d\tau\Big)^{1/2}\Big), \quad \text{for } t \ge 0.$$

Together these estimates show the result stated. □

We now demonstrate an almost $O(h^{2\beta})$ estimate away from the corners.

Theorem 19.7 *Let $\Omega_0 \subset \Omega_1 \subset \Omega$ be such that Ω_1 does not contain any corner of Ω and the distance between $\partial\Omega_1 \cap \Omega$ and $\partial\Omega_0 \cap \Omega$ is positive. Assume that the triangulations associated with S_h are quasiuniform in Ω_1. Then we have, for the solutions of* (19.6) *with $v_h = R_h v$ and* (19.1), *for $\beta < s < 1$, with $C = C_{s,T}$,*

$$\|u_h(t) - u(t)\|_{L_\infty(\Omega_0)} \le Ch^{2\beta}\ell_h^{1/2}\Big(\|u(t)\|_{W_\infty^{2s}(\Omega_1)} + \|\Delta v\| + |g|_1 + \Big(\int_0^t \|f_t\|^2 d\tau\Big)^{1/2}\Big), \quad \text{for } t \le T.$$

Proof. By Lemmas 19.9 and 19.5 we have, with $\beta < s < 1$,

$$\begin{aligned}\|\rho(t)\|_{L_\infty(\Omega_0)} &\le Ch^{2\beta}\big(\|u(t)\|_{W_\infty^{2s}(\Omega_1)} + |u(t)|_{1+s}\big)\\ &\le Ch^{2\beta}\Big(\|u(t)\|_{W_\infty^{2s}(\Omega_1)} + \|\Delta v\| + \|g\| + \int_0^t \|f_t\| d\tau\Big), \quad \text{for } t \le T.\end{aligned}$$

Further, using the superconvergence result of Lemma 19.7,

$$\|\theta(t)\|_{L_\infty} \le C\ell_h^{1/2}\|\nabla\theta(t)\| \le Ch^{2\beta}\ell_h^{1/2}\Big(|g|_1 + \Big(\int_0^t \|f_t\|^2 d\tau\Big)^{1/2}\Big), \quad \text{for } t \ge 0.$$

Together these estimates show the error bound stated. □

We remark that, in the case of a globally quasiuniform mesh, it can be shown that the singularity at the nonconvex corner pollutes the finite element solution of the elliptic problem everywhere in Ω and that therefore the $O(h^{2\beta})$ convergence away from the nonconvex corner is best possible. However, as we shall now see, optimal order $O(h)$ and $O(h^2)$ convergence in H^1 and L_2, respectively, may be obtained, for both the elliptic and the parabolic problem, provided the triangulations are systematically refined towards the nonconvex corner as follows.

For a triangulation $\mathcal{T}_h = \{\tau\}$ of Ω, let h_τ be the diameter of τ so that $h = \max_{\mathcal{T}_h} h_\tau$, and let d_τ denote the distance from τ to the nonconvex corner O. We now assume that the $\mathcal{T}_h$ is graded towards O in such a way that for $d_\tau \ge d_I \approx h^{1/\beta}$ we have $ch\, d_\tau^{1/\beta} \le h_\tau \le Ch\, d_\tau^{1/\beta}$. Near O, for $d_\tau \le d_I$, we assume that the ratio h_τ/d_I is bounded above and below, so that, in particular, $\mathcal{T}_h$ is locally quasiuniform for $|x| \le d_I$.

It can be seen that under these conditions dim $S_h \le Ch^{-2}$, so that, asymptotically, the size of the system that has to be solved is of the same order as for globally quasiuniform triangulations. Construction of families of meshes which fulfil these requirements can be found in the references given at the end of this chapter.

We then have the following result for the elliptic problem.

Lemma 19.10 *With triangulations as above, let u_h and u be the solutions of* (19.5) *and* (19.4). *Then we have*

$$\|u_h - u\| + h\|\nabla(u_h - u)\| \le Ch^2\|\Delta u\| = Ch^2\|f\|.$$

Proof. We start with the $O(h)$ estimate for the gradient. With $u = u_R + u_S$ as in (19.21) we have

$$\|\nabla(u_h - u)\| \le \|\nabla(I_h u - u)\| \le \|I_h u_R - u_R\|_1 + \|I_h u_S - u_S\|_1.$$

If $f \in L_2$, then $u_R \in H^2$ and $I_h u_R$ exists. Further, we have by (19.21),

$$\|I_h u_R - u_R\|_1 \le Ch\|u_R\|_2 \le Ch\|f\|.$$

Since $|\kappa(f)| \le C\|f\|$, it now remains to show

$$\|I_h S - S\|_1 \le Ch. \tag{19.30}$$

For this we first introduce some notation. Let $d(x)$ denote the distance from x to the nonconvex corner O, and let $\bar{d} = \max_\Omega d(x)$. Set $d_j = \bar{d}\,2^{-j}$, and let

$$\Omega_j = \{x \in \Omega : d_{j+1} \le d(x) \le d_j\}, \quad \text{for } j = 0, \dots, J,$$

with J chosen so that $d_J \approx d_I$. Furthermore, let $\Omega_j' = \Omega_{j-1} \cup \Omega_j \cup \Omega_{j+1}$ and $\Omega_I = \{x \in \Omega : d(x) \le d_J/2\}$. The triangulations are then quasiuniform on each Ω_j'. We have $h_j \approx c\,h\,d_j^{1/\beta}$ for the maximal mesh-size on Ω_j and hence, with $\varepsilon = 1/\beta + \beta - 1 > 1$,

$$\|I_h S - S\|_{H^1(\Omega_j)} \le Ch_j\|S\|_{H^2(\Omega_j')} \le Ch_j d_j^{\beta-1} \le Ch\,d_j^{\varepsilon},$$

and

$$\|I_h S - S\|_{H^1(\Omega_I)} \le \|I_h S\|_{H^1(\Omega_I)} + \|S\|_{H^1(\Omega_I)} \le C d_J^{\beta} \le Ch,$$

which implies (19.30) after taking squares and summing.

The L_2-bound now follows by a standard duality argument. □

he optimal order error bounds for the elliptic problem in Lemma 19.10, obtained by refinements of the triangulations towards the nonconvex corner, can be carried over to the parabolic problem.

Theorem 19.8 *Assume that the triangulations underlying the S_h are refined as in Lemma 19.10. We then have, for the solutions of* (19.6) *and* (19.1), *with $v_h = R_h v$ and g as in* (19.8), *for any $\varepsilon \in (0, \frac{1}{2})$, with $C = C_\varepsilon$,*

$$\|u_h(t) - u(t)\| \le Ch^2\Big(\|\Delta v\| + \|g\|_\varepsilon + \int_0^t \|f_t\|_\varepsilon\,d\tau\Big), \quad \textit{for } t \ge 0.$$

Proof. Bounding $u_h - u$ as in (19.25), we have, in view of Lemma 19.10,

$$\|u_h(t) - u(t)\| \le \|\rho(0)\| + 2\int_0^t \|\rho_t\|\,d\tau \le Ch^2\Big(|v|_2 + \int_0^t |u_t|_2\,d\tau\Big).$$

The result stated now follows by Lemma 19.5. □

We shall now give an example of a nonsmooth data error estimate and demonstrate that, for the homogeneous parabolic equation, an $O(h^{2\beta})$ error estimate holds for the semidiscrete approximation for positive time even when the initial data are only assumed to be in L_2, provided the discrete initial data are appropriately chosen.

Theorem 19.9 *Let $u_h(t)$ and $u(t)$ be the solutions of* (19.6) *and* (19.1) *with $f=0$, and let $v_h = P_h v$. Then we have, for $\beta < s < 1$, with $C = C_s$,*

$$\|u_h(t) - u(t)\| \le Ch^{2\beta} t^{-(1+s)/2} \|v\|, \quad \text{for } t > 0.$$

Proof. We recall the inequality (3.14),

$$\|u_h(t) - u(t)\| \le Ct^{-1} \sup_{\tau \le t} \big(\tau^2 \|\rho_t(\tau)\| + \tau\|\rho(\tau)\| + \|\widetilde{\rho}(\tau)\|\big), \quad \text{for } t > 0,$$

where $\widetilde{\rho}(t) = \int_0^t \rho(\tau)d\tau$. This was shown in Chapter 3 for smooth $\partial\Omega$; the smoothness of $\partial\Omega$ is not required for the proof. By (19.23) and (19.15), and using the definition of $\dot{H}^{-1+s}$ we easily obtain that

$$\tau\|\rho(\tau)\| \le C\tau h^{2\beta} |u(\tau)|_{1+s} \le Ch^{2\beta} \tau^{(1-s)/2} \|v\|.$$

Hence, since $s < 1$, we find

$$\|\widetilde{\rho}(\tau)\| \le \int_0^\tau \|\rho(\eta)\| d\eta \le Ch^{2\beta} \int_0^\tau \eta^{(-1-s)/2} \|v\|\, d\eta \le Ch^{2\beta} \tau^{(1-s)/2} \|v\|.$$

In the same way,

$$\tau^2 \|\rho_t(\tau)\| \le Ch^{2\beta} \tau^2 |u_t(\tau)|_{1+s} \le Ch^{2\beta} \tau^{(1-s)/2} \|v\|.$$

Together these inequalities complete the proof. □

As examples for fully discrete methods we will show some error estimates for the application of the *backward Euler* and the *Crank–Nicolson* methods to the discretization in time of the spatially semidiscrete problem (19.6). Letting k denote the constant time step, $U^n = U_h^n$ the approximation in S_h of the exact solution $u(t)$ of (19.1) at $t = t_n = nk$, and setting $\bar{\partial} U^n = (U^n - U^{n-1})/k$, we consider first the backward Euler method

$$(\bar{\partial} U^n, \chi) + (\nabla U^n, \nabla \chi) = (f^n, \chi), \quad \forall \chi \in S_h,\ n \ge 1, \qquad U^0 = v_h = R_h v. \tag{19.31}$$

We first show the following error estimate in L_2-norm.

Theorem 19.10 *Let U^n and $u(t_n)$ be the solutions of* (19.31) *and* (19.1), *respectively, with g as in* (19.8). *Then, for any $\varepsilon \in (0, \frac{1}{2})$ and $T > 0$ we have, with $C = C_{\varepsilon,T}$,*

$$\|U^n - u(t_n)\| \le C(h^{2\beta} + k)\Big(\|\Delta v\| + \|g\|_\varepsilon + \int_0^{t_n} \|f_t\|_\varepsilon \, d\tau\Big), \quad \text{for } t_n \le T.$$

Proof. Analogously to (19.9) we write

$$U^n - u(t_n) = (U^n - R_h u(t_n)) + (R_h u(t_n) - u(t_n)) = \theta^n + \rho^n. \tag{19.32}$$

Here ρ^n is bounded as desired as in the proof of Theorem 19.4. To bound θ^n we note that

$$(\bar\partial\theta^n, \chi) + (\nabla\theta^n, \nabla\chi) = -(\omega^n, \chi), \quad \forall \chi \in S_h, \tag{19.33}$$

where

$$\omega^n = \omega_1^n + \omega_2^n = (R_h - I)\bar\partial u(t_n) + (\bar\partial u(t_n) - u_t(t_n)).$$

Choosing $\chi = \theta^n$ in (19.33) we obtain as in the proof of Theorem 1.5, since $\theta^0 = 0$,

$$\|\theta^n\| \le k\sum_{j=1}^n \|\omega_1^j\| + k\sum_{j=1}^n \|\omega_2^j\| = I + II.$$

Here $k\omega_1^j = \int_{t_{j-1}}^{t_j} \rho_t \, d\tau$, and hence, again as in Theorem 19.4,

$$I \le \int_0^{t_n} \|\rho_t\| \, d\tau \le C_T h^{2\beta}\Big(\|g\| + \int_0^{t_n} \|f_t\| \, d\tau\Big), \quad \text{for } t_n \le T.$$

Further, as in Theorem 1.5, and using Lemma 19.5, we find

$$II \le Ck\int_0^{t_n} \|u_{tt}\| \, d\tau \le C_\varepsilon k\Big(\|g\|_\varepsilon + \int_0^{t_n} \|f_t\|_\varepsilon \, d\tau\Big).$$

Together these estimates complete the proof. □

We note that in this and the following error estimates there is no reduction in the convergence rate in time.

Next, we will show the following estimate for the gradient of the error.

Theorem 19.11 *Let U^n and $u(t_n)$ be as in Theorem 19.10. Then we have, with $C = C_T$,*

$$\|\nabla(U^n - u(t_n))\| \le C(h^\beta + k)\Big(\|\Delta v\| + |g|_1 + \big(\int_0^{t_n} \|f_t\|^2 d\tau\big)^{1/2}\Big), \quad \textit{for } t_n \le T.$$

Proof. Here $\nabla\rho^n$ is bounded as desired by the proof of Theorem 19.5. Further, choosing $\chi = \bar\partial\theta^n$ in (19.33), we have, cf. (1.53),

$$\|\nabla\theta^n\|^2 \le 2k\sum_{j=1}^n \|\omega_1^j\|^2 + 2k\sum_{j=1}^n \|\omega_2^j\|^2 = I' + II'.$$

Here, using (19.29) and Lemma 19.6,

$$I' \le 2\int_0^{t_n} \|\rho_t\|^2\,d\tau \le Ch^{2\beta}\Big(\|g\|^2 + \int_0^{t_n} \|f_t\|_{-1}^2\,d\tau\Big), \quad \text{for } t_n \ge 0.$$

Further, once more by Lemma 19.6,

$$II' \le Ck^2\int_0^{t_n} \|u_{tt}\|^2\,d\tau \le Ck^2\Big(|g|_1^2 + \int_0^{t_n} \|f_t\|^2\,d\tau\Big), \quad \text{for } t_n \ge 0.$$

Together these estimates complete the proof. □

Next we will show the following nonsmooth initial data estimate.

Theorem 19.12 *Let U^n and $u(t_n)$ be the solutions of* (19.31) *and* (19.1) *with $f = 0$, but with $v_h = P_h v$. Then we have, for $\beta < s < 1$, with $C = C_s$,*

$$\|U^n - u(t_n)\| \le C(h^{2\beta}t_n^{-(1+s)/2} + kt_n^{-1})\|v\|, \quad \textit{for } t_n \ge 0.$$

Proof. In view of Theorem 19.9 it suffices to note that

$$\|U^n - u_h(t_n)\| \le Ckt_n^{-1}\|P_h v\| \le Ckt_n^{-1}\|v\|, \quad \text{for } t_n \ge 0.$$

The former inequality is a special case of, e.g., Theorem 7.2. □

We also include a maximum-norm error estimate.

Theorem 19.13 *Assume the family of triangulations underlying S_h is such that $h_{\min} \ge Ch^\gamma$ for some $\gamma > 0$. Then, for any s with $0 \le s < s_1 < \beta$, we have, for the solutions of* (19.31) *and* (19.1), *with $C = C_{s,s_1}$, $n \ge 0$,*

$$\|U^n - u(t_n)\|_{L_\infty} \le C(h^s + \ell_h^{1/2}k)\Big(\|\Delta v\| + |g|_1 + \Big(\int_0^{t_n} \|f_t\|^2 d\tau\Big)^{1/2}\Big).$$

Proof. The term ρ^n is bounded as desired by the argument in the proof of Theorem 19.6, and by Lemma 6.4 the estimate for θ^n follows from that of $\nabla\theta^n$ in the proof of Theorem 19.11. □

As a final example of a fully discrete method we will consider the *Crank–Nicolson* method for the discretization in time of the semidiscrete problem (19.6), combined with such refinement in space that yields an optimal order $O(h^2)$ error estimate in space. With the above notation, and setting $\widehat{U}^n = \frac{1}{2}(U^n + U^{n-1})$, the Crank-Nicolson method is defined by

$$(19.34) \qquad \begin{aligned} (\bar\partial U^n, \chi) + (\nabla \widehat{U}^n, \nabla\chi) &= (f(t_{n-\frac12}), \chi), \quad \forall \chi \in S_h,\ n \ge 1, \\ U^0 &= v_h = R_h v. \end{aligned}$$

We show the following error estimate.

Theorem 19.14 *Let U^n and $u(t_n)$ be the solutions of* (19.34) *and* (19.1), *with $g_1 = u_{tt}(0) = \Delta g + f_t(0)$, and let $\varepsilon \in (0, \frac{1}{2})$. Assume that the triangulations are refined as in Lemma 19.10. Then we have, with $C = C_{\varepsilon,T}$,*

$$\|U^n - u(t_n)\| \le C(h^2+k^2)\Big(\|\Delta v\| + \|g\|_\varepsilon + \|g_1\|_\varepsilon + \int_0^{t_n} (\|f_t\|_\varepsilon + \|f_{tt}\|_\varepsilon)\,d\tau\Big), \quad \textit{for } t_n \le T.$$

Proof. We again represent the error as in (19.32). For ρ^n we have, as in the proof of Theorem 19.8,

$$\|\rho^n\| \le Ch^2\Big(\|\Delta v\| + \|g\|_\varepsilon + \int_0^{t_n} \|f_t\|_\varepsilon\,d\tau\Big).$$

To bound θ^n we write

$$(\bar\partial\theta^n, \chi) + (\nabla\widehat{\theta}^n, \nabla\chi) = -(\omega^n, \chi), \quad \forall \chi \in S_h, \ n \ge 1, \tag{19.35}$$

where, cf. (1.56),

$$\omega^n = (R_h - I)\bar\partial u(t_n) + \big(\bar\partial u(t_n) - u_t(t_{n-\frac12})\big) + \Delta\big(u(t_{n-\frac12}) - \widehat{u}(t_n)\big) = \sum_{j=1}^{3} \omega_j^n.$$

As in the proof of Theorem 1.6 this yields, since $\theta^0 = 0$,

$$\|\theta^n\| \le k\sum_{j=1}^n \|\omega_1^j\| + k\sum_{j=1}^n \|\omega_2^j\| + k\sum_{j=1}^n \|\omega_3^j\|.$$

Here, as in (19.10) and the proof of Theorem 19.8,

$$k\sum_{j=1}^n \|\omega_1^j\| \le \int_0^{t_n} \|\rho_t\|\,d\tau \le Ch^2\Big(\|g\|_\varepsilon + \int_0^{t_n} \|f_t\|_\varepsilon\,d\tau\Big), \quad \text{for } t_n \le T.$$

Further, by Taylor expansion around $t_{n-\frac12}$, as in the proof of Theorem 1.6,

$$k(\|\omega_2^j\| + \|\omega_3^j\|) \le Ck^2 \int_{t_{j-1}}^{t_j} (\|u_{ttt}\| + \|\Delta u_{tt}\|)\,d\tau \le Ck^2 \int_{t_{j-1}}^{t_j} (\|u_{ttt}\| + \|f_{tt}\|)\,d\tau,$$

where we have also used $\Delta u_{tt} = u_{ttt} - f_{tt}$. Hence, applying also (19.27), we obtain

$$k\sum_{j=1}^n (\|\omega_2^j\| + \|\omega_3^j\|) \le Ck^2 \int_0^{t_n} (\|u_{ttt}\| + \|f_{tt}\|)\,d\tau \le C_\varepsilon k^2\Big(\|g_1\|_\varepsilon + \int_0^{t_n} \|f_{tt}\|_\varepsilon\,d\tau\Big), \quad \text{for } t_n \le T,$$

which bounds θ^n as desired. The proof is now complete. □

As mentioned earlier, the classical treatises of elliptic problems in domains with corners are Grisvard [110], [111], see also Dauge [65], Kondratiev [140], Nazarov and Plamenevsky [176] and Kozlov, Mazya and Rossmann [141]. Finite element methods in polygonal domains have been considered in Babuška and Aziz [11], Babuška and Rosenzweig [12], Kellogg [137], Bacuta, Bramble and Xu [14] and Bacuta, Bramble and Pasciak [13]. The use of refinement near the corners was initiated in Babuška [9] and Raugel [202]. Our treatment of the parabolic problem follows essentially Chatzipantelidis, Lazarov, Thomée and Wahlbin [47], where also further references to, e.g., fractional order spaces, may also be found.

20. Time Discretization by Laplace Transformation and Quadrature

In this chapter we consider an alternative to time stepping for the discretization in time of an initial value problem for a parabolic equation. We now use a representation of the solution as an integral along a smooth curve extending into the complex right half plane, with an integrand containing the resolvent of the associated elliptic operator. This integral is then evaluated to high accuracy by a quadrature rule. In this way the problem is reduced to a finite set of elliptic equations, which may be solved in parallel. The procedure is combined with finite element discretization in the spatial variables.

We consider first the approximate solution of the abstract parabolic problem

$$(20.1) \qquad u_t + Au = f(t), \quad \text{for } t > 0, \quad \text{with } u(0) = v,$$

in a complex Banach space $\mathcal{B}$, where v and $f(t)$ are given, and A is a closed operator in $\mathcal{B}$ such that $-A$ generates a bounded analytic semigroup $E(t) = e^{-At}$. More precisely, we assume that the spectrum $\sigma(A)$ of A is contained in a sector of the right half plane, and that the resolvent $R(z;A) = (z\,I - A)^{-1}$ of A satisfies, for some $\delta \in (0, \pi/2)$ and $M \geq 1$ independent of z,

$$(20.2) \quad \|R(z;A)\| \leq M(1+|z|)^{-1}, \quad \text{for } z \in \Sigma_\delta = \{z : \delta \leq |\arg z| \leq \pi\} \cup \{0\}.$$

We note that since $\|A^{-1}\| \leq M$ it follows that $z \in \rho(A)$ for $|z| < 1/M$, and that $\|R(z;A)\| \leq 2M$ for $|z| \leq 1/(2M)$, say.

For our present approach to the solution of (20.1), let $\widehat{u}(z)$ denote the Laplace transform of u, so that for some $x_0 \in \mathbb{R}$,

$$(20.3) \qquad \widehat{u}(z) = \int_0^\infty e^{-zt} u(t)\, dt, \quad \text{for } \operatorname{Re} z \geq x_0.$$

Taking Laplace transforms in (20.1), we then obtain the transformed equation

$$(20.4) \qquad (z\,I + A)\widehat{u}(z) = v + \widehat{f}(z),$$

where we assume that $\widehat{f}(z)$ is analytic for $\operatorname{Re} z \geq x_0$. We then formally have

$$(20.5) \qquad \widehat{u}(z) = R(z; -A)(v + \widehat{f}(z)), \quad \text{for } \operatorname{Re} z \geq x_0.$$

Taking inverse Laplace transforms we find

$$u(t) = \frac{1}{2\pi i}\int_{x_0-\infty}^{x_0+\infty} e^{zt} R(z;-A)(v+\widehat{f}(z))\,dz, \tag{20.6}$$

or, after a change of variables $z \to -z$, and with $g(z) = v + \widehat{f}(-z)$,

$$u(t) = \frac{1}{2\pi i}\int_{\Gamma} e^{-zt} w(z)\,dz, \quad \text{where } w(z) = R(z;A)g(z). \tag{20.7}$$

Initially Γ is the line $\Gamma_0 = -x_0 + i\mathbb{R}$ parallel to the imaginary axis in the complex plane, with $\operatorname{Im} z$ decreasing along Γ_0, but for our purposes, assuming that $w(z)$ may be continued analytically in an appropriate way, we shall want to take for Γ a deformed contour in the set $\Sigma'_\delta = \Sigma_\delta \cup \{z;\ |z| < 1/(2M)\}$, with Σ_δ as in (20.2), which behaves asymptotically as a pair of straight lines in the right half plane, with slopes $\pm\sigma \neq 0$, say, where $\sigma \geq \tan\delta$, so that the factor e^{-zt} decays exponentially as $|z| \to \infty$ along Γ. Since clearly the resolvent $R(z;A)$ is analytic in Σ'_δ, the question of analyticity of $w(z)$ along Γ depends on the forcing term $f(t)$ in (20.1). The reason for the change of sign in z above is that then the representation (20.7) conforms with the formula (6.35) in the case of the homogeneous equation.

For concreteness, we take

$$\Gamma = \{z = z(s) = \varphi(s) - i\sigma s,\ s \in \mathbb{R}\} \subset \Sigma'_\delta,\ \varphi(s) = -\gamma + \sqrt{s^2+\nu^2}, \tag{20.8}$$

for suitable positive parameters γ, ν, and σ . The curve Γ is then the right-hand branch of a hyperbola, which crosses the real axis at $\varphi(0) = -\gamma + \nu < 1/(2M)$. Some of the constants below will depend on the parameters of Γ. With the choice of the minus sign in the imaginary part of $z(s)$, we have that $\operatorname{Im} z$ decreases along Γ as s increases from $-\infty$ to ∞.

Letting $\mathbb{R}_+ = [0,\infty)$, we assume thus that $\widehat{f}(-z)$, and therefore also $g(z)$, has a bounded analytic continuation from the complex half-plane bounded to the right by Γ_0 to the closed subset $G = \Gamma - \mathbb{R}_+$ of the complex plane to the left of Γ, so that all singularities of $g(z)$ lie to the right of Γ. The same property will then apply to $w(z)$ in (20.8).

Examples of such functions are linear combinations of functions of the form $f(t) = t^l e^{-\lambda t} b$, with l a nonnegative integer, λ a complex number, and $b \in \mathcal{B}$. We then have $\widehat{f}(-z) = l!(\lambda - z)^{-l-1} b$ which is analytic for $z \neq \lambda$. In the presence of this function, Γ should be chosen to the left of λ. In the particular case of the homogeneous equation, i.e., when $f(t) = 0$, Γ may be chosen as any curve in Σ'_δ which may be homotopically deformed to Γ_0.

Using our assumptions on A, Γ and $\widehat{f}(z)$, one may use the representation (20.7) of $u(t)$ to show some stability and smoothness estimates. Here and below we write

$$\|g\|_W = \sup_{z\in W} |g(z)|, \quad \text{for } W \subset \mathbb{C}.$$

Theorem 20.1 *Assume that $g(z)$ is bounded and analytic in G, and let $\kappa = -\varphi(0) = \gamma - \nu$. Then we have for the solution $u(t)$ of (20.1),*

$$\|A^j u^{(k)}(t)\| \leq CM(e^{\kappa t} + t^{-j-k})\|g\|_\Gamma, \quad \text{for } t > 0,\ j = 0, 1,\ k \geq 0.$$

Proof. We begin with the stability estimate, the case $j = k = 0$. For $t \geq 1$ we find at once by (20.7), (20.8) and (20.2),

$$\begin{aligned}\|u(t)\| &\leq C \int_\Gamma e^{-t\,\mathrm{Re}\,z}\|R(z;A)\|\,|dz|\ \|g\|_\Gamma \\ &\leq CM \int_{-\infty}^{\infty} e^{-t\varphi(s)}(1+|s|)^{-1}\,ds\ \|g\|_\Gamma.\end{aligned}$$

Here, since $\varphi(s) - \varphi(0) = \sqrt{s^2+\nu^2} - \nu \geq \frac{1}{2}|s| - \frac{1}{2}\nu$, we have

$$-t\varphi(s) = \kappa t - t(\varphi(s) - \varphi(0)) \leq \kappa t - \tfrac{1}{2}|s| + \tfrac{1}{2}\nu. \tag{20.9}$$

Hence

$$\|u(t)\| \leq CMe^{\kappa t}\int_0^\infty e^{-\frac{1}{2}s}\,ds\ \|g\|_\Gamma = CMe^{\kappa t}\,\|g\|_\Gamma.$$

For $0 < t < 1$, since the integrand in (20.7) is analytic in $G = \Gamma - \mathbb{R}_+$, we may replace the part of Γ for which $|s| \leq 1/t$ by the part of the circle $\{z : |z| = \rho_t = |z(1/t)|\}$ that lies in G, and thus integrate over $\Gamma_t \cup \gamma_t \subset \overline{G}$, where $\Gamma_t = \{z \in \Gamma; |z| \geq \rho_t\}$ and $\gamma_t = \{z \in G; |z| = \rho_t\}$, appropriately oriented. Since $|z(s)| \geq \sigma|s|$ we have $\|R(z;A)\| \leq C(1+|s|)^{-1}$ on Γ, and since also $-\varphi(s) \leq \gamma - |s|$, we find

$$\begin{aligned}\int_{\Gamma_t} e^{-t\,\mathrm{Re}\,z}\|R(z;A)\|\,|dz| &\leq CM\int_{1/t}^\infty e^{-t\varphi(s)}s^{-1}\,ds \\ &\leq CMe^{\gamma t}\int_{1/t}^\infty e^{-ts}s^{-1}\,ds \leq CM.\end{aligned}$$

Further, since $|z(s)| \leq C(1+|s|)$ we have $\rho_t \leq Ct^{-1}$ and hence

$$\int_{\gamma_t} e^{-t\,\mathrm{Re}\,z}\|R(z;A)\|\,|dz| \leq CM\int_{-\pi}^{\pi} e^{t\rho_t}\rho_t^{-1}\rho_t\,d\theta \leq CM.$$

It follows that

$$\|u(t)\| \leq CM\|g\|_G.$$

Noting that $\|g\|_G = \|g\|_\Gamma$ by the maximum-principle, since $g(z)$ is analytic in G, and since $e^{\kappa t}$ is bounded below for $t < 1$, this completes the proof.

Turning to the case $j = 0, 1,\ j + k > 0$, we have

$$A^j u^{(k)}(t) = \frac{(-1)^k}{2\pi i}\int_\Gamma z^k e^{-zt}A^j R(z;A)g(z)\,dz,$$

so that

$$\|A^j u^{(k)}(t)\| \le CM \int_0^\infty (1+s)^{j+k-1} e^{-t\varphi(s)}\,ds\, \|g\|_\Gamma.$$

Here $\varphi(s) \ge -\kappa$ for $s \in \mathbb{R}$ and $\varphi(s) \ge \frac{1}{2}s$ for $s \ge s_0$, for some $s_0 > 0$. Hence

$$\begin{aligned}\int_0^\infty (1+s)^{j+k-1} e^{-t\varphi(s)}\,ds &\le C\int_0^{s_0} e^{\kappa t}\,ds + C\int_{s_0}^\infty s^{j+k-1} e^{-\frac{1}{2}ts}\,ds \\ &\le C(e^{\kappa t} + t^{-j-k}),\end{aligned}$$

which completes the proof. □

With the deformed contour represented as in (20.8), the integral (20.7) may be written as

$$u(t) = \int_{-\infty}^\infty v(s,t)\,ds, \quad \text{with } v(s,t) = \frac{1}{2\pi i} e^{-z(s)t} w(z(s)) z'(s). \tag{20.10}$$

We note that the integrand decays exponentially for large $|s|$ when $t > 0$.

Our approximate solution will now be defined by approximating the integral by means of a quadrature scheme,

$$U_N(t) = \sum_{j=-N}^N \omega_j v(s_j, t) = \sum_{j=-N}^N \widetilde{\omega}_j e^{-z_j t} w(z_j), \tag{20.11}$$

with certain quadrature points $s_j \in \mathbb{R}$ and nonnegative weights ω_j, and where $z_j = z(s_j)$, $\widetilde{\omega}_j = z'(s_j)\omega_j/(2\pi i)$. We remark that although the exact solution $u(t)$ does not depend on Γ, this approximate solution does. Below we shall consider in more detail two specific such quadrature formulas.

By the definition in (20.7), the values of $w(z)$ needed in (20.11) satisfy

$$(A - z_j I) w(z_j) = -g(z_j), \quad \text{for } |j| \le N. \tag{20.12}$$

This expresses a central feature of our method, namely that the $2N+1$ values $w(z_j) \in \mathcal{B}$ entering in (20.11) are independent, and hence may be found in parallel. We remark also that the functions $w(z_j)$ determine the approximate solution (20.11) for all $t > 0$.

We shall now consider a first quadrature formula for an integral over the real axis $\mathbb{R}$ with values in $\mathcal{B}$, by applying a truncated trapezoidal rule. Under appropriate conditions this quadrature formula has a high order of accuracy. We shall then apply this formula to our representation (20.10) of the solution of the parabolic problem. More precisely, we shall study the quadrature rule

$$Q_N(v) = k\sum_{j=-N}^N v(s_j) \approx J(v) = \int_{-\infty}^\infty v(s)\,ds, \quad \text{where } s_j = jk, \tag{20.13}$$

where we choose $k = N^{-(1-\varepsilon)}$ with some $\varepsilon \in (0,1)$. If we apply this quadrature rule to our representation (20.10) of the solution of (20.1), this defines the approximation to $u(t)$ as

$$U_N(t) = Q_N(v(\cdot,t)) = \frac{k}{2\pi i}\sum_{j=-N}^{N} e^{-z_j t} w(z_j) z'(s_j), \quad k = N^{-(1-\varepsilon)}, \tag{20.14}$$

where $z_j = z(s_j) = \varphi(s_j) - i\sigma s_j$. Note that $\max_{|j|\le N}|z_j| = O(N^\varepsilon)$.

We begin our analysis with the following stability result. As earlier we set

$$\ell(t) = \max(1, \log(1/t)). \tag{20.15}$$

Lemma 20.1 *Assume that $v : \mathbb{R} \to \mathcal{B}$ satisfies*

$$\|v(s)\| \le V(1+|s|)^{-1}e^{-\mu|s|}, \quad \textit{for } s \in \mathbb{R},\ \mu > 0. \tag{20.16}$$

Then, for $Q_N(v)$ defined in (20.13), *with $k = N^{-(1-\varepsilon)}$, $\varepsilon \in (0,1)$, we have, with $C = C_\varepsilon$,*

$$\|Q_N(v)\| \le CV\ell(\mu), \quad \textit{for } \mu > 0.$$

Proof. We have

$$\|Q_N(v)\| \le Vk\sum_{j=-\infty}^{\infty}(1+|s_j|)^{-1}e^{-\mu|s_j|} \le V\Big(k + 2\int_0^\infty (1+s)^{-1}e^{-\mu s}\,ds\Big).$$

The result now follows from the easily proven fact that

$$\int_0^\infty e^{-\mu s}(1+s)^{-1}ds \le C\ell(\mu), \quad \text{for } \mu > 0. \qquad \square \tag{20.17}$$

Using this lemma we now show the following stability estimate for the time discrete solution $U_N(t)$ of (20.1), as defined by (20.14), demonstrating that the discrete solution is bounded in each closed subinterval of $(0,\infty)$, with a bound that grows logarithmically for t small.

Theorem 20.2 *Under the assumptions of Theorem 20.1, let $U_N(t)$ be defined by* (20.14). *Then we have*

$$\|U_N(t)\| \le CM\, e^{\kappa t}\,\ell(t)\,\|g\|_\Gamma, \quad \textit{for } t > 0, \quad \textit{with } \kappa = -\varphi(0).$$

Proof. Using (20.2) and (20.10), we have, since $|z'(s)|$ is bounded, that

$$\|v(s,t)\| \le Ce^{-t\varphi(s)}\|w(z(s))\| \le CMe^{-t\varphi(s)}(1+|s|)^{-1}\|g\|_\Gamma, \quad \text{for } s \in \mathbb{R}.$$

For $t \le 1$ this shows

$$\|v(s,t)\| \le CMe^{t(\gamma-|s|)}(1+|s|)^{-1}\|g\|_\Gamma \le CMe^{-t|s|}(1+|s|)^{-1}\|g\|_\Gamma,$$

and the bound stated therefore follows from Lemma 20.1. For $t \geq 1$ we have by (20.9)

$$\begin{aligned}\|U_N(t)\| \leq & CMk \sum_{|j|\leq N} e^{-t\varphi(s_j)}\|g\|_\Gamma \\ & \leq CMe^{\kappa t}k \sum_{|j|\leq N} e^{-|s_j|/2}\|g\|_\Gamma \leq CMe^{\kappa t}\|g\|_\Gamma,\end{aligned}$$

which completes the proof. □

We next turn to an estimate for the quadature error in (20.13).

Lemma 20.2 *Let $r \geq 1$ be given and assume that, with $C = C_r$,*

$$\|v^{(j)}(s)\| \leq C(1+|s|)^{-1}e^{-\mu|s|}, \quad \text{for } j \leq r, \quad s \in \mathbb{R}, \quad \mu > 0. \tag{20.18}$$

Then, with $Q_N(v)$ defined in (20.13), with $k = N^{-(1-\varepsilon)}$, $\varepsilon \in (0,1)$, we have, with $C = C_{r,\varepsilon}$,

$$\|Q_N(v) - J(v)\| \leq CV\ell(\mu)\big(N^{-r(1-\varepsilon)} + e^{-\mu N^\varepsilon}\big), \quad \text{for } \mu > 0.$$

Proof. We shall use the following easy consequence of the Euler-Maclaurin summation formula, see, e.g., [66], p. 208. Let $Q_\infty(v) = k\sum_{j=-\infty}^{\infty} v(jk)$. Then, for $r > 1$,

$$\|Q_\infty(v) - J(v)\| \leq \frac{Ck^r}{(2\pi)^r}\int_{-\infty}^{\infty} \|v^{(r)}(s)\|\,ds.$$

Under our assumption (20.18) it follows that

$$\|Q_\infty(v) - J(v)\| \leq CVN^{-r(1-\varepsilon)}\int_0^\infty (1+s)^{-1}e^{-\mu s}\,ds.$$

We also have

$$\|Q_N(v) - Q_\infty(v)\| \leq kV\sum_{|j|>N}(1+|s_j|)^{-1}e^{-\mu|s_j|} \leq CV\int_{Nk}^\infty (1+s)^{-1}e^{-\mu s}\,ds.$$

Here, since $Nk = N^\varepsilon$, we find

$$\int_{Nk}^\infty (1+s)^{-1}e^{-\mu s}\,ds \leq e^{-\mu Nk}\int_0^\infty (1+s)^{-1}e^{-\mu s}\,ds \leq Ce^{-\mu N^\varepsilon}\ell(\mu), \tag{20.19}$$

where in the last step we have used (20.17). □

We now show the following error estimate for the discrete solution.

Theorem 20.3 *Let U_N be defined by (20.14). Then, under the appropriate assumptions on $g(z)$ and Γ, we have, for any $r \geq 1$ and $\widetilde{\kappa} > \kappa$, with $C = C_{r,\varepsilon}$,*

$$\|U_N(t) - u(t)\| \leq CM\,e^{\widetilde{\kappa}t}\,\ell(t)\big(N^{-r(1-\varepsilon)} + e^{-tN^\varepsilon}\big)\max_{k\leq r}\|g^{(k)}\|_\Gamma, \quad \text{for } t > 0.$$

Proof. We recall from (20.10) and (20.14) that

$$U_N(t) - u(t) = Q_N(v(\cdot,t)) - J(v(\cdot,t)).$$

To apply Lemma 20.2 we use (20.2) and the Leibniz rule applied to $w(z)$ as defined in (20.7) to obtain

$$\|w^{(j)}(z)\| \le CM(1+|z|)^{-1} \max_{k\le j} \|g^{(k)}(z)\|, \quad \text{for } z \in \Gamma,$$

and hence, from the definition of $v(\cdot,t)$ in (20.10),

$$\|v^{(j)}(s,t)\| \le CM(1+t^r)e^{-t\varphi(s)}(1+|s|)^{-1} \max_{k\le r} \|g^{(k)}\|_\Gamma, \quad \text{for } j \le r, \ s \in \mathbb{R}.$$

Since $(1+t^r)e^{-t\varphi(s)} \le Ce^{t\tilde{\kappa}}e^{-|s|/2}$ by (20.9), for $t \ge 1$, and $(1+t^r)e^{-t\varphi(s)} \le Ce^{-t|s|}$ for $t \le 1$, the theorem follows by Lemma 20.2. □

Since r is arbitrary, this error bound is of order $O(N^{-q})$ for any $q > 0$, for fixed $t > 0$, but deteriorates as t tends to 0.

Choosing the time step $k = c/\sqrt{N}$ in (20.13) and (20.14), i.e., $\varepsilon = \frac{1}{2}$, and using a different analysis based on a representation of the quadrature error as an integral over the boundary of a strip around Γ in the complex plane, one may improve the error estimate in Theorem 20.3 to $O(e^{-c\sqrt{N}})$ for $t \ge 0$ with $c > 0$. We shall not carry out the details but we apply this alternative approach to our next quadrature scheme.

To define this second quadrature formula we begin with a change of variable in (20.8), and set

$$s = \nu \sinh \xi, \quad \text{for } \xi \in \mathbb{R}, \tag{20.20}$$

The representation (20.7) of the solution may now be thought of as an integral with respect to the real variable ξ,

$$u(t) = \int_{-\infty}^{\infty} v(\xi,t)\,d\xi, \quad \text{where } v(\xi,t) = \frac{1}{2\pi i}\, e^{-z(\xi)t} w\big(z(\xi)\big) z'(\xi). \tag{20.21}$$

where, with $\alpha = \arctan \sigma \in (0, \frac{1}{2}\pi)$, $\lambda = \nu\sqrt{1+\sigma^2}$,

$$z = z(\xi) = -\gamma + \lambda(\cos\alpha \cosh\xi - i \sin\alpha \sinh\xi) = -\gamma + \lambda\cos(\alpha + i\xi). \tag{20.22}$$

This time the time-discretization will be affected by choosing the quadrature rule (20.13), with $k = \log N/N$, whose quadrature points are equally spaced in $[-\log N, \log N]$. Applying this to (20.21), and setting $z_j = z(\xi_j)$, $\xi_j = jk$, our approximate solution to (20.1) becomes

$$U_N(t) = \frac{k}{2\pi i} \sum_{j=-N}^{N} e^{-z_j t} w(z_j) z'(\xi_j), \quad \text{with } k = \log N/N. \tag{20.23}$$

We note that this time, for N large,

$$\max_{|j|\le N} |z_j| = |z_N| = |\gamma - \lambda\cos\alpha\cosh(\log N) + i\lambda\sin\alpha\sinh(\log N)| \approx \tfrac{1}{2}\lambda N.$$

The asymptotic behavior of the function $z = z(s)$ given in (20.8) implies that $|e^{-z(s)t}| \approx e^{-|s|t}$ for $|s|$ large. Using instead the parameter ξ in (20.20), (20.22) shows the "double exponential" behavior $|e^{-z(\xi)t}| \approx e^{-\nu t\cosh\xi}$ for large $|\xi|$, which will lead to an improved error bound. In the analysis of the quadrature rule (20.13) we shall now need the following.

Lemma 20.3 *Let $Q_N(v)$ be defined in (20.13), with $k = \log N/N$, and assume that the integrand v satisfies*

$$\|v(\xi)\| \le Ve^{-\mu\cosh\xi} \quad \text{for } \xi\in\mathbb{R},\ \mu>0.$$

Then we have, with C independent of N, V and μ,

$$\|Q_N(v)\| \le C\ell(\mu)\,V, \quad \text{for } N\ge 1.$$

Proof. We have

$$\|Q_N(v)\| \le k\sum_{j=-\infty}^{\infty} Ve^{-\mu\cosh(jk)} \le kV + 2V\int_0^\infty e^{-\mu\cosh\xi}\,d\xi,$$

and, changing variables by $s = \cosh\xi - 1$ and using (20.17),

$$\begin{aligned}(20.24)\qquad \int_0^\infty e^{-\mu\cosh\xi}\,d\xi &= e^{-\mu}\int_0^\infty \frac{e^{-\mu s}}{\sqrt{s^2+2s}}\,ds \\ &\le \int_0^1 \frac{ds}{\sqrt{2s}} + \sqrt{2}\int_1^\infty \frac{e^{-\mu s}}{1+s}\,ds \le C\,\ell(\mu).\end{aligned}$$

Since k is bounded for $N\ge 1$, this shows the lemma. □

By applying Lemma 20.3 to the integrand $v(\xi,t)$ in (20.21) we obtain the following stability result.

Theorem 20.4 *Let $U_N(t)$ be the approximate solution of (20.1) defined by (20.23). Then, under the appropriate assumpions on $g(z)$ and Γ, we have*

$$\|U_N(t)\| \le CMe^{\kappa t}\ell(t)\|g\|_\Gamma, \quad \text{for } t>0, \quad N\ge 1.$$

Proof. Recalling (20.7) and (20.2), we see that

$$\|w(z(\xi))\| \le \frac{CM}{1+|z(\xi)|}\|g\|_\Gamma, \quad \text{for } \xi\in\mathbb{R}.$$

By (20.22), we have

$$(20.25)\qquad \frac{z'(\xi)}{z(\xi)} = \frac{-i\lambda\sin(\alpha+i\xi)}{-\gamma+\lambda\cos(\alpha+i\xi)}$$

$$= \frac{-\cos\alpha\sinh\xi + i\sin\alpha\cosh\xi}{\gamma\lambda^{-1}-\cos\alpha\cosh\xi+i\sin\alpha\sinh\xi} \to \pm 1, \quad \text{as } \xi\to\pm\infty,$$

so that $|z'(\xi)| \le C(1+|z(\xi)|)$ for $\xi\in\mathbb{R}$, and hence, since $\lambda\cos\alpha = \nu$,

$$(20.26)\qquad \|v(\xi,t)\| \le CMe^{-t\,\mathrm{Re}\,z(\xi)}\|g\|_\Gamma = CMe^{t\gamma}e^{-t\nu\cosh\xi}\|g\|_\Gamma, \quad \text{for } \xi\in\mathbb{R}.$$

It therefore follows, by Lemma 20.3, with $\mu = t\nu$, since $\ell(\nu t)\le C\ell(t)$ and $e^{\gamma t}\le C$, that

$$\|U_N(t)\| = \big\|Q_N(v(\cdot,t))\big\| \le CM\ell(t)\|g\|_\Gamma, \quad \text{for } t\le 1.$$

Since $\mathrm{Re}\,z(\xi) = -\gamma+\nu\cosh\xi = -\kappa+\nu(\cosh\xi-1)$, we have $-t\,\mathrm{Re}\,z(\xi)\le \kappa t-\nu(\cosh\xi-1)$ for $t\ge 1$, and hence

$$\|U_N(t)\| \le CMe^{\kappa t}k\sum_{|j|\le N} e^{-\nu\cosh\xi_j}\|g\|_\Gamma \le CMe^{\kappa t}\|g\|_\Gamma, \quad \text{for } t\ge 1,$$

which completes the proof. □

The analysis of the quadrature error will depend on assuming that the integrand may be extended into a closed strip $Y_r = \{\zeta : |\mathrm{Im}\,\zeta|\le r\}$ around the real axis, and satisfies certain boundedness properties there. The next lemma shows that under appropriate conditions the quadrature error is of order $O(e^{-cN/\log N})$ as $N\to\infty$.

Lemma 20.4 *Let $Q_N(v)$ be defined by (20.13), with $k=\log N/N$, and assume that the integrand $v(\zeta)$ is analytic and bounded in Y_r, and if*

$$\|v(\xi+i\eta)\| \le Ve^{-\mu\cosh\xi} \quad \textit{for } \xi\in\mathbb{R} \quad \textit{and} \quad |\eta|\le r, \quad \textit{with } \mu>0.$$

Then, with $\bar r = 2\pi r$, and with C independent of N, V and μ, we have

$$\|Q_N(v)-J(v)\| \le CV\ell(\mu)\big(e^{-\bar r N/\log N}+e^{-\mu N/2}\big), \quad \textit{for } N\ge 2.$$

Proof. Let $Q_\infty(v) = k\sum_{j=-\infty}^{\infty} v(jk)$. We first show that

$$(20.27)\qquad \|Q_\infty(v)-J(v)\| \le \frac{e^{-\bar r/k}}{1-e^{-\bar r/k}}\int_{-\infty}^{\infty}\big(\|v(\xi+ir)\|+\|v(\xi-ir)\|\big)\,d\xi.$$

For this we observe that

$$Q_\infty(v) = \frac{1}{2\pi i}\int_{\mathcal{C}_r} v(\zeta)\pi\cot\Big(\frac{\pi\zeta}{k}\Big)\,d\zeta,$$

where the contour $\mathcal{C}_r = \mathcal{C}_r^+\cup\mathcal{C}_r^- = \partial Y_r$ consists of the lines $\mathcal{C}_r^\pm : \{\zeta = \mp\xi\pm ir\}$, with $\xi\in\mathbb{R}$ increasing. Since

$$\frac{1}{2i}\cot\Big(\frac{\pi\zeta i}{k}\Big) = \mp\frac{1}{2} \mp \frac{e^{\mp i2\pi\zeta/k}}{1-e^{\mp i2\pi\zeta/k}}, \quad \text{for } \zeta \in \mathcal{C}^{\pm},$$

and, by deforming the contours in the complex plane,

$$\int_{\mathcal{C}_r^{\pm}} v(\zeta)\,d\zeta = \mp\int_{-\infty}^{\infty} v(\xi)\,d\xi = \mp J(v),$$

it follows that

$$Q_\infty(v) = J(v) - \int_{\mathcal{C}_r^+} \frac{v(\zeta)e^{i2\pi\zeta/k}}{1-e^{i2\pi\zeta/k}}\,d\zeta + \int_{\mathcal{C}_r^-} \frac{v(\zeta)e^{-i2\pi\zeta/k}}{1-e^{-i2\pi\zeta/k}}\,d\zeta.$$

Since $\mathrm{Re}(i2\pi\zeta/k) = \mp\bar{r}/k$ on $\mathcal{C}_r^{\pm}$, the inequality (20.27) now follows by obvious estimates.

From (20.27) it follows, using (20.24), that, since $e^{-\bar{r}N/\log N} < 1$ for $N \geq 2$,

$$\|Q_\infty(v) - J(v)\| \leq \frac{4Ve^{-\bar{r}N/\log N}}{1-e^{-\bar{r}N/\log N}} \int_0^\infty e^{-\mu\cosh\xi}\,d\xi \leq C\ell(\mu)Ve^{-\bar{r}N/\log N}.$$

For the remainder of the infinite sum we have

$$\|Q_\infty(v) - Q_N(v)\| \leq 2Vk\sum_{j=N+1}^{\infty} e^{-\mu\cosh(jk)} \leq 2V\int_{Nk}^{\infty} e^{-\mu\cosh\xi}\,d\xi.$$

Here, as in (20.24) we have by (20.19), with $s = \cosh\xi - \cosh(Nk)$,

$$\begin{aligned}\int_{Nk}^{\infty} e^{-\mu\cosh\xi}\,d\xi &= e^{-\mu\cosh(Nk)}\int_0^\infty \frac{e^{-\mu s}}{\sqrt{(s+\cosh(Nk))^2-1}}\,ds \\ &\leq e^{-\mu N/2}\int_0^\infty \frac{e^{-\mu s}}{\sqrt{s^2+2s}}\,ds \leq Ce^{-\mu N/2}\ell(\mu),\end{aligned}$$

where we have used $\cosh(Nk) = \cosh(\log N) \geq N/2$. Together these estimates complete the proof. □

For the purpose of application of Lemma 20.4 to the function $v(\xi,t)$ in (20.21) we define a conformal mapping

$$z = z(\zeta) = -\gamma + \lambda\cos(\alpha + i\zeta), \tag{20.28}$$

from the strip Y_r onto the set

$$Z_r = \{\, z(\zeta) : \zeta \in Y_r \}. \tag{20.29}$$

The contour Γ is then just the image in the z-plane of the real axis in the ζ-plane, and may now be defined as $z = z(\xi)$ for $\xi \in \mathbb{R}$. In the error analysis below the assumptions made above on Γ now have to hold with Γ replaced by Z_r for some r satisfying $0 < r < \alpha$. In particular, we assume that r is so small that the singularities of $g(z)$ are to the right of Z_r.

Writing $z = x + iy$ and $\zeta = \xi + i\eta$, we find that

$$x = -\gamma + \lambda\cos(\alpha - \eta)\cosh\xi, \quad y = -\lambda\sin(\alpha - \eta)\sinh\xi, \tag{20.30}$$

and thus the line η = constant is mapped to the right branch of the hyperbola

$$\left(\frac{x+\gamma}{\lambda\cos(\alpha-\eta)}\right)^2 - \left(\frac{y}{\lambda\sin(\alpha-\eta)}\right)^2 = 1,$$

whose asymptotes are $y = \pm(x+\gamma)\tan(\alpha-\eta)$, with angles $\pm(\alpha-\eta)$ with the positive real axis, and which cuts the real axis at $x = -\gamma + \lambda\cos(\alpha-\eta)$. Hence, sufficient conditions to ensure that $Z_r \subset \Sigma_\delta$ and that $\operatorname{Re} z \to \infty$ whenever $|\operatorname{Im} z| \to \infty$ with $z \in Z_r$ are

$$0 < r < \alpha, \quad \alpha - r > \delta, \quad \gamma > \lambda\cos(\alpha - r). \tag{20.31}$$

In applying Lemma 20.4 to our approximate solution of (20.1) we need to assume that our assumptions hold with Γ replaced by a strip Z_r with $r > 0$. In particular, $g(z)$ now has to be bounded and analytic in $\widetilde{G}_r = Z_r - \mathbb{R}_+$. The contour Γ is thus required to lie more to the left for the error estimate than for the stability bound of Theorem 20.4.

Theorem 20.5 *Let $u(t)$ be the solution of the initial value problem* (20.1), *and assume that $Z_r \subset \Sigma_\delta$. Then, under the above assumptions on $g(z)$, the approximate solution $U_N(t)$ of* (20.1) *defined by* (20.23) *satisfies, with $c = \frac{1}{2}\lambda\cos(\alpha+r)$ and $\bar{r} = 2\pi r$, for any $\widetilde{\kappa} > \gamma - \lambda\cos(\alpha + r)$,*

$$\|U_N(t) - u(t)\| \le CMe^{\widetilde{\kappa}t}\ell(t)\big(e^{-\bar{r}N/\log N} + e^{-ctN}\big)\|g\|_{Z_r}, \quad \textit{for } t > 0.$$

We note that for any given $t > 0$, the first term in the parenthesis is the dominant term, so that this result shows a convergence rate of order $O(e^{-\bar{r}N/\log N})$, which deteriorates as t tends to zero. We also note that a larger r results in a higher convergence rate.

Proof of Theorem 20.5. We have, for $\zeta = \xi + i\eta \in Z_r$,

$$\|v(\xi + i\eta, t)\| \le \frac{1}{2\pi}e^{-\operatorname{Re} z(\xi+i\eta)t}\|w(\sigma + i\eta)\|\,|z'(\xi + i\eta)|,$$

and $\operatorname{Re} z(\xi + i\eta) = -\gamma + \lambda\cos(\alpha - \eta) \ge -\gamma + \lambda\cos(\alpha + r)$. As in (20.7), since $Z_r \subset \Sigma_\delta$, we have

$$\|w(z)\| \le \frac{CM}{1+|z|}\|g\|_{Z_r}, \quad \text{for } z \in Z_r,$$

and (20.25) generalizes to

$$\frac{z'(\xi+i\eta)}{z(\xi+i\eta)} = \frac{-\cos(\alpha-\eta)\sinh\xi + i\sin(\alpha-\eta)\cosh\xi}{\gamma\lambda^{-1} - \cos(\alpha-\eta)\cosh\xi + i\sin(\alpha-\eta)\sinh\xi} \to \pm 1, \tag{20.32}$$

as $\xi \to \pm\infty$, uniformly for $|\eta| \le r$. Hence, for $\xi \in \mathbb{R}$ and $|\eta| \le r$,

$$\|v(\xi+i\eta,t)\| \le CMe^{(\gamma-\lambda\cos(\alpha+r))\,t\cosh\xi}\|g\|_{Z_r} = CMe^{\widetilde{\kappa}t}e^{-\mu t\cosh\xi}\|g\|_{Z_r},$$

where $\mu = \kappa - \gamma + \lambda\cos(\alpha+r) > 0$. The result then follows from Lemma 20.4. □

As we pointed out above, the convergence result of Theorem 20.5 may be described as a nonsmooth data error estimate. We shall now discuss a modification of the above method (20.23) for which we shall be able to show an error bound that holds uniformly down to $t = 0$, but which is only of order $O(e^{-c\sqrt{N}})$. For this purpose we first write the representation (20.6) in a slightly different form. Recall that the solution operator of the homogeneous case of (20.1) may be written as

$$E(t)v = \frac{1}{2\pi i}\int_\Gamma e^{-zt}R(z;A)v\,dz,$$

where Γ is defined by (20.22). Using Duhamel's principle we have

$$\begin{aligned} u(t) &= E(t)v + \int_0^t E(t-\tau)f(\tau)\,d\tau \\ &= \frac{1}{2\pi i}\int_\Gamma e^{-zt}R(z;A)v\,dz + \int_0^t \frac{1}{2\pi i}\int_\Gamma e^{-z(t-\tau)}R(z;A)f(\tau)\,dz\,d\tau, \end{aligned}$$

or, after changing the order of integration,

$$u(t) = \frac{1}{2\pi i}\int_\Gamma R(z;A)\widetilde{g}(z,t)\,dz,$$

where

$$\widetilde{g}(z,t) = e^{-zt}v + \int_0^t e^{-(t-\tau)z}f(\tau)\,d\tau = e^{-zt}\Big(v + \int_0^t e^{\tau z}f(\tau)\,d\tau\Big).$$

We note that the latter integral equals $\widehat{f}(-z)$ if $f(\tau)$ vanishes for $\tau > t$, which connects this formulation with the old representation (20.7). This is reasonable since the value of $u(t)$ is independent of $f(\tau)$ for $\tau > t$.

The main idea in our analysis is now to use the fact that if Γ crosses the real axis at $\varphi(0) = -\gamma + \nu > 0$, but sufficiently close to $z = 0$ so that this point is to the left of Γ, then

$$\frac{1}{2\pi i}\int_\Gamma e^{-zt}\frac{dz}{z} = 0, \quad \text{for } t > 0,$$

and hence, as is easily seen,

$$\frac{1}{2\pi i}\int_\Gamma \frac{\widetilde{g}(z,t)}{z}\,dz = 0.$$

The solution of (20.1) may therefore now be represented as

$$u(t) = \frac{1}{2\pi i}\int_\Gamma (R(z;A)\widetilde{g}(z,t) - z^{-1}\widetilde{g}(z,t))\,dz = \frac{1}{2\pi i}\int_\Gamma \widetilde{w}(z,t)\,dz,$$

where

$$\widetilde{w}(z,t) = \widetilde{R}(z;A)\widetilde{g}(z,t), \quad \text{with } \widetilde{R}(z;A) = R(z;A) - z^{-1}I.$$

The reason for this modification is that $\widetilde{R}(z;A)$ decays more rapidly than $R(z;A)$ as $|z| \to \infty$ on Γ. With the deformed contour represented as in (20.22), we obtain this time

$$u(t) = \int_{-\infty}^{\infty} v(\xi,t)\,d\xi, \text{ with } v(\xi,t) = \frac{1}{2\pi i}\widetilde{w}(z(\xi))z'(\xi). \tag{20.33}$$

Using again the quadrature rule (20.13) we now get an approximate solution of our problem of the form

$$U_N(t) = \frac{k}{2\pi i}\sum_{j=-N}^{N} \widetilde{w}(z_j,t)\,z_j', \tag{20.34}$$

with k to be specified below.

To compute the approximate solution we thus need to find

$$\widetilde{w}(z_j,t) = e^{-z_jt}W(z_j,t) - z_j^{-1}\widetilde{g}(z_j,t), \quad \text{for } |j| \le N,$$

where the $W(z_j,t)$ are the solutions of the $2N+1$ elliptic equations

$$(A - z_j)\,W(z_j,t) = -e^{z_jt}\widetilde{g}(z,t) = -v - \int_0^t e^{\tau z_j} f(\tau)\,d\tau, \quad |j| \le N.$$

Note that for the homogeneous equation this system is independent of t. This means that in this case the solution of the system of elliptic equations yields the discrete solution at all times. For the inhomogeneous equation one system of elliptic equations has to be solved for each time t where the approximate solution is sought.

In our error analysis we shall need some regularity of the data. To express this we define a scale of Banach spaces

$$\mathcal{B}^\sigma := D(A^\sigma) = \{v \in \mathcal{B} : A^\sigma v \in \mathcal{B}\}, \quad \text{for } \sigma > 0,$$

and write the norm in this space as $\|v\|_\sigma = \|A^\sigma v\|$. We note that (cf. [194], Theorem 2.6.10)

$$\|v\|_{1-\sigma} \le C_\sigma \|v\|^\sigma\,\|v\|_1^{1-\sigma}, \quad \text{for } 0 \le \sigma \le 1. \tag{20.35}$$

We begin our error analysis with a bound for the modified resolvent.

Lemma 20.5 *If A satisfies the resolvent estimate* (20.2), *then*

$$\|\widetilde{R}(z;A)v\| \le \frac{C_\sigma M}{|z|(1+|z|)^\sigma}\|v\|_\sigma, \quad \text{for } z \in \Sigma_\delta,\ 0<\sigma\le 1.$$

Proof. We note that

$$\widetilde{R}(z;A)v = z^{-1}R(z;A)Av = z^{-1}R(z;A)A^{1-\sigma}A^\sigma v$$

and, by (20.35),

$$\begin{aligned}&\|R(z;A)A^{1-\sigma}w\| = \|R(z;A)w\|_{1-\sigma}\\ &\le C_\sigma \|R(z;A)w\|^\sigma \|AR(z;A)w\|^{1-\sigma} \le CM(1+|z|)^{-\sigma}\|w\|,\end{aligned}$$

from which the result stated follows by setting $w = A^\sigma v$. □

Our next lemma is an error estimate for quadrature rule using a time step adapted to application with the estimate of Lemma 20.5 for the modified resolvent.

Lemma 20.6 *If the integrand $v(\zeta)$ is analytic and bounded in Y_r, and*

$$\|v(\xi+i\eta)\| \le Ve^{-\sigma|\xi|}, \quad \text{for } \xi\in\mathbb{R} \text{ and } |\eta|\le r,$$

then, with $Q_N(v)$ defined by (20.13) *and $\bar r = 2\pi r$,*

$$\|Q_N(v)-J(v)\| \le C_{r,\sigma}Ve^{-\sqrt{\bar r\sigma N}}, \quad \text{for } k=\sqrt{\bar r/(\sigma N)}.$$

Proof. For the infinite quadrature sum $Q_\infty(v) = k\sum_{j=-\infty}^{\infty} v(jk)$ we have using (20.27)

$$\|Q_\infty(v)-J(v)\| \le \frac{4Ve^{-\bar r/k}}{1-e^{-\bar r/k}}\int_0^\infty e^{-\sigma\xi}\,d\xi \le \frac{4V\sigma^{-1}e^{-\bar r/k}}{1-e^{-\bar r/k}},$$

whereas for the tail of the infinite sum we now have

$$\|Q_\infty(v)-Q_N(v)\| \le 2Vk\sum_{j=N+1}^{\infty} e^{-\sigma\xi_j} \le 2V\int_{Nk}^\infty e^{-\sigma\xi}\,d\xi \le 2V\sigma^{-1}e^{-\sigma Nk}.$$

Hence by the triangle inequality,

$$\|Q_N(v)-J(v)\| \le 2V\sigma^{-1}\Big(\frac{2e^{-\bar r/k}}{1-e^{-\bar r/k}} + e^{-\sigma Nk}\Big).$$

The error bound now follow by choosing k so that $\bar r/k = \sigma Nk$. □

We are now ready for our smooth data error estimate.

Theorem 20.6 *Let $u(t)$ be the solution of the initial value problem (20.1), and assume that $Z_r \subset \Sigma_\delta$. Then, for $0 < \sigma \le 1$, if we choose $k = \sqrt{\bar r/(\sigma N)}$, the approximate solution $U_N(t)$ defined in (20.34) satisfies, with $\widetilde\gamma = \gamma - \lambda\cos(\alpha + r)$ and $\bar r = 2\pi r$,*

$$\|U_N(t) - u(t)\| \le CMe^{\widetilde\gamma t} e^{-\sqrt{\bar r\sigma N}}\Big(\|v\|_\sigma + \int_0^t \|f\|_\sigma d\tau\Big), \quad \text{for } t > 0.$$

Proof. It follows from (20.33), (20.32) and Lemma 20.5 that

$$\begin{aligned}\|v(\xi + i\eta, t)\| &\le C\|\widetilde w(z(\xi + i\eta))\|\,|z'(\xi + i\eta)|\\ &\le CM(1 + |z(\xi + i\eta)|)^{-\sigma}\|\widetilde g(z(\xi + i\eta), t)\|_\sigma.\end{aligned}$$

Since $-\operatorname{Re} z(\xi + i\eta) = \gamma - \lambda\cos(\alpha - \eta)\cosh\xi \le \widetilde\gamma$, for $|\eta| \le r$, we have

$$\|\widetilde g(z(\xi + i\eta), t)\|_\sigma \le e^{\widetilde\gamma t}\Big(\|v\|_\sigma + \int_0^t \|f\|_\sigma d\tau\Big),$$

and, using also $|z(\xi + i\eta)| \ge c\cosh\xi \ge ce^{|\xi|}$, we thus find

$$\|v(\xi + i\eta, t)\| \le CMe^{\widetilde\gamma t}e^{-\sigma|\xi|}\Big(\|v\|_\sigma + \int_0^t \|f\|_\sigma d\tau\Big).$$

The result now follows by Lemma 20.6. □

Note that the higher the regularity assumed, i.e., the bigger the σ, the shorter the time step k and the faster the convergence. We remark that this error bound does not assume $\widehat f(z)$ to have an analytic continuation as required earlier.

We shall now apply our above results to the discretization in both space and time of the initial boundary value problem for the heat equation,

$$\begin{aligned}(20.36)\qquad u_t - \Delta u &= f(t) \quad \text{in } \Omega, \quad \text{with } u(\cdot, t) = 0 \quad \text{on } \partial\Omega, \quad \text{for } t > 0,\\ u(\cdot, 0) &= v \quad \text{in } \Omega,\end{aligned}$$

where Ω is a convex bounded domain in R^2 with smooth boundary $\partial\Omega$. We first consider this problem in the Banach space $C_0(\bar\Omega)$, normed with $\|v\|_{L_\infty} = \sup_{x\in\Omega}|v(x)|$.

Let S_h denote standard piecewise linear finite element spaces defined on a family of quasiuniform triangulations of Ω and vanishing on $\partial\Omega$. We consider the spatially semidiscrete problem corresponding to (20.36) to find $u_h(t) \in S_h$ such that

$$(20.37)\quad (u_{h,t}, \chi) + (\nabla u_h, \nabla\chi) = (f, \chi), \quad \forall \chi \in S_h,\ t > 0, \quad \text{with } v(0) = P_h v,$$

where P_h denotes the orthogonal L_2-projection onto S_h, or with Δ_h the discrete Laplacian defined by (1.33), and $A_h = -\Delta_h$,

$$u_{h,t} + A_h u_h = P_h f, \quad \text{for } t > 0, \quad \text{with } v(0) = P_h v. \tag{20.38}$$

This problem is of the form (20.1) when S_h, equipped with the maximum-norm, is considered as a Banach space.

Recall from Lemma 6.1 that P_h is bounded in maximum-norm, and from Theorem 6.6 that a maximum-norm resolvent estimate for A_h of the form (20.2) holds, uniformly in h, so that for any $\delta \in (0, \pi/2)$ there is a $C \geq 1$ such that

$$\|R(z; A_h)\|_{L_\infty} \leq C(1 + |z|)^{-1}, \quad \text{for } z \in \Sigma_\delta.$$

Before we discuss the fully discrete schemes, we shall establish the following maximum-norm estimate for the error in the semidiscrete solution. In the case of the homogeneous equation, i.e., when $f(t) = 0$, then Γ may be chosen to pass through the origin, so that $\kappa = 0$, and the error bound then reduces to that of Theorem 6.10.

Lemma 20.7 *Assume that $g(z)$ is analytic and bounded in G, and let $u_h(t)$ and $u(t)$ be the solutions of (20.38) and (20.36). Then, with $\|g\|_{L_\infty,\Gamma} = \sup_{z\in\Gamma} \|g(z)\|_{L_\infty}$, we have*

$$\|u_h(t) - u(t)\|_{L_\infty} \leq C h^2 \ell_h^2 t^{-1} e^{\kappa t} \|g\|_{L_\infty,\Gamma}, \quad \text{for } t > 0.$$

Proof. We have the representation

$$u_h(t) - u(t) = \frac{1}{2\pi i} \int_\Gamma e^{-tz} G_h(z) g(z)\, dz, \quad \text{with } G_h(z) = R(z; A_h)P_h - R(z; A).$$

We shall show below that, in operator norm,

$$\|G_h(z)\|_{L_\infty} \leq C h^2 \ell_h^2, \quad \text{for } z \in \Gamma. \tag{20.39}$$

Assuming this for a moment, we find by (20.9) that for $t \geq 1$,

$$\begin{aligned} \|u_h(t) - u(t)\|_{L_\infty} &\leq C\, e^{\kappa t}\, h^2 \ell_h^2 \int_0^\infty e^{-ts/2}\, ds\, \|g\|_{L_\infty,\Gamma} \\ &\leq C h^2 \ell_h^2 t^{-1} e^{\kappa t}\, \|g\|_{L_\infty,\Gamma}. \end{aligned}$$

For $t \leq 1$ we may replace Γ by the curve $\Gamma_t \cup \gamma_t$ used in the proof of Theorem 20.1 and show $\int_{\Gamma_t \cup \gamma_t} e^{-t \operatorname{Re} z}\, |dz| \leq C t^{-1}$, which completes the proof.

To prove (20.39) we write

$$G_h(z) = \big(R(z; A_h)P_h - P_h R(z; A)\big) + (P_h - I)R(z; A) = G_h'(z) + G_h''(z).$$

Here, with $R_h : H_0^1 \to S_h$ the elliptic projection defined by (1.22), and since $P_h A = A_h R_h$,

$$\begin{aligned} G_h'(z) &= R(z; A_h)P_h(zI - A)R(z; A) - R(z; A_h)(zI - A_h)P_h R(z; A) \\ &= R(z; A_h)(A_h P_h - P_h A)R(z; A) = R(z; A_h)A_h(P_h - R_h)R(z; A), \end{aligned}$$

We shall need the L_p–error estimate

$$\|(R_h - I)v\|_{L_p} \le Ch^2\ell_h\|v\|_{W_p^2} \quad \text{if } v = 0 \quad \text{on } \partial\Omega, \quad \text{for } 2 \le p < \infty,$$

easily obtained by interpolation between the cases $p = \infty$ and $p = 2$, and recall the Agmon-Douglis-Nirenberg regularity estimate

$$\text{(20.40)} \qquad \|v\|_{W_p^2} \le Cp\|Av\|_{L_p}, \quad \text{if } v = 0 \quad \text{on } \partial\Omega.$$

Using also the inverse estimate $\|\chi\|_{L_\infty} \le Ch^{-2/p}\|\chi\|_{L_p}$ on S_h (cf. the proof of Lemma 6.4), as well as the maximum-norm boundedness of $AR(z;A) = I - zR(z;A)$ for $z \in \Sigma_\delta$ (note that (20.2) now holds for $A = -\Delta$, see (6.42)) and similarly for $R(z;A_h)A_h = A_hR(z;A_h)$, we find that, for $v \in \mathcal{C}(\bar{\Omega})$,

$$\begin{aligned}\|G_h'(z)\,v\|_{L_\infty} &\le C\|(P_h - R_h)R(z;A)v\|_{L_\infty}\\ &\le Ch^{-2/p}\|P_h(R_h - I)R(z;A)v\|_{L_p} \le Ch^{2-2/p}\ell_h\|R(z;A)v\|_{W_p^2}\\ &\le Ch^{2-2/p}\ell_h\,p\,\|AR(z;A)v\|_{L_p} \le Ch^{2-2/p}\ell_h\,p\,\|v\|_{L_\infty}.\end{aligned}$$

Thus, with $p = \ell_h = \log(1/h)$ for small h,

$$\|G_h'(z)\,v\|_{L_\infty} \le Ch^2\ell_h^2\|v\|_{L_\infty}, \quad \text{for } z \in \Gamma.$$

To bound $G_h''(z)v$ we introduce the piecewise linear interpolant $I_h : C(\bar{\Omega}) \to S_h$ and note that, for any triangle τ of the triangulation, the Bramble–Hilbert lemma implies

$$\|I_hv - v\|_{L_\infty(\tau)} \le Ch^{2-2/p}\|v\|_{W_p^2(\tau)},$$

from which the corresponding estimate follows with τ replaced by Ω. Hence, since P_h is bounded in maximum-norm, and using (20.40),

$$\begin{aligned}\|(P_h - I)v\|_{L_\infty} &= \|(P_h - I)(I_h - I)v\|_{L_\infty}\\ &\le Ch^{2-2/p}\|v\|_{W_p^2} \le Ch^{2-2/p}\,p\,\|Av\|_{L_p}.\end{aligned}$$

Since $\|AR(z;A)\|_{L_\infty} \le C$, it follows that

$$\|G_h''(z)v\|_{L_\infty} \le Ch^{2-2/p}\,p\,\|AR(z;A)v\|_{L_p} \le Ch^{2-2/p}\,p\|v\|_{L_\infty},$$

and, again with $p = \ell_h$,

$$\|G_h''(z)v\|_{L_\infty} \le Ch^2\ell_h\|v\|_{L_\infty},$$

which completes the proof of (20.39). □

The result of Lemma 20.7 is a nonsmooth data error estimate in the sense of Chapter 3. For solutions which are smoother in x, the factor t^{-1} and one of the factors ℓ_h may be removed. When the Banach space is the Hilbert space $L_2(\Omega)$ the factors ℓ_h are superfluous.

The fully discrete solution obtained by application of our first time discretization method (20.14) to the semidiscrete problem (20.38) is thus defined by

$$(20.41)\quad U_{N,h}(t) = k \sum_{j=-N}^{N} e^{-z_j t} w_h(z_j) z'(s_j), \quad w_h(z) = R(z; A_h)^{-1} P_h g(z),$$

and correspondingly for our second method (20.23). In both cases, to find $U_{N,h}(t)$ it is thus required to solve the $2N+1$ discrete elliptic problems

$$(20.42)\quad (\nabla w_h(z_j), \nabla \chi) - z_j (w_h(z_j), \chi) = -(g(z_j), \chi), \quad \forall \chi \in S_h, \quad |j| \leq N.$$

We now establish error estimates for the fully discrete methods defined by our above two choices of quadrature rules. For the first rule we have the following.

Theorem 20.7 *Let $u(t)$ be the solution of* (20.36), *and let $U_{N,h}(t)$ be the approximation defined by* (20.41) *and* (20.14). *Then, under the appropriate assumptions on $g(z)$ and Γ, we have, for any $\widetilde{\kappa} > \kappa$ and $t > 0$, with C independent of N and h,*

$$\|U_{N,h}(t) - u(t)\|_{L_\infty} \leq C e^{\widetilde{\kappa} t} \big(\ell(t) (N^{-r(1-\varepsilon)} + e^{-tN^{\varepsilon}}) + h^2 \ell_h^2 t^{-1} \big) \max_{k \leq r} \|g^{(k)}\|_{L_\infty, \Gamma}.$$

Proof. We write

$$U_{N,h}(t) - u(t) = (U_{N,h}(t) - u_h(t)) + (u_h(t) - u(t)).$$

By Theorem 20.3 we obtain, uniformly in h, with $C = C_r$,

$$\|U_{N,h}(t) - u_h(t)\|_{L_\infty} \leq C e^{\widetilde{\kappa} t} \ell(t) \big(N^{-r(1-\varepsilon)} + e^{-tN^{\varepsilon}} \big) \max_{k \leq r} \|g^{(k)}\|_{L_\infty, \Gamma}, \quad t > 0.$$

In view of Lemma 20.7 this shows the result stated. □

For the second fully discrete method defined by (20.23), the same argument, using Theorem 20.5 instead of Theorem 20.3, yields the following.

Theorem 20.8 *Let $u(t)$ be the solution of* (20.36), *and let $U_{N,h}(t)$ be the result of application of* (20.23) *to the semidiscrete problem* (20.38). *Then, under the appropriate assumptions on $g(z)$, we have, with C and c independent of N and h,*

$$\|U_{N,h}(t) - u(t)\|_{L_\infty} \leq C e^{\widetilde{\kappa} t} \big(\ell(t) (e^{-\bar{r} N / \log N} + e^{-ctN}) + h^2 \ell_h^2 t^{-1} \big) \|g\|_{L_\infty, Z_\delta}.$$

We close by applying the modified second rule (20.34) to the initial-boundary value problem (20.36), which problem we now consider in the framework of the Hilbert space L_2, in which as usual we denote the norm by $\|\cdot\|$. The Banach space $\mathcal{B}^\sigma$ is then the space $\dot{H}^{2\sigma}$ introduced in Chapter 3, with $\|v\|_\sigma = |v|_{2\sigma}$. The fully discrete solution $U_{N,h}(t)$ obtained by application of our method (20.34) to (20.38) is then defined by

$$U_{N,h}(t) = \frac{k}{2\pi i} \sum_{j=-N}^{N} \widetilde{w}_h(z_j, t)\, z_j', \tag{20.43}$$

where

$$\widetilde{w}_h(z,t) = w_h(z,t) - z^{-1} P_h \widetilde{g}(z,t), \quad \text{with } w_h(z,t) = (zI - A_h)^{-1} P_h \widetilde{g}(z,t).$$

Theorem 20.9 *Let $u(t)$ be the solution of (20.36), and let $U_{N,h}(t)$ be the result of application of (20.23) to the semidiscrete problem (20.37). Let $0 < \sigma \le 2$ and $\varepsilon > 0$, and assume that P_h is such that*

$$\|A_h^{\sigma/2} P_h v\| \le C \|A^{\sigma/2} v\| = C|v|_\sigma, \quad \forall v \in \dot{H}^\sigma. \tag{20.44}$$

Then we have, with $k = \sqrt{2\bar{r}/(\sigma N)}$ and $\widetilde{\gamma} = \gamma - \lambda\cos(\alpha + r)$,

$$\|U_{N,h}(t) - u(t)\| \le C'_{\varepsilon,T}(v,f) h^2 + C''_\sigma(v,f) e^{\widetilde{\gamma} t} e^{-\sqrt{\bar{r}\sigma N/2}}, \quad \textit{for } t \le T,$$

where

$$C'_{\varepsilon,T}(v,f) = C\Big(\|\Delta v\|_\varepsilon + \|f(0)\|_\varepsilon + \int_0^t \|f_t\|_\varepsilon d\tau\Big),$$

and

$$C''_\sigma(v,f) = C e^{\widetilde{\gamma} t} e^{-\sqrt{\bar{r}\sigma N/2}} \Big(|v|_\sigma + \int_0^t |f|_\sigma \, d\tau\Big).$$

Proof. The estimate for the error $u_h(t) - u(t)$ in the semidiscrete solution follows from Theorem 19.2 and Lemma 19.1. By Theorem 20.6 we have, using assumption (20.44),

$$\|U_{N,h}(t) - u_h(t)\| \le C e^{\widetilde{\gamma} t} e^{-\sqrt{\bar{r}\sigma N/2}} (\|A_h^\sigma P_h v\| + \int_0^t \|A_h^{\sigma/2} P_h f\| \, d\tau) \le C''_\sigma(v,f).$$

Together these estimates complete the proof. □

We make some remarks on condition (20.44). We first note that in the case that the triangulations $\mathcal{T}_h$ underlying the S_h form a quasiuniform family, then it is easily seen that (20.44) holds with $\sigma = 2$. In fact,

$$A_h P_h v = A_h R_h v - A_h P_h (R_h - I) v = P_h A v - A_h P_h (R_h - I) v,$$

and hence, using the inverse estimate $\|A_h\chi\| \le Ch^{-2}\|\chi\|$ for $\chi \in S_h$,

$$\|A_hP_hv\| \le \|Av\| + Ch^{-2}\|(R_h - I)v\| \le C\|v\|_2 \le C\|Av\|.$$

We further recall that, under certain conditions on the $\mathcal{T}_h$, weaker than quasiuniformity, P_h is stable in H_0^1, see [59]. Under such conditions, (20.44) holds with $\sigma = 1$. In fact,

$$\|A_h^{1/2}P_hv\|^2 = (A_hP_hv, P_hv) = \|\nabla P_hv\|^2 \le C\|\nabla v\|^2 = C\|A^{1/2}v\|^2.$$

By interpolation between this inequality and $\|P_hv\| \le \|v\|$ one finds, under the above conditions on $\mathcal{T}_h$, that (20.44) holds for $0 \le \sigma \le 1$. In particular, if $0 \le \sigma < 1/2$, since then $H^\sigma = \dot{H}^\sigma$, this means that the result of Theorem 20.9 is valid for such σ if $v \in H^\sigma$ and $f(\tau) \in H^\sigma$ for $\tau > 0$, thus not requiring these functions to vanish on $\partial\Omega$.

The approach to discretization in time of parabolic problems described in this chapter was introduced in Sheen, Sloan and Thomée [215], [216]. In [216] the integral representation (20.10) was transformed to an integral over a finite interval which was then approximated by the trapezoidal rule to yield a $O(h^r)$ error estimate for arbitrary r, and $t > 0$. The analysis of the quadrature formulas described here by extension to a strip containing Γ is based on López-Fernándes and Palencia [158] and applied to evolution problems in McLean and Thomée [171] and McLean, Sloan, and Thomée [172]. The modification of the second quadrature method in order to attain uniform convergence town to $t = 0$ is taken from Gavrilyuk and Makarov [106], see also [105], [107], [108].

References

1. R. ADAMS AND J. FOURNIER, *Sobolev Spaces*, Pure and Applied Mathematics, No. 140, Academic Press, 2003.
2. S. AGMON, A. DOUGLIS, AND L. NIRENBERG, *Estimates near the boundary for solutions of elliptic partial differential equations satisfying general boundary conditions*, Comm. Pure. Appl. Math., 12 (1050), pp. 623–727.
3. G. AKRIVIS, M. CROUZEIX, AND C. MAKRIDAKIS, *Implicit-explicit multi-step finite element methods for nonlinear parabolic problems*, Math. Comp., 67 (1998), pp. 457–477.
4. H. AMANN, *Existence and stability of solutions for semi-linear parabolic systems, and applications to some diffusion reaction equations*, Proc. Roy. Soc Edinburgh Sect. A, 81 (1978), pp. 35–47.
5. W. ARENDT, *Semigroups and Evolution Equations: Functional Calculus, Regularity and Kernel Estimates*, in Handbook of Differential Equations: Evolutionary Differential Equations, C.M. Dafermos, E. Feireisl eds, Elsevier/North Holland.
6. A. ASHYRALYEV AND P. E. SOBOLEVSKIĬ, *Well-posedness of parabolic difference equations*, vol. 69 of Operator Theory: Advances and Applications, Birkhäuser Verlag, Basel, 1994. Translated from the Russian by A. Iacob.
7. J.-P. AUBIN, *Behavior of the error of the approximate solutions of boundary value problems for linear elliptic operators by Galerkin's and finite difference methods*, Ann. Scuola Norm. Sup. Pisa (3), 21 (1967), pp. 599–637.
8. A. K. AZIZ AND P. MONK, *Continuous finite elements in space and time for the heat equation*, Math. Comp., 52 (1989), pp. 255–274.
9. I. BABUŠKA, *Finite element method for domains with corners*, Computing 6 (1970), 264–273.
10. I. BABUŠKA, *The finite element method with Lagrangian multipliers*, Numer. Math., 20 (1973), pp. 179–192.
11. I. BABUŠKA AND A. K. AZIZ, *Part I. Survey Lectures on the Mathmatical Foundation of the Finite Element Method*, in The Mathematical Foundations of the Finite element Method with Applications to Partial Differential Equations, Academic Press, New York,1972, pp. 3-359.
12. I. BABUŠKA AND M. ROSENZWEIG, *A finite element scheme for domains with corners*, Numer. Math. 20 (1972), pp. 1-21.
13. C. BACUTA, J. H. BRAMBLE AND J. PASCIAK *New interpolation results and application to finite element methods for elliptic boundary value problems*, in Numerical Linear Algebra with Applications???
14. C. BACUTA, J. H. BRAMBLE, AND J. XU, *Regularity estimates for elliptic boundary value problems in Besov spaces*, Math. Comp. 72 (2003), 1577–1595.

15. C. Baiocchi and F. Brezzi, *Optimal error estimates for linear parabolic problems under minimal regularity assumptions*, Calcolo, 20 (1983), pp. 143–176.
16. N. Y. Bakaev, *On the bounds of approximations of holomorphic semigroups*, BIT, 35 (1995), pp. 605–608.
17. N. Bakaev, *Maximum norm resolvent estimates for elliptic finite element operators*, BIT, 41 (2001), pp. 215–239.
18. N. Bakaev, *Linear Discrete Parabolic Problems*, North-Holland Mathematics Studies, No. 203, 2006.
19. N. Y. Bakaev, M. Crouzeix, and V. Thomée *Maximum-norm resolvent estimates for elliptic finite element operators on nonuniform triangulations*, (2006), under preparation.
20. N. Y. Bakaev, V. Thomée, and L.B. Wahlbin *Maximum-norm estimates for resolvents of elliptic finite element operators*, Math. Comp., 72 (2002), pp. 1597–1610.
21. G. A. Baker, J. H. Bramble, and V. Thomée, *Single step Galerkin approximations for parabolic problems*, Math. Comp., 31 (1977), pp. 818–847.
22. R. E. Bank and T. Dupont, *An optimal order process for solving finite element equations*, Math. Comp., 36 (1981), pp. 35–51.
23. J. Becker, *A second order backward difference method with variable steps for a parabolic problem*, BIT, 38 (2002), pp. 644-664.
24. A. Berger, R. Scott, and G. Strang, *Approximate boundary conditions in the finite element method*, in Symposia Mathematica X, Academic Press, 1972, pp. 259–313.
25. J. Bergh and J. Löfström, *Interpolation spaces. An introduction*, Springer-Verlag, Berlin, 1976. Grundlehren der Mathematischen Wissenschaften, No. 223.
26. C. Bernardi, *Numerical apprxoimation of a periodic linear parabolic problem*, SIAM J. Numer. Anal., 19 (1982), pp. 1196–1207.
27. G. Birkhoff, H. H. Schultz, and R. S. Varga, *Piecewise Hermite interpolation in one and two variables with applications to partial differential equations*, Numer. Math., 11 (1968), pp. 232–256.
28. J. Blair, *Approximate solution of elliptic and parabolic boundary value problems*, PhD thesis, Univ. of California, Berkeley, 1970.
29. J. H. Bramble, *Discrete methods for parabolic equations with time-dependent coefficients*, in Numerical Methods for PDE's, Academic Press, 1979, pp. 41–52.
30. ——, *Multigrid Methods*, Pitman, New York, 1993.
31. *The Analysis of Multigrid Methods*, in Handbook of Numerical Analysis vol VII. P. G. Ciarlet and J. L. Lions, eds., North-Holland, Amsterdam, 2000, pp. 173–412.
32. J. H. Bramble, T. Dupont, and V. Thomée, *Projection methods for Dirichlet's problem in approximating polygonal domains with boundary value corrections*, Math.Comp., 26 (1972), pp. 869–879.
33. J. H. Bramble, J. A. Nitsche, and A. H. Schatz, *Maximum norm interior estimates for Ritz-Galerkin methods*, Math.Comp., 29 (1975), pp. 677–688.
34. J. H. Bramble, J. E. Pasciak, P. H. Sammon, and V. Thomée, *Incomplete iterations in multistep backward difference methods for parabolic problems with smooth and nonsmooth data*, Math. Comp., 52 (1989), pp. 339–367.

35. J. H. Bramble and P. Sammon, *Efficient higher order single step methods for parabolic problems: Part I*, Math. Comp., 35 (1980), pp. 655–677.
36. J. H. Bramble and A. H. Schatz, *Higher order local accuracy by averaging in the finite element method*, Math. Comp., (1977), pp. 94–111.
37. J. H. Bramble, A. H. Schatz, V. Thomée, and L. B. Wahlbin, *Some convergence estimates for semidiscrete Galerkin type approximations for parabolic equations*, SIAM J. Numer. Anal., 14 (1977), pp. 218–241.
38. J. H. Bramble and V. Thomée, *Semi-discrete-least squares methods for a parabolic boundary value problem*, Math.Comp., 26 (1972), pp. 633–648.
39. ———, *Discrete time Galerkin methods for a parabolic boundary value problem*, Mat. Pura Appl., 101 (1974), pp. 115–152.
40. P. Brenner, M. Crouzeix, and V. Thomée, *Single step methods for inhomogeneous linear differential equations in Banach space*, RAIRO Anal. Numér., 16 (1982), pp. 5–26.
41. P. Brenner and V. Thomée, *On rational approximation of semigroups*, SIAM J. Numer. Anal., 16 (1979), pp. 683–694.
42. S. C. Brenner and L. R. Scott, *The Mathematical Theory of Finite Element Methods*, Springer-Verlag, New York, 1994.
43. A. Calderon and A. Zygmund, *On the existence of certain singular integrals*, Acta Math., 88 (1952), pp. 85–139.
44. A. Carasso, *On least square methods for parabolic equations and the computation of time-periodic solutions*, SIAM J. Numer. Anal., 11 (1974), pp. 1181–1192.
45. J. Céa, *Approximation variationelle des problèmes aux limites*, Ann. Inst. Fourier (Grenoble), 14 (1964), pp. 345–444.
46. L. Čermák and M. Zlámal, *Transformation of dependent variables and the finite element solution of nonlinear evolution equations*, Internat. J. Numer. Methods Engrg., 15 (1980), pp. 31–40.
47. *Parabolic finite element equations in nonconvex polygonal domains*, to appear.
48. C.-M. Chen, S. Larsson, and N.-Y. Zhang, *Error estimates of optimal order for finite element methods with interpolated coefficients for the nonlinear heat equation*, IMA J. Numer. Anal., 9 (1989), pp. 507–524.
49. C.-M. Chen and V. Thomée, *The lumped mass finite element method for a parabolic problem*, J. Austral. Math. Soc. Ser. B, 26 (1985), pp. 329–354.
50. H. Chen, *An L^2 and L^∞-Error Analysis for Parabolic Finite Element Equations with Applications by Superconvergence and Error Expansion*, PhD thesis, Universität Heidelberg, 1993.
51. P. G. Ciarlet, *The Finite Element Method for Elliptic Problems*, North-Holland, Amsterdam, 1978.
52. R. Courant, K. Friedrichs, and H. Lewy, *Über die Partiellen Differenzengleichungen der Mathematischen Physik*, Math. Ann., 100 (1928), pp. 32–74.
53. M. Crouzeix, *Sur l'approximation des équations différentielles opérationelles linéaires par des méthods de Runge-Kutta*, PhD thesis, Université Paris VI, 1975.
54. ———, *Une méthode multipas implicite-explicite pour l'approximation des équations d'évolution parabolique*, Numer. Math., 35 (1980), pp. 257–276.
55. ———, *On multistep approximation of semigroups in Banach spaces*, J. Comput. Appl. Math., 20 (1987), pp. 25–35.

56. M. Crouzeix and A. Mignot, *Analyse Numérique des Équations Différentielles*, Masson, Paris, 1984.
57. M. Crouzeix and P. A. Raviart, *Approximation d'équations d'évolution linéaires par des méthodes multipas*, vol. 7 of Méthodes Math. de l'Informatique, Dunod, Paris, 1978, pp. 133–150. Proc. Sympos., Novosibirsk, 1978.
58. M. Crouzeix and V. Thomée, *On the discretization in time of semilinear parabolic equations with nonsmooth initial data*, Math. Comp., 49 (1987), pp. 359–377.
59. ——, *The stability in L_p and W_p^1 of the L_2-projection onto finite element function spaces*, Math. Comp., 48 (1987), pp. 521–532.
60. ——, *Resolvent estimates in l_p for discrete Laplacians on irregular meshes and maximum-norm stability of parabolic finite difference schemes*, Comput. Meth. Appl. Math, 1 (2001), pp. 3–17.
61. M. Crouzeix, S. Larsson, and V. Thomée, *Resolvent estimates for elliptic finite element operators in one dimension*, Math. Comp., 63 (1994), pp. 121–140.
62. M. Crouzeix, V. Thomée, and L. B. Wahlbin, *Error estimates for spatially discrete approximations of semilinear parabolic equations with initial data of low regularity*, Math. Comp., 53 (1989), pp. 25–41.
63. M. Crouzeix, S. Larsson, S. Piskarev, and V. Thomée, *The stability of rational approximations of analytic semigroups*, BIT, 33 (1993), pp. 74–84.
64. G. Da Prato and E. Sinestrari, *Differential operators with non dense domain*, Ann. Scoula Norm. Sup. Pisa, 14 (1987), pp. 285–344.
65. M. Dauge, *Elliptic Boundary Value Problems on Corner Domains*, Vol 1341 of Lecture Notes in Mathematics, Springer-Verlag, Berlin, 1988.
66. P. J. Davis and P. Rabinowitz, *Methods of Numerical Integration*, second edition, Academic Press, London, 1984.
67. M. Delfour, W. Hager, and F. Trochu, *Discontinuous Galerkin methods for ordinary differential equations*, Math. Comp., 36 (1981), pp. 453–473.
68. J. K. Demjanovič, *The net method for some problems in mathematical physics*, Dokl. Akad. Nauk SSSR, (1964), pp. 250–253. (In Russian).
69. J. E. Dendy, Jr., *Galerkin's method for some highly nonlinear problems*, SIAM J. Numer. Anal., 14 (1977), pp. 327–347.
70. J. Descloux, *On finite element matrices*, SIAM J. Numer. Anal., 9 (1972), pp. 260–265.
71. M. Dobrowolski, *L^∞-convergence of linear finite element approximation to quasilinear initial boundary value problems*, RAIRO Anal. Numér., 12 (1978), pp. 247–266.
72. ——, *L^∞-convergence of linear finite element approximation to nonlinear parabolic problems*, SIAM J. Numer. Anal., 17 (1980), pp. 663–674.
73. J. Douglas, Jr., *Effective time-stepping methods for the numerical solution of nonlinear parabolic problems*, in Mathematics of Finite Elements and Applications, III (Proc. Third MAFELAP Conf., Brunel Univ., Uxbridge, 1978, Academic Press, London, 1979, pp. 289–304.
74. J. Douglas, Jr. and T. F. Dupont, *Galerkin methods for parabolic equations*, SIAM J. Numer. Anal., 7 (1970), pp. 575–626.
75. ——, *Galerkin methods for parabolic equations with nonlinear boundary conditions*, Numer. Math., 20 (1973), pp. 213–237.

76. ——, *The effect of interpolating the coefficients in nonlinear parabolic Galerkin procedures*,
77. J. Douglas, Jr. and T. Dupont, *H^{-1} Galerkin methods for problems involving several space variables*, in Topics in Numerical Analysis III, J. J. H. Miller, ed., Academic Press, London, 1977, pp. 125–141.
78. J. Douglas, Jr., T. Dupont, and R. E. Ewing, *Incomplete iterations for time-stepping a Galerkin method for a quasilinear parabolic problem*, SIAM J. Numer. Anal., 16 (1979), pp. 503–522.
79. J. Douglas, Jr., T. Dupont, and M. Wheeler, *H^1-Galerkin methods for the Laplace and heat equations*, in Mathematical Aspects of Finite Elements in Partial Differential Equations, C. de Boor, ed., Academic Press, New York, 1974, pp. 383–416.
80. J. Douglas, Jr., T. Dupont, and M. F. Wheeler, *Some superconvergence results for an H^1-Galerkin procedure for the heat equation*, in Computing methods in applied sciences and engineering (Proc. Internat. Sympos., Versailles, 1973), Part 1, vol. 10 of Lecture Notes in Comput. Sci., Springer-Verlag, Berlin, Heidelberg, New York, 1974, pp. 288–311.
81. J. Douglas, Jr., T. Dupont, and M. Wheeler, *A quasi-projection analysis of Galerkin methods for parabolic and hyperbolic equations*, Math. Comp., 32 (1978), pp. 345–362.
82. N. Dunford and J. T. Schwartz, *Linear Operators: Part I. General Theory*, Interscience Publ., New York, 1958.
83. T. Dupont, *L_2 error estimates for projection methods for parabolic equations in approximating domains*, in Mathematical Aspects of Finite Elements in Partial Differential Equations, Academic Press, New York, San Fransisco, London, 1974, pp. 313–352.
84. T. Dupont, G. Fairweather, and J. P. Johnson, *Three-level Galerkin methods for parabolic equations*, SIAM J. Numer. Anal., 11 (1974), pp. 392–410.
85. C. M. Elliott and S. Larsson, *Error estimates with smooth and nonsmooth data for a finite element method for the Cahn-Hilliard equation*, Math. Comp., 58 (1992), pp. 603–630.
86. ——, *A finite element model for the time-dependent Joule heating problem*, Math. Comp., 64 (1995), pp. 1433–1453.
87. C. M. Elliott and A. M. Stuart, *The global dynamics of discrete semilinear parabolic equations*, SIAM J. Numer. Anal., 30 (1993), pp. 1622–1663.
88. K. Eriksson and C. Johnson, *Adaptive finite element methods for parabolic problems. I. A linear model problem*, SIAM J. Numer. Anal., 28 (1991), pp. 43–77.
89. ——, *Adaptive finite element methods for parabolic problems. II. Optimal error estimates in $L_\infty(L_2)$ and $L_\infty(L_\infty)$*, SIAM J. Numer. Anal., 32 (1995), pp. 706–740.
90. ——, *Adaptive finite element methods for parabolic problems. IV. Nonlinear problems*, SIAM J. Numer. Anal., 32 (1995), pp. 1729–1749.
91. ——, *Adaptive finite element methods for parabolic problems. V. Long-time integration*, SIAM J. Numer. Anal., 32 (1995), pp. 1750–1763.
92. K. Eriksson, C. Johnson, and S. Larsson, *Adaptive finite element methods for parabolic problems. VI. Analytic semigroups*, SIAM J. Numer. Anal., 35 (1998), pp. 1315–1325.

93. K. Eriksson, C. Johnson, and V. Thomée, *Time discretization of parabolic problems by the discontinuous Galerkin method*, RAIRO Modél. Math. Anal. Numér., 19 (1985), pp. 611–643.
94. K. Eriksson and V. Thomée, *Galerkin methods for singular boundary value problems in one space dimension*, Math. Comp., 42 (1984), pp. 345–367.
95. D. Estep and S. Larsson, *The discontinuous Galerkin method for semilinear parabolic equations*, RAIRO Modél. Math. Anal. Numér., 27 (1993), pp. 35–54.
96. L.C. Evans, *Partial Differential Equations*, Graduate Studies in Mathematics, Vol. 19, American Mathematical Society, Rhode Island, 1998.
97. R. Falk and J. Osborn, *Error estimates for mixed methods*, RAIRO Anal. Numér., 14 (1980), pp. 249–277.
98. K. Feng, *Finite difference schemes based on variational principles*, Appl. Math. Comput. Math., 2 (1965), pp. 238–262. (In Chinese).
99. G. Fix and N. Nassif, *On finite element approximations in time dependent problems*, Numer. Math., 19 (1972), pp. 127–135.
100. A. Friedman, *Partial Differential Equations of Parabolic Type*, Prentice-Hall, Englewood Cliffs, N.J., 1964.
101. K. O. Friedrichs and H. B. Keller, *A finite difference scheme for generalized Neumann problems*, in Numerical Solution of Partial Differential Equations, Academic Press, New York, London, 1966, pp. 1–19.
102. H. Fujii, *Some remarks on finite element analysis of time-dependent field problems*, in Theory and Practice in Finite Element Structural Analysis, University of Tokyo Press, Tokyo, 1973, pp. 91–106.
103. H. Fujita and A. Mizutani, *On the finite element method for parabolic equations, I; approximation of holomorphic semigroups*, J. Math. Soc. Japan, 28 (1976), pp. 749–771.
104. H. Fujita and T. Suzuki, *Evolution Problems*, in Handbook of Numerical Analysis, Vol. II. Finite Element Methods (Part 1), P. Ciarlet and J. Lions, eds., North-Holland, New York, 1991, pp. 789–928.
105. I. P. Gavrilyuk and V. L. Makarov, *Exponentially convergent parallel discretization methods for the first order evolution equation,* Computational Methods in Applied Mathematics, 1 (2001), 333-355.
106. ———, *Exponentially convergent algorithms for the operator exponential with applications to inhomogeneous problems in Banach spaces,* SIAM J. Numer. Anal. 43 (2005), pp. 2144–2171.
107. ———, *An exponentially convergent algorithm for nonlinear differential equations in Banach spaces,* Reports on Numerical Maathematics, Jenaer Schriften zur Mathematik und Informatik, Friedrich-Schiller-Universität, Jena, No. 02/05, 2005.
108. ———, *Algorithms without accuracy saturation for evolution equations in Helbert and Banach spaces,* Math. Comp. 74 (2005), pp. 555–583.
109. R. Grigorieff, *Stability of multistep-methods on variable grids*, Numer. Math., 42 (1985), pp. 359–377.
110. P. Grisvard, *Elliptic Problems in Nonsmooth Domains*, Pitman, Massachusetts, 1985.
111. P. Grisvard, *Singularities in Boundary Value Problems,* Masson, 1992.
112. W. Hackbusch, *Parabolic multi-grid methods*, in Computing Methods in Applied Sciences and Engineering VI, R. Glowinski and J. L. Lions, eds., North-Holland, Amsterdam, 1984, pp. 189–197.

113. E. Hairer and G. Wanner, *Stiff and differential–algebraic problems* in Solving ordinary differential equations. II, vol. 14 of Springer Series in Computational Mathematics, Springer-Verlag, Berlin, 1991.
114. A. Hansbo, *Error estimates for the numerical solution of a time-periodic linear parabolic problem*, BIT, 31 (1991), pp. 664–685.
115. A. Hansbo, *Strong stability and non-smooth data error estimates for discretizations of linear parabolic problems*, BIT, 42 (2002), pp. 351–379.
116. R. Haverkamp, *Eine Aussage zur L_∞-Stabilität und zur genauen Konvergenzordnung der H_0^1-Projektionen*, Numer. Math., 44 (1984), pp. 393–405.
117. H.-P. Helfrich, *Fehlerabschätzungen für das Galerkinverfahren zur Lösung von Evolutionsgleichungen*, Manuscripta Math., 13 (1974), pp. 219–235.
118. ——, *Error estimates for semidiscrete Galerkin type approximations for semilinear evolution equations with nonsmooth initial data*, Numer. Math., 51 (1987), pp. 559–569.
119. D. Henry, *Geometric Theory of Semilinear Parabolic Equations*, vol. 840 of Lecture Notes in Math., Springer-Verlag, Berlin, 1981.
120. J. G. Heywood and R. Rannacher, *Finite element approximation of the nonstationary Navier-Stokes problem I. Regularity of solutions and second-order spatial discretization*, SIAM J. Numer. Anal., 19 (1982), pp. 275–311.
121. ——, *Finite element approximation of the nonstationary Navier-Stokes problem II. Stability of solutions and error estimates uniform in time*, SIAM J. Numer. Anal., 23 (1986), pp. 750–777.
122. ——, *Finite element approximation of the nonstationary Navier-Stokes problem III. Smoothing property and higher order estimates for spatial discretization*, SIAM J. Numer. Anal., 25 (1988), pp. 489–512.
123. ——, *Finite element approximation of the nonstationary Navier-Stokes problem IV. Error analysis for second-order time discretization*, SIAM J. Numer. Anal., 27 (1990), pp. 353–384.
124. E. Hille and R. Phillips, *Functional Analysis and Semigroups*, American Mathematical Society, Providence, R.I., 1957.
125. M. Huang and V. Thomée, *Some convergence estimates for semidiscrete type schemes for time-dependent nonselfadjoint parabolic equations*, Math. Comp., 37 (1981), pp. 327–346.
126. ——, *An error estimate for the H^{-1} Galerkin method for a parabolic problem with non-smooth initial data*, Calcolo, 19 (1982), pp. 115–124.
127. ——, *On the backward Euler method for parabolic equations with rough initial data*, SIAM J. Numer. Anal., 19 (1982), pp. 599–603.
128. P. Jamet, *Galerkin-type approximations which are discontinuous in time for parabolic equations in a variable domain*, SIAM J. Numer. Anal., 15 (1978), pp. 912–928.
129. D. Jespersen, *Ritz-Galerkin methods for singular boundary value problems*, SIAM J. Numer. Anal., 15 (1978), pp. 813–834.
130. C. Johnson, S. Larsson, V. Thomée, and L. B. Wahlbin, *Error estimates for spatially discrete approximations of semilinear parabolic equations with nonsmooth initial data*, Math. Comp., 49 (1987), pp. 331–357.
131. C. Johnson, Y.-Y. Nie, and V. Thomée, *An a posteriori error estimate and adaptive timestep control for a backward Euler discretization of a parabolic problem*, SIAM J. Numer. Anal., 27 (1990), pp. 277–291.

132. C. JOHNSON AND V. THOMÉE, *Error estimates for some mixed finite element methods for parabolic type problems*, RAIRO Modél. Math. Anal. Numér., 15 (1981), pp. 41–78.
133. O. KARAKASHIAN, *On Runge-Kutta methods for parabolic problems with time-dependent coefficients*, Math.Comp., 47 (1986), pp. 77–101.
134. T. KATO, *Abstract evolution equations of parabolic type in Banach and Hilbert spaces*, Nagoya Math. J., 5 (1961), pp. 93–125.
135. T. KATO AND H. TANABE, *On the abstract evolution equation*, Osaka Math. J., 14 (1962), pp. 107–133.
136. S. L. KEELING, *Galerkin/Runge-Kutta discretizations for semilinear parabolic equations*, SIAM J. Numer. Anal., 27 (1990), pp. 394–418.
137. B. R. KELLOGG, *Interpolation between subspaces of a Hilbert space*, Technical note BN–719, Institute for Fluid Dynamics and Applied Mathematics, College Park, 1971.
138. R. P. KENDALL AND M. F. WHEELER, *A Crank-Nicolson-H^{-1}-Galerkin procedure for parabolic problems in a single-space variable*, SIAM J. Numer. Anal., 13 (1976), pp. 861–876.
139. S. N. S. KHALSA, *Finite element approximation of a reaction-diffusion equation. part i: Application of topological techniques to the analysis of asymptotic behavior of the semidiscrete approximations*, Quart. Appl. Math., 44 (1986), pp. 375–386.
140. V. KONDRATIEV, *Boundary value problems for elliptic equations in domains with conical or angular points*, 16 (1967), pp. 227–313. Trans. Moscow Math. Soc.,
141. V.A. KOZLOV, V.G. MAZYA AND J. ROSSMAN, *Elliptic Boundary Value Problems in Domains with Point Singularities*, American Mathematical Society, Mathematical Surveys and Monographs, vol. 52, 1997.
142. M. KRIŠEK AND P. NEITTAANMÄKI, *On superconvergence techniques*, Acta Appl. Math., 9 (1987), pp. 175–198.
143. S. LARSSON, *The long-time behavior of finite element approximations of solutions to semilinear parabolic problems*, SIAM J. Numer. Anal., 26 (1989), pp. 348–365.
144. ——, *Nonsmooth data error estimates with applications to the study of the long-time behavior of finite element solutions of semilinear parabolic problems*, Math. Comp., 68 (1999), pp. 55-72.
145. S. LARSSON AND J.-M. SANZ-SERNA, *The behavior of finite element solutions of semilinear parabolic problems near stationary points*, SIAM J. Numer. Anal., 31 (1994), pp. 1000–1018.
146. ——, *A shadowing result with applications to finite element approximation of reaction-diffusion equations*, Math. Comp. 68 (1999), pp. 55–72.
147. S. LARSSON, V. THOMÉE, AND L. B. WAHLBIN, *Finite-element methods for a strongly damped wave equation*, IMA J. Numer. Anal., 11 (1991), pp. 115–142.
148. S. LARSSON, V. THOMÉE, AND N.-Y. ZHANG, *Interpolation of coefficients and transformation of the dependent variable in finite element methods for the nonlinear heat equation*, Math. Methods Appl. Sci., 11 (1989), pp. 105–124.
149. S. LARSSON, V. THOMÉE, AND S. Z. ZHOU, *On multigrid methods for parabolic problems*, J. Comput. Math., 13 (1995), pp. 193–205.

150. I. LASIECKA, *Convergence estimates for semidiscrete approximations of non-selfadjoint parabolic equations*, SIAM J. Numer. Anal., 21 (1984), pp. 894–909.
151. ——, *Galerkin approximations of abstract parabolic boundary value problems with rough boundary data; L_p theory*, Math. Comp., 47 (1986), pp. 55–75.
152. M.-N. LEROUX, *Semi-discrétisation en temps pour les équations d'évolution paraboliques lorsque l'operateur dépend du temps*, RAIRO Anal. Numér., 13 (1979), pp. 119–137.
153. ——, *Semidiscretization in time for parabolic problems*, Math. Comp., 33 (1979), pp. 919–931.
154. ——, *Variable stepsize multistep methods for parabolic problems*, SIAM J. Numer. Anal., 19 (1982), pp. 725–741.
155. P. LESAINT AND P. RAVIART, *On a finite element method for solving the neutron transport equation*, in Mathematical Aspects of Finite Elements in Partial Differential Equations, C. de Boor, ed., Academic Press, New York, 1974, pp. 89–123.
156. J. L. LIONS AND E. MAGENES, *Problèmes aux Limites Non Homogènes et Applications, I*, Dunod, Paris, 1968.
157. G. LIPPOLD, *Error estimates and step-size control for the approximate solution of a first order evolution equation*, RAIRO Modél. Math. Anal. Numér, 25 (1991), pp. 111–128.
158. M. LÓPEZ-FERNÁNDEZ AND C. PALENCIA, *On the numerical inversion of the Laplace transform of certain holomorphic functions.*
159. A. LOUIS, *Acceleration of convergence for finite element solutions of the Poisson equation*, Numer. Math., 33 (1979), pp. 43–53.
160. C. LUBICH AND A. OSTERMANN, *Multi-grid dynamic iteration for parabolic equations*, BIT, 27 (1987), pp. 216–234.
161. ——, *Runge-Kutta methods for parabolic equations and convolution quadrature*, Math. Comp., 60 (1993), pp. 105–131.
162. ——, *Interior estimates for time discretizations of parabolic equations*, Appl. Numer. Math., 18 (1995), pp. 241–251.
163. ——, *Linearly implicit time discretization of nonlinear parabolic equations*, IMA J. Numer. Anal., 15 (1995), pp. 555–583.
164. ——, *Runge-Kutta approximation of quasi-linear parabolic equations*, Math. Comp., 64 (1995), pp. 601–627.
165. M. LUSKIN, *A Galerkin method for nonlinear parabolic equations with nonlinear boundary conditions*, SIAM J. Numer. Anal., 16 (1979), pp. 284–299.
166. M. LUSKIN AND R. RANNACHER, *On the smoothing property of the Crank-Nicolson scheme*, Appl. Anal., 14 (1982), pp. 117–135.
167. ——, *On the smoothing property of the Galerkin method for parabolic equations*, SIAM J. Numer. Anal., 19 (1982), pp. 93–113.
168. M. MARION AND R. TEMAM, *Nonlinear Galerkin methods: the finite elements case*, Numer. Math., 57 (1990), pp. 205–226.
169. M. MARION AND J. XU, *Error estimates on a new nonlinear Galerkin method based on two-grid finite elements*, SIAM J. Numer. Anal., 32 (1995), pp. 1170–1184.
170. W. MCLEAN AND V. THOMÉE, *Numerical solution of an evolution equation with a positive type memory term*, J. Austral. Math. Soc. Ser. B., 35 (1993), pp. 23–70.

171. W. McLean and V. Thomée, *Time discretization of an evolution equation via Laplace transforms*, IMA J. Numer. Anal. 24 (2004), pp. 439–463.
172. W. McLean, I. H. Sloan, and V. Thomée, *Time discretization via Laplace transformation of an integro-differential equation of parabolic type*, Numer. Math. 102 (2006), pp. 497–522.
173. M. A. Murad, V. Thomée, and A. F. D. Loula, *Asymptotic behavior of semidiscrete finite element approximations of Biot's consolidation problem*, SIAM J. Numer. Anal., 33 (1996), pp. 1065–1083.
174. T. Nakagawa and T. Ushijima, *Finite element analysis of the semi-linear heat equation of blow-up type*, in Topics in Numerical Analysis III, Proceedings of the Royal Irish Academy Conference on Numerical Analysis, 1976, J. J. H. Miller, ed., Academic Press, New York, 1977, pp. 275–291.
175. F. Natterer, *Über die punktweise Konvergenz finiter Elemente*, Numer. Math., 25 (1975), pp. 67–77.
176. S.A. Nazarov and B.A. Plamenevsky, *Elliptic Problems in Domains with Piecewise Smooth Boundaries*, Expositions in Mathematics, vol. 13, de Gruyter, New York, 1994.
177. Y.-Y. Nie and V. Thomée, *A lumped mass finite element method with quadrature for a nonlinear parabolic problem*, IMA J. Numer. Anal., 5 (1985), pp. 371–396.
178. J. Nitsche, *Ein Kriterium für die quasioptimalität des Ritzchen Verfahrens* Numer. Math., 11 (1968), pp. 346–348.
179. J. A. Nitsche, *Lineare Spline-Funktionen und die Methoden von Ritz für elliptische Randwertprobleme*, Arch. Rational Mech. Anal., 36 (1970), pp. 348–355.
180. J. Nitsche, *Über ein Variationsprinzip zur Lösung von Dirichlet-problemen bei Verwendung von Teilräumen, die keinen Randbedingungen unterworfen sind*, Abh. Math. Semin. Univ. Hamb., 36 (1971), pp. 9–15.
181. ——, *On Dirichlet problems using subspaces with nearly zero boundary conditions*, in The Mathematical Foundations of the Finite Element Method with Applications to Partial Differential Equations, A. K. Aziz, ed., Academic Press, New York, 1972, pp. 603–627.
182. ——, *L_∞-convergence of finite element approximation*, in Second conference on finite elements, Rennes, France, 1975.
183. ——, *L_∞-convergence of finite element approximations on parabolic problems*, RAIRO Numer. Anal., 13 (1979), pp. 31–54.
184. ——, *Interior error estimates for semidiscrete Galerkin approximations for parabolic equations*, RAIRO Numer. Anal., 15 (1981), pp. 171–176.
185. J. Nitsche and A. H. Schatz, *Interior estimates for Ritz-Galerkin methods*, Math. Comp., 28 (1974), pp. 937–958.
186. J. A. Nitsche and M. F. Wheeler, *L_∞ convergence of the finite element Galerkin operator for parabolic problems*, Numer. Funct. Anal. Optimization, 4 (1981-82), pp. 325–353.
187. L. A. Oganesjan and L. Ruhovec, *Convergence of diofference schemes in case of improved approximation of the boundary*, Zh. Vychisl. Mat. i Mat. Fiz., 6 (1966), pp. 1029–1042. (In Russian).
188. ——, *An investigation of the rate of convergence of variational-difference schemes for second order elliptic equations in a two-dimensional region with smooth boundary*, Zh. Vychisl. Mat. i Mat. Fiz., 9 (1969), pp. 1102–1120. (In Russian).

189. A. Ostermann and M. Roche, *Runge-Kutta methods for partial differential equations and fractional orders of convergence*, Math.Comp., 59 (1992), pp. 403–420.
190. C. Palencia, *A stability result for sectorial operators in Banach spaces*, SIAM J. Numer. Anal., 30 (1993), pp. 1373–1384.
191. ———, *Stability of rational multistep approximations of holomorphic semigroups*, Math. Comp., 64 (1995), pp. 591–599.
192. ———, *Maximum-norm analysis of completely discrete finite element methods for parabolic problems*, SIAM J. Numer. Anal., 33 (1996), pp. 1654–1668.
193. C. Palencia and B. Garcia-Archilla, *Stability of linear multistep methods for sectorial operators in Banach spaces*, Appl. Numer. Math., 12 (1993), pp. 503–520.
194. A. Pazy, *Semigroups of Linear Operators and Applications to Partial Differential Equations*, Springer-Verlag, New York, 1983.
195. S. Piskarev, *Error estimates for approximation of semigroups of operators by padé fractions*, Soviet Math. (Iz. VUZ), 23 (1979), pp. 31–36.
196. H. Price and R. Varga, *Error bounds for semi-discrete Galerkin approximations of parabolic problems with applications to petroleum reservoir mechanics*, in Numerical Solution of Field Problems in Continuum Physics, American Mathematical Society, Providence, R.I., 1970, pp. 74–94.
197. M. H. Protter and H. F. Weinberger, *Maximum Principles in Differential Equations*, Prentice-Hall, Englewood Cliffs, N.J., 1967.
198. H. H. Rachford, Jr., *Two-level discrete-time Galerkin apprxoimations for second order nnonlinear parabolic partial differential equations*, SIAM J. Numer. Anal., 10 (1973), pp. 1010–1026.
199. H. H. Rachford,Jr. and M. M. Wheeler, *An H^{-1}-Galerkin procedure for the two point boundary value problem*, in Mathematical Aspects of Finite Elements in Partial Differential Equations, Academic Press, New York, San Fransisco, London, 1974, pp. 353–382.
200. R. Rannacher, *L^∞-stability and asymptotic error expansion for parabolic finite element equations*, Bonner Math. Schriften, 228 (1991).
201. R. Rannacher and R Scott, *Some optimal error estimates for piecewise linear finite element approximations*, Math. Comp., 38 (1982), pp. 437-445.
202. G. Raugel, *Résolution numérique par une méthode d'élements finis du problème de Dirichlet pour le laplacien dans un polygone*, C.R. Acad. Sc. Paris, Série A 286 (1977), 791–794.
203. P. Raviart, *The use of numerical integration in finite element methods for solving parabolic equations*, in Topics in Numerical Analysis, J. Miller, ed., Academic Press, New York, 1973, pp. 263–264.
204. P. Raviart and J. Thomas, *A mixed finite element method for 2nd order elliptic problems*, in Proc. of the Symposium on the Mathematical Aspects of the Finite Element Method, Rome, December, 1975, Springer Lecture Notes in Mathematics, Springer-Verlag, Berlin, Heidelberg, New York, 1977, pp. 292–315.
205. P. Sammon, *Convergence estimates for semidiscrete parabolic equation approximations*, SIAM J. Numer. Anal., 19 (1983), pp. 68–92.
206. P. Sammon, *Fully discrete approximation methods for parabolic problems with nonsmooth initial data*, SIAM J. Numer. Anal., 20 (1983), pp. 437–470.

207. G. SAVARE, *A(θ)-stable approximations of abstract Cauchy problems*, Numer. Math., 65 (1993), pp. 319–335.
208. A. SCHATZ AND L. B. WAHLBIN, *On the quasi-optimality in L_∞ of the H^1-projection into finite element spaces*, Math. Comp., 38 (1982), pp. 1–22.
209. A. H. SCHATZ, V. THOMÉE, AND L. B. WAHLBIN, *Maximum norm stability and error estimates in parabolic finite element equations*, Comm. Pure Appl. Math., 33 (1980), pp. 265–304.
210. ——, *Stability, analyticity and almost best approximation in maximum-norm for parabolic finite element equations*, Comm. Pure Appl. Math., 51 (1998), pp. 1349-1385.
211. R. SCHOLZ, *Optimal L_∞-estimates for a mixed finite element method for second order elliptic and parabolic problems*, Calcolo, 20 (1983), pp. 354–373.
212. R. SCHREIBER AND S. EISENSTAT, *Finite element methods for spherically symmetric elliptic equations*, SIAM J. Numer. Anal., 18 (1981), pp. 546–558.
213. L. R. SCOTT, *Interpolated boundary conditions in the finite element method*, SIAM J. Numer. Anal., 12 (1975), pp. 404–427.
214. ——, *Optimal L^∞ estimates for the finite element method on irregular meshes*, Math. Comp., 30 (1976), pp. 681–697.
215. D. SHEEN, I. H. SLOAN, AND V. THOMÉE, *A parallel method for time-discretization of parabolic problems based on contour integral representation and quadrature*, Math. Comp. 69 (1999), pp. 177–195.
216. D. SHEEN, I. H. SLOAN, AND V. THOMÉE, *A parallel method for time-discretization of parabolic equations based on Laplace transformation and quadrature*, IMA J. Numer. Anal. 23 (2004), pp. 269–299.
217. P. E. SOBOLEVSKII, *Equations of parabolic type in a Banach space*, Trudy Moscov. Math. Obšč., 10 (1961), pp. 297–350. translated in Amer. Math. Soc. Transl. 49(1966), pp. 1-62.
218. J. SQUEFF, *Superconvergence of mixed finite element methods for parabolic equations*, RAIRO Modél. Math.Anal.Numér, 21 (1987), pp. 327–352.
219. E. M. STEIN, *Singular Integrals and Differentiability Properties of Functions*, Princeton University Press, Princeton, New Jersey, 1970.
220. B. STEWART, *Generation of analytic semigroups by strongly elliptic operators*, Trans. Amer. Math. Soc., 199 (1974), pp. 141–161.
221. G. STRANG AND G. J. FIX, *An Analysis of the Finite Element Method*, Prentice-Hall, Englewood Cliffs, N. J., 1973.
222. T. SUZUKI, *On the rate of convergence of the difference finite element approximation for parabolic equations*, Proc. Japan Acad. Ser. A Math. Sci., 54 (1978), pp. 326–331.
223. R. TEMAM, *Stability analysis of the nonlinear Galerkin method*, Math. Comp., 57 (1991), pp. 477–505.
224. V. THOMÉE, *Spline approximation and difference schemes for the heat equation*, in The Mathematical Foundations of the Finite Element Method with Applications to Partial Differential Equations, A. K. Aziz, ed., Academic Press, New York and London, 1972, pp. 711–746.
225. ——, *Some convergence results for Galerkin methods for parabolic boundary value problem*, in Mathematical Aspects of Finite Elements, C. de Boor, ed., Academic Press, New York, 1974, pp. 55–88.
226. ——, *High order local approximations to derivatives in the finite element method*, Math. Comp., 31 (1977), pp. 652–660.

227. ——, *Some interior estimates for semidescrete Galerkin approximations for parabolic equations*, Math. Comp., 33 (1979), pp. 37–62.
228. ——, *Negative norm estimates and superconvergence in Galerkin methods for parabolic problems*, Math. Comp., 34 (1980), pp. 93–113.
229. ——, *Galerkin Finite Element Methods for Parabolic Problems*, vol. 1054 of Lecture Notes in Math., Springer-Verlag, Berlin and New York, 1984.
230. ——, *Finite difference methods for linear parabolic equations*, in Handbook of Numerical Analysis vol I. Finite Difference Methods 1, P. G. Ciarlet and J. L. Lions, eds., North-Holland, Amsterdam, 1990, pp. 5–196.
231. V. Thomée and L. B. Wahlbin, *On Galerkin methods in semilinear parabolic problems*, SIAM J. Numer. Anal., 12 (1975), pp. 378–389.
232. ——, *Maximum-norm stability and error estimates in Galerkin methods for parabolic equations in one space variable*, Numer. Math., 41 (1983), pp. 345–371.
233. ——, *Stability and analyticity in maximum-norm for simplical Lagrange finite element semidiscretizations of parabolic equations with Dirichlet boundary conditions*, Numer. Math., 87 (2000), pp. 373–389.
234. V. Thomée, J.-C. Xu, and N.-Y. Zhang, *Superconvergence of the gradient in piecewise linear finite-element approximation to a parabolic problem*, SIAM J. Numer. Anal., 26 (1989), pp. 553–573.
235. F. Tomarelli, *Regularity theorems and optimal order error estimates for linear parabolic Cauchy problem*, Numer. Math., 45 (1984), pp. 23–50.
236. H. Triebel, *Interpolation Theory, Function Spaces, Differential Operators*, VEB Deutscher Verlag, Berlin, 1978.
237. T. Ushijima, *On the uniform convergence for the lumped mass approximation to the heat equation*, J. Fac. Sci. Univ. Tokyo, 24 (1977), pp. 477–490.
238. ——, *Error estimates for the lumped mass approximation of the heat equation*, Mem. Numer. Math., 6 (1979), pp. 65–82.
239. L. B. Wahlbin, *On maximum norm error estimates for Galerkin approximations to one-dimensional second order parabolic boundary value problems*, SIAM J. Numer. Anal., 12 (1975), pp. 177–182.
240. ——, *A remark on parabolic smoothing and the finite element method*, SIAM J. Numer. Anal., 17 (1980), pp. 33–38.
241. ——, *A quasioptimal estimate in piecewise polynomial Galerkin approximation of parabolic problems*, in Numerical Methods, Proceedings, Dundee 1981, G. Watson, ed., Springer-Verlag, New York, 1981, pp. 230–245.
242. ——, *On the sharpness of certain local estimates for H_0^1-projections into finite element spaces: Influence of a reentrant corner*, Math. Comp., 42 (1984), pp. 1–8.
243. ——, *Superconvergence in Galerkin Finite Element Methods*, vol. 1605 of Lecture Notes in Math., Springer-Verlag, Berlin and New York, 1995.
244. L. B. Wahlbin, *Local Behavior in Finite Element Methods*, Handbook of Numerical Analysis, vol. II, Finite element Methods (Part 1), P.G. Ciarlet and J.L. Lions, Eds, Elsevier, 1991, pp. 353–522.
245. M. F. Wheeler, *L_∞ estimates of optimal order for Galerkin methods for one dimensional second order parabolic and hyperbolic equations*, SIAM J. Numer. Anal., 10 (1973), pp. 908–913.
246. ——, *A priori L_2 error estimates for Galerkin approximations to parabolic partial differential equations*, SIAM J. Numer. Anal., 10 (1973), pp. 723–759.

247. ——, *An H^{-1} Galerkin method for a parabolic problem in a single space variable*, SIAM J. Numer. Anal., 12 (1975), pp. 803–817.
248. K. Yosida, *Functional Analysis*, Springer-Verlag, Berlin, 1964.
249. O. C. Zienkiewicz, *The Finite Element Method in Engineering Science*, McGraw-Hill, London, New York, 1977. Third edition.
250. M. Zlàmal, *On the finite element metod*, Numer. Math., 12 (1968), pp. 394–405.
251. ——, *Finite element multistep discretizations of parabolic boundary value problems*, Math. Comp., 29 (1975), pp. 350–359.
252. ——, *Finite element methods for nonlinear parabolic equations*, RAIRO Modél. Math. Anal. Numér., 11 (1977), pp. 93–107.
253. ——, *A finite element solution of the nonlinear heat equation*, RAIRO Modél. Math. Anal. Numér., 14 (1980), pp. 203–216.

Index

MIX
Papier aus verantwortungsvollen Quellen
Paper from responsible sources
FSC® C105338

If you have any concerns about our products,
you can contact us on
ProductSafety@springernature.com

In case Publisher is established outside the EU,
the EU authorized representative is:
Springer Nature Customer Service Center GmbH
Europaplatz 3, 69115 Heidelberg, Germany

Printed by Libri Plureos GmbH
in Hamburg, Germany